3-98

WIT ☑ P9-DBI-843

	DATE DUE		
APR 1 '98			
	SEP 24 '98		
JUL 17 '98			
AUG 07 '98	DEC 17 '99		
AUG 28 '98			
JAN 22 '99			
APR 9 '99			
APR 30 '99			
JUL 9 '99			
SEP 3 '99			

International Encyclopedia of Horse Breeds

International Encyclopedia of Horse Breeds

By Bonnie L. Hendricks

Foreword by Anthony A. Dent

University of Oklahoma Press : NORMAN AND LONDON

Library of Congress Cataloging-in-Publication Data

Hendricks, Bonnie L. (Bonnie Lou), 1940–
    International encyclopedia of horse breeds / by Bonnie L. Henricks ; foreword by Anthony A. Dent. — 1st ed.
        p. cm.
    Includes bibliographical references (p.   ) and index.
    ISBN 0–8061–2753–8 (cloth)
    1. Horse breeds—Encyclopedias. 2. Horses—Encyclopedias. 3. Horses—Breeding—Encyclopedias.
I. Title.
SF291.H37   1995   95–11430
636.1'003—dc20   CIP

BOOK DESIGN BY BILL CASON

The paper in this book meets the guidelines for permanence and durability of the Committee on Production Guidelines for Book Longevity of the Council on Library Resources, Inc. ∞

1 2 3 4 5 6 7 8 9 10

In loving memory,
To Gary
Whose encouragement and support spurred me on

Contents

Foreword

By Anthony A. Dent

DURING THE MONTH WHEN I BEGAN TO WRITE this prefatory note, a mild fuss was created in the press of France, where I live, concerning a horse presented to President Mitterand on the occasion of his state visit to Turkmenistan and the manner in which this goodwill gift was disposed of. I cannot refrain from commenting on the inadvisability of presenting horses to people who do not ride—people such as the president (who is nearly my age) and the current prime minister of the United Kingdom (little more than half my age but not a rider), who received a similar present when he visited the same country. Both horses were the same breed, the Akhal-Teke strain of Turcoman horse.

The question has arisen as to what will happen to the horse now occupying luxurious quarters in the barracks of the Garde Républicaine when Mitterand's term of office comes to an end in the spring of 1995. According to the constitution of the Fifth Republic, such gifts are made to him not as an individual but solely by virtue of his office. Consequently when he retires he must turn it over to . . . whom? Ten to one his successor will not be a riding man either. Most people assumed that the obvious destination was the National Stud Service, which nowadays is a department of the Ministry of Agriculture. But a high official of that ministry, in the course of an interview, said that is impossible because, "This stallion not having been born in France, his progeny cannot be registered in our Stud Book." However, the Stud Book of the French Saddle Horse is replete with stallions imported from other countries—Trakehners, Han-overians, Andalusians, Lusitanians, Barbs, Wielkopolski—you name it, they have it. The French Saddle Horse becomes less French every day.

As for the Stud Book of Racehorses (other than trotting horses) in France—the only such registry in France not directly subject to the Ministry of Agriculture—the animals whose ancestry it records are known as Pursangs Anglais (Psa for short) because the foundation stock was imported from England at the end of the eighteenth century. And whence came the Founding Fathers whose written memorial is the General Stud Book, the holy writ of Newmarket? For long every harness room in England held as an article of faith that the holy equine trinity consisted of the Darley Arabian, the Byerley Turk, and the Godolphin Barb. The names are not entirely what they seem.

The Darley Arabian, bought at Aleppo in Syria, was described thus by the Yorkshireman who bought him: "The name of the race is manaka." That is an accurate enough rendering, by the linguistic standards of the day, of the Arabic word *muniqui,* a principal strain of Arab horse not used for warfare or for hunting but exclusively for racing, coming from the far north of Iraq and reputed among the Bedouin to carry Turcoman blood.

As for the Byerley Turk, its derivation sounds straightforward enough. In 1689, Captain Byerley brought it back to England from the siege of Budapest, where it had been captured from a Turkish officer. At that time Turkey, though it filled the whole of Asia Minor (Anatolia) and all the Balkans up to the

Danube, was much more closely tied to Turkestan (the homeland of the Ottomans) than it is now. And this great sea of grass served as an inexhaustible reservoir of troop horses and officers' chargers for the armies of the sultan. So Captain Byerley's acquisition need not necessarily have been foaled in any part of what is now, or was in the seventeenth century, part of the Turkish dominions.

That leaves the Godolphin Barb, or as some call it, the Godolphin Arabian, which after twenty years in England died and was buried on the Gogmagog Hills in Cambridgeshire. Lord Godolphin was its second owner. It came by way of Paris and no doubt was called *Barb* because the majority of Oriental horses in France were Barbs. (After Spain, France was the European country with the easiest access to North Africa.) Quite recently, after painstaking, accurate, and prolonged research, the greatest American hippologue of our time, Alexander Mackay Smith of Virginia, revealed the true provenance of the Godolphin. He successfully disproved the story (swallowed in succession by Edwin Coke, the first English owner, and Lord Godolphin) of its having been given by the Bey of Tunis to the King of France, who gave it to his cook, who set it to pulling a cart in the streets of Paris. It is now clear that—like the Byerley horse—it was a Turk, captured by the army of Prince Eugene in its victorious advance down the Danube. I will not spoil this intriguing story but rather will let you read it in Mackay Smith's own work. Of the three original Orientals, the Godolphin is the only one whose name figures in the pedigree of Thoroughbreds living today.

To conclude, given that two of the horses were of Turcoman origin and the other of a strain believed to carry some Turcoman blood, it is hard to see what objection there can be to the admission of another such horse into a register derived from the General Stud Book. Comparable gifts have been exchanged by crowned heads or the equivalent since the days of the Hittite empire, when King Solomon was not yet born, and have been at the root of many "new" breeds in a number of countries listed in the foregoing work.

Not, however, the one I remember best.

Quite early in his years as premier, Khrushchev embarked on a state visit to Queen Elizabeth accompanied by his minister of defense, A. N. Bulganin, and bringing a gift of two horses of an ancient breed from Karabakh, that mountainous Caucasian region that has lately been the scene of an armed dispute between Armenian and Azerbaijani forces. In contrast to the gift last mentioned, this was a really sensible offering. The monarch of the United Kingdom was a keen horsewoman, her husband a polo player. Her children also rode, and her daughter, the Princess Royal, since has achieved great distinction in that field. Certainly, these two Caucasian newcomers would have been most welcome.

You will be surprised, as you read this book, to find the vast number of breeds described that come from what was the Soviet Union. And you will be rather more surprised to learn that more than a lifetime ago there were almost twice as many breeds in the Russian Empire, which comprised roughly the same territory. At the turn of the century the Imperial (now Russian) Academy of Sciences in St. Petersburg published in French a comprehensive book, by Simonov and Moerder, on the horse breeds of the Russian Empire. I saw a copy only once but was most impressed. I seem to recall that there were about a hundred entries. Even more surprising to me is the number of Chinese horse breeds described in Bonnie Hendricks's book. I cannot sufficiently admire her industry and perseverance in obtaining information that must have been most difficult to come by, illustrations and all.

Do not be disappointed to find that a great many countries are not treated in this volume. There are two reasons for this. First, in the Pacific and Indian oceans are many states, recently colonies, in which there has never been a native race of horse, and such horses as are to be found are a legacy of the late colonial power—in Madagascar, for instance. In the same quarter of the world are countries well on the way to independence, such as New Caledonia and Tahiti, in which the situation is similar. Second, there are countries in Central Africa that simply are not viable for horses because of the sickness spread by the tsetse fly. These include what is now the

Central African Republic, once the domain of the infamous and Napoleon-struck Emperor Bokassa. Having proclaimed himself emperor and learning that the great Bonaparte had gone to his coronation in a carriage drawn by white horses, Bokassa ordered a half-dozen white horses from France. But when they arrived, his Lord Chamberlain tremblingly announced that a black team had been sent in error. Nothing daunted, Bokassa ordered, "Paint them white. All over!" This was done, and—not surprising, because they could not sweat—they all died before the day of ceremony dawned, instead of dying a few days later from the equine equivalent of sleeping sickness.

In general, national pride plays havoc with classification of breeds. Look under Israel, and you will find a local breed listed as *Native Horse* that in fact is predominantly an Arab, but *Arab* having been so very recently a dirty word in the Zionist state, it cannot officially figure as such. Again, the two breeds Andalusian and Lusitanian really are nearly identical, but as the Spaniards and Portuguese respectively insist theirs is the only genuine article, foreigners have invented the term *Iberian Riding Horse* and even have produced crossbreeds from these two identical groups that were the preferred mount of every monarch in Europe from the time of William the Conqueror (who rode a black Andalusian at Hastings) until the eighteenth century.

I suppose that what the reader looks for in a reference book of this kind is above all accuracy, comprehensiveness, and up-to-dateness. I believe this book possesses these qualities insofar as it is possible for one author living in one country to produce one. Hendricks's book is a very comprehensive survey of the universal equine scene *today*, which inevitably implies that it will not be *quite* right tomorrow. It is in no sense a history, though many historical facts are adduced to explain particular features of some races. Breeds come and go, dwindle almost to the point of extinction, and are revived more quickly than one would think. Despite all the care, knowledge, and skill that have gone into the book's making, the day after its publication the author or the publisher will read a letter bearing the heading of this legation or that embassy and the signature of the military, agricultural, commercial, or even cultural attaché, deploring the omission of all mention of the Patagonian Prancer or the Coromandel Cob or the Abominable Snow Horse of the Himalayas, "which has played such a notable part in the development of my country."

Never mind. I am convinced this is the best that can be done in an imperfect world, and I wish it every success.

Preface

THE HORSE! WE ARE ENCHANTED AND FASCInated by this wonderful creature. It must have been a subject of intrigue even for early peoples who hunted it for food and painted it on the walls of caves, this animal of such admirable speed and strength combined with the timidity of a rabbit. Human use of the horse can hardly be overstated. In looking backward at our history one must wonder how far we would have progressed by this time without use of the horse. It has been our food; its hide has provided clothing, tent coverings, blankets, and bags for carrying various objects. Horses have drawn our light chariots and carriages in splendor and stateliness—and have trudged along patiently and unperturbed, pulling our heavy burdens. They have been our sacrifices to obscure gods and our war machines, vehicles, and pleasure horses.

The horse became our mode of transportation when our ancestors yearned to explore the world around them or expand their horizons—as they assaulted, victimized, and plundered their neighbors. No other animal has worked so diligently for humankind, giving us the benefit of its great strength and ability. This noble creature has voluntarily and obediently died for us when, in our need or ignorance, we required more than its great heart could accomplish. The horse has suffered outrageous abuse at our hands—and enjoyed luxuriant pampering and good care. Truly, the horse has experienced our lives and evolution with us.

This remarkable animal has survived from its beginning as tiny *Eohippus* due to its incredible ability to adapt. Many other creatures became extinct when their environment experienced a sudden change. The horse, like a giant chameleon, simply altered to fit the environment or migrated on sturdy legs, using an unerring sense of direction, to a more suitable place. It survived the Ice Age. It can live in the desert. It can live in the Alps. It can live in marshes. Steppes or mountains, lowlands or high, intense heat or bone-chilling cold—the horse flourishes and survives. In cold regions, horses developed a shorter stature, thicker skin, and a long, thick winter coat. Horses adapted to cold or wet regions have special whorls in their coats that divert water from sensitive parts of the body. A horse can paw through several feet of snow to find withered grass and, if necessary, roots for food. In hot, arid deserts, horses have developed longer legs to raise their bodies above the scorching sand and a thin skin to help them cool more quickly; they have developed the ability to exist on minimal water and on hard, dry feed.

Left to its own devices the horse, suspiciously appearing as a natural layabout, favors a relaxed existence on lush pastures and perhaps a playful romp in the cool of the day. For defense it used speed to avoid predators, including humans. Once captured, it revealed a submissive nature that must have suggested possible usefulness far beyond that of a mere food source. For yet to be determined ages the horse served as a beast of burden and meat source before being ridden. For all our pride over being the most intelligent creatures on earth, we humans do not always think particularly fast.

Normally somewhat placid in demeanor, once mounted, the horse arched its neck and stepped out in a lively manner or even pranced and quickly learned to obey human commands. Imagine the thrill of revelation the first experimental riders must have felt!

Both man and horse experienced a profound transformation with the advent of horseback riding. Each found a new pride, arrogance, and aggressiveness never before possessed once people learned to ride and not simply eat the horse. Suddenly, the rider could look down upon the world from the superior loft provided by the back of the snorting, mane-tossing, powerful horse. The horse transformed a benign and plodding pedestrian — trudging along laboriously dragging or carrying his meager possessions — into the fearsome mounted warrior of the steppes, who could strike and retreat with equal speed, covering great distances in a short time. Uninitiated neighbors suddenly had to contend with unprecedented problems. No longer did they merely align themselves along their respective borders and throw rocks or spears or perhaps shout insults. Instead, great hordes of mounted warriors crossed the borders of neighboring tribes, killing, looting, terrorizing — some, as an ultimate gesture of contempt, making gold-lined drinking cups of the skulls of their enemies.

Early hunter-gatherers must have lived a life of constant fear, concerned with collecting and catching enough to eat and avoiding themselves becoming the victims of predators. For thousands of years their life-style had changed only in hesitant degrees. Oxen were used for some time prior to domestication of the horse, but they were slow — another plodder. The horse, on the other hand, revolutionized the development of human culture, virtually overnight.

The horse, appearing proud to be in possession of the rider on its back, took on an arrogant attitude formerly unseen in a creature so benign. The ancient war horse with flashing eyes, bowed neck, and flaring nostrils, charging into battle with striking forefeet, scarcely resembled the frightened rabbit of former days.

Hast thou given the horse strength? Hast thou clothed his neck with thunder? Canst thou make him afraid as a grasshopper? The glory of his nostrils is terrible. He paweth in the valley, and rejoiceth in his strength. He goeth on to meet the armed men. He mocketh at fear, and is not affrighted; neither turneth he back from the sword. The quiver rattleth against him, the glittering spear and the shield. He swalloweth the ground with fierceness and rage. *(Job 39:21–24)*

Becoming a warrior's mount was one more situation to which the horse needed to adapt. And adapt it did, gloriously and to perfection.

We love our horses and we want them to love us. It is obvious that most of them feel an affectionate friendliness toward us. I suspect, however, in admiration for this truly independent and adaptable species, that we deal continually with its keen instinct for survival and preoccupation with food. Take down the fence and offer them liberty, and horses will depart for places unknown and put owners out of their minds. We need them — these proven survivors do not need us.

Of the animals domesticated by humankind, none has resisted as determinedly as the horse. It works for us amiably and obediently, but possibly only because this is the safest reaction to the moment. The horse's preferences are clearly shown when it is given a choice. In open country with no fences, where freedom is theirs, the best trained and gentlest horses will often turn and run like the wind when approached by someone with halter or bridle in hand. Usually it is the horse that associates a human visitor with a morsel of grain that meets its owner halfway in the field. Such independence of spirit is delightful, especially combined with *willingness* to be tractable. There are several other animals that could be useful to people, but *will* not.

The horse is not, as one local county agricultural official told me, a dying species. Again adapting and meeting the challenge of a new situation now that they are no longer necessary on farms in most areas, horses shine today in sports and recreation as never before. They thrill us with their speed and stamina in races, agility at equestrian games, jumping expertise, and all-around athletic ability. A horse can master precision of movement like a ballet dancer, controlled and steady—or can throw its head down between its front legs and behave like an outlaw in the rodeo arena: anything to please. I had a gelding at one time that would buck only under certain circumstances. This horse was used a few times to fill in as a bareback bronc in a rodeo, and no one ever managed to finish a ride on him. After the rodeo, we saddled and rode him home with no misgivings—much to the surprise of wide-eyed onlookers.

In a very real sense the last century has seen a complete change in the relationship between people and horses. Today horses enjoy the well-earned position of being almost *our masters,* while we have taken the role of servants. We carry their food and water—whereas they used to carry ours! We clean their stalls and bed these with fresh straw or shavings. We groom their coats to a glossy shimmer and go to great expense in equipment, training, shoeing, veterinary care, hay, grain, and whatever else is necessary for the privilege of enjoying their company.

I believe there is a space within the human heart reserved especially for a relationship with the horse. Horses were among the first animals to appear on earth, I am told, while we, they say, are the latest; yet we have this great affinity for each other (or at least, we for the horse). We love its smell; we love to look into those huge eyes and wonder what wise thoughts might reside behind them; we love the feel of that silky coat; the strange and fascinating combination of gentleness and great power and fire residing within the same entity. We admire the horse's great strength and appreciate its willingness to be our friend.

While the dog was domesticated earlier than the horse and has served us well as a companion, protector, and even beast of burden, the title given the dog of "man's best friend" seems misapplied when compared to the great service we have received throughout the centuries from the horse.

Certainly I am not the first to attempt to document accurately all of the horse and pony breeds on earth. And—after twenty-three months of persistent letter writing—I too have not succeeded in including them all in one volume. Authorities in some countries simply did not respond. In addition to the breeds and types presented here in main articles, many subtypes are also described. There remain many breeds not contained in this volume that I hope to include in the first revision.

I have made every effort to present a unique and accurate reference book on equine breeds, including much of their history. Information presented was obtained from breed authorities around the world and not "borrowed" from other books. Consequently, the reader will find a good deal of new information and many breeds never previously presented in *any* book.

In writing this volume I have gained great respect for fellow authors who have ventured down the path to produce a book on world horse breeds. By following their steps down this perilous trail I have gained a feeling of kinship and understanding of the overwhelming task it is to write about horses in an objective manner.

Beyond aspiring to present all the respective varieties of our friend, *Equus caballus,* in the world, I have labored in fear and trembling to give objective and useful information. Fear and trembling, because there are many "authorities" quick to challenge any who dare to venture into statements of fact if they make a mistake. Few subjects are as controversial as the horse, whether one is addressing origins, abilities, methods of feeding, training, breeding, or any other matter. Even the question of what constitutes a breed as opposed to a type is food for argument.

At first my reasons for attempting a volume on world horse breeds stemmed from the fact that I knew of several kinds of horses never presented in other books. Many years ago I set out to attempt this project but was unable to proceed. My late husband, Gary, believed in my ability and encouraged me to begin again. I discarded the outdated information and photos I had collected and launched out once again, in January 1989. Once I began to work, it became obvious that many relatively recent writers had not contacted breed authorities for current information. A book was needed, I felt, that included the addresses of breed associations around the world, especially today when there is so much activity in import and export of horses. The sport horse reigns in this age and has joined the "jet set." But how do you arrange to purchase a European Warmblood or Russian Akhal-Teke or South American Mangalarga Marchador if you have no idea whom to contact? Others may have omitted the addresses of breed organizations to avoid having their books become obsolete when addresses change. Even as I write this volume, it is becoming dated. The world changes so quickly. The reunification of Germany and unraveling of the Soviet Union required revisions in the emerging manuscript. The solution is regularly revised editions with current information. I believe the address list in appendix C, most difficult to obtain, will be useful to readers.

Few of my personal opinions will be found within the covers of this book. I found many of my most cherished beliefs worthless and my store of "reliable" horse lore sadly lacking when I began in earnest to ask the world for truths.

Occasionally, when response from breed authorities was brief or nonexistent, I consulted other knowledgeable sources, and I feel confident in their opinions. The bibliography is scant for a volume of this scope because I consulted few books, finding them too often repeating misconceptions from earlier books. The list of addresses in appendix C should be considered my main source of information.

It seems redundant to state in nearly every article that this breed or that has a good temperament. The horse, with few exceptions, is an amiable, submissive creature by nature. Some are more excitable than others; some have been spoiled or taught bad habits—the fault of handlers and not horses.

I have omitted an extended treatment of the origins of the horse or a chapter on early domestication, for I am not an archaeologist, paleontologist, or anthropologist and there are many books available by such qualified people. Also omitted is information regarding breeding, feeding, foaling, training, riding, grooming, and other aspects of handling. Such subjects are considered general, controversial, and usually a matter of personal opinion and preference. Our adaptable equine friend seems able to cope in spite of all our various and sundry methods.

This is a reference book of horse breeds and types; their origins, history, physical aptitudes, descriptions, and uses and interesting information about distinctive aspects of the various types. I have faithfully reported what breed authorities claim about their respective horses and have not judged their merits of good conformation (or lack of same) or their right to exist. An expensive hotblood beauty in a stable likely would not survive one night in Siberia at $-70°F$, nor one day's trek in the high altitude and cold, rarefied air of the mountains of China or the high Andes of South America carrying a 200- to 300-pound pack. Many of the horses in the world have their special jobs at which they excel daily when others would not survive a day trying to follow their tracks. An attempt to "improve" or "Arabize" these hardy animals could only result in loss of ability to survive in their inhospitable environment. The old saying, "if it ain't broke, don't fix it," surely applies to horses that are perfectly adapted for their task. Amazingly, the horse, seemingly created especially for our use, has adapted to nearly every inhabited area on the earth.

Missing from this volume is the mythical creature commonly called "Chinese pony" by apparently all but the Chinese people. In its stead, I am delighted to include over thirty distinct horse breeds of China through the gratefully received assistance of Professor Youchun Chen in Beijing, director of the Chinese Institute of Animal Science, and his associate, Professor Tiequan Wang. To furnish information for this work, Professor Chen translated the book *Horse and Ass Breeds in China,* written in part by Professor Tiequan Wang, who also furnished some photographs. Professor Chen supplied information on additional Chinese breeds not found in *any* book, so some are being introduced to the world for the first time in this volume. Because the horses of China are hardly known outside of their homeland, I have included a section, "Chinese Horse," which contains some general information.

Also missing herein is the Falkland Island pony, another mythical entity found in some books. Authorities in the Falklands informed me that while a wide variety of horses and ponies have been imported, they are mixed without discrimination according to the current fad; therefore, no consistent or distinct type has ever emerged. The Haiti pony also does not exist as a distinct breed or type, being a mixed variety descended from imported animals and bred with no specific goal in mind.

The Karacabey of Turkey is included for edification only, though this breed is found in nearly every other current book on world horse breeds. This breed, usually presented as the only horse breed of Turkey, has not existed since 1979. Turkey does, however, have over a dozen other breeds, mostly of ancient origin, as worthy of inclusion in a book on world horse breeds as any others.

The Barb, intriguing to horse chroniclers of both past and present, is included in this volume under the heading, "Moroccan Barb." While the Barb horse originated across all of north Africa and not only in Morocco, this is the major country breeding the Barb today.

Individual treatment is given to each of the various Spanish Mustang associations due to the fact that there is some difference among the horses of these groups or among their breeding intentions. Confusion has evolved around the term *mustang,* which many people believe applies to any feral, scrub horse. The feral horses of the United States are included in this volume under the heading "Mustang," which refers to a horse that has little or no connection with the Spanish Mustangs that descended from the horses of the conquistadores.

I have set aside the usual format in books on horse breeds—of grouping them into separate categories as draft, light horse, and pony. While the standard generally followed states that any equine standing less than 14.2 hands is a pony, I found many small horses that simply are *not* ponies. The Caspian is an excellent example. Skeletal remains examined by scientists found the Caspian to be a very fine, small horse and not a pony, one determining factor being length of the cannon bone. Yet, the Caspian averages only 11 hands high. The Hucul (Carpathian) of Romania is considered a small horse and not a pony by Romanian authorities yet is included in the pony section of most breed books. Conversely, the Bardigiano of Italy, similar in size to the Hucul, is classified by the Italians as "a pony type and not a small horse." The Miniature horse is not considered a pony, and in Iceland, the national equine is known as the Icelandic Horse. Not so many years ago many Arab horses stood lower than 14.2 hands. But who in his or her right mind would dare call the purebred Arab standing at 14.1 hands a pony?

The term "pony," in addition to referring to an equine standing under 14.2 hands high, also usually relates to a small equine with decided draft-type qualities—a shorter leg, squarish or boxy type of head, massive muscling—as opposed to a light horse type with a longer cannon and more refined head and lighter body muscling. Even the subject of what distinguishes a pony from a horse is one fraught with controversy. One paleontologist told me that ponies like those of the British Isles are exactly the same equine as the large draft horses of today. Island living causes reduction in size, I was told, whereas selective breeding on the continent produced the "great horse" and subsequent large draft animals. Interestingly, blood group testing such as that being done at the University of Kentucky reveals differences between the genetic markers of draft animals and those of ponies. Also, one must ask why the Caspian of Iran, the Guoxia of China, the Rahvan of Turkey, the Carpathian, and some others have remained small, although they are not found on islands.

Regarding the blood testing mentioned, Professor E. Gus Cothran of the University of Kentucky kindly rendered this brief explanation in a personal letter.

The need for a form of permanent identification and parentage verification has led to widespread use of blood typing for many horse breeds. Blood typing employs modern biochemical techniques to determine the genetic characteristics of individuals. Differences in these characteristics, or genetic markers, among individuals represent genetic differences. Within a breed, individuals that share the greatest number of genetic markers are likely to be more related to each other than to individuals that share fewer markers. This principle also applies to comparisons among breeds. Frequency of occurrence of each genetic marker within a breed can be compared to the frequencies within other breeds. The comparison gives a measure called genetic similarity. Genetic similarity can generally be considered an indication of relationships. By comparing genetic marker data of a large number of breeds, it may be possible to decipher much of the evolutionary history of modern horse breeds. In addition, genetic information of this type can provide information about the demographic history of a breed. Continued monitoring of the genetics of a breed can indicate how current breeding practices affect the genetic makeup of a breed and may be useful for determining strategies for the management of rare breeds.

There is some controversy relating to the classification of draft horses from "light draft." In view of the various opinions and out of respect for those who do not consider their small horse a pony, or their chunky riding horse a draft animal, I have followed the encyclopedic style and presented all of the breeds in alphabetical order. An appendix is included listing the breeds under the country from which they originate.

Our generation is seeing the disappearance of many breeds as they are blended into others defined as "sport horse" or "warmblood." Within a few years many of these older breeds could cease to exist as pure entities within the horse kingdom. Breeds being blended to develop the sport horse are included in the section dealing with the new breed of which they are now a part and can also be found in the index. Consequently, the number of breeds represented in this volume is much greater than the number suggested by the headings. Beneath the title of each entry I have included the native spelling of the breed name when possible, as well as common synonyms.

I wonder, with all of the blending in and phasing out of many of our present breeds to bring to fulfillment the creature called "sport horse," how many breeds will exist individually in another hundred years? Will we have come full circle, in a sense, and reduced the horse to but a few main types, as was the case at the beginning of its domestication? Will we ever wish we could have some of those ancient breeds back, to broaden the gene base again and replace some of what has been lost?

One should consider that, in our quest to breed the perfect horse, we have drastically reduced the available gene pool. Many genes that imparted hardiness and strength through natural selection have been lost forever. As the gene pool continues to narrow, what do we sacrifice? Will this proven survivor lose the ability to persist without human pampering? The rare breeds left today should be carefully preserved and guarded for that day when we may realize we have nearly lost something extremely precious: the genetic perfection that enabled the horse to adapt and survive almost anywhere in the world. In the desire to breed a better horse, we have often created a weaker one. We have horses that are beautiful to observe but suffer with splints, sidebone, ringbone, bowed tendons, navicular and numerous other ailments. In the remaining more primitive types, such problems of the legs and feet are nearly *unknown*.

Representatives of horse breeds around the world are invited to contact me (in care of the University of Oklahoma Press) with additional or updated information or corrections (please provide references) so that these can be included in the first revision of this volume. *International Encyclopedia of Horse Breeds* will be an ongoing endeavor with revisions and updates. It is my desire to add breeds omitted from this first edition and to update, enlarge, and when necessary correct information presently herein. I would also like to include in the first revision photographs of more recent champions and good examples of the breeds.

I suspect there is much more yet to be learned about the horse. Let us unite as many kinds of this adapting, persisting, enduring, and endearing survivor, *Equus caballus*, as we can in one encyclopedia. In the process, we and the horse will find many new friends in other lands.

BONNIE L. HENDRICKS
Vacaville, California

Acknowledgments

MANY PEOPLE ASSISTED ME IN VARIOUS WAYS in writing this encyclopedia, and to name all of them and their deeds individually would make a book in itself. This volume is the result of the combined effort of many people around the world. I am grateful to the United States Department of State for furnishing embassy and diplomatic lists and to the many people in American embassies in every area of the world and the foreign embassies in the United States for their cheerful assistance in helping to furnish addresses imperative to this project. They made every effort to direct me in my pilgrimage for accurate information.

I wish also to express my deep appreciation to all of the horse breed organizations and authorities who so abundantly supplied information, photographs, stud books, and often long letters to describe their respective horse or pony breeds. The address of each breed organization that responded to my request for information is included in appendix C for the reader's convenience and for promotion of the horse breeds.

Some people went that extra mile to assist me and their efforts merit a very special and heartfelt thanks. I have especially appreciated the efforts of Marye Ann Thompson, Willcox, Arizona, who proofread the manuscript and has been a gold mine of assistance and information; Evelyn Simak, Munich, Germany, for her work in translating information on German horses and especially the rare breeds, and for the many photographs from her personal archives; John A. Williams and Marie Carroll, American Embassy, Jeddah,

Saudi Arabia, for their tireless efforts in obtaining information on the Arab breed; Luciano Miguens, Buenos Aires, Argentina; Dr. Peter Chakarov, Sofia, Bulgaria; Ebangi Achenduh, Garoua, Cameroon; Dr. Rafael Mancilla, Santiago, Chile; Professor Youchun Chen, director of the Institute of Animal Science in Beijing, China, who spent many hours translating Chinese text and answering a myriad of questions; Feng Weiqi of Beijing, who kindly supplied conformation photographs of the Chinese horses; Professor Wang Tiequan of the Institute of Animal Science in Beijing, China, for photographs and information; Dr. David W. Cantero, Havana, Cuba; Professor Harold Barclay of Edmonton, Alberta, Canada; veterinarian Dr. Alzbeta (Betty) Kovacova-Pecarova of Kosice, Czechoslovakia, for many hours spent translating Russian text and information on horse breeds in her own country; François De La Sayette of the Union Nationale Interprofessionnelle du Cheval (U.N.I.C.) in Paris, France; the French Ministry of Agriculture; Professor H. C. Schwark, Karl-Marx University, Leipzig, Germany; HRH Princess Alia Al-Hussein, Amman, Jordan; Professor Dominicus C. Chong, Cheju National University, Korea; Professor E. Pern, Ryazan, former USSR; Dr. Frans van der Merwe, Pretoria, South Africa; Kurt Graaf, Jönköping, Sweden; Ertuğrul Güleç, Ankara, Turkey; Louise Firouz, Teheran, Iran for photographs and a wealth of information, and for her inclusion on the Oriental horse; Caren Firouz of Teheran, Iran, for his beautiful photographs of Iranian horse breeds; Dr. Jorge De Paiva Benites, the

Azores, Portugal; Peter Fischer, Dromana, Australia; Brenda Dalton, British Caspian Society; Alex McIntosh, Edinburgh, Scotland; Marit Jonsson and her daughter, Kristin Jonsson, of Hillerød, Denmark, for translations on the horses of Denmark; Raja Bhupat Singh, Jodhpur, India; Rex and Marge Moyle, Star, Idaho; Mary Gharagglon, Teheran, Iran; Barbara J. Christie, Halifax, Nova Scotia, Canada; Zoe Lucas, Halifax, Nova Scotia, Canada; Gilbert Jones, Finley, Oklahoma; Jim and Retha Carter, Anderson, California; Susan Paulton, Hot Springs, South Dakota; Mariko Hendricks, Valley Center, California, for translating Japanese text; Professor E. Gus Cothran, Lexington, Kentucky; William P. Huth of the American Embassy in Moscow, who assisted greatly in directing me to knowledgeable people in the (then) Soviet Union; Roger Thorp of Custer, South Dakota, for his photographic expertise; Ian Mason, Edinburgh, Scotland, for his enthusiastic review, and last—but most certainly not least—Anthony Dent in St. Pierre-de-Chignac, France, whose knowledge of history, eye for detail, moral support, and friendship I very keenly appreciate.

My most grateful thanks to all of you. By your response you have confirmed that all of the diverse breeds of horses and ponies have their special niche; that they are of interest to horse lovers everywhere; and that reliable information should be available within the covers of one reference volume. Your enthusiasm for this project was not only gratifying, but expressed tremendous desire worldwide for such a volume.

International Encyclopedia of Horse Breeds

Abtenauer

ORIGIN: *Austria*
APTITUDES: *Heavy draft work*
AVERAGE HEIGHT: *14.3 to 15 h.h.*
POPULATION STATUS: *Rare*

Abtenauer mare and foal. Many foals have a curly coat such as this one. Photo: Evelyn Simak

The Abtenauer is a rare breed with an estimated population today of under one hundred. It is the smallest variant of the Noriker breed. Bred in the isolated valley of Abtenau, south of Salzburg, it is a consolidated population in itself. Of lighter bone than the Noriker, the Abtenauer breeds true to type and is well adapted to the poor soil on which it lives.

This breed is known for its easy, flowing trotting action. There is a saying in the Austrian army that a soldier who excels in marching travels like an Abtenauer. The Abtenauer is an elegantly built draft horse with a well-shaped head and strong legs. The most common colors are black, blue roan, and chestnut. All colors are accepted except leopard spotting, which is undesired by breeders.

Mares and foals spend the summers in pasture high in the mountains and are fed salt once a week to prevent them from becoming feral. They spend the summer freely roaming the alpine meadows with cattle.

Often foals are born with peculiar curly coats that are lost when the baby hair is shed.

Because of its agility and good sense of balance, this breed is valuable for work in mountain forests. The Abtenauer has a quiet, willing disposition.

Adaev

(Adaevskaya)

ORIGIN: *Kazakhstan (former Soviet Union)*
APTITUDES: *Riding, light draft*
AVERAGE HEIGHT: *13 to 14.1 h.h.*
POPULATION STATUS: *Common*

A horse of the steppes, the Adaev is one type found within the Kazakh breed, which dates to the fifth century B.C. Researchers have distinguished two widespread types within the breed; the other is the Jabe (see breed account, this volume). The Adaev is a riding type, while the Jabe is a heavyweight type more suited to draft work.

The Adaev nearly disappeared in the past due to extensive outcrossing, while the Jabe was improved and increased in number, partly due to Adaev blood.

Early in 1985, 27,000 head of Adaev horses were bred. The task of breeders is not only to save but to increase the numbers of this breed, which is well adapted to the severe conditions of the Prikaspi desert (Caspian Depression). The plan in the last decade has been improvement of breeding and productive qualities of the Adaev horse through the use of *taboon* raising (see below), recording of pedigrees of individual horses, establishment of additional food resources, and increase in size of the horses.

The modern Adaev was produced by crossbreeding the Kazakh with Don, Thoroughbred, and Orlov Trotter, creating a horse with more pronounced saddle characteristics and clean-cut conformation. The head is light, with a long neck, well-defined withers, and a straight back. Some Adaev horses have a narrow chest and insufficient bone, the result of primitive management conditions and lack of proper breeding selection.

The Adaev is bay, grey, palomino, or chestnut. This strain of the Kazakh is not as well suited to harsh environmental conditions as other Kazakh horses.

The horses are raised in herds to be used by herdsmen and shepherds as work horses and also for meat production. Mares of the breed are remarkable for their high yield in milk, and are highly prized on farms specializing in the production of mare's milk. The milk is usually fermented into a drink called *koumiss*, which is very healthful and used to cure many ailments.

Within the Adaev breed there are three types—massive, medium, and light. Production of the massive type has been of the utmost importance. This type is well balanced and proportional with a massive body and skeleton. These horses have the best endurance, mobility, and adaptability to environmental conditions. They are used under saddle and as meat animals and this dual usefulness gives them outstanding value.

Taboon, or *tabun*, is a word that describes a method of livestock keeping used throughout Asia. Territories are huge and unfenced. A taboon is a herd of horses numbering up to 200 head, attended continually by mounted herdsmen.

Medium and light Adaev horses are specifically of the saddle type, having highly set heads and easy gaits with a noble appearance. Both the medium and light types have emerged through the use of Akhal-Teke blood. The goal is to preserve all three types. A great practical value is the increase of meat and milk production in Adaev horses previously giving low production while retaining their adaptability to climate. The solution to the problem came through crossing with the Kushum and Jabe breeds, which have great live weight and good milking ability.

In 1978 an experiment was initiated including 240 horses, offspring of Adaev mares and Kushum, Jabe, and Adaev stallions. Each of the young was raised under taboon-*tebenevka* conditions *(tebenevka* means year-round pasture raising without supplemental feed). The results showed that the crossbreds withstood severe conditions well and gave higher weight gains than the purebreds.

The typical Adaev horse raised in the southern regions is of taller growth than other Kazakh horses. The head is large, lean, and often Roman-nosed; the neck is of medium length and high set; the trunk is deep; the legs are clean with clearly defined tendons and joints; the skin is thin with visible subcutaneous veins; the coat is well developed in winter with long mane and tail and some feather on the legs.

Akhal-Teke

(Akhaltekinskaya)

ORIGIN: *Turkmenistan (former Soviet Union)*
APTITUDES: *Riding, racing*
AVERAGE HEIGHT: *14.2 to 16 h.h.*
POPULATION STATUS: *Rare*

Akhal-Teke stallion in Germany. Photo: Evelyn Simak

Akhal-Teke stallion in Ryazane, USSR. Photo: Nikiphorov Veniamin Maksimovich

The Akhal-Teke stallion, Sarsgyn, eighteen years old. Photo: Evelyn Simak

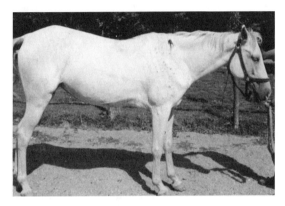

Oriental mare showing blood sweating phenomenon. Photo: Louise Firouz

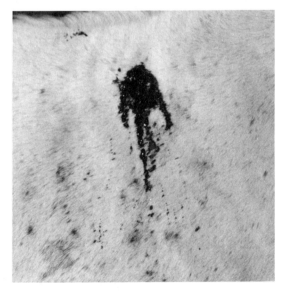

Close-up of "blood sweating" mare. Photo: Louise Firouz

The Akhal-Teke is an ancient breed that emerged in its present form by the turn of the eighth century, at a time when Turkmenian horses were regarded as the best to be found in Central Asia. It developed in the oases southeast of the Kara Kum desert and the foothills of the Kopet Mountains, on the border of Iran and the former Soviet Union. Turkmene horses were exported to both near and far countries, causing a stir of admiration.

While the original, ancient Turkmenian horse is considered extinct, horses bred in the area of Turkmenistan are still referred to by that name. (Turkmenistan, Turkistan, and Turkmenia refer to the same geographic area and are used interchangeably.) This breed was well known as early as 1000 B.C. as a race horse, and 30,000 Bactrian horsemen made up the cavalry of King Darius of Persia. The Turkmene breed was also ridden by the army of Alexander the Great. Philip, Alexander's father, obtained great numbers of horses (20,000) from the area of Fergana in what is now Turkmenistan for his troops (Bökönyi, 1968). Bucephalus, the favored horse of Alexander, while generally said to have been a Thessalonian horse, was more likely of Turkmenian blood. Most light-breed horses in Greece had been brought from Fergana or had descended from stock brought from there earlier. The horses of Fergana were taller and swifter than most other breeds of that time.

Alexander took a spoil of 50,000 eastern horses from the Persians (Bökönyi, 1968). These horses, when crossed with the local European animals, produced larger horses that were used by the Roman cavalry. With the Romans, Turkmenian blood was spread throughout the known world, adding refinement to the heavier, shorter draft types found throughout Europe.

Modern representatives of the ancient breed are the Akhal-Teke and Iomud in the former Soviet Union. In Iran, descendants of the Turkmenian are represented in the Turkoman, a name that includes both the Akhal-Teke and Yamud (see Turkoman).

The Turkmenian, an original, indigenous Oriental horse, had great influence on the light breeds of Europe and North Africa and, indirectly, the world, although this is not generally acknowledged. This horse was commonly called "Turk" in Europe, and subsequent writers assumed the animals were of Arabian extraction. An example is found regarding the outstanding sire Turkmen Atti (sometimes written as Turk main Atti), without a doubt a Turkmene horse, but written into the Trakehner stud book as an Arab. The Darley Arabian (see English Thoroughbred) is said to have had many Turkmenian crosses in his pedigree.

The Eurasian steppe was dominated by several different tribes and peoples over the centuries, and a great interchange of horses took place. During the Hsiung-nu (better known as Hun) empire, which mirrored the Chinese, organized warfare across the Gobi Desert alternated with periods of peace. Formalized exchanges of tribute-gifts allowed the rulers of each side to strengthen themselves by acquiring rare and valuable goods. The Chinese obtained horses for the army and other imperial uses, while the Hsiung-nu rulers acquired grain, silks, and other luxuries.

The Hsiung-nu displaced other peoples by a series of migrations into what is now Iran and across the Hindu Kush into India. Various Iranian tribes — Śakas (a Persian name for Scythian) and Kushāns chief among them — were the protagonists of these displacements. As they vacated their grazing lands, they came under control of Turkish tribes, and the frontier of Indo-European languages began to shrink as the Turks advanced. By the end of the second century B.C., the patterns of migration altered. At that time the Persian border was effectually defended by the new guardians, the Parthians, whose horses were of Turkmenian descent.

The key change was the introduction of alfalfa (lucerne) as a cultivated crop that proved to be fine fodder for horses. It was discovered that horses fed on alfalfa with some additional grain and provided with shelter from the harsh elements could be bred larger and stronger than the steppe ponies that had only grass to eat. Large horses could support armored men and even carry armor to protect their own bodies.

The fame of the Parthian horses soon reached the Chinese imperial court. Emperor Wu Ti began his reign in 141 B.C. during the Han dynasty. He was known as a ruthless leader who had little regard for the welfare of his troops. Driven by a passion to possess some of the "blood sweaters," as the Fergana horses were called by the Chinese, in 104 B.C. he sent an envoy carrying much gold, including some valuable golden horse statues, to the kingdom known to the Chinese as Dawan, or Wan, for exchange. The capital, Fergana, was called Haohan. The king of Wan was influenced by his premier and refused to accept the gifts; instead the entire embassy was killed and the treasure confiscated. Refusing to accept defeat, Wu Ti sent another expedition under the charge of General Li Guang in the year 101 B.C. The expedition was supported by 60,000 soldiers, 30,000 horses, and 10,000 cattle. On this occasion a coup d'etat was accomplished, the Wan king, Wugua, was killed, and the surviving aristocrats made peace with Han. Ten "elite" horses and 3,000 horses of average quality were accepted, although only 1,000 of the animals survived to reach the Yumenguan gate-garrison at the Great Wall. An agreement was made that each year the Dawan kingdom would present two elite horses to the Han emperor.

There has been a great deal of speculation as to the true identity of the ancient horse the Chinese called the "blood-sweating" horse. Some have theorized that the horses suffered from insect bites that bled; others have thought the blood sweater to be Appaloosa-colored horses with red markings. In a personal communication from Professor Youchun Chen of Beijing, China, director of the Chinese Academy of Animal Science, it has been confirmed that the ancient blood-sweating horse was none other than the breed known today as Akhal-Teke, the famous Turkmenian horse. Fine examples of this breed have always been called elite horses.

Emperor Wu Ti wanted the blood-sweating horses because he believed this special breed held significance and possession of them was considered a mark of Heaven's grace. The old empire of Dawan covered the area from the western part of Xinjiang Uygur Autonomous Region to Turkmenia, including Fergana. In ancient times, the Dawan horse was slightly heavier in build than the modern-day Akhal-Teke breed. Many were a golden color with metallic sheen, still seen today in many Akhal-Teke horses.

Descendants of the Fergana breed are still known to "sweat blood" today when bred in their ancient homeland. One source reported that the "blood sweating" is due to the extremely thin skin of the Akhal-Teke horses, and results when a blood vessel is broken in the skin during exertion. Louise Firouz in Tehran, Iran, a well-known breeder of both Caspian and Akhal-Teke horses, reported that the "blood sweating" is due to a parasite and not an insect. Bleeding through the skin need not be accompanied by exertion. According to Mrs. Firouz, there is only one place in the world, the general area of Fergana and northern Iran, where this parasite exists in the river water. The guilty rivers are, respectively, the Gorgan and the Fergana. Animals drinking from the river are infected. At a certain point in the cycle of the parasite in late spring, the organism breaks through the horse's skin, causing a small amount of unsightly bleeding. The Firouz horses are given rain water or water from other sources to prevent the infestation. The bleeding phenomenon also occurs in other animals drinking the water—mules, donkeys, and cattle—but not in people. It takes approximately four years for an animal to be cleansed of this parasite once removed from the area.

As the native breed of the area since ancient times is today known as Akhal-Teke, and considering information provided by Professor Youchun Chen in Beijing, it is safe to say that the ancient and revered blood sweater was none other than the direct ancestor of the Akhal-Teke breed. While the ancient Caspian is also native to this area, the small size of the Caspian eliminates all possibility of its having been called the elite blood sweater, as the elite horses were quite tall.

The Turkmenian steppe was one of the earliest areas of horse domestication (preceeded only by the Ukraine) and as horses in this area became taller and swifter than the smaller types found on the steppes, the breed was much sought after and spread over vast distances. It should be considered that the ancient Turkmenian and Caspian horses may have been the first true hotbloods. If one considers the world map and the area of early domestication of the light horse, and the subsequent spread by nomadic tribes as they invaded toward the south to Syria and Jordan into what is now Saudi Arabia then across the north of Africa, it is not difficult to believe that the Turkmenian breed, perhaps with some crosses to Tarpan types and the ancient Caspian, could easily have fathered the Arab and Barb and contributed greatly through them to the Andalusian and other Iberian breeds. According to Sandor Bökönyi (1968), the foundation of eastern stock was formed by the Scythian horses that spread from northern Iran and present southern Russia by Scythian expansion and trade into central Europe, North Africa, and eastward to the Altai Mountains and eventually, with the Yakuts, to the Arctic Ocean. When viewing a herd of purebred Akhal-Teke horses, one is struck by the strong similarity to the Arab breed—yet the Turkmenian is centuries older than the Arab.

This certainly indicates that Turkmenian blood would have been present very early to form a base for development of the North African Barb, believed a descendant in part of the indigenous Iberian horse (Sorraia).

Many authorities believe the Byerley Turk, a foundation sire of the Thoroughbred breed, was indeed a "Turk" and that the Godolphin Barb *was* a Barb and not an Arab horse as has often been stated. The Thoroughbred, although much larger, bears strong resemblance to the Turkmenian in general morphology. Many Turkmenian horses (referred to as Turks) were imported into Great Britain about the time of development of the English Thoroughbred. Additionally, many horses acquired in Syria were termed "Arab" in England because they had been obtained from Arab peoples, when in fact they descended from the Turkmenian breed, widely used in that area. While not firmly established at this time, it is anticipated that genetic blood-group testing now being conducted will eventually reveal to which breed the Thoroughbred is more closely related. Early testing of this kind has revealed little relationship between the Thoroughbred and Arab breeds.

For many centuries the Akhal-Teke breed has not changed in conformation. Throughout its entire history the Akhal-Teke has been carefully bred pure. The Turkmene people accurately kept pedigrees through oral tradition, and selection was always strict and purposeful. Like the Exmoor pony, which resembles the purest type of Celtic (now extinct) or Northern pony, the Akhal-Teke is probably the purest descendant of the ancient Scythian horse. Perfectly preserved horses of this type were found frozen in Siberia (Pazyryk) in ancient burial mounds that are dated from 520 to 212 B.C. (there is some disagreement on the dates). This horse has been bred from time immemorial by nomadic tribes indigenous to the deserts of Turkmenistan. The name *Akhal* means "pure," and *Teke* is the name of the tribe that has bred these horses since earliest recorded time.

These were wary and unsettled people like their ancestors, the Scythians. Horses with the ability to run long distances without fatigue were necessary, as the people lived by raiding other nomadic tribes. A tall, heavy horse could not survive where feed consisted of scant mouthfuls of dry grass or where water was scarce. The Turkmenian horse has always been known as a "meatless" type, very lean and without excess muscle. To condition their horses, the nomads wrapped them in numerous felt blankets to sweat out all surplus fat. They have trained their horses for centuries using peculiar methods that have proven to be effective. Youngsters at about twenty-one months of age are started in training for races. They are worked at a walk and gallop in the morning and evening, and sweating under blankets is usual during the day. This is done to remove all fat from the body. They are worked and fed so that ribs show. The day before a race the horses are fed wheat, and the night prior to a race they are not allowed to sleep but are walked and sometimes given a bit of alfalfa. They are fed a mixture of cultivated fodder consisting of limited but highly nutritive feed such as corn, alfalfa, bread, and animal protein (often boiled chicken) when available, as grass is scarce in the desert areas. Horses for breeding were always selected according to results of races and the quality of their offspring. Mares have often been taken great distances to be mated with the best stallions. As a result of wide use of the best stallions and their sons, various types were developed within the pure breed. The Turkmene breeders refused to use inbreeding except in very special cases. Horses with great speed were always selected.

From the Akhal and Tejen oases in Turkmenia, the breed was introduced to many places. During the eighth to tenth centuries the guards of the khalif of Baghdad were Turkmene horsemen mounted on Turkmenian horses. Persian, Bukhara, and Turkmenian stallions were used extensively in the stables of Russian tsars in the fourteenth to seventeenth centuries. In the fifteenth to nineteenth centuries the Turkmenian penetrated into western Europe and England (see also Oriental).

In the former Soviet Union, the Akhal-Teke is now found in Turkmenia, southern Kazakhstan, and North Caucasus. Many of the breeds found in Turkey, Iran, Iraq, and Afghanistan are directly descended from this ancient breed.

The Akhal-Teke is a slim animal with a fine coat and thin skin. At first glance the horse appears to be made up of angles and is too narrow to be pleasing to the eye of one unaccustomed to this type. Mane and tail hair is often sparse. The head is straight and fine, long in the muzzle and very wedge-shaped, often attached to the neck at an angle rather than a curve; the neck is long and slim, fitting into the shoulders at another angle, and with high withers this horse often appears "ewe-necked"; the back is long and often "soft," or a bit dipped; the croup is usually quite flat but the tail is usually carried low; the chest is narrow and in some individuals not very deep; the shoulders are long and very well sloped; the legs are straight, long, and slim with wonderful bone, well-defined tendons, and hard joints; the feet are well formed, often large, and very hard.

The Akhal-Teke has peculiar, soft gaits, most of them seeming to glide over the ground. This breed has been used for thousands of years in deserts on sandy soil, and this is reflected in its gaits. Horses of other breeds have slightly longer pasterns in the front and shorter, straighter pasterns in the rear. The Akhal-Teke has long, slanted hind pasterns, perhaps an adaptation to walking on sand.

In color, the Akhal-Teke is often a striking golden dun with a metallic sheen. Most breeds closely related to the ancient Turkmenian are found to have the same bright, metallic sheen or glint to their coats. Grey, bay, and black are also seen. White markings are common on the face and legs.

This breed is known to have phenomenal powers of endurance and is an ideal mount in desert conditions. Akhal-Tekes took part in a famous trek from Ashkhabad to Moscow in 1935, a distance of over 1,800 miles including over 300 miles of desert. The latter distance was covered in three days by these extraordinary horses traveling *without water* and carrying riders.

The age of maturity for Akhal-Tekes is five to six years, which is relatively late. They show excellent speed and are used for racing in the former Soviet Union. Their action is distinguished by strength, smoothness, and elegant carriage, which is most appreciated in modern classic events and especially in dressage. The jump is very soft and elastic. Insufficient height (when compared to the larger European warmbloods) prevents this breed from fully matching the requirements of modern competitions; nevertheless, Absent, Muar, and Penteli of the Akhal-Teke breed are well-known names in the world of equestrian sports as winners in international competitions and Olympic games. The unique sporting assets of this breed are employed by crossbreeding with the Trakehner, Hanoverian, and Latvian horses, giving increased size while retaining the desirable qualities of the breed.

The Akhal-Teke combines average fertility with extended longevity. It is not uncommon that stallions are widely used after the age of twenty. The record longevity of thirty years was attained by the mare Elan, who produced seventeen foals in Russia.

Due to intensive agricultural development and poor competitiveness, the Akhal-Teke purebred population has suffered a sharp decline in its homeland. Thus, the gene pool of the breed is strictly limited. Probably the most difficult period in its history was in the last century. After Turkmenia and Russia were joined, the nomadic raids ended and the economic basis for breeding of the pure Akhal horse was finished. The war horse, used for centuries, was no longer needed. The raising of such horses was expensive. Outstanding Akhal-Teke horses went in large numbers to Iran, Turkey, Afghanistan, and India, and were replaced by cheaper working horses. The breed was supported then by only a few enthusiasts.

During 1988, the monthly journal *Konevodstvo*, Horse Breeding and Equestrian Sport, published in Moscow, devoted nearly all issues to the future and past of the Akhal-Teke breed. Many people had expressed concern because the Akhal-Teke was on the brink of extinction. There was a time when, due to the fact that the Akhal-Teke does not match the size of large European warmbloods and Thoroughbreds, the breed lost favor. This is curious when it is known that despite its smaller size, the Akhal horse is very competitive and able. Many purebred Akhal-Teke horses were sent to slaughter. In the homeland, feral herds of Akhal-Teke horses were created when horses were abandoned. By November of 1989, another article in *Konevodstvo* appeared, titled "No Reason to Worry about the Fate of the Akhal-Teke," referring to measures that had been taken to save the breed. A special commission traveled to stud farms to see the conditions of raising and feeding of the horses. The number of pure Akhal-Teke horses increased noticeably from 1980 to 1989.

Another danger to the Akhal-Teke has been the tendency to cross the breed with Thoroughbred to increase the size. These crossings were stopped previously, as the offspring did not reach the quality of the purebred and because it destroyed the breed. Unknown to many, in 1973 there were only eighteen pure mares and three purebred stallions remaining in Russia, due to intense crossbreeding. Since that time and due to some concerned horse breeders, the remnant Akhal horse has been bred strictly in purity. Had it not been for a few concerned people, this ancient and valuable horse would have been lost.

In March of 1988, plans were made to repeat the famous ride from Ashkhabad to Moscow in order to popularize the breed. Horses were chosen from various breeding farms, twenty-seven Akhal-Teke and two Iomud. All of the horses were over six years of age. Riders weighed between 110 and 133 pounds and were aged from seventeen to sixty-four. The ride began on June 1, 1988, and was completed in sixty days. The riders passed through the republics of Turkmenia, Uzbekistan, Kazakhstan, and Russia with sharply differing climatic conditions. The most difficult part of the ride was through the Kara Kum desert, a distance of about 370 miles in scorching heat, and then nearly the same distance across the almost waterless Kazakh steppes. Nearly every day they faced strong wind and sometimes hail. The most important care given to the horses was washing and massaging their legs. Daily feeding consisted of alfalfa hay, barley, wheat bran, and fish oil. The horses moved at a walk and trot, changing back and forth between these gaits. In sixty days the horses traveled more than 1,800 miles. All but one finished the ride. Perhaps, with the increasing popularity of this ancient breed and the realization that it is impossible to reestablish a breed that becomes extinct, those in control will take care to preserve a healthy gene pool of this distinctive and historic breed. The former Soviet Union plans to raise the breeding nucleus to 700 mares. The breed has also begun to spread internationally.

Overcoming what seemed insurmountable difficulties, several Akhal-Teke horses were imported to the United States in recent years by Phil Case of Shenandoah Farm, Staunton, Virginia. Another herd of forty-five animals at Wellington, Colorado, was dispersed on August 4, 1990. As Americans become acquainted with this outstanding and ancient breed, the animals will gain respect and interest. An excellent mount for endurance riding, well able to take heat and exertion with less need of water than most other breeds, the Akhal-Teke is sure to do well in such events.

Albanian
(Myzeqea)

ORIGIN: *Albania*
APTITUDES: *Riding, light draft*
AVERAGE HEIGHT: *12.3 to 14.1 h.h.*
POPULATION STATUS: *Common*

Albanian or Myzeqea mare and foal. Photo: Petrit Dizdari, Ministry of Agriculture, Tirana, Albania

Albanian horses. Photo courtesy Ministry of Agriculture, Tirana, Albania

The Albanian is a very small horse belonging to the Balkan group that includes the small horses of Greece, Bulgaria, Yugoslavia, Romania, and Turkey.

The native horse of Albania has characteristics in common with the horses of neighboring countries, particularly with those of the mountainous regions of the Rodope range and Bosnia-Herzegovina, one of six republics of the former Yugoslavia. The ancient inhabitants of this region were Illyrians, Indo-Europeans who overran the northwest part of the

Balkan peninsula around the fifth century B.C. The Serbians settled here during the seventh century A.D. and were overpowered by the Turks in 1386. During the Ottoman Empire a great deal of Arab blood was infused into the local horses, which were likely various combinations of Tarpan, Turkmenian, and Mongolian stock.

The management of horses is an old passion of the Albanian people, going back many centuries. Albanian horses have always had a good reputation among other Balkan peoples. The Albanian knighthood of Skanderbeg was a rear force for this national hero, terrifying the Osman invaders (Ottoman Empire) and bringing honor and glory to the country. The small local horse of Albania has always been distinguished for its freedom of movement, agility in difficult terrain, resistance to disease, lively temperament, and notable endurance.

Social and economic factors combined with the land and climate have played important roles in determining the numbers of Albanian horses as well as the types seen. In the past horses were used more for transport and riding than for agricultural purposes.

In recent years, measures have been taken to promote and increase the number of Albanian horses and to improve them for agricultural work. The improved breeds are concentrated in large breeding centers such as the Zootechnic Station at Shkodra and at specialized farms where stallions are produced for improving local horses. Since 1980 there has been a great increase in the number of horses in Albania.

There are two types of native Albanian horse, which are referred to as "Mountain" and "Plain," but today the distinction between the two is not as clear as in the past due to a good deal of exchange between the two. Essentially, the Plain type of Myzeqea is distinguished by larger size and greater weight and development. The Plain type stands about 13.2 hands and the Mountain type on average about 12.2 to 12.3 hands.

This small horse is very similar to the Hucul of Romania and other small horses collectively known as Carpathian mountain ponies. The Albanian Myzeqea is very good for long distance use and is exceptionally strong for its size. These horses convert food efficiently and can live on small amounts while exerting great physical effort. They are often used as carriage horses as well as for riding and light draft. Among the horses of the plains, many have a natural ambling gait which is easy on the rider. Turkey also has small horses highly esteemed for their easy ambling gait (see Rahvan).

The goals of Albanian horse breeders today are concentrated on improvement and increase in numbers. Purebred Arab, Haflinger, and Nonius horses are crossed with the native Albanian and several improved types are emerging. The native horse of Albania has few disadvantages of quality, the main need being increased size for heavier agricultural work. The Haflinger breed was imported from Austria to improve the working abilities of horses in the hilly regions of the country. Nonius horses, originating in Hungary, were imported beginning in 1947. They are bred pure at the Zootechnical Station in Shkodra, and each year stallions and mares are distributed to various areas for breed improvement, mainly in the plains areas. Nonius blood improves the growth rate when crossed to local horses. Arab horses represent excellent genetic value to add to the gene pool.

Each year national shows are organized in Albania, where the best individuals of pure breeds and crosses are represented. Recently, shows have been organized for national riding competitions.

Albino

ORIGIN: *United States*
APTITUDES: *Riding Horse*
AVERAGE HEIGHT: *15 h.h.*
POPULATION STATUS: *Common*

While the "albino" is a color and not a breed, the Albino, or white horse, has been given a registry in America.

In days of old, riding a pure white horse was a sign of affluence or high position. White horses were reserved for kings and queens, the bravest knights, Indian chiefs, or

The Albino stallion, Snowman. Photo: Ruth Thompson White

commanding generals and were widely associated with heroes. White horses were described in folklore and portrayed in early art as animals of tremendous endurance and beauty. Painters of the Middle Ages and Renaissance portrayed their subjects on white horses. During the centuries of mounted warfare, the white horse of the commander became a morale builder and rallying point for the troops in time of battle.

El Cid rode a white horse named Babieca into most of his battles. Babieca also carried the embalmed body of El Cid, sitting erect in the saddle, in a last and victorious battle against the Saracens at Valencia. Napoleon owned sixty or more white horses. They were trained to stand steady during the heat of battle (and it is said that this training was necessary as Napoleon was a notoriously poor horseman).

In more recent times the white horse developed a false reputation of being a weak individual with poor eyesight—false because the Albino is not an albino. The true condition of albinism does not exist in the living horse.

When lethal white genes are combined from both sire and dam, the fetus usually dies early in the pregnancy. Live foals may be pure white but they are not true albinos.

The American Albino Horse Association represents horses that are white. The association was started to recognize offspring of a single white stallion named Old King, who had the reputation of producing many pink-skinned, white foals from solid-colored mares.

Old King was purchased by Caleb and Hudson Thompson in 1918 and taken to their ranch in Naper, Nebraska. Although no authentic record had been kept of Old King's ancestry, it was generally assumed that he was of Arab-Morgan ancestry, due to his conformation.

With a herd of Morgan-bred mares, the Thompsons started their venture to produce a breed of all-white horses. They spent the years from 1918 to 1936 building their herd. During this time the horses were not registered with any breed registry. In 1936, Caleb and Ruth Thompson realized the need for a recording system wherein breeding records could be maintained for future guidance in their breeding program. They felt this snow-white strain should have a name synonymous with its most outstanding characteristic, the beautiful white color. The name American Albino was adopted because "albino" is synonymous with "white," and the Latin word *albus* also means white.

In 1937 the American Albino Horse Club was formed at the White Horse Ranch in Naper, Nebraska. The object of the club was to collect, preserve, and verify the pedigrees of the American Albino and to promote interest in the various types of the American Albino.

The American Albino is described as beautiful, intelligent, and good in every instance where the horse is used to entertain the public. It has an excellent disposition and shows a willingness to tackle any job. Generally the white horse exists in nearly all breeds, so many breeds and types are represented in this stud book.

While it is true that a pink-skinned horse may sunburn more easily than a dark-skinned horse, there is no truth in the beliefs that a white horse is a weak individual or that one with blue eyes has poor eyesight. White horses do not always produce white offspring.

Altai
(Altaiskaya)

ORIGIN: *Altai Mountains of Siberia (former Soviet Union)*
APTITUDES: *Riding, packing*
AVERAGE HEIGHT: *13 to 14 h.h.*
POPULATION STATUS: *Common*

An Altai horse of common color pattern. Photo: Nikiphorov Veniamin Maksimovich

Raised in the Altai Mountains for many centuries, this breed is well adapted to its harsh environment. Horses have always been important to the nomads and tribesmen of this mountainous region, requiring horses with a strong heart, lungs, muscles, and tendons along with very hard feet. A surefooted horse is of vital importance, as they must travel over steep mountain trails cut from the rock and cross fast-moving streams and rivers. The Altai horse was developed over a long period and the harsh continental climate created a hardy animal, indispensable to the people who depend on it.

This is one of the most ancient breeds of Siberia. At the beginning of the twentieth century many were crossed with other breeds from Russia—trotters, Dons, and halfbreds. Such crossings were made after the revolution as well as under the Soviet government, and then the crossbred horses were bred "in purity." As a result, the local horse became larger with greater working ability. In the middle of the 1970s a survey was conducted of horse resources in the Altai region and the best stock was collected at breeding farms, numbering 680 mares.

The head of the typical native Altai is average in length, large, and somewhat coarse. The neck is fleshy; the back is long and tends to be slightly dipped; the croup is well developed and muscular; the legs are short and properly set. Occasional defects in conformation include overly sloping pasterns and bowed hocks. The usual colors are chestnut, bay, black, grey, and sometimes *chubary* (leopard spotted).

The Altai exhibits extremely high adaptability to year-round pasture grazing. Altai crossed on other pure breeds produce offspring that show good performance, being larger, more massive, and stronger than the Altai yet retaining their sound health and being undemanding as regards management. Activities are presently under way to develop a new breed by crossing the Altai with the Lithuanian, Russian, and Soviet Heavy Draft horse. While this activity is improving the breed through crossbreeding, a portion of the Altai breed is wisely bred pure to maintain the valuable gene pool.

Breeding work with the Altai is aimed at three goals: the local horse is propagated in purity; good riding horses are being developed for cattle breeders by crossing with the Budyonny breed; and a new meat type is being developed with draft horses and massive Don stallions. Crossbreeding has one poor result in the Altai breed: the loss of unique color patterns such as the Appaloosa. A very low percentage of crossbred Altai horses have the spotted color pattern, which is popular. As a result, a few stud farms working with the Altai breed are specially selecting horses with the Appaloosa color pattern.

In the Altai district most of the horses are crossbred, but in some regions of the Upper Altai there are several thousand aboriginal horses. The breed has retained many of the qualities of its ancient ancestors. At Pazyryk, where kurgans (burial mounds) dated to between the fifth to third centuries B.C. were excavated, horse remains similar in measurement and appearance to the modern Altai horse were found, along with the taller, slimmer, ancient Turkmenian type.

Alter Real

ORIGIN: *Portugal*
APTITUDES: *Riding horse*
AVERAGE HEIGHT: *15.1 to 16.1 h.h.*
POPULATION STATUS: *Uncommon*

Alter Real of Portugal. Photo: Sally Anne Thompson

This breed received its name from Alter do Chão, a small town lying in the Alentejo province of Portugal. This is also the location of the national stud for the Alter Real horse.

In Portuguese, *real* is the word for "royal." The breeding farm was founded by the Braganza royal family to supply suitable high school and carriage horses in Lisbon. The family selected and imported 300 Andalusian mares in 1747. The Alter breed was seriously contaminated during the Napoleonic invasions when Arab, English Thoroughbred, Norman, and Hanoverian blood was introduced with poor results. There was a later attempt to improve the breed with the infusion of Arab blood, but it was not until new Andalusian blood was introduced by government inter-

vention that the breed began to improve and regain its former classic excellence.

The Alter Real breed faced extinction again at the beginning of the twentieth century, when Portugal renounced its monarchy. Records at the Alter stud were nearly all destroyed, and when the government decided to discontinue the stud, stallions were taken away for gelding. At the last moment, Dr. Ruy d'Andrade arrived at Alter in time to save two stallions and a few mares. By 1942, he was able to turn a small but thriving herd over to the Ministry of Agriculture.

The Alter Real is always bay in color. Immensely strong and powerfully built with high breast and broad, muscular back, it has long pasterns and strong hocks that give it an impressive high-stepping action.

An Alter Real stallion named Sublime, taken to Brazil prior to Napoleon's invasion of Portugal, was directly responsible for formation of the Mangalarga Marchador, the most popular and widespread breed in Brazil.

American Bashkir Curly

ORIGIN: *United States*
APTITUDES: *Riding horse*
AVERAGE HEIGHT: *14.3 to 15 h.h.*
POPULATION STATUS: *Common*

American Bashkir Curly stallion, Colonel's Fiesty Fella. Photo courtesy Carolyn Joy

The American Bashkir Curly dates to 1898, when Peter Damele and his father were riding horseback in the Peter Hanson mountain range in the remote high country of central Nevada, near Austin. Peter, who passed away in 1981 at the age of ninety, vividly recalled finding three horses with tight, curly ringlets covering their entire bodies. Since then, curly horses have been found on the Damele range and many Curlies in the United States can be traced to that herd.

The Curly horses have been thought to be descended from the Bashkir, an ancient Russian breed, but the American horses may have been incorrectly named. The Russian breed most often found with a curly coat is the Lokai. Bashkir horses are occasionally found to have a curly coat but this may be due to crossing with the Lokai.

There have been various theories proposed about how the Russian horses appeared ranging in the state of Nevada. Some have theorized that they came across the Bering Strait on the ice with the Mongols in ancient times, but the writer found no evidence to support this. It has been scientifically established that domestic horses did not inhabit the Americas until they were brought by the Spanish conquistadores (see Spanish-American Horse).

It is possible that curly-coated horses were imported while the Russians occupied Alaska and parts of the North American coast as far south as Fort Ross in California (see Kuznet). Some may have been traded to miners or farmers. This could account for feral horses with curly coats being found in Nevada.

Regarding the use of Russian horses in Alaska, a letter from Katrina Foster, park volunteer, United States Department of the Interior, Sitka, Alaska, revealed the following information, partly gleaned from the book, *Imperial Russia in Frontier America*, by James R. Gibson.

Horses were used only in a limited way during the late 1700s and early 1800s during the Russian experimentation with farming. With the inadequate equipment and knowledge of the time, accompanied by severe climatic conditions, farming proved to be a great disappointment to the Russians. Vegetable gardening was more successful.

Stockbreeding was not very successful, with most settlements only able to keep a few cattle, sheep, pigs, and perhaps chickens. In 1817 there were only sixteen horses in Russian America.

Pack horses were used in Siberia for the transportation of goods to Okhotsk on the Siberian coast off the Bering Sea, where the supplies were loaded aboard ships bound for New Archangel (Sitka).

The trip across Siberia was extremely rigorous, and nearly half of the horses died each year from various causes. Anthrax took a heavy toll, and many died from famine and fatigue. The trail was littered with horse carcasses and skeletons. As a result Yakut horses were sometimes scarce, and the Yakut people were reluctant to part with them. The governor of Siberia, Michael Speransky, recommended around 1820 that the Russian American Company use the oceanic route from Cronstadt instead of the Siberian route.

It has been suggested by some that Russian horses in Alaska may have escaped and made their way south to the open range in Nevada. This notion stretches the imagination to the breaking point. Such horses would have run a gauntlet of rugged terrain and predation by wolves, grizzly, and cougar in hundreds of miles of heavily forested country. For many months of the year the area is gripped by deep snow and bitter cold. Thick forest offers little feed or appeal for horses. Several treacherous, wide and swift rivers would have had to be crossed.

It seems that any horses brought across the Bering Strait to America would have been the hardy Yakut and not the Bashkir or Lokai, whose habitation is much further south and west of the Yakut. It is doubtful that even this horse could have reached Nevada.

The mystery surrounding the origin of horses with curly coats in America is an intriguing one. Horses with curly coats have been reported occasionally for many years in South America as well.

The American Bashkir Curly transmits the curly characteristics to offspring approximately half of the time even when mated to horses without the curly coat. Curly horses are hardy and able to survive extreme winter conditions. In the winter of 1951–52, the Curly horses were the only ones to survive without being given supplemental feed on the range in Nevada.

The American Bashkir Curly Registry was established in 1971 in an attempt to preserve this breed. Curly horses may be found in nearly all areas of the country today. One odd feature of Curlies is that they often completely shed out the mane hair and sometimes the tail in the summer, growing it back in the winter. The hair of the mane and tail is fine and silky but often quite kinky. The summer coat is often wavy or rather straight with the curls returning in the winter coat.

Curly horses appear in all common horse colors including Appaloosa and Pinto. A typical Curly is of medium size, resembling the early-day Morgan in conformation. Many individuals have been found without ergots. Some have small, soft chestnuts. The wide-set eyes (characteristic of Oriental horses) are said to give this breed a wider range of vision to the rear. They are alert, have a proud carriage, and most move at a running walk or foxtrot. The hooves are black and hard, almost perfectly round in shape. Curly horses have an exceptionally high concentration of red blood cells, stout, round cannon bones, and straight legs. The knees are flat. They have strong hocks and short, strong backs; the rump is round without a crease; shoulders are powerful and rounded; and the chest is wide and deep. Foals arrive with thick, curly coats, curls inside their ears, and curly eyelashes.

The American Bashkir Curly has a gentle nature and is easy to train. It has excelled in many events, including barrel racing, pole bending, Western riding, gymkhana, hunter, jumper, roping, cutting, English equitation, English pleasure, Western equitation, Western pleasure, gaited pleasure, and competitive and endurance trail riding.

American Indian Horse

ORIGIN: *United States*
APTITUDES: *Riding horse*
AVERAGE HEIGHT: *13 to 15 h.h.*
POPULATION STATUS: *Rare*

The American Indian Horse Registry (AIHR) was established in 1961 for the purpose of

American Indian Horse. Photo: Nancy Falley

collecting, recording, and preserving the pedigrees of American Indian horses.

Horses are registered in five classifications. Class O horses are not bred to conform to popular current standards but to preserve original bloodlines of American Indian tribes. Horses registered since 1979 have pedigrees tracing to various tribes. Horses registered with the Spanish Mustang registry and Southwest Spanish Mustang Association qualify for O classification.

Class AA horses are at least one-half O in breeding. Some Bureau of Land Management (BLM) horses may qualify for AA classification.

Class A horses are those with unknown pedigrees. BLM horses qualify as Class A.

Class M horses include breeding of modern type. Sires or dams may be registered Quarter Horse, Appaloosa, and so forth.

Class P is for ponies of Indian horse type. Eligible ponies include those with Galiceño, POA, Welsh, or Shetland in their pedigrees. Ponies of unknown ancestry may also qualify.

What is an American Indian Horse? This little horse has been known by a multitude of names: cow pony or buffalo horse, mustang or Indian pony, cayuse or Spanish pony. Virtually every color known to the horse appears in this breed.

The history of the American Indian Horse is a long and colorful one. Frederick Remington, famous artist and writer, said of the horse in 1888 ("Horses of the Plains," *Century*

Magazine): "One thing is certain; of all the monuments which the Spaniard has left to glorify his reign in America there will be none more worthy than his horse. The Spaniards' horses may be found today in countless thousands, from the city of the Montezumas to the regions of perpetual snow; they are grafted into our equine wealth and make an important impression on the horse of the country. He has borne the Moor, the Spanish conqueror, the Indian, the mountain man and the vaquero through all the glories of their careers."

For additional information see Spanish-American Horse.

American Miniature stallion, JR's Red Colonel. Photo: J. Burchill

American Miniature

ORIGIN: *United States*
APTITUDES: *Riding, light draft*
AVERAGE HEIGHT: *Up to 34"*
POPULATION STATUS: *Common*

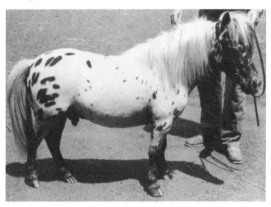

American Miniature (Falabella) stallion. Photo: J. Burchill

American Miniature. Photo: Tom Nebbia

The American Miniature horse is a unique breed. The limiting characteristic is size, which may not exceed thirty-four inches in height at the withers. It must be a sound, well-balanced horse with the proper conformation features common to most of the larger breeds. If there were no size reference, the Miniature horse might well give the illusion of being a full-sized horse. The American Miniature gives the impression of agility, strength, and alert intelligence and comes in all possible horse colors.

This breed descended from many sources. Undeniably, the blood of the English and Dutch mine horses—brought to this country in the nineteenth century for use in some Appalachian coal mines until the 1950s—played a part in development of the breed. Blood of the Shetland pony appears in the pedigrees of many registered Miniatures. In the past decade several breeders have imported Miniatures from England, Holland, Belgium, and West Germany, and the famous Falabella from Argentina is also present. Other breeders have selectively bred Miniatures from larger breeds of horses.

In 1978 the American Miniature Horse Association was formed. It is one of the only registries dealing exclusively with true Miniatures, 34 inches and under. As indicated, ponies over 34 inches are not eligible for registration.

American Miniature horses thrive on attention and display a curiosity and intelligence that make them delightful companions, enjoyed by people of all ages. Those who can no longer mount or handle the 1,000-pound horse do not have to give up their passion for horses; they can simply switch to the smaller animal. Those who have never experienced the special thrill of ownership, yet always wanted to, are finding the "mini" a wonderful opportunity. These little horses have already proven their worth in therapeutic programs for disabled children and adults as well as with the aged. People in high-pressure jobs find them to be a wonderful aid for the relief of stress and a delightful avenue of relaxation.

American Mustang

ORIGIN: *United States*
APTITUDES: *Riding horse*
AVERAGE HEIGHT: *13.2 to 15 h.h.*
POPULATION STATUS: *Common*

American Mustang stallion. Photo: Bernie and Lenore Negri

The American Mustang Association (AMA) was formed in 1962 in San Diego County, California, to collect, record, and preserve the pedigrees of American Mustang horses.

While most other registry offices for the Spanish horse in America register only on bloodlines, the AMA has taken a slightly different approach, registering horses that meet morphological standards in measurements. In this way it is believed that the original characteristics of the horse brought to this continent by the Spanish conquistadores will be preserved. According to the American Mustang Association, the size, conformation, and skeletal structure of a horse of Mustang breeding will not permit that horse to be classified with any other limited group of horses, just as an Arab is not mistaken for a Quarter Horse.

Within the boundaries set by the AMA, three basic types will be found: Andalusian, Barb, and Jennet.

Some Bureau of Land Management feral horses are accepted for registration in the AMA if they meet measurement requirements and exhibit other clearly recognizable Spanish characteristics.

In general the conformation of the American Mustang should be symmetrical, smooth muscled, well proportioned, and endowed with the beauty, vigor, and general characteristics of the breed. Long, smooth muscling is desirable. The head is carried high and proudly; it shows refinement and breeding with clean lines, moderate length, and a wide forehead which tapers to a small muzzle. The nostrils are full and flared when the Mustang is exerting itself. The mouth is firm and the eyes alert and showing intelligence. A straight profile is preferred but a slight tendency to be either concave or convex is acceptable. The neck of the stallion is very crested, set high, and arched and flows smoothly into moderate withers well covered with muscle. Mutton (low) withers or high, sharp withers are equally undesirable. The shoulder is long and sloping. The exceptionally short back and the loins are smoothly muscular, strong, and well-coupled; the girth is deep and well-sprung, the croup is rounded with tail set low. A flat croup with a high tail-set and a goose-rump are equally undesirable. Legs should be straight and sound, fine-boned without coarseness, giving the overall appearance of substance with refinement. Hooves are wider at the base than at the coronet band, round, dense, and in proportional size to the horse. Quarters should be strong but smooth in muscling without excessive mass in the gaskin.

For additional information see Spanish-American Horse.

American Paint Horse

ORIGIN: *United States*
APTITUDES: *Riding horse*
AVERAGE HEIGHT: *15 to 16 h.h.*
POPULATION STATUS: *Common*

American Paint Horse. Photo: Catherine van der Goes

The American Paint Horse is a performance type. It should exhibit intelligence, refinement, and good balance, travel true and straight, and smoothly perform any stock horse maneuver. Breeding of the American Paint Horse is based upon bloodlines of horses registered in the American Paint Horse Association (APHA) and the American Quarter Horse Association and upon bloodlines of Thoroughbred horses registered by the Jockey Club of New York or other registries recognized by the Jockey Club.

Paint horses have been favorites since the days when our ancestors drew colorful animals on the walls of caves. The Egyptians left murals, mosaics, paintings, and sculpture showing the spotted horse. In China it was often represented in memorials that have since been unearthed in burial mounds.

The paint horse can be traced not only in the art of ancient civilizations but in the spotted Oriental horse brought from the plains of Eurasia by the Huns and other barbaric tribes during invasions of the Roman Empire abut 500 A.D. The influence of spotted horses in Mediterranean countries was clearly shown in Spain. Breeders crossed the Oriental paint horse with the horses of their homeland. By 700 A.D. Spain had many multicolored horses with the *overo* and *tobiano* patterns.

Glynn W. Haynes, in *The American Paint Horse*, maintains that paint horses were among the shipments of horses brought to the Americas by the Spanish explorers (see Spanish-American Horse). Descendants of these first paint horses in America were prized by the Indians. Paints were among the early Spanish cow ponies and western stock horses of the eighteenth and nineteenth centuries. The modern American Paint horse received great infusions of Thoroughbred and Quarter Horse blood, resulting in a breed of great versatility.

Paints are a color breed, based upon and regulated by restricted bloodlines. The APHA does not recognize crosses to gaited breeds, pony breeds, or draft horses. Outcrosses are permitted but are limited to Quarter Horse and Thoroughbred.

The two major American Paint Horse color patterns, *overo* and *tobiano*, are distinguished by the location of white on the horse. The *overo* rarely has white extending across the back between the withers and tail. Generally at least one and often all four legs are dark in color. Head markings are often bald, apron, or bonnet-faced. Irregular, scattered, or splashy white markings appear on the body, often called calico patterns. Usually, the tail is one color. The animal may be predominantly white or dark.

The *tobiano* usually has a head marked like a solid-colored horse and may have a blaze, stripe, star, or snip. Usually all four legs are white, at least below the hocks and knees.

Spots are usually regular and distinct, often coming in oval or round patterns that extend down over the neck and chest, giving the appearance of a shield. *Tobianos* usually have the dark color on one or both flanks. They may be predominantly either dark or white.

The American Paint Horse Association offers a wide range of events for competitors, including amateur classes, a large youth program, racing, horse shows, and incentive programs. This large organization is the result of two preexisting bodies, the American Paint Stock Horse Association and the American Paint Quarter Horse Association, which joined in 1965. The APHA is dedicated to the continuance of registration and promotion of the performance type American Paint Horse. The association is entrusted with the recording and maintenance of show and race records.

When the APHA organized in 1965, there were 3,900 American Paint Horses in the registry. By 1984 more than 100,000 animals were registered in the United States and Canada. Today the APHA is also represented in Australia, Denmark, England, Switzerland, Holland, South Africa, Mexico, Guatemala, Ecuador, and Japan.

The American Paint Horse is a solidly built stock horse, compactly made but refined. The head has a straight profile, large eyes, and pricked, average-sized ears; the neck should be muscular and nicely formed; the shoulder is sloped; the back is short and strong; the legs are solid and well muscled with strong joints and tendons.

American Quarter Horse

ORIGIN: *United States*
APTITUDES: *Riding, racing*
AVERAGE HEIGHT: *14.3 to 16 h.h.*
POPULATION STATUS: *Common*

The American Quarter Horse might well be referred to as the most successful horse-breeding venture in the world. The American Quarter Horse Association (AQHA) is the largest horse breed organization on earth and in all of history, with affiliate associations in fifteen countries. More than 2.7 million registered Quarter Horses grace the stud books of the AQHA, and approximately 900,000 members support and participate in AQHA programs. The AQHA approves more than 2,400 horse shows each year, with winners accumulating points toward year-end awards.

The American Quarter Horse's quick burst of speed and tremendous agility make it the ideal choice for barrel racing. Photo: American Quarter Horse Association

The American Quarter Horse is the breed best known for "cow sense," the ability to work cattle. Photo: American Quarter Horse Association

Known as the sprinter of horse racing, the American Quarter Horse provides plenty of racing excitement with close races. The popular 440-yard, quarter-mile race usually is run in less than twenty-two seconds. Photo: American Quarter Horse Association

"The American Quarter Horse" as painted by noted equine artist Orren Mixer is recognized as the breed ideal for American Quarter Horse breeders worldwide. Photo: American Quarter Horse Association.

This "wonder horse" came by its talents honestly, and the breed has proven over the years to be consistent and versatile. Originally a race horse by profession, it branched out as a cow horse, where it still excels, and on to any job a horse may be asked to do, including pleasure riding, show jumping, dressage, and a number of other uses.

The Quarter Horse is one of America's oldest and most popular breeds, tracing its family tree to the horses of the Spanish conquistadores on one side and to the colonial horses brought from England and Europe on the other.

Early stock of this breed were called Chickasaw (see Chickasaw), a small, agile type which the planters obtained from the Chickasaw Indians. Planters brought the animals north out of Florida, known at that time as Spanish Guale. In Spanish Guale there were a number of breeding establishments founded earlier by the Spanish admiral, Pedro Menéndez de Avilés. He entered and named the bay of St. Augustine on August 28, 1565, having with him 11 ships, 2,000 men, 100 horses, 400 pigs, 200 sheep, and 400 lambs. Some of the horses he had obtained from Cadiz; the rest were imported from the royal Spanish ranches in Hispaniola. Although it is difficult to say exactly when the Indians first began their traffic in Spanish horses, the breeding of the Menéndez original stock is considered a matter of record.

It used to be taken for granted that the colonial planters got their Spanish horses from strays and runaways from de Soto's expeditions. Others have supposed Cortés responsible. Spanish Guale, however, is the most likely source. Horses ridden by the Chickasaws were not scrubs, as some have imagined, but were among the very finest horses that have ever, even to this day, set foot on American soil.

Men by the names of Sandy, Wood, and Gookin were the first to concentrate seriously on importing English horses to Virginia, about 1610 to 1620. Soon after, horse racing was legalized by Governor Nicholson and became immediately popular. By 1690, large purses were being offered. Lack of race tracks or even straight stretches of road made it necessary to run short races, frequently on town streets. These races were usually staged over a quarter of a mile and often were matches between two horses. Some of the best early running horses in the colonies were raised in Rhode Island, and many of the best horses in Rhode Island were bred by William Robinson.

His original stallion, Old Snipe, was found running in a herd of wild horses. It is likely that Old Snipe descended from the horses in Andalusia or Cordoba in Spain. The get of Old Snipe were well known for their speed.

When intercolonial match races were held

between horses of Rhode Island and Virginia, the success of the Spanish horses was so great that the Virginians obtained some of Old Snipe's get to improve their own stock. The Spanish blood was being crossed into the Anglo-American horses, and the American Quarter Horse was in the making. (The Virginia horses probably already had some Spanish blood, as there were many Spanish horses running by this time in the backwoods of the colony.)

The English Thoroughbred had no influence on the Quarter Horse in the beginning, as Quarter Horses were well established before the Thoroughbred set foot in the North American colonies. Later, when raced against Quarter Horses, Thoroughbreds were beaten on short distances because of the lightning-speed Quarter Horse start. It surely was a great affront to the owners of the sleek, greyhound-like Thoroughbreds to be left at the post by the short, chunky horses of the colonies.

Imported in 1752 by Mordecai Booth, Janus seems to have been the only Thoroughbred to influence the Quarter Horse at that time. His offspring were noted more for their early speed than for their stamina.

Distance racing in America did not become common until the nineteenth century, and did not reach its peak until about 1850. In the colonies, the "short-horses" were the specialty. The short-horses, so called because of the short races, or quarter-mile races, were later called Quarter Horses.

In the beginning of the nineteenth century, breeding of the Quarter Horse dropped to a low level in favor of the English horse, which could run longer distances. The breed could have died out except that another talent, always present but until then not used a great deal, was discovered. As more and more people came to America and began to move west, the Quarter Horse once again became indispensable. It pulled the wagons, buggies, and plows of the pioneers, carried preachers and their Bibles to the furthermost points of worship, and raced country doctors to the bedsides of their patients. And when the great herds of wild cattle spelled fortune for anyone with the gumption to round them up and drive

them to markets, the Quarter Horse revealed its talent as a cow horse, inherited from its Spanish ancestors, long used to work and fight the formidable bulls of Spain and Portugal.

The Quarter Horse, along with being a superb race horse, excels in work with cattle, and much has been written about its "cow sense." It is fast and agile and can "turn on a dime." It also makes a fine pleasure horse, having an easy gait and kind, easy disposition.

From time to time over the years Thoroughbred blood has been used to refine and upgrade the Quarter Horse. This breed tends to become too "short" in its running ability without some Thoroughbred blood, due to the blood of the chunky, well-muscled colonial horses brought from England, Holland, and some other European countries. In addition to the original Spanish blood of the Chickasaw horses, later Quarter Horses obtained some infusions of draft blood. The addition of Thoroughbred blood refines these qualities and generally improves the breed. After infusion of a good deal of Thoroughbred blood, the stud books were closed to all but those horses whose sires and dams were already registered in the AQHA.

In conformation, the Quarter Horse has a short, fine head with small, alert ears and lively eyes set wide apart. The profile is usually straight. The neck is muscular, well formed, and somewhat arched; withers are well defined and usually prominent. The back is short and straight; the croup long, muscular, and rounded, dropping gently to the haunches; the chest wide and deep; the shoulder long, usually well sloped, and muscular. The legs are solid and well formed with broad, clean joints and very muscular thighs, gaskins, and forearms. The feet are generally good, although often considered too small for a horse of this size and massiveness.

The Quarter Horse has thirteen recognized colors ranging from shades of brown to black, sorrel, and grey. Sorrel is the most common color, and about one-third of all American Quarter Horses are registered as sorrel. Appaloosa and Pinto markings are not allowed, but white markings on the face and legs are common.

American Saddlebred

ORIGIN: *United States*
APTITUDES: *Gaited riding horse*
AVERAGE HEIGHT: *15 to 16 h.h.*
POPULATION STATUS: *Common*

An American Saddlebred three-gaited pleasure horse. Donaldson photo

American Saddlebred shown in fine harness. Donaldson photo

If you had the power of creation and chose to create the perfect all-around saddle horse, where would you begin? First, you would want a horse with beauty and style. It should have easy gaits and a good temperament with willingness to learn and obey. This horse should have substance, quality, and stamina; it should be strong and well built with good legs and a reputation of soundness. Termed the "world's most beautiful horse" by its host of admirers, the American Saddlebred may be considered the "peacock of the horse world." It is also one of the most versatile horses.

Commonly known as the Kentucky Saddler, this breed was developed by the pioneers of Kentucky, who desired a utility horse of style, one that had the qualities set forth above. The American Saddlebred met the challenge to the utmost.

The horse carried pioneers over hazardous trails through the mountains from the coastal country of Virginia and the Carolinas. People arrived riding Morgans, Narragansett Pacers, Canadians, Spanish horses, and trotters, all of which were the foundation stock used in Kentucky to develop the American Saddlebred.

From the Morgans and Canadians, American Saddlebreds inherited substance, compact bodies with short, strong backs, and much ruggedness and stamina. The trotters endowed them with gameness and a speedy gait for driving, which became the popular means of travel with the development of roads. The Narragansett Pacer provided the free and easy natural gaits which made the new breed stand apart from other horses. These qualities made the American Saddlebred the most popular horse for riding and driving with wealthy sportsmen of the day.

The Narragansett Pacers were a breed from the state of Rhode Island. They were small, usually sorrel in color, and distinctive in their habit of pacing or ambling. Horses of this type were much in demand as saddle horses because of their ease of movement, hardiness, and surefootedness.

There is some controversy as to the exact origin of the Narragansett. One theory that is widely accepted is that Governor Robinson of Rhode Island imported a Spanish Jennet stallion which became the progenitor of the pacing horses. Others refute this possibility, maintaining that the Jennet is an Oriental breed and could not have produced *pacing* offspring. One should consider that often, horses traveling at a lateral, four-beat gait have been erroneously called pacers. While the pace is not a comfortable gait to ride, the "amble" is, and possibly the Narragansett Pacer was not a pacer at all. However, the mares used could have been the source of the pacing gait. Another story is that a stallion of unknown parentage swam in from the sea and became the founder of the Narragansetts. It should be taken into account that the Narragansett originated in an area saturated with Canadian blood, and this old breed always possessed lines of horses that paced. In any event, the pacing horses were well liked and well known.

The Narragansetts were not especially flashy and the mood changed for a more stylish horse when roads were developed and driving horses became fashionable. Consequently, the little Narragansetts slowly disappeared, their bloodlines absorbed into the new breeds being developed.

Another breed developed in Canada that contributed greatly to development of the American Saddlebred was the Canadian Pacer. These horses derived from a cross of the French-Canadian mares with Narragansett Pacer stallions obtained in New York and New England. The most famous of the Canadian Pacers was a stallion named Tom Hal, a blue roan foaled around 1806. Imported to Philadelphia from Canada and then taken to Kentucky, he was entered in the American Saddlebred Horse Register and became the founder of the Hal family of Tennessee, the Tom Hal Saddlers of Kentucky, and the Blue Bull family of Indiana. The blood of this horse was also a strong influence in development of the Tennessee Walking Horse and the American Standardbred.

The ideal American Saddlebred is a beautiful animal with much fineness and quality and ample substance to make it capable of being a good horse to ride or drive under any conditions. Average height is from 15 to 16 hands and weight is 1,000 to 1,200 pounds. Colors are bay, chestnut, black, or grey, with an occasional roan. Its manners are gentle and pleasant and it is spirited enough to be stylish under saddle or in harness. Many American Saddlebreds have extreme speed and action, which good trainers develop to make them the

champion three- and five-gaited horses of the show ring. Others have the action required to make fine harness champions.

American Saddlebreds, versatile to a great degree, are popular for driving, riding, and as parade horses or hunters. They also make good cow horses and show horses. During the Civil War they were popular as officers' chargers due to their regal, lofty way of going, powers of endurance, smooth action, and easy gaits.

General John Hunt Morgan used Gaines Denmark and John Dillard in his command during the Civil War, and they were perfect examples of loyalty, stamina, and courage. General Basil W. Duke, one of the most distinguished cavalry officers, rode a Saddlebred while riding with Morgan's Raiders.

When Morgan marched from Sparta, Tennessee, in August of 1862 to surprise a garrison at Gallatin, he made the distance of ninety miles (including detours to conceal his route) in about twenty-five hours. On the Ohio raid, after more than two weeks of severe marching, his command, then about 2,100 strong, marched without stopping from Summansville, Indiana, to a point twenty-eight miles east of Cincinnati, a distance that is fairly estimated as ninety-four miles. This march was made in about twenty-five hours. The greatest percentage of the Kentucky Saddlebred horses that had started on the raid performed the march without flinching, and many of them went on to Buffington, some even bearing their riders across the Ohio River and returning to the Confederacy.

Other horses that had been impressed in Ohio and Indiana failed in such an ordeal, never lasting more than a day or two and often succumbing to exhaustion after a ride of eight to ten hours. This is certainly evidence of the great stamina of the American Saddlebred horse.

Judges of this breed agree a better saddle horse is a beautiful horse with graceful style and intelligent bearing. It must have good-sized, well-formed feet; clean, flat-boned legs; a short back with smooth loin; a compact body deep through the heart and barrel-ribbed close to the hips, which should be well muscled with full quarters; and a high, level croup with a flowing tail coming out high and carried straight. It should have a well-shaped, finely chiseled head with a lean, smooth jaw, large bright eyes set wide apart, and small ears well set upon the head, preferably sharp and dainty and used alertly.

The neck should be medium to long, nicely arched, fitting onto the head correctly, and with a fine, small throat latch. The neck should also fit properly into a well-sloped shoulder. The withers should be prominent and not beefy; this breed should have a wide breast with the legs coming out of the corners, with plenty of width between them. The legs should be set on the feet straight and have true, high, straight, smooth action. The pasterns should be long and sloping with a good deal of spring.

A good saddle horse should have a friendly, gentle nature, amiable to education and responsive to a trainer's commands or signals. For a five-gaited show horse, the animal must possess the animation, gameness, and leg motion to learn to do five gaits properly. The trot should be square and bold, with natural high action, and speed is desirable if done in form. The canter should be slow and rhythmic with rocking-chair smoothness and motion. The rack should be rapid and free from any lateral motion or pacing. It should be a single-foot with the exaggerated, showy action learned from a good trainer that can be recognized by ear even if one is blindfolded. The knee and hock action should be snappy and sufficient for this beautiful gait. The slow gait should be a high, methodical, showy stepping pace, done very slowly, with restrained speed; if done properly it is the most thrilling of the five gaits. The walk should be prompt, in an elastic step, "primpy," with head well set and an alert manner.

In the three-gaited horse the walk, trot, and canter are the three required gaits. These must be done cleanly and distinctly. The trot is the gait most emphasized and must be true, high in action, with excessive motion of knees and hocks. The walk is prompt, showy, and done cheerfully in correct form. A showy walk of about five miles per hour without any dancing or fretting is desired by most judges. The canter should be slow and rhythmic.

Although the original list of foundation sires contained seventeen stallions, this list was gradually reduced, and at a meeting of the association on April 10, 1908, it was decided to list Denmark, by Imported Hedgeford, as the sole Foundation Sire. The others on earlier lists were given numbers and placed on the Noted Deceased Sire List. The designation of Denmark alone as Foundation Sire was made because in number of registered descendants no other sire could compare with him.

Denmark was a brown horse foaled in 1839 and is reported to have been a horse of great beauty, a game, consistent four-mile race horse. His son, Gaines Denmark 61, was the real founder of the Denmark family of Saddlebred horses. He was the product of mating the fine Cockspur pacing mare known as the Stevenson mare to Denmark. Gaines Denmark was a beautiful black stallion with both hind feet white, a horse of wonderful style and finish. Foaled in 1851, he was such an outstanding horse that he was selected by General Hunt Morgan to use in his command in the Civil War.

Gaines Denmark was made famous by his four sons that he produced prior to his entry in the Civil War. He produced little after his return. These were Washington Denmark 64, Diamond Denmark 68, Star Denmark 71, and Sumpter Denmark 65.

The American Saddlebred is a versatile, beautiful horse that can easily meet the needs of almost any owner.

American Shetland

ORIGIN: *United States*
APTITUDES: *Riding, driving pony*
AVERAGE HEIGHT: *Maximum of 46"*
POPULATION STATUS: *Common*

The American Shetland pony was developed from the original Scottish Shetland and crosses to other pony breeds such as registered Hackney and Welsh. This has resulted in a slightly taller, more refined type of pony than the original, which is referred to as the Island type.

The American Shetland Pony Club maintains two stud books: Division A, which is for pure Shetlands, and Division B, which is for ponies from one Division A parent with the other being Hackney, Welsh, or Harness Show Pony. All Division B animals must carry at least 50 percent Shetland blood.

Registered Shetland ponies are judged in most of the nation's leading state fairs and in major horse shows and expositions. Classes are of two types, breeding (shown at halter) and performance (in harness or roadster).

The American type of Shetland is more delicate and refined than the Island type, with a higher head and slimmer lines. Both make excellent children's ponies, but by and large the American public has shown preference for the American type. This does not imply a large pony, but rather a small, well-made, squarely built animal with short ears, wide-apart eyes, small muzzle, and fine neck.

The American Shetland Pony Club also maintains a stud book for Miniature horses. For additional information regarding the history of this breed, see Shetland.

American Walking Pony

ORIGIN: *United States*
APTITUDES: *Riding pony*
AVERAGE HEIGHT: *Maximum of 14 h.h.*
POPULATION STATUS: *Common*

American Walking Pony stallion, BT Golden Splendor. Photo: Joan Brown

American Walking Pony. Photo: Joan Brown

The American Walking Pony Registry was founded in 1968 by Joan Hudson Brown, serving as executive secretary. After several years of experimental breeding, this new breed was developed, the offspring being a large Arab-type pony about 14 hands tall with smooth saddle gaits.

The foundation cross that produced the American Walking Pony was the registered Tennessee Walking Horse and the registered Welsh pony. The Walking Horse contributed the smooth saddle gait and the Welsh pony the beautiful Arab-type head and long, arched neck. Ponies of various bloodlines of this cross are accepted for registration. When a large enough breeding nucleus exists, the stud book will be closed and only ponies with both sire and dam registered will be allowed.

The unique gaits of the American Walking Pony are the Pleasure Walk, the Merry Walk, and the canter. American Walking Ponies also trot. The breed is actually seven-gaited, comparable to Roan Allen, a champion Walking Horse of yesterday who also exhibited and won in both the fine harness division and five-gaited division.

The Walking Pony, a most versatile breed, has good jumping ability inherited from the Welsh and is very successful in open shows as a pony hunter.

The gaits of this pony breed are inherited and the action is light and flowing. They are shown at the Pleasure Walk, a slow, free and easy four-beat gait faster than a regular walk and with some head action; and the Merry Walk, a smooth but faster four-beat gait with slight head motion. The pony should be well collected and flexed at the poll with a good head-set. The canter should be smooth, unhurried, and straight on both leads with moderate collection. The breed is also shown in formal and pleasure driving classes.

The Walking Pony is described as having a neatly chiseled and comparatively small head. The eyes are set wide apart and bold with white (sclera) often showing; the ears are pointed and well shaped; the neck, which is long, arched, and well muscled, is crested in stallions but leaner in mares. The neck is set on high and well back into moderately high withers; the back is short; the shoulder is long and sloping; the ribs are well sprung and the girth deep; the chest is of good width. The croup should be moderate in slope and length with hips that are muscular. The tail is carried straight and in a gay manner. The front legs are straight and rear hocks are either straight or inclined to turn in slightly. The pasterns should have medium length and slope.

Due to the light and flowing gaits of the American Walking Pony, the breed is known as "a dream walking."

The Anadolu, an ancient breed of Turkey, in its working clothes. Photo: Ertuğrul Güleç

Anadolu

(Anadolu Ati)

ORIGIN: *Turkey*
APTITUDES: *Riding, pack horse*
AVERAGE HEIGHT: *12.1 to 13.3 h.h.*
POPULATION STATUS: *Common*

Descended from crosses of Turkoman, Arab, Persian, Karabakh, Akhal-Teke, Kabarda, Deliboz, Mongolian, and the ancient Anatolia horse, the Anatolian breed, better named Anadolu, is itself an ancient breed in Turkey. Some books refer to this small horse as Native Turkish Pony or Turk, but Professors Selâhattin Batu and M. Nurettin Aral made a distinction in types between the horses in Anatolia as Anadolu and East and Southeast Anadolu.

A good example of the Anadolu horse. Photo: Ertuğrul Güleç

The Anadolu horse has been known for one thousand years in Turkey. *Anadolu* can be translated as "Turkey on the Asia." This is the most populous Turkish horse breed.

The Anadolu is a small horse. The head is small and shows refinement. Both convex and concave profiles are found within the breed. The mouth is small; nostrils are open and flexible; withers are somewhat low; the chest is narrow; the croup is sloped.

The Anadolu is a very strong horse with superb endurance and good speed. It is able to live in extremely poor conditions and is found throughout Turkey. There are approximately 930,000 Anadolu horses.

Andalusian
(Purebred Spanish Horse)

ORIGIN: *Spain (Andalusia)*
APTITUDES: *Riding horse*
AVERAGE HEIGHT: *15.1 to 15.3 h.h.*
POPULATION STATUS: *Common*

Andalusian stallion, Seville, Spain. Photo courtesy Asociación de Criadores de Caballos de Pura Raza Española

Capricoso S II, a thirteen-year-old imported Andalusian stallion owned by El Canto Andalusians. Photo courtesy Ann B. Ohrel

Andalusian stallion, Seville, Spain. Photo courtesy Asociación de Criadores de Caballos de Pura Raza Española

Andalusian stallion. Photo courtesy Asociación de Criadores de Caballos de Pura Raza Española

The Andalusian, officially known as the Purebred Spanish Horse, reigned for several centuries throughout the known world as the embodiment of perfection in horseflesh. There is hardly a breed in existence that has not felt the dynamic impact of its influence and been greatly enhanced.

So widespread was the use of this horse that it became known by many different names, resulting in a good deal of confusion. The Andalusian is represented by the names Iberian Saddle Horse, Iberian War Horse, Jennet, Ginete, Lusitano, Alter Real, Carthusian, Spanish Horse, Portuguese, Peninsular, Castilian, Extremeño, Villanos, Zapata, and Zamoranos. The famous Jennet of ancient times is unfortunately extinct. While the word *Jennet* applied to a specific type of Andalusian famous for its smooth, fast, ambling gait, the term *ginete* referred to a style of riding (with shortened stirrup) and only indirectly to the horse ridden. As the Spanish horse was ridden in this manner after the invasion by the Moors, it became widely known as Ginete or Jennet.

Those wishing to know the Andalusian should drop some of the widely spread misconceptions about the breed. The Andalusian did not appear suddenly with the Muslim invasion of the Iberian Peninsula in 711 A.D., when Berber horses were crossed on the native horse of Spain. Often we see history more clearly when it is more distant in time. Archaeology, anthropology, paleontology, and other sciences have rewritten history as new facts have been revealed.

Another misconception is that the Andalusian obtained its convex profile from the North African Barb. An observation of the probable origin of the Barb will dispel this notion and reveal the reverse to be true.

Invasion by the Arabs brought fresh infusions of Oriental (not Arab) blood to the already renowned horses of Andalusia. Prior to the Muslim invasion, however, the Iberian Peninsula had been invaded and occupied by many others, and the indigenous horses of Spain and Portugal had already received infusions of hot eastern blood as well as cold northern blood. Great quantities of Oriental blood were introduced into Spain centuries prior to the birth of Christ. Periods of civilization and/or invasion of the peninsula include those of the Iberians (originally from the north of Africa), peoples of the Alamanni, Basques (province of Navarre), Carthaginians, Celts, Cimbrians, Franks, Greeks, the Moorish invasion of 172–175 A.D., the Muslim invasion of 711 A.D., Ostragoths, Phoenicians, Romans, Suebi, Teutons, Vandals, Visigoths, and perhaps some others (and not in the order given). Each of these civilizations

brought horses that had an influence on the native horses of Spain.

Some claim the Iberians domesticated the "Iberian" horse in prehistoric times, and often the impression is given that this means indigenous horses on the peninsula. It is interesting to note, however, that the Iberian people (Ibero in Spanish) who first inhabited the southern coast of Spain are believed to have migrated from "white Africa." Some migrated about the eleventh century B.C., so it is conceivable that the ancient Iberians took with them into Spain in very early times horses that crossed with the indigenous horse. The true origin of their horses is unknown. Possibly they were descendants of early Equidae migrations from Spain, crossing the once existing land-bridge at Gibraltar; possibly they had already been mixed with horses from the east.

The Hyksos, Semitic Asiatics who gradually infiltrated the area around the Nile Valley, seized power in Lower Egypt in the seventeenth century B.C.—600 years prior to the migration of Iberians into Spain. The Hyksos ruled in Egypt during the fifteenth dynasty (ca. 1674–1567 B.C.). The name Hyksos, given by the Egyptain historian Manetto, was translated by the Jewish historian Josephus (first century A.D.) to mean "king shepherds" or "captive shepherds." It was probably derived from the Egyptain word *hequ-khase*, meaning "rulers of foreign lands." The Hyksos introduced many things to Egypt—until then a "backward" civilization—such as the horse, chariots, the compound bow, improved battle axes, and advanced fortification. They were expelled from Egypt by Ahmose I, who reigned about 1570 to 1546 B.C. It is not difficult to believe that horses spread across the north of Africa once they were established in Egypt. Horses of the Hyksos came from western Asia.

History records that the early Iberian horses improved the existing stock in Spain and that, as early as the eleventh century B.C., the horses of southern Spain were considered better than stock in other parts of the country.

The native or indigenous Iberian horse was the ancient *Equus stenonius*, still represented by a small remnant of the Sorraia breed.

While most historians claim domestication of the horse first took place in the Ukraine, possibly 5,000 B.C., in China about 4,000 B.C., and in Mesopotamia about 2,500 B.C., evidence has been found that could indicate horse domestication as early as 25,000 B.C. (Loch, 1986) on the Iberian Peninsula. Cave paintings were discovered in 1879 in Altamira in northern Spain, and these were linked to previous discoveries in the Dordogne in France. In cave paintings dated approximately 5,000 B.C., in Canforos de Peñarubia in the northwest of Spain, Mesolithic horses are portrayed being led by men and women—and Magdalenian horse paintings dated about 15,000 B.C. are shown with what appear to be rope halters on their heads. Some scientists dispute this claim, suggesting that the artist was only showing the border of the "mealy mouth" and the lateral lines running down the side of the skull, which together give the impression of a halter (personal communication, Anthony A. Dent).

The horses depicted bear an undeniable likeness to the Sorraia breed, with so-called "Barb" head clearly in evidence. It should be noted that drawings of horses being led does not prove horses were *ridden* during that period. Probably they were kept as meat and milk animals long before they were used as beasts of burden in any capacity.

Equus stenonius was one of the types of original horse believed to have inhabited Spain in prehistoric times. Research indicates this horse migrated into North Africa several millennia prior to domestication of the horse. The Arab invasion in 711 A.D. brought horses from the east which crossed with descendants of Iberian horses in North Africa, producing the Barb. When they penetrated Spain it is probably true to say that more original Spanish blood was returned to the peninsula than any other type. The horses of Spain had become quite mixed by that time with many breeds and were heavier than those of North Africa. The Barb horse did not impress its convex profile on the Andalusian; rather, the ancient Iberian was in full possession of a "Barb" head centuries before there was a horse known as Barb. Infusions at various times of European and Oriental blood influ-enced the native horses of Spain, and the Andalusian developed.

The Celtic people, migrating in waves from the eighth to sixth centuries B.C. onward, settled heavily in north and central Spain, penetrated Portugal and Galicia, but left the indigenous Bronze Age Iberian people intact. The name Celtiberian was given to peoples where the two cultures overlapped.

One breed usually overlooked when considering the equine bloods taken into Spain and Portugal is the Camargue horse, highly thought of by Julius Caesar and introduced to the peninsula by the Romans. One interesting characteristic of this ancient breed is the color; all of the foals are born dark, black or bay, and turn grey with maturity—a well-known characteristic in the Lipizzan, a descendant of the Andalusian.

The Goths (Visigoths) from the island of Gotland (in Sweden) also made a contribution to the horses of Spain. About 200 B.C., they migrated through the region now constituting Germany and Poland and into western Russia, settling near the Black Sea for approximately one hundred years. Sacking Rome in 410 A.D., they invaded the Spanish Peninsula in 414 A.D. and ruled until the Muslim conquest. This points toward influence of the Gotland horse and undoubtedly to horses of the Central Asian steppes such as the Turkmenian.

This great melting pot simmered with varying degrees of infusion and influence of different horse breeds and types as they crossed with the various native horses of Spain and Portugal—and there were several. It is a great testimony to the dominant character of one particular native horse, the Sorraia, that throughout all the admixture of foreign blood the emerging result still retained the so-called Barb head, sloping croup, powerful quarters, exceptional action, and extremely calm, kind nature.

When we speak of the original Spanish horses we should acknowledge that none is more "Spanish" than another. Those still exist today in relative purity (excepting the Galician), yet their blending added to the ingredients of the Andalusian. Perhaps the most prevalent and dominant was the Sorraia.

The Asturian from the northwestern province of Asturia in Spain is another ancient equine native that contributed to development of the Andalusian. In the Asturian we may find the origin of the famous Jennet, for the Asturian was (and still is) an ambler. The Galician was the ancient small horse of Galicia, a small Spanish province west of Asturias in the north of Spain. The pure Galician is extinct today but was likely also among the horses that made up the Andalusian. The Garrano is another early Spanish and Portuguese horse, thought to be one ancestor to the Asturian and surely an ingredient of the Andalusian. The Pottok or Basque pony, a breed of great antiquity, is also a native of Spain and a possible part of the Andalusian recipe. It is unrealistic to believe, as many do, that the North African Barb was taken into Spain and crossed to the Sorraia alone to produce the Andalusian. It was a much more involved process than that.

While the Andalusian was greatly admired prior to the Arab invasion in 711 A.D. and used as a war horse by the Romans and others, the horses of the Berbers brought refinement and refreshment to the heavier breed, as well as a new style of riding to the country; *a la jineta* (riding with a shortened stirrup). The inhabitants of Spain, riding their heavier Andalusians in the old style, *a la brida*, and encumbered by their heavy armor, were no match for the fast Berber warriors, who literally ran circles around them.

The addition of Berber blood was frosting on the cake in the recipe that produced the world's greatest horse breed, the Andalusian. From this time forward there would never be a war horse to equal it. Every army desired it. This horse was the perfect combination of desirable characteristics in agility, strength, and beauty and in addition possessed great docility, an obedient nature, and strong loyalty to its master.

Interest was focused on the Andalusian with the decline of chivalry. The breed had been widely distributed as far east as Cypress, during the time when the coldblooded heavy chargers were being bred to their maximum size, and was highly esteemed also in the northern countries. The first Andalusians recorded in England were two black chargers ridden by William the Conqueror at Hastings. The Andalusian was the perfect blend of both cold and Oriental blood: fast, agile, pleasant to ride, and able to carry great weight. This horse was the weapon of the Reconquest against the Arabs and of the conquistadores against the Incas; the prized mount of El Cid and of Pizarro; the charger of the Christian armies of the Danube when confronting the Turks.

The first real error in the breeding of Andalusians seems to have come during the reign of King Philip III (1598–1621). In 1600 he put the royal stud into the hands of a Neapolitan horseman, Juan Jeronimo Tiuti. This man imported slow-moving, huge Norman, Danish, Flemish, and Neapolitan stallions to the stud in Cordoba and crossed them with the light, agile Spanish mares. The resulting catastrophe revealed itself in a short time as the Andalusians lost their speed and refinement, gaining muscle and slow, lethargic movements.

Two types emerged later in the Andalusian breed due to crossbreeding. After the Peninsular Wars, the old Spanish studs suffered a shortage of good breeding stock. For improvement, a Spanish delegation purchased twenty-six Arab stallions, twelve mares, and three foals and put them in the royal stud at Aranjuez. King Alfonso XII imported three Arab stallions from France between 1884 and 1885, and in 1893 a royal order called for the breeding of Arab horses by the state (Loch, 1986). Military involvement occurred as well as that of private breeders. Favor had turned toward an Andalusian with a more "Oriental" shape to the head, and the old Iberian blood was threatened with extinction. For a number of years registration was refused to purebred Andalusian horses displaying the classic Iberian head. The urge to put an Arab head on nearly every breed of horse in the world seems universal. The Arab has a lovely head. But one of the wonderful things about the horse is its great diversity of size, type, and use.

With the loss of the ancient Iberian type with convex profile, agility and the powerful forward-going thrust of the hocks were also sacrificed. Fortunately, a few individuals continued to breed the classic horse and the bloodlines were preserved. The most important remaining breeders were the Carthusian monks (see Carthusian) of Jerez de la Frontera. When the monks broke up their stud, these fine horses passed into the hands of a few famous breeders. Among these were the owners of the famous Miura fighting bulls, and they, along with a few other breeders, have kept the bloodlines pure to the present day. While some contend that the Andalusian is no longer pure in its original form due to the years of Arab infusion, blood-group testing being done at the University of Kentucky has shown none of the genetic markers unique to the Arab, proving that at least some of the lines were kept pure.

It was in the hands of the bull owners that the Andalusian earned its reputation as the greatest stock-working animal in the equine world. The Quarter Horse and other breeds noted for their "cow sense" inherited this ability from their Andalusian ancestors. In the valley of the Guadalquivir River, Spanish cowboys have long used their Andalusian horses in handling the bulls, considered exceedingly temperamental stock. Few horses would feel comfortable working these dangerous animals, yet Andalusians appear to delight in the work. With incredible speed and handiness, they can maneuver an angry bull, dodging in and out and barely missing the hooking horns when the bull charges.

The very best Andalusians are used in a daring spectacle that takes place in the bull ring. A skilled bullfighter on horseback, called a *rejoreador*, fights and kills a *toro bravo* (fierce bull) in a spectacular display which combines intricate high school movements with curving dashes, coming within inches of the dangerous horns. It is here that the obviously superior qualities of the Andalusian as a stock-working horse are readily apparent to all. The natural calm temperament is underscored by the fact that one moment the horse is perilously close to death and the next, he turns to doing intricate high school movements in a perfectly calm state of mind.

The process of eliminating the old type and selecting for the new continued unchecked until 1975; then, slowly, a change began to take place as breeders turned once again to their original, classic horse. Had it not been for the few breeders who had clung to the old style of horse for which Spain had been famous for centuries, none would have survived in purity to the present.

Blood of the Andalusian had a strong influence on almost every breed of the known world in ancient times. Agile and powerful, the Andalusian made the best type of war horse. Besides agility and strength, this horse has always had a regal carriage and high step fit for any king or knight. Far be it from any wealthy knight (for the Andalusian has always been expensive) to be seen plodding along on a bored, low-headed mount!

According to traditional fables, the Spanish horse was bred by Zephyr, the golden or gentle west wind. Also known as Pegasus since ancient times, this horse was the reigning symbol in Olympus of all the contemporary horses of the Punic-Roman world until the last years of the eighteenth century.

The Spanish horse, bred in Andalusia (formerly Roman Betica), made up the Iberian and Celtiberian squadrons of the famous Carthaginian horse troops that carried the Roman army in its conquests throughout the ancient world. Breeds were spawned by the spread of this horse from the lukewarm waters of the Betis River (now called Guadalquivir) to the frozen banks of the Volga. One of the most famous chargers of old was the Friesian breed, developed when Andalusians were taken into Holland during the Crusades and crossed with the native Friesland horses. The Friesian has been bred pure for centuries, yet the Spanish influence is still evident.

Mounted on Spanish Andalusian horses like El Cid's horse Babieca, Christian horsemen battled the Arabs of the Reconquest, and because of its well-earned reputation, Andalusia, the land of their birth, was called by Cervantes, "the origin of the best horses in the world."

This same horse, along with the beloved Spanish Jennet, carried the conquistadores on their forays into the Americas, and both North and South America owe a great deal to Spain for the quality found in American breeds. In addition to the Andalusian and Jennet, the Spaniards brought the Garrano, Sorraia, Asturian, and Galician to the Americas, possibly as pack animals. Without a doubt, these small horses contributed to the various types found throughout South, Central, and North America.

The famous Kladruby stud, created by Maximilian II of Austria, the Imperial de Lipizza stud, and that of Mezöhegyes were founded on Spanish horses and mares. The Kladruby stud has retained the ancient, Baroque head shape of the Iberian, resisting thoughts of "Arabizing" the head of their elegant breed. The truly *old* Spanish blood has vanished almost everywhere else in the world.

The aesthetics of pure Spanish blood stand out because of this breed's incomparable elegance, harmony, and distinction. Spanish blood, once introduced into any other breed, leaves a distinctive stamp that may still be seen centuries later. Famous for their fire and elegance combined with majestic peacefulness, great endurance, and liveliness, Andalusians are a perfect example of controlled power. It was the Andalusian that gave the Lipizzan breed its great strength and high school ability.

The Andalusian possesses excellent muscular action; a finely sculptured head with subconvex profile; short, active ears; vivacious eyes with a quiet and kind expression; elegant, arched neck; and profuse and often wavy mane and tail. It has well-defined withers and strong, sloping back; wide, well-developed chest; rounded quarters with low tail-set; well-sloped shoulders; strong legs with broad joints; and the hoof is well formed and strong.

Grey is the predominant color today, with chestnut next, then bay. In early times all colors were found within the breed, including spotted.

In studying the history of almost every horse breed, blood of the Andalusian will be found and with it, an elegance and quality that are unmistakable. It seems a cruel twist of past allegiance that the breeders of the world, having drawn so heavily on the horse of Spain for many centuries to create and improve native breeds elsewhere, now tend to overlook the Spanish blood inherent in their stock and look only to the Arab for improvement. This is not meant to detract from the high quality of the beautiful Arab, universally used to improve other breeds. It simply is the case that the Spanish horse *was* the original improver of many of those breeds and it is curious that for refreshment breeders should look elsewhere.

Andean
(Andino, Morochuco, Peruvian Criollo)

ORIGIN: *Peru*
APTITUDES: *Riding, pack horse*
AVERAGE HEIGHT: *12 to 13.2 h.h.*
POPULATION STATUS: *Common*

A distinct type of the Peruvian national horse (see Peruvian Paso), the Andean is a small, strong animal. Habitat for this horse is in the Andes at an altitude of over 9,000 feet. As a result, the Andean horse has developed huge lung capacity and a dense coat, products of its environment. The forehead of the Andean often protrudes around the eyes. The Morochuco is a type of Andean horse.

This breed adapted to the harsh environment and high altitude of the Andes and has many characteristics reminiscent of the Tibetan breed. A strong, able climber, the Andean exists and indeed thrives where other breeds would perish. This horse has also learned to distinguish toxic plants, avoiding poisoning or death.

The Andean has a somewhat heavy head with subconvex profile; the neck is short and muscular; the skin and coat are usually heavier than in related Spanish breeds. The color is generally dark, predominantly chestnut brown. The conformation is usually more angular than that of the Peruvian Paso or Costeño. The pasterns are short and straight and the hooves are small, strong, and dark colored with high heels. Like the cousins mentioned, however, the Andean horse possesses the same smooth, lateral gait.

The Andean has great resistance to fatigue and easily climbs steep hills at high altitude.

This horse often carries a pack weighing more than 200 pounds over mountainous regions of over 16,000 feet in elevation, with no ill effects.

Races for the Andean are often held on holidays on the plains of the high country, where visiting people and animals can only walk at a slow pace due to the altitude.

Anglo-Kabarda
(Anglo-Kabardinskaya porodnaya-gruppa)

ORIGIN: *Caucasus (former Soviet Union)*
APTITUDES: *Riding, sport horse*
AVERAGE HEIGHT: *15.2 to 16 h.h.*
POPULATION STATUS: *Common*

This new breed was created by crossing Kabarda (also called Kabardin) mares with Thoroughbred stallions at the Malokarachaevski and Malkin studs. The bay stallions Lestorik (1939) and Lukki (1939) and the dark bay stallion Lok-Sen (1923) were of particular importance in development of the breed. Today there is from 25 to 75 percent Thoroughbred blood in well-bred Anglo-Kabarda horses.

In the new breed it is hoped that the best qualities of both breeds will be combined. Anglo-Kabarda horses are well suited to the climate of the Caucasus, thriving at pasture the year round, and are able to negotiate difficult mountain terrain skillfully. At the same time, they are much larger and faster than the purebred Kabarda, and conformation is more like that of the Thoroughbred.

The Anglo-Kabarda has inherited the Kabarda head with a Roman-nosed profile, the straight back, and a slightly sloping croup. From the Thoroughbred it has inherited comparatively long legs with well-developed joints and a long neck and good set to the head, with a well-sloped shoulder.

Anglo-Kabarda horses participate at national and Olympic events and are used for work under saddle on farms of the northern Caucasus. The total number of the breed is about 6,300 head.

For additional information see Kabarda.

Appaloosa

ORIGIN: *United States*
APTITUDES: *Riding horse*
AVERAGE HEIGHT: *14.2 to 16 h.h.*
POPULATION STATUS: *Common*

Doc's A Rollin, Appaloosa Horse Club Champion owned by James Moxley, Bolingbrook, Illinois

Dreamfinder, owned by Kenard Farms. Photo courtesy Appaloosa Horse Club

Appaloosa. Photo: D. Moors

Appaloosa. Photo: Appaloosa Horse Club

The Appaloosa has a heritage as distinctive and colorful as its coat pattern. While the Appaloosa as a breed originated in the United States, spotted horses have existed since prehistoric times when great ice flows were part of the landscape and people lived in caves. Twenty thousand years ago ancient artists drew detailed images of spotted horses on cave walls in what is now France, perhaps as part of special rites, hoping for good luck on the hunt. The drawings indicate the Appaloosa as one of the oldest identifiable horses known to humanity.

The term "spotted" has been used in generalizations referring to various markings on horses, but the spots of an Appaloosa are unmistakable and unique to the breed. Typical Appaloosa spot patterns are depicted in cave art and the later art of Asia and China. Thus, a horse with an Appaloosa coat pattern is clearly distinguished from horses having pinto or paint markings.

Establishment of the spotted breed is further evidenced in history of other parts of the world. Today's Appaloosa most likely traces its origins to the hardy and fleet horses used by nomadic tribes in the treeless, old steppes of Central Asia some 3,500 years ago. Domestication of the horse changed the lives of these people dramatically. In addition to providing hides and food, the horse allowed them to wander as never before, pulling or carrying their belongings as they sought better living conditions. Horses became widely known in the Mediterranean region because of the forays of these Asian tribes. The spotted horse was the subject of art in Egypt and Greece in 1400 B.C. and of Italy and Austria in 800 B.C.

In ancient Persia, spotted horses were worshipped as the sacred horses of Nisaea, and the great hero of Persian literature, Rustam, rode a spotted horse named Rakush. Rakush was chosen from among thousands of horses brought from the region that is now Iran and Afghanistan. He was known as a great war horse and sire of beautiful spotted foals.

The spotted horse was introduced to China around 100 B.C. Prior to this time the Chinese had found their small mounts no match for the Central Asian horses. As the western horses became coveted, the spotted horse was especially prized and often given in tribute to conquering rulers. Because of their value and beauty, spotted horses became common in Chinese art starting in the seventh century A.D.

Spotted horses have appeared in western Europe throughout recorded history but were specifically bred only in a few places, such as Denmark and Austria. The famous Lipizzan horses often exhibited spots during the six-teenth through eighteenth centuries.

Spotted horses were common among the sturdy Spanish Andalusians of the sixteenth century and spread with Spanish exploration of the New World. As Spanish horses were caught or stolen by American Indians and traded to other tribes, they spread northward; most tribes even of the Pacific Northwest were mounted by around 1710 (see Spanish-American Horse for additional information).

The Nez Perce Indians of the areas now called Washington and Oregon were partic-ularly sophisticated in their use of horses and their mounts were prized and envied by other tribes. Unlike other Indians, the Nez Perce carefully chose their breeding stock. Only the best animals were allowed to reproduce. If a horse was not strong or handsome it was traded away. The result was a group of strong, swift animals that could travel great distances in rugged terrain on sturdy legs and hard hooves.

The horse was an important part of life for the American Indian, and records tell of Ap-paloosas pulling plows as well as running the races so popular with the Nez Perce. No other breed of horse can claim such a close associa-tion with a particular group of American Indians, and no other group of Indians so carefully selected and cultivated horses with specific, desirable characteristics.

Nez Perce horses performed various tasks according to their value. The most precious were ridden in war. The war horse had to have strength, speed, courage, and intelligence. A horse with these qualities that was also highly marked had even greater value. Spots were especially prized as they helped to camou-flage the horse as well as to decorate it.

The spotted mounts of the Indians were referred to by white settlers as "Palouse horses," for the Palouse River runs through the area. Later referred to as "a Palouse," the horse's name eventually was slurred to "Ap-paloosey." Today's spelling was officially adopted by the Appaloosa Horse Club when it was founded in 1938.

During the Nez Perce War of the late 1800s the stamina of the Appaloosa thwarted the U.S. Cavalry for months as the Nez Perce fled over 1,300 miles of rugged, punishing terrain.

The Indians were never defeated. Chief Jo-seph simply stopped running due to the hard-ships suffered by his people, for there were women and young children on this long trek. All were suffering from the winter cold.

Some of the prized horses of the Nez Perce were abandoned, while others were confis-cated by the whites. Hundreds were destroyed by army rifle fire to assure the Indians' cap-ture and eliminate the possibility of their escape on the indefatigable horses. Some of the Spanish mares were bred to heavy horses to produce working farm animals, and the Nez Perce stallions spared by the army were gelded. Almost overnight an entire breed of superior horses was nearly destroyed.

A small group of dedicated breeders joined together in 1938 to protect and promote this distinctive breed. At first some Arab blood was allowed, to bring refinement back into the strain, and later a great deal of registered Quarter Horse blood was used.

Today the Appaloosa Horse Club (ApHC) is one of the largest breed organizations in the United States, with some 20,000 members and nearly half a million registered horses.

The Appaloosa has several characteristic color patterns, known as snowflake, leopard, marble, frost, spotted blanket, and white blanket. The head is small and well shaped, usually with a straight profile; the ears are pointed and the eyes large with obvious white sclera; the neck is long and muscular, often slightly arched; the withers are moderately pronounced; the back is short and straight; the chest is deep; the shoulder is long and sloping; the croup is usually slightly sloping. The legs are solid and well muscled with good bone structure, and the hooves are charac-terized by black and white stripes. The skin of the nose, lips, and genitals is mottled.

AraAppaloosa

ORIGIN: *United States*
APTITUDES: *Riding horse*
AVERAGE HEIGHT: *14 to 15 h.h.*
POPULATION STATUS: *Common*

The AraAppaloosa, while not new in type, is the horse represented by the AraAppaloosa and Foundation Breeder's International, a new horse registry in the United States.

Breeding what they consider to be the original type of Appaloosa horse, the AAFBI supports breeders who choose to incorporate Arab bloodlines into their foundation Appaloosa breeding programs. The AAFBI viewpoint is that the backgrounds of the spotted horse (or Appaloosa) and the Arab have much in common. History records the spotted horse as one of the oldest identifiable distinct breeds, and in reality the horse of Asia Minor was influential in the foundation of the Arab breed. The Arab dates back many centuries, and often early types were particolored, as depicted in Middle Eastern and Egyptian art.

Meriwether Lewis, a well-known explorer and horseman, was one of the first white men to visit the Nez Perce Indian tribe of the northwestern United States and see their premodern Appaloosa horses. In his journal entry dated Saturday, February 15, 1806, he wrote, "Their horses appear to be of an excellent race; they are lofty, elegantly formed, active and durable."

Claude Thompson, founder of the Appaloosa Horse Club (ApHC) in 1938, remembered seeing the beautiful Nez Perce Indian Appaloosas in his youth. He felt that Arab blood was the only way to recreate the true Appaloosa. Arab blood was infused into his Appaloosa breeding program. Indeed, the only outcross originally permitted in the ApHC at that time was the Arab, so many of the foundation Appaloosas had this blood.

The AraAppaloosa is a fine Appaloosa of great quality; one with color, elegance, performance ability, soundness, and stamina. The AraAppaloosa combined the color, personality, and good temperament of the foundation-bred Appaloosa with the refined bloodlines and color patterns of the old Arab strains.

Years ago, the ApHC certified the first 4,932 horses to attain permanent status by production and through inspection as "Foundation Appaloosas." These horses were given an "F" prefix on their registration number. Many of these were refined, excellent examples of well-planned breeding programs and carried a high percentage of Arab blood along with the Appaloosa blood tracing to the horses of the Nez Perce, Palouse, and other Indian tribes.

Many of the foundation horses had distinct characteristics and qualities worthy of passing on. Also, many were genotypes capable of reproducing their color and other attributes several generations down the line. The AAFBI wants to preserve the early foundation lines of the Appaloosa breed.

Arab
(Arabian)

ORIGIN: *Middle East*
APTITUDES: *Riding horse*
AVERAGE HEIGHT: *14 to 15 h.h.*
POPULATION STATUS: *Common*

In writing of horse breeds there is no ground so hazardous as that one treads when speaking of origins of the Arab horse. Many avid lovers of the beautiful Arab would almost rather the breed had no earthly lineage whatsoever—hesitating to admit to parentage any less mysterious than the very wind itself. Unfortunately for those who cling to the fables of Arab origin in the southern parts of the Arabian Peninsula, the modern sciences of archaeology, paleontology, anthropology, and archaeozoology have worked against the old beliefs much like a battering ram against an ancient fortress gate. History is being rewritten daily as the spade of the archaeologist and modern scientific techniques enable us to have a sharper perspective on the past.

Fossil evidence was discovered early in this century that proved beyond any doubt that Equidae first appeared on the North American continent. Prior to this archaeological discovery, many held to the belief that the Arab horse originated and developed independently of all other breeds in the southern part of the Arabian Peninsula, as a distinct subspecies. Some have erroneously gone so far as to call the half-Arab a "hybrid," so completely did they separate the Arabian horse from all others.

The Arab has been claimed to be the first "true" breed, the main root from which *all* others sprang. Try as one will, it is equally difficult to imagine either the Shire or the Shetland emerging from the pure Arab. And one must ask, What, then, can be said of Przewalski's Horse? This clearly ancient species is at the far end of the scale from the Arabian, and it stretches the imagination to think this horse sprang from the Arab–or that the Arab could have sprung from the Przewalski.

Other opinions are based on the belief that the Arab originated in southern Saudi Arabia while another, coldblooded northern horse developed in Siberia, and the mixture of the two in varying degrees caused all other breeds to develop.

An important truth in matters pertaining to the Arab horse is that this breed has been embraced by the entire world as a proven enhancement to most other breeds or types of horses. There are few areas of the globe that have not used Arab blood to refine and upgrade native horses. In this regard the Arabian horse reigns as undisputed king, regard-

International Champion Arab stallion, *Ralvon Elijah. Photo: Elizabeth M. Salmon, Elijah Enterprise

Carol Shows of Vacaville, California, on her Arab mare, Seyn Tara, at the end of their first thirty-mile endurance ride. Photo: Hughes Photography

Princess Alia Al Hussein at the Royal Stables in Jordan with two lovely Arab mares. Photo courtesy Princess Alia Al Hussein of Jordan

The Arabian mare, Sarra, who has represented the Royal Stables of Jordan at shows in the United Kingdom, Switzerland, and Holland, as well as at home. Photo courtesy Princess Alia Al Hussein of Jordan

Mares of the royal stud of Jordan at the winter Arab show on the grounds of the Royal Stables. Photo: Gabrielle Boiselle, West Germany

The unforgettable Arab stallion, *Morafic. Photo courtesy Gleannloch Farms

The Arab stallion, Mashour, of the UMM ARGUB strain. Photo courtesy Princess Alia Al Hussein of Jordan

less of its heritage. People, however, are universally fascinated by the subject of origins; many have spent lifetimes on hands and knees carefully removing fossilized bone, both human and animal, from earth or rock and subjecting it to numerous tests and years of scrutiny and speculation. We are driven to know about origins, not only our own but those of the universe and everything it contains.

The revelations of modern science have exploded old beliefs regarding evolution of the horse, including the Arabian, which is acknowledged as distinctively different yet a legitimate member of the family after all. It shares the heritage of all horses, emerging at first not as a "perfect" horse but as a tiny animal about the size of a fox terrier, with several toes on each foot rather than a hoof and given the unlikely sounding name of *Eohippus*—a relative of the hippopotamus, no less.

It is not the purpose of this volume to go into detail regarding the evolution of the horse. The subject has been dealt with fully by far more knowledgeable authors. But suffice it to say that over a period of millions of years, *Eohippus* became our present friend, *Equus caballus*, the horse. Wave after wave of prehistoric *E. caballus*, fully developed to its present state, migrated over the land-bridge that existed at what is now the Bering Strait into Siberia. The horses remaining in North and South America died out about the same period for an unknown reason, and the species continued its development in Asia and Europe as it spread to the west and south.

Spreading and adapting to various climates, elevations, and food resources, horses developed into several distinct types over time, undoubtedly all differing in varying degrees from the original models that trekked across the ancient land-bridge. Controversy has roared over theories regarding the types of prehistoric horses that developed and the exact number of those types. Many scientists believe there were two main waves of migration from North America, and in those two waves variances perhaps occurred from beginning to end of each. Many authorities believe that four distinct prehistoric horses existed at the time that the land-bridge became covered with water and all Equidae died out in the Americas. The four types, called Pony Type I, Pony Type II, Horse Type III, and Horse Type IV, are roughly described below.

Pony Type I was small and chunky, the ancestor of the Celtic pony and all European ponies. Averaging about 13 hands, it was adapted to a cold, damp climate, possessed a dorsal stripe, and most likely was some shade of brown with a black mane and tail. Pony Type II, resembling Przewalski's Horse or Taki (see Przewalski) and called the Tundra Pony, was small and blocky in build but

slightly larger than Pony Type I, averaging about 13.2 hands. It was adapted to dry cold and inhabited all of Asia north of the Himalayas and all of Europe north of the Alps, perhaps into the Iberian Peninsula. This pony was pale in color, probably sandy dun with a black mane, tail, and legs. It is believed that this pony turned white in winter. Horse Type III, averaging about 15 hands, migrated from North America during a warm climatic phase. This horse had a large head, long neck, long back, long skull with narrow forehead, and a sloped croup, and was the probable prototype of the Turkmenian horse. Horse Type IV, the smallest of the wild races, averaging about 12 hands, had high withers, slender legs, a straight or concave profile, broad forehead, and high tail carriage. This horse could be called the prototype of the modern Arab. Little is known about the colors of the latter two types, but there seems to have been more variety, including golden dun, black, grey, bay, or chestnut. These did not have the dorsal stripe.

Current scientific indications are that all modern horses are descendants of those prehistoric types, some of which began their long journey to the Arabian Peninsula from North America through Siberia and Mongolia.

Various aspects of history relating to the Arab appear in the articles in this volume for the Akhal-Teke, Andalusian, Barb, Caspian, Oriental, Persian Arab, Turkoman, and, collectively, the South American breeds. The Arab has touched the history of nearly every horse breed on earth. So universal is this breed that it is hardly possible to speak of any horse breed without referencing the Arabian.

Every book I have consulted on horse breeds states that origin of the Arab breed was Saudi Arabia. In keeping with the policy for this volume to collect data directly from breed authorities rather than borrowing from other books, Saudi Arabia was contacted. The response came not in a letter but in a phone call from Jeddah. Despite the legends, I was told, the Arab horse did *not* originate on the Arabian Peninsula, but with the "Arab" peoples of Iran (Persia), Iraq, Syria, and Turkey—countries which were all at one time referred to as "Arab" before Saudi Arabia existed as a country. Some of the best Arab horses, I was told, are to be found in Iran and Jordan. Further, I was informed that Saudi Arabia had never had many good Arab horses, and just that week breeding stock was being imported from England. The only wild horses that existed on the African continent or Arabian Peninsula were zebras and asses, both members of the Equidae. Even they are descendents of migrants from Siberia.

The first people to introduce the domestic horse into Africa were the Hyksos (whose name means "ruler of foreign lands"), who came from Anatolia in the seventeenth century B.C., seizing power in Lower Egypt. The earliest skeleton of a horse in Africa was found in the debris of a fortress at Buhen, dated at the seventeenth century B.C.

Solomon, King of Israel, supplied the Egyptians with horses imported from Cilicia (in ancient Anatolia, now Turkey), where he also obtained horses for his army. Greater numbers of horses were introduced into North Africa from Asia in the seventh to eleventh centuries A.D. These horses were all of the Oriental type, coming from the countries then termed "Arab," because they were heavily populated by Arab people.

One of the most misunderstood terms when dealing with history of the horse is the word *Oriental*. Oriental blood does not always imply the Arabian horse—which at the time of emergence of the old Andalusian horse (see Andalusian) did not exist. Peoples called "Arab" have historically lived in regions including those that today comprise Syria, Turkey, Iraq, Iran, Jordan, and more, and Saudi Arabia did not exist as a country until the twentieth century. Nor were there horses on the Arabian Peninsula. Observers of yesteryear whose words have been brushed aside stated that the Arab horse descended from Oriental horses. It is for this reason that the Arab is referred to as an Oriental horse. Usually, because the Oriental horses had been obtained from Arab peoples, by the time they reached Europe they were called Arabs.

A very different—occidental—type of horse was introduced into North Africa from Spain and the islands of the Aegean Sea. From Spain it reached Morocco, Algeria, and Tunisia, while from the Aegean islands it was taken in boats to Libya by the Sea Peoples. This occidental horse differed from the Oriental type in conformation of head and body. The presence of the Iberian type of horse was recorded by the Egyptians about 1230 B.C., among spoils of war taken by Merneptah from western invaders, and also some decades later, when Rameses III captured 180 horses and asses from a Libyan tribe.

Horses were not present in Arabia at the time of the Hyksos invasion of Egypt. While nearly all the peoples of western Asia and northern Africa possessed horses, the Arabs (of the Arabian Peninsula) were found to be camel riders. Shortly before the rise of Mohammed, the Arabs began to breed horses on a very small scale, the scale determined by lack of feed for large herds of horses in their desert homeland. Some of the first breeding stock in the Arabian Peninsula was obtained in raids to Persia (Iran) in the sixth century A.D.

The domesticated horse was introduced into Asia Minor, northern Syria, and Mesopotamia shortly after 2,500 B.C. by invaders or migrants from the north. Prior to this time, the Syrians and Mesopotamians had relied on use of the wild ass known as the onager to pull their solid-wheeled wagons and carts.

The search for evidence of Arab horse origins resulted in more information than was collected on any breed or type in the world, much of it controversial, all of it interesting. A booklet from the Arabian Horse Registry of America and the Arabian Horse Trust, both of Westminster, Colorado, acknowledged recent evidence regarding the Arab breed, stating that there are certain arguments for the ancestral Arabian's having been a wild horse in northern Syria, southern Turkey, and possibly to the east as well. This area, which includes part of the Fertile Crescent comprising part of Iraq and running along the Euphrates and west across Syria and along the coast to Egypt, offered a climate with enough rain and sufficiently mild temperatures to provide an ideal environment for horses.

Enough evidence has been revealed to state flatly that the Arab horse did not develop alone in the desert without selective breeding into a distinct "subspecies." To be a sub-species an animal must be identifiable by a distinctive number of chromosomes. The Arab has the same number of chromosomes as all other domestic horses. Genetic blood-group testing done on Arabian horses, while showing genetic markers unique to the breed (as all relatively pure breeds show unique genetic markers) does not show anything that would indicate a subspecies. While many Arab horses have one less lumbar vertebra than horses of other breeds, several breeds not related to the Arab are often found to have one less lumbar vertebra than the usual six. Przewalski's Horse is found to have only five lumbar vertebrae with a frequency compara-ble to that of the Arab. Studies have revealed that the number of lumbar vertebre varies from five to six in all equine breeds and species—horses, zebras, asses, and Przewal-ski's Horse.

Few would dispute the long-standing purity of the Arab breed. The dispute, if there must be one, should lie in the ancient breeds that contributed to the Arabian horse, for this breed, like all other modern horse breeds, descended from prehistoric ancestors that crossed the Bering Strait.

The Egyptian Agricultural Organization in Cairo reported that the first known pure Arabs in Egypt were "owned by King El Nasir Mohamed, Ruler of Egypt (1300 A.D.), impor-ted from Nejd, El Ishaa, Iraq, Koteif and Bahrain." The first stud book kept in Egypt was begun by the Royal Agricultural Society in 1900. The first stallion to be registered was Saklawi, foaled on November 27, 1886.

Princess Alia Al-Hussein of Jordan kindly sent photographs of some of the outstanding Arab horses bred at the Royal Stables in Am-man. The stud in Jordan is founded on the horses of her great-grandfather, King Ab-dallah. Some he brought from the Hejaz dur-ing the Arab revolt and others he obtained in Jordan when he settled there. The mare lines are based on the original stock, although from time to time a few mares and stallions are imported for an outcross. However, the per-centage of "desert-bred" blood is kept high to maintain the particular characteristics of the royal horses.

"In Arabian horses," the Princess stated, "the mare's strain is the one the foals take, and it is noticeable how strongly the mare lines influence looks and temperament in the off-spring. Of course, the sires pass theirs on too, but one can tell which female line the foals belong to far more easily on the whole." The aim of the Royal Stud is to preserve the Arab horse in the Arab world, and now most Arab countries are awakening to the importance of the horses as an integral part of Islamic and Arab culture and history. Some of the Arab countries have allowed their horses to die out and must now import breeding stock from the West, while the Royal Stables of Jordan have carefully maintained their bloodlines.

As regards the origin of the Arabian horse, Princess Alia said:

There are many theories and it is difficult to speculate with any accuracy. There are engravings in Egypt of the Pharaohs with horses showing Arab characteristics such as the dished face and high tail carriage. Others believe them to have come from the Yemen, or Nejd. Certainly, by the time of the Prophet Mohammed they were valued as a precious part of the Arab household. They are mentioned in the Koran several times, and we are told there that every grain one feeds a horse is counted as a good deed. The Prophet said, "Good fortune is knotted in the horse's forelock until the Day of Judgement, and [the horse's] people [own-ers] are assisted in their care of them, so stroke their forelocks and pray that they be blessed.

Princess Alia related that horses of the Royal Stables are all used for riding, and even those used primarily for breeding are broken to saddle first. They have good tempera-ments, and Arab horses enjoy work.

The reputation some people have given them of being difficult or "high-strung" is more often the result of insensitive handl-ing—the Arab horse does *think* more than most breeds and likes to be treated with re-spect. Intelligent, interested, and usually willing, Arab horses can easily be aroused by abuse or mistreatment. I have known Arabs that seemed nervous and flighty and others that were calm and easily handled—and the difference was always found to be in the handlers and their methods.

The purebred Arab is now found all over the world and the very active breed societies in most countries, affiliated with the World Arabian Horse Organization (WAHO), assure the future of the breed in its pure form. The United States currently possesses more Arab horses than any other country.

In the desert, there have long been different families and strains within the Arab breed. One story relates that there were five original strains from the famous mares of Mohammed: Kuhaylan, Ubayyah, Saglaviyah, Dahmah, and Shuwaymah. Historians differ regarding the origin of the different strains (and the proper spelling of their names) and none of the theories can be verified. In modern breeding there are three main types recognized: Kuhaylan (or Koheilan), the stockier, more thickset type; Saglawi, the refined type; and Muniqi, the racing type. Due to interbreeding of these types it is increasingly difficult to distinguish them in western horses. In the Middle East, strain breeding is still an important factor in the Arab horse (see Persian Arab).

There is no disputing the facts: The Arab is one of the oldest pure breeds in the world; it is the most influential breed, having been used to improve nearly every other horse breed; it is the most widespread horse on earth, with breed societies in nearly every country; it is one of the most beautiful and elegant of all horses. The Arab horse has held the lead worldwide in endurance rides since this sport saw an increase in popularity.

An elegant horse, the Arab may be grey, bay, chestnut, black, or more rarely, roan. White markings on the face and legs are common. The head is small and delicate with a straight or slightly dished (concave) profile, a wide forehead, small muzzle, and large, expressive eyes. The ears are small and curved inward at the tip. The nostrils are flared and flexible and the lips are firm. The neck is long and crested with a clean throat, broad at the base; the withers are pronounced and long; the back is short and straight; the croup is broad and usually quite level; the tail is set high and carried gaily. The chest is muscular, deep, and broad; the shoulder long and sloping; the legs muscular with broad,

strong joints and clearly defined tendons; and the hooves small with very tough horn, wide at the heel. The coat is fine and silky and the skin is invariably black. The mane and tail are full.

Araba

ORIGIN: *Turkey*
APTITUDES: *Riding, carriage use*
AVERAGE HEIGHT: *12 to 13.3 h.h.*
POPULATION STATUS: *Common*

Bred from the Anadolu in Turkey, the Araba is a special type within that breed. The word *araba* means "horse wheel carriage," and this special type has been selectively bred for use with carriages. Araba horses are beautiful and enduring with a lot of speed. This horse is more refined and classy in appearance than the Anadolu.

Horse carriages are much more economical to use than automobiles in Turkey, although their use is forbidden in large cities. In small villages the Araba is very popular. It can easily pull one ton of weight in places where motorized vehicles cannot travel. The breed is economical to feed and has great endurance and strength, especially considering its small size.

Argentine Criollo

ORIGIN: *Argentina*
APTITUDES: *Riding horse*
AVERAGE HEIGHT: *14 h.h.*
POPULATION STATUS: *Common*

The Argentine Criollo is the native horse of Argentina, descending from Barb, Arab, and Andalusian horses of the Spanish conquest. Horses were shipped to the River Plate from Spain when parties went to explore and conquer South America. They brought the toughest, hardiest horses they could, and while it is delightful to speculate and romanticize over the highly prized Andalusian, the fact is that most of the early Spanish horses to set foot in the Americas were more likely mixtures of

A good example of the Argentine Criollo. Photo: Luciano Miguens

Argentine Criollo mare and foal. Photo: Luciano Miguens

Argentine Criollos, dun in color. Photo: Luciano Miguens

Argentine Criollo. Photo: Luciano Miguens

Andalusian, Garrano, and—likely a majority—the Sorraia breed.

Conditions were tough on such a long voyage. On board the small vessels there was almost no room to move about; water supplies often ran out; the horses were weakened by the constant motion of the ship and lack of proper food. Many horses died on those voyages and those reaching land had often declined to the point that they needed time to recover before carrying conquering Spaniards. It is doubtful to me that Spain sent its finest horses on such early excursions, knowing their survival would be in peril.

In the long campaigns and skirmishes with Indians, many horses escaped or were intentionally turned loose. After destruction of Buenos Aires by the Indians, many horses were driven into a wild environment. Natural selection resulted in physical hardiness because of the severe struggle to survive. The weak and unsound died. The survivors became the progenitors of the Criollo breed.

In Argentina the summers are very hot but winters are severely cold. In the early days before the Indians were ousted, the Pampas area was known among white settlers as the "desert," although it does have tall grass. Only an occasional lagoon or sweeping undulation breaks the monotony of the landscape. During seasonal migrations or when drought forced the wild horses to travel hundreds of miles, weaklings perished miserably, to be devoured by armadillos, foxes, birds, or wild dogs. Foals lacking the stamina and speed to keep up with their dams fell behind to be struck down by pumas. Stallions fought to the death for supremacy of the herds. The Criollo was forged in the furnace of adversity and emerged a horse of steel, fit for any task.

Several explorers who ventured far out into the Pampas left records of the tremendous herds of wild horses and cattle that were to be seen there some two centuries after their ancestors had gained their freedom. The wild horses were known as *baguals*, and some writers recorded seeing herds numbering in the thousands.

Many long rides have been taken on Criollo horses—astonishing distances that only the hardiest horse could withstand. The most famous ride in this century was that of A. F. Tschiffely, who, alternately riding and packing two Criollo geldings, rode from Buenos Aires, Argentina, to Washington, D.C., in 1925–28, a distance of some 10,000 miles. He crossed plains, cold, barren mountains, steaming jungles, hot deserts, and pestilent swamps. Both of these geldings made the trip in fine shape, and both lived to an old age back in their Argentina home.

In 1767, an order arrived in Buenos Aires to expel the Jesuits from the Americas. The governor, Francisco de Paula Bucarelli, selected a man named Merlo to take the written news to the viceroy in Lima, Peru, a distance of at least 3,000 miles. An average of seventy-five miles per day was achieved on this ride. Fresh horses were used on the trip, but very long distances had to be traveled between fresh mounts. Starting out from Buenos Aires, the lone messenger headed across the Pampas in a north-northwesterly direction. He traveled through hostile Indian territory; vast, waterless regions; shimmering salt beds; grotesque cactus forests; the rocky desolation of mighty valleys; roaring mountain torrents; and high regions with cold and rarified air. Up into the high regions near the snow line, braving icy blasts so strong that they caused the horse to stagger, he rode across range after range through unmapped passes, 16,000 feet and more above sea level. The rider urged his seemingly tireless mount on and on, from regions covered with blinding snow and down break-neck slopes into dense and steaming tropical jungles to battle with mosquitoes and flies. Merlo delivered the governor's message to the viceroy in Lima forty days after he had set out from Buenos Aires. Thirty-nine years later, Merlo's amazing equestrian feat was almost equaled when an unknown rider covered the same distance in forty-one days.

Another outstanding ride was made by Major Corvalan in 1810. He left Buenos Aires and five days later reached the town of Mendoza, which is situated at the foot of the Andes. This meant he averaged 133 miles per day—and he had only one mount. There are many such stories, all authenticated, of the amazing endurance of the Criollo horse.

Today, the Criollo is mainly a working cow horse. It is also used for pleasure riding and rodeo events as it is easy to handle, agile, and quite fast.

For many years endurance rides have been held in Argentina to prove and even increase the remarkable endurance of the Criollo horse. These are not ordinary trail rides, but test the real stamina of each horse. The rides, called "raids," are organized by the Criollo Breeders Association and are open only to purebred Criollo horses. The primary purpose of the rides is to select animals for breeding purposes—ones that will pass on their unusual stamina to their offspring. The distance of the ride is 750 kilometers (about 465 miles) and the maximum time allowed to complete it is seventy-five hours; minimum is fifty-six hours. The ride is completed in fourteen daily stints, with the distances per day being shorter the first few days to assist horses that have been *idle*. The shorter distance is eighty kilometers, and is covered in four hours. The only feed the horses are allowed during these fourteen days is the grass they snatch along the roadside. They are checked by a veterinarian at the end of each day and any doubt regarding their fitness causes elimination.

The Criollo is a hardworking horse that thrives on the grass it can find without supplemental feeding. Even under hard work, this horse seems to bloom and grow healthier. The bone structure is said to be made like steel, and problems in the legs and feet are extremely uncommon.

While the purebred Criollo horse is not used for polo, the cross of this horse to the English Thoroughbred has made Argentina famous as a breeder of outstanding polo horses.

The Criollo is found in a great variety of colors, with the only colors excluded for registration being *pintado* (painted) and *tobiano* (see American Paint). These colors do exist in the Criollo. Many Argentine breeders favor the *gateado* (dun, translating as "cat colored"), which is said to be the color of the very toughest horses. A great majority of Criollo horses are either dun or *lobuno* (grullo), with dorsal stripe and zebra stripes on the legs.

This is a sturdy, thickset horse with heavy muscling. The neck is short and strong, the croup is sloped in Barb fashion, and the tail is carried close to the buttocks whether at work or rest. The mane and tail are thick. The head of the modern Criollo has either a straight or a slightly concave profile, whereas some twenty years ago this horse was always found to have the typical convex-shaped Barb head. The Criollo breed bears a considerable resemblance to the Sorraia breed of Portugal and Spain.

It is difficult to imagine a horse with more endurance than the famous Criollo breed. For additional information see Spanish-American Horse.

Argentine Polo Pony. Photo: Evelyn Simak

Argentine Polo Pony. Photo: Evelyn Simak

Argentine Polo Pony

ORIGIN: *Argentina*
APTITUDES: *Riding, sport horse*
AVERAGE HEIGHT: *14.2 to 15 h.h.*
POPULATION STATUS: *Common*

While the Argentine Polo Pony is not technically a breed, and polo is played all over the world on many types of horses, Argentina has been recognized as the best breeder of polo ponies on earth due to the blood of the hardy, strong Criollo horse. Although the term "pony" is used when referring to polo horses, these do not possess pony blood. The traditional game of polo was originally played on ponies or small horses.

Polo is an ancient game, originating in the Orient over 2,000 years ago. The earliest references to it are found in writings concerning Alexander the Great and Darius, King of Persia. It is thought that the game originated in Persia but was played in various forms throughout the East, from China and Mongolia to Japan.

Muslim invaders from the northwest and Chinese from the northeast took the game to India, and in the middle of the nineteenth century the British saw it and introduced it to England. In 1876 the height limit in India for polo ponies was set at 13.2 hands high, and in

England at 14 hands. In 1919 the height limit was abolished. Manipuri "ponies" in India were the first to be used extensively in the game of polo. Many small Thoroughbreds are used for polo, but it has been the crossbreds that have proven the best in competition.

In Argentina, realizing that the Criollo horse possessed the vital qualities of toughness, calmness, and stamina for polo but lacked the speed of the Thoroughbred, breeders set out to produce the ideal polo horse. Since 1900, the Argentines have continued to upgrade their native stock while importing good Thoroughbred stallions. By 1930 Argentina had a polo pony as attractive as any other, but tougher and with more bone.

A polo pony must be fast, able to gallop, turn, and stop very quickly. The horse must be bold and athletic to the highest degree. The long neck, good shoulders, short, strong back, depth of girth, and exceptionally strong quarters are essential. Argentine Polo Ponies are bred for these characteristics. Additionally, they must be sensitive and responsive to the rider.

There is hardly a horse with more stamina than the Criollo breed, and the strong bone and joints it has contributed to the Argentine Polo Pony have produced an animal not only fast and agile, but one whose legs will stand up under the strenuous activity needed for the game.

The high quality of the breed is carefully monitored by Argentine breeders. Only horses that have proven their ability on the polo field are used for breeding.

Assateague/Chincoteague

ORIGIN: *United States*
APTITUDES: *Riding pony*
AVERAGE HEIGHT: *12 to 14 h.h.*
POPULATION STATUS: *Rare*

The Assateague and Chincoteague ponies are combined in this volume as they are essentially the same. They inhabit the island of Assateague off the coast of Virginia and Maryland, and are divided only by a fence which is an extension of the border between

A sign in Assateague State Park. Photo: Kevin Farley, park ranger

Assateague stallion. Photo: Kevin Farley, park ranger

Ponies on Assateague Island. Photo courtesy Assateague Island National Seashore

Assateague Island stallion and mare. Photo by the author

the two states. Each herd is managed by the state on the side of the fence on which it lives.

While many enjoy the romantic idea that these ponies, or small horses, are descendants of Spanish horses surviving from shipwreck off the coast during early exploration of the Americas, a serious study of this possibility may dispel the notion (see Sable Island). Assateague Island National Seashore provided material regarding the famous ponies of Assateague Island.

One popular tradition is that the first settlers found large numbers of wild horses in the meadows of Assateague Island, the parent stock having come ashore from a shipwrecked Spanish galleon. There is no documented foundation for this theory. Had there been horses on the island, some record would have been left by the first settlers. When the island was granted to one of the colonists in 1670, no mention of horses was made. When Colonel Henry Norwood was shipwrecked on the nearby coast in 1649, he spent some time in the neighborhood as a guest of a hospitable Indian chieftain named Kickotanke. He wrote of the presence of large numbers of hogs in the marshes near Chincoteague but made no mention of horses.

It has also been claimed that the wild ponies of Assateague were descended from horses left by pirates, but this is improbable. Dr. Philip H. Bruce in his *Economic History of Virginia in the 17th Century* pointed out that the number of horses in the colony in 1631 was very small, and prior to 1649 references in the records of Virginia to horses are rare.

In an effort to increase the number of horses, the Quarter Court convened at Jamestown in March 1639 and granted Thomas Stegge and Jeremy Blackman the right to import horses to the colony. A few years later the Assembly passed laws to encourage their further importation. In 1649 there were only 300 horses in the colony, but by 1669 so many had been imported and natural increase had been so rapid that horses became a burden. Consequently, their further importation was prohibited.

After a tax on horses was imposed in 1669, requiring owners to pen them between July 20 and October 20, some of the planters on the peninsula transported stock to the nearby barrier islands to avoid the expense of fencing or taxes. Thus, the horses of Assateague and Chincoteague came to be there.

Some have said the coarse provender of the salt marshes and continual exposure to the elements stunted the growth of the horses trapped on the island. In the 1660s, however, *all* of the equine stock in the colonies was of small stature, and attempts were constantly being made to increase their size. From all reports, the horses taken to the island were not "ponies," but small horses, evidenced by a longer cannon bone than is usually found on true ponies.

By 1671 "horses, geldings, mares and colts" had increased to the point that a great deal of destruction was done to cornfields, and importation of horses to the province was forbidden by law for three years.

Unwanted horses of Somerset County were turned loose in large numbers to range over the island marches. The unpatented outlying beaches belonged to the county, and as such they were leased to the occupants of bayside farms to be "held by the Manor of Somerset." These beaches were long used as a sort of pasture commons, to which livestock was transported from the mainland on barges or scows; occasionally in narrow places, such as the Sinepuxent Neck, and during low waters, wading was possible. Domestic animals permitted to run at large were generally branded for identification by their owners.

Origin of the equine population of Assateague Island seems quite clear in view of the above documented history. The Maryland herd is managed by the National Park Service. The Virginia herd is owned by the Chincoteague Volunteer Fire Company and allowed by permit to graze on Chincoteague National Wildlife Refuge.

The Chincoteague ponies were made famous by Marguerite Henry's book for children, *Misty of Chincoteague*, and the ponies have been a major attraction for several years. Children and their parents travel to the island to see firsthand what they have read about. The ponies roam wild and free and are a year-round delight.

Each year during the last week of July, the

ponies are rounded up for the famous swim to Chincoteague for "Pony Penning," one of the best-publicized carnivals on the Atlantic Seaboard. On Tuesday, riders enter the Chincoteague National Wildlife Refuge on saddle horses and round up the ponies. On Wednesday, the ponies are swum across to Chincoteague and spend the night in a pound in the center of the carnival grounds. On Thursday, some of the ponies are auctioned off to the highest bidders and the proceeds are used to help finance the activities of the Chincoteague Volunteer Fire Company. On Friday, the old stock and ponies that are not sold are returned to Assateague. This annual practice helps to keep the pony population at a level where they can survive without facing starvation. This event has occurred so many times that it is not difficult to convince the ponies to return to the island. They plunge enthusiastically into the water, anxious to be free once again. Pony Penning had its origin in colonial days when livestock owners were required to mark their colts in the presence of neighbors.

In 1989 I visited Assateague Island to photograph the ponies, and it was a most interesting venture. Though feral, the ponies are accustomed to seeing tourists, for many people use the beaches. Calmly maintaining their independence among the swarms of visitors, the ponies often walk past people within arm's reach, seeking out tasty morsels of tender grass, and behave as though they are totally unaware there is a human in the area. I could have run my hand along the back of a pinto stallion as he walked past following a mare and foal, and do not ever recall feeling so totally ignored. He seemed resolved not even to glance at me, and the mare with her foal exhibited the same behavior.

It was also interesting, having visited many U.S. state and national parks that have signs warning against feeding bears or other animals, to see signs saying: "Wild ponies bite and kick. Keep your distance. Do not feed."

In recent years, some Chincoteague ponies sold at public auction have been returned to the refuge when severe Atlantic storms or other situations reduced the population on the island. At least one Spanish Mustang stallion was donated to the herd, and possibly Shetland blood has also found its way to Assateague when pony populations have been threatened. Thus, today, the Chincoteague and Assateague ponies are not the pure descendants of the early colonial horses, as they once were. The ponies viewed while visiting the island seemed more like small horses, being longer in the leg and not having a typical squarish pony head.

Genetic blood testing done by the University of Kentucky revealed that the horses of Assateague and Chincoteague are related to common American stock and do not show more Spanish descent than is normally seen in many American breeds.

The Assateague Island pony is independent and hardy, taking care of itself and well adapted to its island home. There are approximately 300 ponies, 150 to each herd, and as described above numbers are maintained at this level to assure ecological balance on the island. The ponies have divided themselves into small groups, many accompanied by a herd stallion who jealously keeps his band apart from the others.

The diet of Assateague Island ponies consists mostly (80 percent) of coarse salt marsh cordgrass, supplemented by American beach grass, thorny greenbrier stems, rose hips, bayberry twigs, seaweed, and even poison ivy. Because they eat so much salty food, the ponies drink twice as much fresh water as most domestic horses and have even been seen to take a sip of salt water.

Each "harem" consists of a stallion, his mares, and their offspring. The size of each individual band is determined by the ability of the herd stallion to hold his mares together. Spectacular displays of rearing, biting, and kicking occur when stallions battle over their mares. Once dominance is established, the stallion controls his band with typical body postures and vocal signals. Young stock (both colts and fillies) are usually forced out of the band as yearlings by the herd sire. This is nature's way of preventing inbreeding in wild or feral herds. The youngsters band together until about three years of age before forming their own groups. Those who believe there is rampant inbreeding in feral herds should spend some time observing horses left completely to their own devices, as this behavior is true of all horses left to nature. It is also true that half-siblings may end up in the same group at a later time, but usually the bachelor groups break up and go their separate ways.

Assateague ponies are found in all common colors, with many pintos among them. The head is small with a refined look and usually a small muzzle, the eyes are widely spaced, and the ears are small; morphology is generally light; the withers are prominent; the croup is usually sloped; the shoulder generally also has a good slope. The legs are strong, although slim, with good joints and hard feet.

A most reliable and expert breeder, nature has carefully eliminated weak individuals from the Assateague herds, leaving in their wake small, tough animals, which, when tamed and broken to ride, give many years of usefulness and joy to children who have a friend of Chincoteague's famous Misty.

An Asturian mare and foal. Photo: Ontañon

Asturian
(Asturcon)

ORIGIN:	*Northern Spain (Asturias and Galicia)*
APTITUDES:	*Riding, packing*
AVERAGE HEIGHT:	*11.2 to 12.2 h.h.*
POPULATION STATUS:	*Rare*

Centuries ago the existence of a small horse breed originating in the northwest of Spain was recorded. The Romans referred to these horses as *asturcones* and thought well of them—and they were popular with the French during the Middle Ages. Pliny (23–79 A.D.) described them as a small breed that did not trot, but moved in an easy gait by alternately moving both legs on one side.

The ambling gait was natural for this small horse, and done in such a way that it gave a

Asturian horses in the north of Spain. Photo: Onieva

comfortable ride. As a result, they became popular as ladies' mounts. Known as *palfreys* in England, they were called *haubini* in France, a word which later became *hobbye* and eventually *hobby horse*. Much of this blood was taken to Ireland, where the "Irish hobby" was greatly admired.

It is thought by some that the Asturian developed as a cross between the Garrano pony of northern Portugal and Spain—a direct descendant of the Celtic pony—and the Sorraia, the original saddle horse of Iberia, which gave the breed its calm temperament. Some other blood must have been present in the Asturian's lineage, however, because the ambling gait is not present in either the Sorraia or Garrano. Suspected by the author is a stronger and more direct link to the ancient Celtic pony, of which some strains at least must have been amblers. The Celtic pony was taken to Iceland centuries ago, and the Icelandic Horse has always traveled at the amble, or tølt. There is a narrow but clear trail of ambling horses to be found in Turkey, China, Mongolia, and Siberia, tracing the route of the prehistoric horse to the now submerged land-bridge at the Bering Straits.

The Asturian is found high in the Asturian mountains such as the Sierra de Sueve, but the largest and most important group is now concentrated in the western part of Asturia.

Living in a feral state for the most part, under difficult conditions, the breed was facing extinction, but recently breed associations have been formed to protect it. Most commonly, the Asturian is black or bay with no white markings. At one time the breed was combined with its close relative, the Galician (see Andalusian), but presently that breed is nearly extinct in its pure form. The inhabitants of Spain's Pontevedra district round the ponies up at certain times of the year for branding and culling.

The Asturian has a small although sometimes rather heavy head, with a straight profile, small ears, and large eyes; the neck is long and quite thin with a flowing mane; the withers are moderately high; the back straight and strong; the croup is sloping with a low tail-set; the shoulder is well sloped. The feet of this pony are well shaped and very tough.

History does not seem to record much about this little horse, so used by the Romans, English, Irish, and other peoples, but it is obvious that its blood was spread far beyond its Spanish home and no doubt had a strong influence on many other breeds, for which it receives no credit.

The Spanish Jennet, a favorite horse of old Spain, was a confirmed ambler. It is probable that the Jennet carried blood from the Asturian horse, and this was the source of its gait.

Australian Pony

ORIGIN: *Australia*
APTITUDES: *Riding pony*
AVERAGE HEIGHT: *11 to 14 h.h.*
POPULATION STATUS: *Common*

One of the most beautiful and versatile pony breeds in the world developed in Australia. This unique pony evolved from a somewhat complex beginning, drawing on the most suitable characteristics of several pony breeds for over a century.

History of the breed dates to the early 1800s. Australia had no indigenous equine breeds, and early settlers to this vast land brought horses and ponies to work the land and provide transportation. When the ship, *First Fleet*, arrived at Sydney Cove in 1788, horses were among the livestock aboard.

Early in the twentieth century many good pony types were being produced as a result of crossbreeding the various horses and ponies that had been imported. Main contributors to the breeding of Australian ponies during the early years were English Thoroughbred, Hackney and Hackney pony, Welsh Mountain and Cob type, Arab, Timor, Hungarian, and Exmoor ponies. Additional imports arrived in the early 1900s, and these, crossed with the Australian-bred stock, produced many ponies worthy of a stud book.

In 1931 the Australian Pony Stud Book Society was formed to record pedigrees and to establish a standard of type. Initially, the

Supreme Champion Australian Pony. Photo: Marion Costello

Champion Australian Pony stallion, Moorooduc Park Sebastion. Photo: Teare

Australian Pony in dressage. Photo: Marion Costello

and the one mare in the reference section were imported to Australia during the late 1800s and early 1900s. The only Australian-bred pony included in the reference section was Tam O'Shanter, foaled in 1882, by the imported Exmoor pony Sir Thomas, one of only two purebred Exmoors to come to Australia, the other being Dennington Court.

Imports of Timor ponies date to early in the 1800s, with the first pony landing at Sydney Cove in 1803. A large draft of Timor ponies landed at Port Essington in the Northern Territory in 1849, and over the years many of the breed ran wild and found their way down the West Australian coastline. Undoubtedly, descendants of those are included in the mixtures found in the feral Brumby.

The early imports of Exmoor and Timor ponies had a strong influence on the breeding of the Australian Pony, as did the imported Hungarian pony sire, Bonnie Charlie, who, with some mares, was reportedly brought to Australia with a circus in the mid-1800s. Several studs were formed on the bloodlines of these three pony breeds prior to the founding of the A.P.S.B. Society, and their ancestors were recorded in the early volumes of the stud book.

Today all states in Australia have a flourishing branch of the Australian Pony Stud Book Society, with eleven pony breeds registered: Australian, Shetland, Hackney, Connemara, Dartmoor, Highland, New Forest, and Welsh A, B, C, and D. Since the beginning of the stud book, over 27,000 ponies have been registered.

Coat color of the Australian Pony is usually grey, although any color is admitted with the exception of piebald or skewbald.

The standard of excellence of the Australian Pony is as follows: Head should show quality, with alert, well-proportioned ears, flat forehead, dark, well-filled eyes, and open nostrils. The head should be well set into a well-defined gullet. The neck should be slightly crested with good length and no sign of coarseness. Shoulders should slope back to well-defined withers with no trace of heaviness or coarseness and the chest should be neither too narrow nor too wide. The back should be strong and the loins well coupled,

stud book contained three sections: Shetland, Hackney Pony, and Australian. The Australian section incorporated all of the imported British mountain and moorland pony breeds, along with the Australian-bred stock. In 1950 a section was included in the stud book for Welsh Mountain ponies, and later volumes included sections for other imported pony breeds such as Welsh B, Connemara, and New Forest. Entry into the Australian section was upon inspection for type, which remained the case until 1960. At that time, the section was closed to all but stock from registered sires and dams. The height limit for the Australian pony was set at 14 hands.

A reference section for the pedigrees of prominent ponies used prior to the founding of the stud book in 1931 was included in volume one. All but one of the eight stallions girth deep and well ribbed. Hindquarters should be well rounded and show good proportionate length of croup. Tail should be well set on and gaily carried. Leg bones should be quite flat, showing strength without coarseness; joints should be well shaped and proportionate to the pony; cannon bones should be short and straight, and the pasterns of moderate slope with proportionate length. Hooves should be strong, neat, and well shaped. In gait, action should be smooth, showing free flexion of joints without exaggeration; stride should be of good length, straight and true. The pony should have good presence and should show agility, character, alertness, and good ground coverage. The feet should be well placed, standing square and true.

Australian Stock Horse

ORIGIN: *Australia*
APTITUDES: *Riding horse*
AVERAGE HEIGHT: *14.2 to 16 h.h.*
POPULATION STATUS: *Common*

Australian Stock Horse. Photo: Mrs. G. Kay Ritchie

Australian Stock Horse. Photo: Mrs. G. Kay Ritchie

Australian Stock Horse in the campdrafting event. Photo courtesy Mike Rushbrook

History of the Australian Stock Horse cannot be separated from that of the Waler, for in the beginning they were one and the same (see Waler). The Australian Stock Horse Society was formed in 1971 to preserve the Australian working horse.

After the First World War, despite the recognition Australian horses had earned and although the Waler was known as a distinctive type, there was no stud book or registry. Mechanization of primary industries reduced the need for working horses and it was not until the 1960s that interest in horses was revived in Australia. Additional leisure time led to increasing use of the horse for recreation.

At the 1971 Sydney Royal Show, Alex Braid of Wellington, New South Wales, and Bert Griffith of Scone gathered a group of enthusiastic people together to discuss formation of a society. In June 1971, about 100 people met at Tamworth to launch the new breed society, which at last gave the home-bred horses recognition and a formal organization.

Today we live in a mechanized world, but there are still tasks on the farm best performed on horseback. With increasing leisure time, people have come more and more to appreciate that those attributes making the Australian Stock Horse so outstanding at work also make it a fine mount for sports and general pleasure riding.

The list of champion show jumpers in Australia is dominated by Australian Stock Horses. The fastest and best polo and polocross "ponies" are invariably of the Australian Stock Horse strain.

Nowhere, however, is this brilliant horse seen to better advantage than in the hands of an able rider in that special Australian event, campdrafting. This sport calls for agility, speed, and surefootedness as well as intelligence on the part of the horse.

Campdrafting is a sport where all the attributes so much appreciated in the breed are best illustrated. Nearly every weekend in New South Wales and Queensland (and increasingly in other states), spectators thrill to the sight of a horse and rider cutting a steer out of a "camp," holding him for a few seconds, and then running him around a course consisting of two pegs so that the steer is driven in a large figure eight, then around a third peg and through a gate. The golden rule for success is that the horse must have "cow sense." It is a demanding, thrilling exhibition of the skill, intelligence, strength, and ability of both horse and rider.

As Australia developed rapidly with the explorations of the 1830s, knowledgeable horse breeders imported a steady stream of Thoroughbreds to improve the local horse strains. Use of Thoroughbred stallions over the condition-hardened mares produced the beautiful strain of tough but stylish animal seen today in the Australian Stock Horse. Conspicuous in the pedigrees of Stock Horse stallions are winners of the English Derby. The mighty Hyperion appears in many pedigrees, while his own sire, Gainsborough, appears in even more. And there are the important Barham, as well as Spion Cop, Spearmint, and others.

While the English Thoroughbred undoubtedly played the major part in formation of the modern Australian Stock Horse, another influence strongly felt since the 1950s has been the American Quarter Horse. In 1954 the King Ranch of Texas introduced Santa Gertrudis cattle into Australia, and it was not long before talk turned to horses. President of the King Ranch, Robert Kleburg, Jr., well known throughout the world for his Quarter Horses, selected four stallions in 1953 to send to Australia. Two of them, Vaquero and Jackeroo, went to the King Ranch stud "Risbon" at Warwick in Queensland; Gold Standard went to another King Ranch property at Clermont in Queensland. These three were imported by King Ranch (Australia) Pty., Ltd. The fourth stallion, Mescal, was imported by the late Samuel Horden and went to his property at Bowral. The influence of these and other Quarter Horse importations in the ensuing years is also making a mark in Australian Stock Horse breeding.

The ideal Australian Stock Horse has an alert, intelligent head with broad forehead, full eye, and wide nostril; the neck is of good length and well set into sloped shoulders. Withers are well defined and slightly higher than the croup; the chest is deep, not too wide in proportion; the ribs are well sprung; and the back is strong and of medium length. The forearms are well developed; the cannon slightly flat; the pasterns not too long and slightly sloping; the quarters are powerful, well muscled, and nicely rounded, wide and deep in the thigh and gaskin; the legs are straight, do not toe-in or toe-out, and are not based wide or too narrow. The whole of the horse should be in balance according to its size.

All colors are accepted in the Australian Stock Horse.

Auxois

ORIGIN: *France*
APTITUDES: *Heavy draft and farm work*
AVERAGE HEIGHT: *15.2 to 16.2 h.h.*
POPULATION STATUS: *Rare*

Auxois of France. Photo courtesy UNIC, Paris, France

Closely related to the Ardennais type, the Auxois is the result of crossbreeding between local mares of Bourguignon type and Ardennais sires, as well as having had Northern Ardennais, Percheron, and Boulonnais influence during the nineteenth century. The large Ardennais and Northern Ardennais are the only breeds used since the beginning of this century. The Ardennes, an indigenous breed having evolved in the mountain range by that name, was used by Belgium, France, and Sweden to form their own distinct breeds from this important root stock.

The Auxois is larger than the Ardennais. Coat colors are mainly bay and roan, with red (chestnut) roan and chestnut occurring occasionally. The head is short with a wide forehead and small, mobile ears. The neck is muscular and well set; the body is massive with prominent withers, a wide chest, and wide and short back and loins. Hind quarters are long with very muscular croup; the tail is carried low. The shoulder of the Auxois is well slanted; the forearms are very muscular with powerful knees and hocks on short, clean cannon bones. The limbs are hardy with little feather. The gaits are ample and supple in spite of the great bulk. The Auxois has a calm and gentle temperament.

The breeding area of the Auxois is around the Cluny stud, including all of the southwestern part of the Côte-d'Or with an extension to the Yonne and the Saône-et-Loire. This area is slightly hilly and fertile with rich pastures.

Although population of the Auxois breed has remained quite small, the Ardennais horse (see French Ardennais) from the Auxois has long been very appreciated for agriculture. It is currently directed mostly toward meat production, and yields a heavy carcass.

The stud book for this breed has been kept since 1913.

Avelignese

ORIGIN: *Italy*
APTITUDES: *Riding, light draft*
AVERAGE HEIGHT: *13.3 to 14 h.h.*
POPULATION STATUS: *Common*

The Avelignese is the most numerous and widespread breed in Italy. Some claim this to be exactly the same as the Austrian Haflinger; others say it is closely related but bred slightly differently. In any event, the two share a common ancestry. The Avelignese appears slightly heavier than the Haflinger. No official records were kept for the breed until

Avelignese, close cousin of the Haflinger. Photo: Dr. Paolo Piccolino Boniforti

Avelignese. Photo: Dr. Paolo Piccolino Boniforti

Azerbaijan
(Azerbaidzhanskaya)

ORIGIN: *Azerbaijan (former Soviet Union)*
APTITUDES: *Riding horse*
AVERAGE HEIGHT: *13.1 to 14 h.h.*
POPULATION STATUS: *Rare*

1874, but it is known to be of ancient origin and has many of the same foundation sires as the Haflinger.

In the Alps and Apennines the Avelignese is popular as a pack horse and is used extensively for light agricultural draft work. This breed is extremely hardy and surefooted, an excellent horse for use in the rugged mountainous terrain. The Avelignese is capable of carrying or pulling great weight and shows marked endurance.

Like the Haflinger, the Avelignese is always a golden chestnut with mane and tail white or flaxen and very thick and heavy. Usually there are white markings on the face, most often a full blaze. The head shows a great deal of Arab influence while the body tends to be chunky and muscular, broad and deep with short, strong legs. This horse has extremely hard hooves.

The name Avelignese originates from Avelengo (Hafling in German), a small area on the Alto Adige, an Italian region since 1918. In Italy, the Avelignese is not crossed with the Arab or any other breed but is kept strictly pure.

Ten to fifteen years ago some Avelignese were still found to have a dorsal stripe and striping on the legs, but this is considered undesirable by breeders and is seen much less frequently today.

This breed belongs to the saddle-pack group and is of ancient origin. Influence by the Persian and Karabakh breeds is probable. The economic and political conditions of the ancient Caucasus were the cause of development of a saddle-pack horse, able to move quickly and travel long distances. Arab authors of the eighth to tenth centuries spoke of the outstanding qualities of Azerbaijan horses, which were in great demand in areas now constituting Iran, Iraq, and Syria. The history of this breed is connected with many wars, as the war horse was indispensable.

The Azerbaijan horse is taller than the mountain breeds of the Caucasus but smaller than the Kabarda and Lokai. It is possible to increase the size of the breed with selective breeding and improved conditions, but imperative, on the other hand, to remember that the saddle-pack horse should be smaller than saddle horses raised on the steppes. One-sided selection for larger size would result in a loss of balance over steep mountain slopes.

The head of the Azerbaijan horse is well proportioned and wedge shaped; the ears are thin and short and the eyes large and expressive; the neck is set high, of medium length, and muscular; the head is carried high. The chest is well developed and wide, a typical characteristic of the breed. Withers are moderately pronounced; the back is short, firm, straight, and moderately wide; the croup is long, wide, and sloping; and the legs are strong with firm hooves. The hair of this horse is thin and fine with a sparse mane and tail. Nearly 10 percent of Azerbaijan horses have no ergots on the hind legs. The prevailing colors are bay and grey, with sorrel, black, grullo, and buckskin. Palomino is rarer.

This breed is healthy, long-lived, and fertile. Under *taboon* conditions the Azerbaijan completes its growth at four or five years.

Winters are spent on pastures at high altitude. The Azerbaijan is a good pack horse, having a natural sense of balance and skill in movement with either pack or rider on steep mountain slopes. It is able to move at speed in mountainous terrain. Many individuals of this breed are natural pacers.

The Deliboz is a direct descendant of the Azerbaijan horse and inherited its pacing gait. This breed is considered the best type of present Azerbaijan horse.

Azores

(Garrano da Ilha Terceira)

ORIGIN: *Azores, Portugal*
APTITUDES: *Riding pony*
AVERAGE HEIGHT: *11.3 to 12.3 h.h.*
POPULATION STATUS: *Rare*

Azores pony (right). Photo: Jorge De Paiva Benites

Little written material is available regarding the small horse of the Azores, but the breed has existed since very early times. According to Dr. Jorge De Paiva Benites of Angra Do Heroismo, the Azores pony is virtually extinct, with only twenty to twenty-five pure individuals remaining. He reports this pony to be very attractive, perfectly balanced and symmetrical in conformation, and very spirited, "reminding one of pure blood." The

The pony of the Azores, spirited and beautiful. Photo courtesy Jorge De Paiva Benites

Azores pony is completely different in type from the Iberian ponies and the Lusitano type of horse. Body conformation is more that of a small horse and not a pony.

In the distant past when the islands were colonized, several horse breeds were introduced, mainly by nobles, including the Lusitano and Alter Real. Horses on the islands suffered the effects of poor management and nutrition, and it is believed that this diminished the size of the Azores breed.

Domingos Augusto Alves da Costa Oliveira, a lieutenant of the cavalry, in his book *Raças Cavalares da Peninsula e Marcas a Ferro* (1906), referred to a stallion named Califa, purchased in Morocco sometime prior to 1872, being introduced to the island. Presumably there were others. The horses from Morocco were said to be small.

Most horses of the Azores are not presently of clearly defined breeds but are based on the Lusitano race. Being of a completely different type than the Portuguese breeds, we must assume that the beautiful Azores pony is a descendant primarily influenced by the horses brought from Morocco, possibly of Arab extraction, but they were likely Barbs, the main breed of that country. The word

Garrano, when used referring to the Azores pony, does not infer a relationship to the Garrano of northern Portugal.

The main colors found in Azores ponies are brown and bay. Remaining individuals are used for agricultural work. They are quite spirited and hotblooded for use as children's mounts.

Azteca

ORIGIN: *Mexico*
APTITUDES: *Riding horse*
AVERAGE HEIGHT: *14.3 to 15 h.h.*
POPULATION STATUS: *Common*

The Azteca is the first breed to be developed in Mexico. The majority of horses in Mexico are descendants of Spanish horses, such as the Galiceño, or recent imports from Spain or Portugal. The Spanish horse has always been favored in Mexico and the Andalusian was the foundation for the new breed, the Azteca.

Breeding of the Azteca began in 1972 by crossing the Andalusian and American Quarter Horse, and some crosses were made between Andalusian stallions and Criollo mares. Azteca horses are increasingly in demand in Mexico, and breeders face the problem of insufficient foals to satisfy buyers.

The breed registry for the Azteca in the United States was formed in June of 1989.

Azteca. Photo: Dolores Dougherty

The Azteca mare, Lady Luck. Photo: Dolores Dougherty

Bloodlines may be crossed back and forth in many combinations, as long as the resulting offspring are always six-eighths or less of any one breed. The purpose of producing this breed is to combine the qualities of the Andalusian with those of the Quarter Horse, which was also founded on Spanish blood.

The Azteca is an attractive warmblood horse, found in all horse colors with the exception of piebald, skewbald, or spotted. The head is lean with a straight or slightly convex profile, small, pricked ears, and large, expressive eyes; the neck is well muscled and slightly arched. The withers are rather high; the back is straight and short; the croup broad and rounded; the chest deep and broad; and the shoulder long and sloping. The legs are well muscled with good joints, long, narrow cannons, and clearly defined tendons and joints. The hooves are well shaped.

Baise
(Guangxi)

ORIGIN: *China (Western Guangxi Autonomous Region)*
APTITUDES: *Riding, pack horse*
AVERAGE HEIGHT: *11 to 11.2 h.h.*
POPULATION STATUS: *Common*

Guangxi region is adjacent to the provinces of Guizhou and Yunnan, and horses originating from these areas have been related since early times. As the climate is good there, horse breeding has long been a common practice

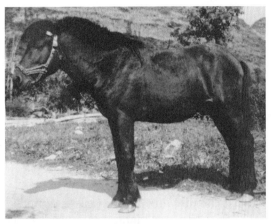

Baise stallion. Photo: Weigi Feng and Youchun Chen

Baise horse of China, loaded with goods. Photo: Youchun Chen and Tiequan Wang

Baise mare. Photo: Weigi Feng and Youchun Chen

The Baise horse at work in the mountains of China. Photo: Youchun Chen and Tiequan Wang

among the inhabitants. Horse riding was included in the marriage ceremony as early as 20 B.C.

The earliest records regarding this horse were found in Xilin and Guixian Counties in the Guangxi region. Two buried bronze statues covered with gold of horses with riders were found in 1972 and 1980. They are believed to be cultural relics belonging to the period of 206 to 25 B.C. The body structure and conformation of the horses depicted are identical to those of the modern Baise horse. In the record of Tiandong County it is said that in the Han dynasty there was a horse market at the town of Pingma, where a lively trade developed. The records of Tianxi County state that in the Song dynasty horses were important in the marriage ceremony and reception.

The Baise horse has a dry, solid conformation. The head is slightly heavy with a straight profile, wide at the jaw, with small, lively ears; the neck is medium in length; withers are moderately pronounced; ribs are rounded; the back and loin are level; the croup is sloped; the shoulder is short and straight; tendons and joints are well developed; the hoof is solid. The legs are strong; hind legs may be sickle-hocked. Hair in the mane and tail is thick. This horse is a native breed of China.

Baise horses are good pasture animals, raised in the open mountains at high altitude. They are caught when needed for work and allowed to roam freely in the mountains at other times. During the farming season they are kept in barns.

The Baise horse is mainly bay with some grey, chestnut, and black individuals. This is a fast, strong horse with a willing temperament. It is used for riding, packing, and as a carriage and draft animal. Horse meat is a common food in this region.

Balearic

ORIGIN: *Spain (Island of Majorca)*
APTITUDES: *Riding pony*
AVERAGE HEIGHT: *14 h.h.*
POPULATION STATUS: *Rare*

Among the rarest of equine breeds, the Balearic island horse or pony is of ancient origin and little known. Because the breed is considered unimportant by officials, it is seemingly impossible to obtain information about it, and barely more than acknowledgment of its existence is available.

Some report this to be a delicate, beautiful pony with primitive features of upright mane and slender legs. Others say the Balearic is a scrubby individual, roughly made.

Summerhays' Encyclopedia for Horsemen states that the Balearic is an ancient and distinctive type most abundant in the Palma district; differing from other breeds in slender limbs, free, graceful carriage, a short, thick, arched neck, and delicate head with Roman nose. Summerhays claimed the Balearic may

be descended from horses shown on ancient Greek coins and vases.

The ponies are used on small farms for agricultural work and by peasants for transportation and in harness. They are usually bay or chestnut in color, sometimes grey. Possibly there is a link to the tiny Skyros of Greece. Nothing is known to the writer regarding the gaits of the Balearic.

Bali

ORIGIN: *Indonesia (Bali)*
APTITUDES: *Riding and light draft*
AVERAGE HEIGHT: *11.3 to 13 h.h.*
POPULATION STATUS: *Common*

The origin of Indonesian ponies is very ancient. Groenveld (1916) and Van der Poel (1904) attempted to trace the origin of these ponies or small horses. Groenveld cited a great deal of literature regarding China in his book. During the T'ang dynasty, (627 to 649 A.D.) China presented several horses to the king of Java as gifts. Later, during the Yuan or Mongol dynasty (1280 to 1367), Chinese cavalry landed near Tuban, in Java. From these sources it was concluded that the ponies of Indonesia were descended directly from the Mongolian, from which most Chinese horses are in part descended. (Information was received from the Ministry of Agriculture in Indonesia.) It is also known that China had obtained horses centuries prior to this time from Fergana (Turkmenistan) and other areas of western Asia to cross with their native horses. Exact identification of the breeds introduced to Indonesia by the Chinese is not possible.

Indian people took horses to Indonesia also, but the race of horses is unknown. According to Groenveld they had Asian and African blood. During the eighteenth century the Dutch took horses to Indonesia that were native to Persia and Arabia.

The Indonesian ponies can be defined as mixtures of Mongolian and western Asian or Oriental blood. Groenveld reported that the highest percentage of Arab blood was found in Sandel (Sandalwood) horses and the least amount was evident in the Batak animals.

The Bali pony, from the island of that name, is predominantly some shade of dun with dorsal stripe, black mane and tail, and dark points, but other colors also occur. Often the mane is naturally upright.

The Bali is very strong and is used as a pack animal to transport coral for the building trade. It also makes a good mount for tourists who wish to view the natural beauty of the island.

The head of the Bali pony is large with a straight profile, small ears, and almond-shaped eyes; the neck is usually short and straight with bristly, upright mane. Withers are often quite low; the back is short and may be slightly curved upward; the croup is sloped; the chest deep; the shoulder quite straight; and the legs are strong with good joints and very hard hooves. As is true of all Indonesian horses, ailments of the legs or feet are extremely uncommon.

Balikun

ORIGIN: *China (Xingjiang Autonomous Region)*
APTITUDES: *Riding, packing, draft*
AVERAGE HEIGHT: *12.2 to 13.1 h.h.*
POPULATION STATUS: *Common*

Balikun horse of China. Photo: Weigi Feng and Youchun Chen

Balikun County, origin of the breed by that name, was possessed by Kazakh and Mongolian tribes many times, according to records kept by the Qing dynasty, so this breed descended from Kazakh and Mongolian horses. In selective breeding for over two hundred

years, the Balikun became a special breed. This horse is very tough and well adapted to its habitat, which is frequently harsh. It is used extensively in the area for transport.

Balikun horses have a heavy head with small ears; the neck is thick, short, and well muscled; withers may be somewhat low; the back is short, flat, and very strong; the croup is sloped with a low tail-set. The shoulder is somewhat straight but muscular. This breed has powerful quarters and strong legs with good feet. The coat is very dense and thick, and the Balikun is able to live on steppe pasture, even at temperatures under $-40°F$. The most common colors are bay and chestnut.

The Balikun can easily carry a pack weighing 220 pounds up to fifty miles in a day. It is considered a native breed of China.

Baluchi

(Balouch or Banamba)

ORIGIN: *Pakistan (Baluchistan)*
APTITUDES: *Riding horse*
AVERAGE HEIGHT: *14 h.h.*
POPULATION STATUS: *Common*

Baluchi horse of Pakistan. Photo: Abdul Wahab Khan

The Baluchi horse is bred in parts of Baluchistan and Sind Provinces and the districts of Bahawlpur, Dera Ghazi Khan, Muzaffargarh, and Multan in Punjab Province.

In Pakistan, the majority of horses are light in build and larger than ponies. Reminiscent of the Kathiawari of India, the Baluchi has very turned-in ears. Predominant colors are bay, black, brown, or grey.

Reports state that the Baluchi horse is related to the West African Barb through horses of Mali known as Bélédougou, or Banamba. It shares the same ancestors as the Waziri, which is a smaller animal. Used for pleasure riding and sports, the Baluchi is of medium size with a fine, dry head. The profile is straight or slightly convex and wide between the eyes, which are large and expressive. The ears are rather long, curving inward to touch at the tip, a typical characteristic of the breed. The neck is long and well muscled; withers are pronounced; the back is short and strong; the croup is sloping with a moderately high tail-set. The quarters are smoothly muscled and the legs are fine and strong with good joints, clearly defined tendons, and very hard, strong feet.

The Waziri breed is well known for its strength, good conformation, and stamina at hard, fast work in the hills. Nowadays it is scarce and said to be mostly impure. Information about it was not available.

Ban-ei horse racing in Japan. Photos courtesy Hirohumi Kageyama

Ban-ei Race Horse

ORIGIN: *Japan*
APTITUDES: *Racing*
AVERAGE HEIGHT: *14.3 to 16.1 h.h.*
POPULATION STATUS: *Common*

The Ban-ei race horse may not be considered a breed by experts, yet the blend of Percheron and Breton has been bred in Japan for many years specifically for a unique type of racing.

In 1948, draft racing, *ban-ei keiba*, was added to regional public racing in Japan. *Ban-ei keiba* is a race in which the horse pulls a heavy sleigh (on bare ground). This event has been traditional to Hokkaido ever since that region was opened for habitation. The horses are the same breed used for reclamation work in the region: heavy draft horses of Percheron and Breton lineage.

The draft races held in Hokkaido comprise steeplechase races over 200 meters. The horses pull a heavy sleigh made of iron upon which iron slabs are placed for added weight. The weighted sleigh weighs half a ton to one ton, and is pulled over two hills in the 200-meter distance.

In Japan there are four draft racetracks, Iwamizawa, Kitami, Obihiro, and Asahikawa. The horses are bred in Hokkaido, especially in the Tokachi area near Obihiro racetrack. Some Percheron and Breton stallions are imported from France each year, and their sons and daughters are used for this type of racing.

Ban-ei keiba had its origin in a popular game enjoyed by residents in some communities in Hokkaido and Tohoku where many horse breeders are located. These dynamic draft races have a large following and their highest stake race is the Minister for Agriculture, Forestry and Fisheries Cup (for three-year-olds and up), with the prize money totaling 18 million yen (U.S. $129,000).

The driver of the race weighs about 160 pounds (with added weights if necessary) and rides the sleigh along with the iron slabs placed there for the weight handicap. The weight handicap system is based on the winning money. There are different weight-pulling requirements depending on the age of the horses.

Banker Horse

ORIGIN: *United States (North Carolina Outer Banks)*
APTITUDES: *Riding horse*
AVERAGE HEIGHT: *13 to 14.3 h.h.*
POPULATION STATUS: *Rare*

Feral stallions on the Rachel Carson Estuarine Sanctuary, an island located off the coast of Beaufort, North Carolina (Banker horses). Photo: Dr. Elizabeth Franke Stevens

A (Banker) feral mare from the island of Corolla, off the coast of North Carolina. Photo: Dale Burrus

Banker ponies. Photo: Gene W. Wood

Horses inhabiting the Outer Banks of the Carolinas have characteristically been called Banker horses, Shackleford ponies, or Banker ponies. Controversy boils around this small group of feral horses, not only regarding their origin but also whether or not they should be allowed to remain on the Outer Banks. Conservationists claim the horses are destroying the fragile ecological balance of the area, which is made up primarily of sand dunes. Supporters of the horses hold a different view—that the horses have "always been there and should be allowed to remain" and that it is other animals released on the banks and not the horses that are causing the problems.

In order to comprehend the Banker horses one must know something of the location and environment of their habitat. Until recent years the Outer Banks of North Carolina were some of the most isolated and undeveloped areas of the United States. The banks consist of a string of sand dunes that protect the mainland of North Carolina from the Atlantic Ocean. Separated from the mainland by large bodies of water called sounds, the Outer Banks are actually a series of islands separated from each other by inlets. They are approximately 175 miles long from the Virginia line to just south of Cape Lookout. The best description is probably that of one of the Indian tribes that once lived there: *hatteras*, an Algonquian word meaning "there is less vegetation."

Generally, the Banker horses are believed to be descended from Spanish horses brought by the first explorers to the area. Lucas Vasquez de Allyon received from the Spanish king a charter that gave him the right to explore and colonize much of the eastern seaboard. In 1521 he sent one of his captains, Cordilk, on an expedition which landed at

River John the Baptist (thought to have been Cape Fear). Other of Allyon's explorers spent considerable time at a place called Chicora, thought to have been in the same area.

The Spaniards had problems with the Indians. Reports say they took Indian children as slaves and sent them to the West Indies. There was an uprising and the Spaniards were forced to flee to stronger Spanish holdings, leaving their livestock and other belongings behind.

During the period from 1584 to 1590 other livestock was taken to the Outer Banks by the expeditions of Richard Greenville. Sir Walter Raleigh's ships maintained a steady stream of traffic between the Outer Banks and England for four years, and during the period from 1584 until 1711 there were many comings and goings by both the Spanish and English. According to Englishman John Lawson, who explored and documented southeastern North Carolina from 1700 until his death at the hands of the Tuscarora Indians in 1711, these swift, well-shaped horses were very hardy and had few leg problems.

The horses of the Outer Banks have undoubtedly been mixed with various strains of both English and Spanish horses, yet they were of the original stock brought to America and have not been greatly mixed with modern breeds.

In 1856 Edmund Ruffin, famous as an authority on agriculture and also credited with firing the first shot in the Civil War, visited the Outer Banks. He wrote of the horse penning that was carried on twice a year on the banks, at which time all of the horses on the island were corralled and the colts branded. The horse pennings were much attended and very interesting festivals for all the residents of the neighboring mainland. He further reported that all of the horses in use on the reef and many on the nearest farms on the mainland were those previously wild "banks ponies." They were described as of small size, with rough, shaggy coats and long manes, capable of great endurance of labor and hardship.

Modern residents of the North Carolina coast may think it odd that the long, narrow strip of barrier islands separating the Atlantic Ocean and the Carolina sounds was once considered the state's most desirable pasture land. However, for over two centuries the raising of livestock was the most important economic use made of the sand dunes and marshes that make up the Outer Banks. Early records testify to the fact that the coastal islands were first settled for the sole purpose of raising livestock.

The colonial residents had good reasons for selecting the Outer Banks for stock raising. For one thing, pastures surrounded by water removed the danger of predation by wolves and losses to Indian marauders. Also, there was no expense for fencing.

The Banker horses feed on the coarse salt grasses of the marshes and supply their needs for fresh water by pawing away the sand so that water oozes into the excavation, just as do the horses of Sable Island. The horses sniff the ground and seem to know the best places to dig for water.

A 1939 book on North Carolina that was compiled by the Federal Writers Project of the Federal Works Agency Projects Administration reported that on Cape Hatteras there was abundant wildlife consisting of wild ponies, cattle, and hogs, which ranged at will until the federal program of sand fixation by grass planting necessitated a strict stock law. In 1938 the county placed a bounty on the few remaining wild ponies, traditionally considered descendants of Barbary horses brought by the Raleigh colonists or saved from wrecked Portuguese ships.

Presently, small numbers of the horses of the Outer Banks remain, distributed on various islands. Some were captured years ago and have been carefully preserved by the Burrus family. Dale Burrus has done a great deal to promote interest in the remaining horses and to protect them.

In a letter dated June 13, 1990, Thomas L. Hartman, superintendent of Cape Hatteras National Seashore, states:

Oral tradition has it that the horses are descended variously from Spanish Mustang stock, early colonial settlement era animals and more recent escapees. Study and testing of the animals has not been exhaustive, but at this time there is not definitive evidence to suggest these horses represent any specific breed. The Ocracoke horses are nearly as much a symbol of Ocracoke tradition as is the maritime history of the island. Farming, transportation and the Life Saving Service depended on the horses. In recent times the horses have been used by a mounted Boy Scout troop and by the National Park Service for mounted patrol. There are plans currently to resume use of some of the horses in mounted patrol activities. Irrespective of their true origins, the horses are now an integral element of Ocracoke Island.

Another source of information was Dr. Elizabeth Franke Stevens of the Atlanta/Fulton County Zoo in Atlanta, Georgia. She has spent time studying the Banker horses and wrote:

I have enclosed a photograph of two stallions from the Rachel Carson Estuarine Sanctuary, an island located off the coast of Beaufort, North Carolina. This island lies between the Beaufort waterfront and the barrier island, Shackleford Banks. The Rachel Carson sanctuary is managed by the North Carolina Dept. of Natural Resources and Community Development. It is believed that this population originated from a small group of horses which swam over from Shackleford Banks sometime in the 1940s. This is not, however, documented in any official way. The horses on Shackleford Banks are believed to be descendants of horses which arrived on shipwrecked Spanish galleons. This fact is also not set in stone.

We do know however, that on both islands horses that are owned by local residents have been taken on and off the island. In the case of the Shackleford Banks there was once a village which was wiped out by a hurricane in the early 1900s. In the case of the Rachel Carson site there was a resident who took horses back and forth from the island for grazing. Because of all of this mixing I do not know whether these horses are all true "Banker" ponies.

Colors in the Banker horses are buckskin, dun, bay, chestnut, and brown. On Shackleford there are some pintos. Interestingly, neither island is known ever to have had a grey.

Population on the Outer Banks varies according to available feed and environmental changes, and a severe drought in 1987 reduced numbers drastically when many horses starved to death on the Rachel Carson site. The population is now managed by removal of some horses every few years to keep the numbers down to twenty or thirty animals. A good deal of controversy surrounds the horses, with some people desiring to take feed to them in hard times and others maintaining that nature should be allowed to "take its course." The future of the hardy Banker horse is thus uncertain. It was people and not nature who placed the horses on the island and abandoned them, trapped with scarcity of food.

The Banker horse displays many characteristics of the Spanish horse that contributed so greatly to the breeds of America. Dale Burrus, who has raised Banker horses for many years, claims that the Banker horse has a disposition that is different from that of most other feral horses—docile and friendly and very willing to learn. This is a tough, enduring small horse, typical of the Spanish mustang.

Bardigiano

ORIGIN: *Italy (Bardi)*
APTITUDES: *Riding, light draft*
AVERAGE HEIGHT: *13.2 to 14.1 h.h.*
POPULATION STATUS: *Common*

The robust Bardigiano of Italy. Photo courtesy Dr. Ugo Mutti

The Bardigiano. Photo courtesy Dr. Ugo Mutti

Origin of the Bardigiano breed can be traced to the horse of Belgian Gaul. It arrived in the Apennines of Parma with Germanic warriors who came to Italy on various occasions, often stopping in the area surrounding the Po River after the fall of the Western Roman Empire (470 A.D.). For decades the horses lived and reproduced in this area, generating a homogeneous population adapted to the uses of light burden and draft. Spurious characteristics disappeared and typical and distinct traits emerged. In this manner, in the silence of mountain pastures, the Bardigiano was formed. The name comes from the village named Bardi, home of a medieval castle.

During the years between the two World Wars the army used Bardigiano mares to produce robust mules. This was the beginning of decline for this old breed, as the number of purebred animals being produced was reduced alarmingly. At the end of World War II, poor decisions made by those in charge of zootechnical activities caused additional problems. Stallions from several different breeds were introduced into the area, and consequently variations were introduced to the breed which led to the dispersion of traits characteristic to the local population. This was the situation until 1972. That year the Comunità Montana dell'Appennino Parmense (Mountain Community of the Parmesan Apennines) initiated a significant cooperative effort with the APA of Parma, and the idea that the Bardigiano race could be recovered was born.

Initially a great deal of difficulty was experienced, as the breed stayed in mountain pastures for nine months of the year. Surveys of the horses' favored haunts and adjacent pastures, however, revealed that the customs and fervor of mountain agriculturists had resulted in preservation of various groups of animals that bore the traditional characteristics of the Bardigiano breed. From these groups selection was begun.

In 1975 the first Bardigiano horse show was held. Shortly thereafter, with the assistance of the Centro Regionale di Incremento Ippico di Reggio Emilia (Regional Center for Equine Expansion of Reggio Emilia), with headquarters in Ferrara, indigenous stallions were once again used systematically with local mares. These efforts were recognized on August 1, 1977, when the Ministry of Agriculture and Forests approved regulations concerning selection of horses and the *Genealogical Book* of the Bardigiano horse. With this document, the Bardigiano was officially recognized. Recognition made it possible for both a central office and an association representing the breed to be established. Also specified were the precise characteristics of the breed, defined by a ministerial decree.

The typical Bardigiano horse is fairly small. The constitution of the breed is hardy and enduring. The head is light with a slightly concave profile, wide mouth, and jutting upper lip; the eyes are large and vivacious. The head of this horse has a particular aesthetic value, since breeders can determine information about the horse's temperament and aptitude just by looking at its expression. The neck is of proportional length with a wide base, preferably arched, with a thick crest that is often double; the thorax is wide and well formed with a wide, deep, muscular chest; the back has medium length, with short yet ample and well-shaped loins; the croup is well developed, appropriate for animals that are not very fast, like the working horse.

The Bardigiano's legs are of rather slim but strong bone structure; the pasterns are short; the hooves are large and extremely hard. The coat is bay, ranging from ordinary to dark bay. A small star on the forehead is tolerated in the breed, as are white markings on the fetlocks, provided that they are not very high.

It is apparent from the distinctive characteristics that the Bardigiano horse is the product of a long process of adaptation to a hill and mountain environment. This is a frugal race, rustic, resistant to drought conditions, and very adaptable to climatic changes. It is capable of utilizing any type of forage and pasture. The type of breeding that is usually followed is semiwild, the animals staying up in the mountains for eight or nine months of the year.

The grasses in mountain pastures are a complete source of nutrition and an important element in the breeding of Bardigiano horses. Even lactating mares do well on the natural pastures. In addition to being a balanced and complete source of energy, protein, and vitamins and minerals, natural pastures in the mountains have a positive effect on the health of the animals because of the sun and exercise. It has been found that mares in the Apennine pastures produce so much milk that foals gain an average of 2.2 to 2.4 pounds per day.

Supplemental feeding is done only in drought years or when there is an excessive number of horses on the range. The amount of pasture needed per adult varies according to the productivity and nature of the pasture. This is especially true in the Apennines, where double the usual amount of pasture is typically required. In fact, the Apennine regions almost always present serious problems that restrict use of the pastures by other animal species. The most prevalent problem in the pastures is the high rate of ingestion of gastroenteric parasites, due to overstocking. Since such infections reduce the growth rate, it is best to deworm all subjects.

Bardigianos are used for tourism, horse therapy, work, and meat. Because of their docile nature and robust physique, they make good animals for nature hikes. That same docile nature and the small size of the breed make the Bardigiano a good horse for working with handicapped children. This horse is easily accessible and easy to work with.

The Bardigiano bears strong resemblance to the Dales pony of Great Britain and the Mérens pony of France.

Bashkir
(Bashkirskaya)

ORIGIN: *Ural Mountains (former Soviet Union)*
APTITUDES: *Riding, light draft*
AVERAGE HEIGHT: *13.1 to 14 h.h.*
POPULATION STATUS: *Common*

The hardy Bashkir horse. Photo: Nikiphorov Veniamin Maksimovich

Bashkir mares near Moscow in milking stanchions. Photo: Nikiphorov Veniamin Maksimovich

This ancient horse of the mountains and steppe region adjacent to the Volga and Urals has been bred for centuries by the Bashkiri people. A versatile and useful animal, the Bashkir has served as a riding and draft horse and is also used for milk and meat production. *Koumiss*, made from mare's milk, has long been a staple of the native people and is also used as a health product throughout Russia. The mares have always been milked, giving large quantities each day.

Milking of mares is widespread in Bashkiria. In the past it was the practice from April through August to tether the foals all day (up to eighteen hours) away from their dams without nursing; they received their mothers' milk only at night. This had an adverse affect on the growth and development of the youngsters, and subsequently on the entire breed. It was necessary to improve the conditions for raising foals.

The Bashkir people moved to the Bashkiria area in the seventh and eighth centuries. Historians have described them as a nation having knowledge of horse breeding, which was widely developed prior to the appearance of Mongols in the area. Therefore, it is incorrect to include the Bashkir breed in the Mongolian group.

The Bashkir horse derives from the steppe horses of western Asia, related to horses of other nations of Turkish origin. Remains of this breed have been found in ancient tombs from the Krasnodar region to the Volga. As a result of crossing with the forest horses widespread north of Bashkiria and the influence of the forest-steppe climate, the Bashkir gained some of the characteristics of the forest horse. This breed differed from other steppe horses in having a coarser constitution with a build close to the draft type. The breed is strong and sensible with a good disposition.

The Bashkir is a small, wide-bodied horse with a massive head and short, fleshy neck. The withers are low; the back broad; the croup nicely rounded; the ribs well sprung; the chest broad and deep. The legs are short and bony. Mane and tail hair is thick and grows quite long in winter. Colors are usually dun, bay, chestnut, roan, and mouse grey. The duns have a dorsal stripe and zebra stripes on the legs.

There are two types found within this breed: the mountain type, small and more cobby and massive, of draft type but also suitable under saddle; and the steppe type, taller, with a longer body of lighter build, of saddle-harness type.

Bashkir horses were used by the army in 1789, and in 1845 army stud farms were established for their breeding. During the second part of the eighteenth century, increasing poverty and lack of standards by the government brought about the decay in breeding of this horse. After the Revolution standards were accepted and improvements were made in breeding. Some stud farms were established in Bashkiria.

In the past, Bashkir horses were bred in *taboons* year round. Some of the breed's characteristics were influenced by climatic conditions. The climate of Bashkiria is severe and continental. Winters have severe frosts, deep snow, and massive storms. Snow covers the countryside from early November until the end of April. Spring comes late; summers are hot but short.

This horse has a high work tolerance and is remarkably healthy and well adapted to the severe weather conditions in which it has lived for centuries.

The Bashkir horse was famous long ago for its hardiness both under saddle and in harness. Soldiers riding these as well as Don horses reached France in 1812 during the war with Napoleon and successfully returned home to Moscow, a very great distance. Also famous in the past were the Bashkir troikas, able to cover seventy-five to ninety miles a day without feeding. Especially prized were pacers among the breed, appreciated for both speed and endurance.

The Bashkir is being improved by selective pure breeding and by crossing with the Russian Heavy Draft. Experimentally, the Bashkir has been crossed with the Kazakh and Yakut horses. In addition to pure breeding, Bashkir horses have also been crossed to Don, trotter, Thoroughbred, and other breeds. The best results have been crosses with Dons and trotters. The Don-Bashkir cross is larger than the pure Bashkir, retaining the massiveness, and is best adapted to *taboon* raising. Without supplemental feeding, however, this cross does not reach full potential in development. Trotter-Bashkir crosses are small, but have great endurance in the harness. Crosses to Thoroughbred are poorly adapted to severe conditions, and these crosses were canceled. Bashkir stock is mainly concentrated at Ufa stud, the leading center for the breed.

Basuto

ORIGIN: *Lesotho (former Basutoland)*
APTITUDES: *Riding, light draft*
AVERAGE HEIGHT: *14.2 h.h.*
POPULATION STATUS: *Rare*

A typical Basuto. Photo: D. R. Phororo, Ministry of Agriculture, Lesotho

Basuto. Photo: D. R. Phororo, Ministry of Agriculture, Lesotho

The Basuto descended from the original Cape horse of South Africa, which in turn developed when horses were first taken to the Cape in 1652 by the Dutch East India Company. The first imports were "ponies from Java," and while most observers reported these were entirely of Persian and Arab blood, it should be noted that the Java pony was predominantly of Mongolian origin. Horses in Java were crossed to Arab and Persian horses. One source stated that the original ponies from Java were upgraded later, at the Cape, with Arab, Thoroughbred, and Persian blood. In any event, the Basuto bears a great deal of similarity to the hardy Mongolian horse and never at any time in history attained the size of the Arab or Thoroughbred. That the first horses at the Cape were of Oriental descent does not eliminate the likelihood of Mongolian blood, which is indeed Oriental.

Cape horses belonging to farmers on the Orange River border were raided repeatedly by the Griqua and Koranna peoples, who in turn lost many of them to tribes living in what is now Lesotho.

The Basuto people found horses a great asset as soon as they were introduced, and apart from the horses raided from other tribes, they also bought horses from European traders. After 1835, many horses were acquired as payment for work on settlers' farms. By 1870, the entire Basuto nation was mounted. While the Cape horse, through continual infusions of good Arab and Thoroughbred blood, averaged about 15 to 15.2 hands, the Basuto remained about 14 hands in height.

There is no doubt that the rugged terrain and severe mountain climate in Lesotho exerted a strong influence on the type of horse that emerged after some generations. The inhospitable climate eliminated weak animals and the high altitude and steep terrain resulted in development of strong heart and large lung capacity. The Basuto was compact, sturdy, and extremely surefooted.

By the turn of the century, the Basuto was well established as a distinctive type, its excellence acclaimed throughout South Africa. The Basuto people were daring riders who raced up and down the steepest mountain slopes at breakneck speed. Their animals were always at the mercy of the elements as they were given no shelter and had to find their own food.

The hard life molded a type of horse which retained such worthy characteristics of its ancestors as spiritedness, intelligence, hardiness, powers of endurance, and a likeable and amiable nature. Through natural selection the Basuto became even hardier than its ancestors.

During the nineteenth century the Basuto's fame spread throughout South Africa and many were exported to India for military purposes. By the end of the century the Anglo-Boer War had spread the fame of this small horse throughout the British Empire. Unfortunately, the price of this fame was rapid decline of the breed.

British forces in South Africa soon discovered that the Basuto was a better war horse than their own, and during the course of the war they bought at least 30,000 horses from the Basuto nation. This has immortalized the Basuto, and many war stories tell of the courage, loyalty, and hardiness of this horse in the service of Boer and Briton. It was through the war, however, that the Basuto lost its best and largest numbers. The breed never recovered from this loss, and additionally, during World War I, more than 2,000 Basutos were exported to German South West Africa (now Namibia) and an unknown number to East Africa. The remaining animals in Basutoland were inferior, and droughts, overgrazing, and neglect caused further deterioration.

Gordon Naysmith rode from Lesotho to the Austrian border, a journey of 16,000 kilometers (nearly 10,000 miles), on Basuto horses, en route to the Olympic Games in Munich in 1972. The performance of the Basuto on the twenty-one-month journey is an indication of the ability of the breed to endure varying and arduous conditions.

Serious work was begun in 1977 to save the Basuto from extinction with the Basotho Pony Project, assisted by the Irish government. A centrally located elite stud at Thaba-Tseka was established to breed colts from which sires are selected to be mated with the mares registered in the stud book. The mares are selected from the best of the general pony population; the sires bred at the elite stud are rotated to other centers. Eventually the stud book will be closed and selected ponies for export will be registered.

As of September 1989, it was reported by Thomas F. Ryan, project manager of the Basotho Pony Project, that while there is not yet a breed association for the Basuto, it is hoped that one will be formed in the near future. The Basuto, he reported, is far from extinction, with a figure of 10,000 counted in a census in 1980.

In other attempts to revive the failing Basuto breed, the South Africans' idea of preserving it was to do so in lowlands, away from the native mountains and harsh climate that are responsible for its finest characteristics of stamina and durability. They produced the Nooitgedacht, preserving all other traits of the original Basuto including the trippling gait.

Lesotho obtained some of the improved breeding stock from South Africa. Interestingly, when brought into the mountains with the horses involved in the Basotho Pony Project, these horses suffered considerably in the first winter and had to be nursed and given extra feed while their mountain cousins looked on in amazement. By the second winter, however, they were ready for anything nature could throw at them. This is a good indication that even when raised in the lowlands, the Basuto does not lose its inherent ability to adapt. In addition to indigenous stock, Connemara blood has been used in the Basotho Pony Project.

The Basuto, according to Lesotho authorities, is a small horse—not a pony, despite its small size. The Basuto rides like a horse, with a long stride, and is noted for its inherent trippling gait. The tripple is a four-beat gait much like the tølt seen in the Icelandic horse. The Basuto while trippling has only one foot on the ground at any given moment and this permits it to extend deeply, which leads to high speeds, nearly as fast as the gallop. The great advantage of this gait is not merely speed but the ability to maintain a reasonably fast speed for long periods over mountainous terrain. The rider sits perfectly, not being required to post and therefore not tiring as quickly as with other gaits. This is a decided advantage in a country where farmers can easily spend up to ten hours in the saddle each day.

The Basuto is generally chestnut, bay, brown, or grey. The head is of medium size but with a rather heavy jaw and straight profile; the eyes are small but expressive. The neck is long and not very muscular; the withers are pronounced; the back is long and straight; the croup sloping; the tail set low; the chest deep; and the shoulder somewhat straight. The legs are slender but strong, with clean joints and very hard hooves.

Batak stallion. Photo courtesy Dr. Soehadji, Department of Agriculture, Indonesia

Batak

ORIGIN: *Indonesia (North Sumatra)*
APTITUDES: *Riding pony*
AVERAGE HEIGHT: *12 to 13 h.h.*
POPULATION STATUS: *Common*

The Batak, some say, is the most highly regarded pony of the Indonesian group next to the Sandel (or Sandalwood), due to great infusions of Arab blood. Indonesian authorities, however, reported this breed to have the *least* amount of Arab blood. In any event, the Batak is bred selectively on Sumatra and it is known that Arab stallions have been used to upgrade the quality of this pony. The best are taken to other Indonesian islands to improve the stock.

The coat of this pony may be of any common horse color. In appearance the Batak has a light head with slightly convex profile; the neck is arched and well shaped; the withers are quite prominent; the back short and strong; the croup sloped; the chest deep; and the shoulder well sloped. The legs are sturdy, slender, and well-muscled with good, hard feet.

The Batak has a kind nature and is easily managed and economical to keep. The Arab blood has added elegance to the breed.

On Sumatra another breed is also raised called the Gayoe. This pony is not as lively as the Batak and is heavier in build. No further information was received regarding the Gayoe breed.

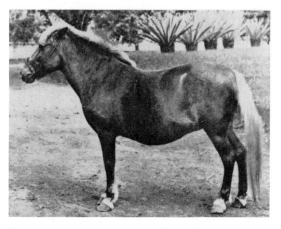

Batak mare. Photo courtesy Dr. Soehadji, Department of Agriculture, Indonesia

For further general information regarding Indonesian ponies, see Bali.

Bavarian Warmblood
(Bayerisches Warmblut)

ORIGIN: *West Germany (Bavaria)*
APTITUDES: *Riding, sport horse*
AVERAGE HEIGHT: *16 to 17 h.h.*
POPULATION STATUS: *Common*

The Bavarian Warmblood was developed in the 1960s based on the ancient Bavarian breed, the Rottal, which is nearly extinct. Using the blood of the outstanding Rottal breed, Hanoverian, Trakehner, and Thoroughbred blood were added to form the new

Bavarian Warmblood. Photo: Evelyn Simak

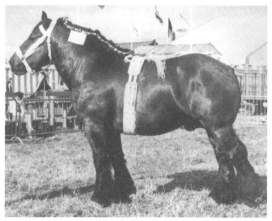

Belgian Ardennais stallion. Photo courtesy P. Laurant, director, Société Royale, Le Cheval de Trait Ardennais

Bavarian Warmbloods in Germany. Photo: Evelyn Simak

Belgian Ardennais mare and foal. Photo courtesy P. Laurant, director, Société Royale, Le Cheval de trait Ardennais

breed. Cleveland Bay, Norman, and Oldenburg horses have also influenced the breed. The old Norman breed is extinct.

The Bavarian Warmblood represents a fine modern riding horse for dressage or show jumping. All colors except piebald and spotted are allowed.

A large and elegant riding horse of heavyweight type with correct body and legs, the Bavarian Warmblood has elastic movements and willing, docile temperament. Most Bavarian Warmblood horses are dark chestnut in color.

Belgian Ardennais

ORIGIN: *Belgium*
APTITUDES: *Heavy draft and farm work*
AVERAGE HEIGHT: *15.1 to 16.1 h.h.*
POPULATION STATUS: *Common*

The Belgian Ardennais originated in the Ardennes Mountains bordering both Belgium and France. A distinction has been made between the Belgian and French breeds, and Sweden also has a version of this breed, established when Ardennais horses were taken into Sweden and crossed with the North Swedish horse. (See Auxois, French Ardennais, Northern Ardennais.)

The modern Ardennais is thought to be descended from the heavy horses praised by Julius Caesar in *De Bello Gallico*, and ancestors to this heavy, sober breed have been bred on the Ardennes plateau for approximately two thousand years. Napoleon valued this breed very highly for its endurance, using the smaller version of the breed.

The climate and vegetation of the Ardennes plateau are perfect for the development of a well-built, energetic large horse. The Ardennais is considered to be among the earliest known of the heavy draft breeds. Arab blood was added to the breed in the eighteenth century and, more recently, Belgian Draft (Brabant) has been used to increase both size and strength.

The Belgian Ardennais is well known for its endurance, ability to withstand harsh climate, willingness to work, and easy management. For its size, the Ardennais is economical to feed.

The breed has a well-shaped, distinctive head with a straight profile, small active ears, and prominent eyes; the neck is crested, short, and muscular and thick at the base; the withers are low but well defined; the back is short, strong, and wide. The loins are broad; the shoulder muscled and well sloped; the croup rounded and sloping. The legs are sturdy, short, and muscular with feather on the lower legs; the hoof is broad and strong.

The Belgian Ardennais may be bay, roan, chestnut, grey, or palomino. Black is excluded from registration and very uncommon.

The Ardennais has a calm, tolerant attitude and is praised for use in hilly and rough terrain, where it is surefooted and agile.

Belgian stallion in Belgium. Photo: L. Rottiers

Belgian Draft. Photo: L. Rottiers

Belgian Draft

(Brabant, Brabançon, Belgian Heavy Draft)

ORIGIN: *Belgium*
APTITUDES: *Heavy draft work*
AVERAGE HEIGHT: *15.3 to 16.3 h.h.*
POPULATION STATUS: *Common*

Origin of this breed is very ancient. The Belgian is most likely descended from the heavy prehistoric horse of the alluvial period, fossilized bones of which have been found from Liège to Dinant along the bank of the Meuse River. Belgium lies in the heart of the area in western Europe that gave rise to the large horses, usually black, known as Flemish or "great horses" by medieval writers.

Until the beginning of this century there were three distinct types known in the Belgian breed; the Big Horse of the Dendre, the Colossal Horse of Mehaigne, and the Grey

Belgians in Germany. Photo: Evelyn Simak

Woodland's Commander, a good example of the American Belgian. Photo: Jim and Retha Carter

Horse of Nivelles. Today, they vary little in conformation or color.

Belgians are large, powerful horses with willing temperaments. They are quick to mature and long-lived, and prior to mechanization of farms and transport were indispensable for heavy draft work. Many Belgians weigh as much as 2,200 pounds. The coat is usually red or chestnut roan, but other colors occur.

The head of the Belgian is light and expressive, somewhat squarish in shape. The profile is straight or slightly concave, the ears small. The neck is short and heavily muscled; the back is broad and short; the flanks are often slightly hollow; the croup is muscular, rounded, and massive; the body is deep and short; the shoulder is sloped; the legs are strong, lean, and sound, with some feather at the fetlock.

The stud book for the Belgian was started in 1885 and is published by the Société Royale le Cheval de Trait Belge. The American stud book was started in Wabash, Indiana, in 1887. Belgians were slower to catch on in America than were the Percheron, Clydesdale, and Shire until into the twentieth century.

Prior to mechanization of farms, Belgian draft horses were exported to many other parts of the world. The government of Belgium worked to fix the characteristics of the breed, and this was accomplished by the early twentieth century. In 1903 the Belgian government sent an exhibit of horses to the St. Louis World's Fair and the International Livestock Exposition in Chicago. This generated a great interest in the breed in America. Many horses were imported to the United States until World War I, when importation ceased, but luckily, enough Belgians had come to America by that time for breeders to continue raising this breed.

The decline of interest in large draft horses after World War II nearly meant the extinction of all draft breeds, but numbers are increasing steadily again as interest in owning and driving heavy horses is reawakening.

The Belgian in America has changed considerably in type from the original Belgian breed. American Belgians are predominantly chestnut-sorrel in color, with flaxen-colored manes and tails. The American horse is taller, more stylish, and has more slope to the shoulder than does the horse of Europe.

In Belgium this horse is primarily directed toward meat production.

Belgian Halfblood

ORIGIN: *Belgium*
APTITUDES: *Riding horse*
AVERAGE HEIGHT: *16 h.h.*
POPULATION STATUS: *Common*

A model Belgian Halfblood. Photo courtesy Hugo Cooreman

Belgian Halfblood. Photo courtesy Hugo Cooreman

The Royal Belgian Halfblood Society was founded in 1920. Until World War II, the breed was primarily used by the army. This breed was based on the indigenous, well-known Belgian Brabant, crossed to English Thoroughbred and Selle Français mares.

After the Second World War Belgian breeders concentrated on breeding an all-around sport horse, combining their best bloodlines with those of surrounding countries. They acquired Selle Français (French Saddle Horse) mares and stallions, German Hanoverians, Dutch Warmbloods, and Thoroughbreds with proven ability such as Furioso xx, Ibrahim, Rantzau xx, and Lugano.

Since the 1960s efforts of these dedicated breeders have shown results, and several Belgian Halfbloods have appeared in international show jumping competitions. The breakthrough came when Gai Luron (de Monnikenhof), under the saddle of François Mathy, won a bronze medal during the Montreal Olympics in 1976.

During the following years many Belgian Halfbloods were purchased by top international riders, and their jumping performance raised the Belgian Halfblood to the same level as the Selle Français, Hanoverian, Holstein, and other well-known European sport horses.

The Royal Belgian Halfblood Society represents 20,000 horses, 2,000 regular and occasional breeders, and 3,000 breeding mares and registers approximately 1,000 foals annually.

Belgian Warmblood

ORIGIN: *Belgium*
APTITUDES: *Riding, sport horse*
AVERAGE HEIGHT: *16 h.h.*
POPULATION STATUS: *Common*

The famous Belgian Warmblood, El Futuro, ridden by Ian Millar of Canada. Photo courtesy E. Bolmain

Stallion testing for the Belgian Warmblood is a very difficult examination. Photo courtesy E. Bolmain

Breeding of the Belgian Warmblood is crucially connected to the Rural Cavalry founded in 1937 in Boezinge (West Flanders) by Canon André De Mey. This priest founded the Rural Cavalry as both a pursuit and entertainment appropriate for country youths and a celebration of the skills of country people.

Predominantly an agricultural country with more than 500,000 people active in agriculture and horticulture, Belgium had an extensive stock of draft horses. In 1935 approximately 40,000 foals were born and more than 11,300 draft animals exported.

The "Canon's horsemen" had to ride the horse they owned, which resulted in representation of several types. The first riding tournament occurred in Boezinge in 1938. During that event, 4,250 spectators were present and thirty-nine horses entered the competition. The program included a parade, a "merry-go-round" in walking and trotting, dressage demonstration by a quartet, riding performances, and jumping. In the competition, twenty-five heavy and fourteen light horses entered. None of the types specifically resembled the Belgian Draft horse or the half-bred horse. All were the breeding result of Thoroughbred or Trotter stallions with heavier mares. This situation clearly obstructed development of a homogenous rural cavalry such as those existing in several surrounding countries. A double objective was determined: By way of riding associations and courses, the country lads had to learn the art of horse riding; and the breeding of a proper type of horse had to be achieved.

Wartime hindered realization of the dream, but when the war was over a base formed for the Belgian Warmblood breed, using horses of Gelderland (Netherlands), Normandy (France), and Hanover (Germany).

The first inspections for stallions to be registered occurred on June 9, 1953, in Waregem, and on July 7 the Minister of Agriculture, Charles Héger, gave approval for experimentation on the condition that all breeding results be noted for five years. In 1954 twenty-seven foals were accepted for registration in the foals stud book.

Founding a stud book became necessary to allow stallions to be officially available for public service. This was not a simple matter, for organizers had to justify a stud book for agricultural-riding horses precisely when mechanization was dramatically altering farming practices and the number of horses used for agriculture was decreasing rapidly. In spite of resistance, demonstrations were held in 1954 to seek recognition of the agricultural-riding horse, and on March 25, 1955, the National Breeding Association of Agricultural-Riding Horses was founded in Louvain, Belgium.

In 1958 competitions replaced demonstrations and on September 1, 1960, the new association received official recognition along with the Belgian Draft horse, Ardennais, Thoroughbred Jockey Club, and Belgian Halfblood.

On February 18, 1970, responding to the modified use of the new type of horse, the name changed to the National Breeding Association of Warmbloods.

Immediately after the jumping contest in Brussels in 1982, the Belgian journalist Ludo Van Bouwel began to pay serious attention to the growing role of Belgian horses in international horse shows and helped spread word of them in the press.

The Belgian Warmblood is an outstanding sport horse and is certain to remain in the news whenever international competitions occur.

Bhutia

ORIGIN: *India (Himalayas)*
APTITUDES: *Riding, pack horse*
AVERAGE HEIGHT: *13 to 13.2 h.h.*
POPULATION STATUS: *Common*

The home of the Bhutia breed lies along the Tibetan border, extending throughout the sub-Himalayan tract from Himachal Pradesh to Darjeeling. Thought to be closely related to the Tibetan and a larger version of the Spiti, the Bhutia is a popular animal in hilly regions for both packing and riding. The breed is perfectly suited to the high mountain slopes which are inaccessible to breeds not native to the area.

The predominant coat of the Bhutia is grey or iron grey, occasionally chestnut or roan. The body is compact with a short, thick neck, strong back, round, muscular quarters, and coarse, hairy legs which are short and strong. The hooves are quite open at the heels. The mane and tail are usually very long and heavy.

Black Sea Horse
(Chernomorskaya)

ORIGIN: *Krasnodar-Rostov area (former Soviet Union)*
APTITUDES: *Riding, light draft*
AVERAGE HEIGHT: *15 to 15.2 h.h.*
POPULATION STATUS: *Rare*

The Black Sea horse (also called Chernomor) survives today in very small numbers. This horse was bred in the coastal areas of the Krasnodar region and in some parts of the Rostov region. Essentially a saddle horse, the Black Sea breed was also used in agriculture and transport.

Development of the breed began at the end of the eighteenth century when Kazakh people moved to Kuban. The aboriginal horses of the region were the Nogai, the old Kuban horse (now extinct), and various types of mountain horses. Most famous was the Nogai, bred by Tartars living along the Kuban River. The Nogai horses were very numerous prior to the arrival of the Kazakhs and widely sold at horse sales in Moscow and other places. Nogai horses were also bought by state-owned horse farms for breeding. Many went to the Don steppes and had a major influence in development of the Don breed. The Nogai was small, lean, of riding type with some characteristics of the mountain horses, and was used to improve those horses.

Kazakh people from Zaporoz settled in Kuban in 1792, bringing their horses with them. The horses were of saddle type, improved by Arab, Turkish, and Persian blood. They were crossed with the Nogai, Kabarda, and other mountain breeds and also with Don and Thoroughbred. The Black Sea horse, produced from these crossings, was a better saddle type and its size was increased.

During the second half of the nineteenth century the people began to raise grain crops and fine-wooled sheep. Shepherds moved to Kuban from Cherson, Poltav, and Ekaterina Gubernias, bringing with them agricultural horses with little saddle blood. During this time large private studs were established to raise saddle horses and remounts for the army. Sires used were Thoroughbred, Don, and Orlov-Rostopchin (Russian Saddle Horse, now technically extinct), crossed to Black Sea and Karabakh mares. To develop a more massive and taller horse, some stud farms also used Orlov Trotters. *Taboons* continued to raise the local type of Black Sea horses, and these animals were quite different from the stud farm animals, being smaller and better adapted to local conditions, having more traits of the native mountain horses.

The Black Sea horse was able to cover sixty miles per day under saddle and had a wonderful, calm temperament and attractive appearance.

Prior to the First World War, a large number of Black Sea horses went to Romania, causing a severe shortage of good breeding stock. Only a few remained after the war at Achtar farm and some other private farms. As a result of various breeding methods, horses from these farms differed in appearance. Some were of draft type, others leaner and more like the Don or light halfbred horse.

All of the surviving Black Sea horses were collected from 1921 to 1923 at the Primor-Achtar farm. After liquidation of this farm, the horses were moved into herds of stud farms in the Don region producing horses for the army. They were not bred pure, but were crossed with Don and Thoroughbred. With their outstanding qualities of appearance, constitution, and working ability, the Black Sea horses of all types had a strong influence in development of the Don and Budyonny breeds. Although breeding at stud farms was discontinued, country breeders continued the breeding, carefully using the blood of Don and Thoroughbred horses.

There were three types within the Black Sea horse breed—massive, having the blood of light draft breeds; light saddle, having influence of Arab, Don, and Thoroughbred;

and local, preserving the traits of the aboriginal mountain horses. During its entire existence the Black Sea horse has been improved with Don, Arab, and Thoroughbred blood. The basic method for breeding was crossing with those three breeds but only to such a degree that the original type of the Black Sea horse was preserved.

In appearance, the Black Sea Horse has a head of medium size with a straight or slightly Roman-nosed profile; the neck is of medium length and straight; the withers are not too well defined; the back and croup are nearly in a straight line; and the body is well muscled. Colors found in the breed were predominantly black and dark brown. Some were sorrel with a golden glint or sheen.

Extremely rare today in its pure form, the Black Sea Horse has mainly been absorbed into the Don, Budyonny, and Ukrainian Saddle Horse breeds, exerting a dominant influence in formation of those esteemed breeds.

Bosnian

(Bosniak, Bosnian Mountain Pony)

ORIGIN: *Yugoslavia*
APTITUDES: *Riding, pack, driving*
AVERAGE HEIGHT: *12.3 to 14.3 h.h.*
POPULATION STATUS: *Common*

Bosnian mare. Photo: Evelyn Simak

Originating in the Balkans and closely related to the Hutsul or Hucul of Romania and the Myzeqea of Albania, the Bosnian is a small, compactly built horse. The small horse breeds found in the Balkans are all thought to be close descendants of the Tarpan with varying degrees of Mongolian influence. Horses of the area were influenced during the Roman occupation of the Balkan peninsula, and when the Turkish cavalry of the Ottoman Empire overtook the area, the blood of eastern horses was infused. By the seventeenth and eighteenth centuries there were several types found within the Bosnian breed, but not all were considered to be improvements.

Since 1900 the breeding of Bosnian horses has been selective and controlled and the rustic, hardy qualities of this small horse have been preserved. Bosnian horses are bay, brown, black, grey, chestnut, palomino, or dun in color. This is a strongly made horse. The head is somewhat heavy with a straight profile and small ears. The eye is large and expressive. The neck is short and muscular with a full mane; withers are moderately pronounced; the back short and straight; the croup slightly sloping; the chest deep and wide; and the shoulder long and sloping. The legs are short and strong, well muscled, with broad joints and clean, strong tendons. The hoof is well formed and very hard.

Bosnian horses are bred in large numbers in Yugoslavia and are still very much in use on farms and for transport. Stallions are strictly controlled by the state, while ownership of mares is left to private breeders. The breed has been used for centuries for transport in the mountains and is surefooted and hardy.

This breed is quite popular in Germany, where numbers are increasing steadily. A Bosnian stallion has been introduced to the Dülmen herd in recent years.

Boulonnais

ORIGIN: *France (Boulogne District)*
APTITUDES: *Draft and farm work*
AVERAGE HEIGHT: *15.1 to 15.2 h.h.*
POPULATION STATUS: *Common*

Boulonnais. Photo courtesy UNIC, Paris, France

The cavalry of the legions of Julius Caesar stayed on the cliffs near Cape Gris-Nez before leaving to invade what is now Great Britain. Much later came the Crusades and the Spanish occupation of Flanders. These events brought a great deal of Oriental and Andalusian blood to the Boulonnais area, and this influence is still seen in the Boulonnais breed today. Later, additional crosses to Andalusian blood and that of Mecklenburg (see Elegant Warmblood) stock from Germany further shaped the breed.

As far back as the seventeenth century the Boulonnais horse enjoyed a fine reputation, and dealers from Picardie and Haute-Normandie came to buy good animals for their business and use.

In former times fish delivery carts were well known to the Parisians. They were drawn by light Boulonnais mares that hauled fish from Boulogne to Paris. Known as Petite Boulonnais or Mareyeuses, the mares were used in relays, the small carts filled with fish and ice.

The Boulonnais is a heavy horse with elegance and distinction. There are two types today: the small-sized "fish cart horse," now very rare, standing 15 to 15.3 hands; and the large Boulonnais, which often weighs over a ton.

The head is elegant and short with a flat and wide forehead, very keen, proud eyes, well-opened nostrils, and small, mobile ears. The neck is thick and muscular; the mane double, thick, and not very long; the chest wide; the rib cage well rounded; and the shoulder perfectly set. Withers are well placed but often set deep into the musculature; the back is straight; and

the limbs are strong and solid with clean joints. There is very little hair on the legs. The color of the Boulonnais is grey, ranging from very light to a dark dappled shade. An occasional chestnut is seen in the breed.

Boulonnais breeding stock is recorded in five departments of the northwest: Pas-de-Calais, North (Dunkerque and Hazebroucq), Somme, Seine-Maritime, and Oise. A large show specializing in this breed is held every summer at Wimmereux in the Pas-de-Calais, bringing together all the best breeding stock.

Energetic, easy, active, and very gentle in nature, the Boulonnais is an excellent draft horse. The breed also has a reputation for improving other breeds of draft horses in spite of very reduced numbers.

The breed is above all an excellent meat animal today, giving a very refined meat and a light bone structure.

Brandenburg
(Brandenburger)

ORIGIN: *Germany (Brandenburg Province)*
APTITUDES: *Riding horse*
AVERAGE HEIGHT: *16 h.h.*
POPULATION STATUS: *Common*

Brandenburg. Photo: Prof. H. C. Schwark

The first Brandenburg stud farm was founded in 1787 in Neustadt/Dosse. During the nineteenth century farmers wanted a heavy warmblood working horse, so many Oldenburgs were used for breeding. There were about one hundred English halfbred (Thoroughbred × warmblood) mares that were mated to Oriental and Thoroughbred stallions. Thoroughbred mares were imported to enlarge the mare herd, and from 1834 there was a pure Thoroughbred herd as well at the stud. In 1845 there were sixty Thoroughbred and sixty halfbred mares and both pure and halfbred stallions. This stud was dissolved in 1866. The Thoroughbreds were sent to Graditz and the halfbreds to Berberbeck.

In 1894 the stud was reestablished using ten Arab mares, which were combined with Thoroughbred mares to form a breeding herd. From that time, Anglo-Arabs were bred. In 1896 a halfbred herd was started from two horses from Trakehnen, two from Graditz, one from Berberbeck, thirteen East Prussians (Trakehner), six West Prussians, ten Hanoverians, three horses from Ireland, and one from England.

The warmblood breeding society was formed in 1922 to produce horses for farmers. Hanoverians were greatly favored. Due to the influence of the climate, feeding, and the breeding stock available, the cross of Hanoverian and East Prussian was desirable. The aim was for a strong-boned, short-legged warmblood horse with a deep chest, good movements, and quiet temperament.

Since 1925, no Oldenburg stallions have been used, and the breeding of the Elegant Warmblood was accomplished by using Hanoverian stallions. Since the 1960s the "riding horse of the GDR" has been bred. Basis for this breeding was the Brandenburg warmblood crossed to Trakehner and Hanoverian stallions and Thoroughbreds.

According to Professor H. C. Schwark at Karl-Marx University, the Brandenburg is historically regarded as a subpopulation of the East German saddle horse.

Brazilian Sport Horse

(Brasileiro de Hipismo, Brazilian Horse)

ORIGIN: *Brazil*
APTITUDES: *Riding horse*
AVERAGE HEIGHT: *over 16 h.h.*
POPULATION STATUS: *Common*

The Brazilian Sport Horse, Best Friend. Photo courtesy Caio Luiz Figueiredo

The Brazilian Sport Horse or Brazilian Horse is the product of crossing stallions from specialized sporting breeds with Brazilian (Criollo) mares specially selected and registered as the basic mares. The principal breeds selected for breeding are the Andalusian, Thoroughbred, Trakehner, Hanoverian, Holstein, Oldenburg, Westphalian, Irish Hunter, Selle Français (French Saddle Horse), Anglo-Argentine, and Belgian Warmblood.

Despite the fact that Brazil has some of the best and largest horse herds in the world, it did not develop horses for sporting events because the demand of the consumer market was supplied by imports from Argentina, Uruguay, and Chile in addition to the Thoroughbred, taken from the tracks. It was not until 1970 that the breeder Enio Monte, of the breeding farm Haras Itapuá, began systematically developing the new sport horse, introducing the breeding of selected animals from the Anglo-Argentine and Trakehner breeds.

Other breeds were later introduced to Brazil and firmed up the formation of the Brazilian sporting horse. On July 9, 1977, the Brazilian Association of Sport Horse Breeders was founded.

The Brazilian horse is an athletic type, versatile and courageous, and has great stature, good structure, and strong muscle with great flexibility and expansive moments. The gaits are cadenced, frisky, and elastic. The character is noble, docile, and balanced, and the temperament is lively without rebellion. This animal has a great disposition for work and excels at show jumping, dressage, or cross-country events.

Currently there are about 6,000 registered horses in the stud book, and as this new breed is developed it is displaying increasing improvement. Many horses of the breed are now participating in various ranks of competition both in Brazil and abroad.

Breton
(Trait Breton)

ORIGIN: *France (Brittany)*
APTITUDES: *Heavy draft and farm work*
AVERAGE HEIGHT: *15 to 16 h.h.*
POPULATION STATUS: *Common*

The Breton horse has a long history, although there are some variances in opinions. According to some, the breed dates back four thousand years or more to the time when it was brought into Europe by Aryans migrating from Asia. Others believe the breed stems from smaller horses bred and improved by Celtic warriors on their conquest of what is now Great Britain.

A population of horses, allegedly descendants of the steppe horses ridden by the Celts, existed for many years in the Breton mountains. At the time of the Crusades, these horses were bred to Oriental stallions and mares, which led to the Bidit Breton. At the end of the Middle Ages two types of Breton horse existed: the Sommier, a northern Brittany pack horse; and the Roussin (meaning cob), originating from the Mountain Bidet, finer and more slender than the former.

During the Middle Ages the Bidet Breton was sought after by military leaders due to its comfortable gait, described as being between a brisk trot and an amble. At this time, the breed stood about 14 hands. This breed was among the horses sent to New France (Canada) by the king of France during the seventeenth century, and the Canadian breed still shows its influence.

During the following centuries and until the nineteenth century, many crossbreedings

Breton. Photo: y. de la Fosse-David, courtesy UNIC, Paris, France

Breton. Photo: y. de la Fosse-David, courtesy UNIC, Paris, France

were made from foreign or even French horses—Boulonnais, Percheron, and Ardennais—in order to adapt the production to the economic needs of various periods.

The crossbreeding of Norfolk stallions at the close of the nineteenth century gave excellent results, producing the "Postier Breton" that made Brittany famous. Since 1930 crossbreeding has been abandoned and selection has become the essential basis of genetic improvement.

At present the Breton horse has these standard characteristics: Coat color is chestnut or chestnut roan, often with flaxen mane and tail, rarely bay or bay roan; the head is square and of medium volume with a wide forehead; the nose is straight but sometimes dished, the nostrils wide; the eye is lively; the ear is small. The neck is strong, slightly short but set well into the withers; the back is wide, short, and muscular; the croup is wide and double; the rib cage is rounded. The shoulder is long and muscular; the limbs are very muscular and short with sound cannon bones.

There are three types of Breton horse. The small Breton draft horse (Center Mountain), considered the real descendant of the ancient Breton horse, has the same general features as the Breton draft horse but is smaller with a more dished face. Very hardy and enduring, this horse is easy to keep and today is gaining once again in popularity. The Breton draft horse, heavier with more bulk, is a strong, muscular, and compact horse. The Postier Breton, having remarkably airy and easy gaits, is very close to the draft horse and is of the same size. This is a more beautiful, distinguished type.

Breeding grounds for the Breton include the Hennebont and Lamballe studs and part of La Roche-sur-Yon stud areas.

Due to quality and popularity, the Breton is the most numerous of the draft horses in France. This breed has been widely exported around the world.

Given its power, hardiness, and energy, the Breton is still used as a work horse on small farms by market gardeners and to gather seaweed. The Breton is mainly bred today for meat production in France.

British Appaloosa
(British Spotted Horse)

ORIGIN: *Great Britain*
APTITUDES: *Riding, sport horse*
AVERAGE HEIGHT: *14.2 h.h. and over*
POPULATION STATUS: *Common*

The name "Appaloosa" is misleading to many when considering the spotted horse in Great Britain. Upon finding the name British Appaloosa Society, I assumed that Appaloosa horses from the United States had been taken to Great Britain and a society formed for them there. The spotted horse in Great Britain has a long history and undoubtedly the same roots as the Appaloosa horse in America; however, the British Appaloosa breed was not based on

British Appaloosa. Photo: Chris Cook

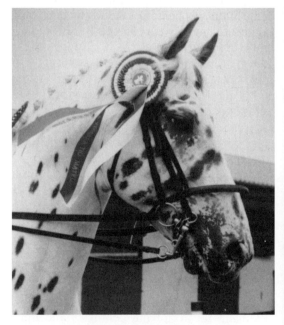

British Appaloosa. Photo: Joanna Prestloich

The British Appaloosa, Cockley Cley Weatherman. Photo: Joanna Prestloich

the Appaloosa of the Nez Perce of North America.

Spotted horses were known in Spain from 100 A.D. An outstanding type of riding horse was bred there in the following centuries, and many individuals were spotted. Spotted horses from Spain were sent to Austria and many other countries. In a manuscript dated about 1397 A.D. one of the first pictures of a spotted horse in Great Britain appears; that of King Richard II and a party of riders reveals one of his followers mounted on a pale chestnut-roan with chestnut spots. Several blanket-marked horses are seen in a Persian painting dating to approximately 1440 A.D.

Likely, the spotted horse came in great numbers into Great Britain with the Spanish Jennet. There is also a very old strain of spotted horses and ponies found in Wales.

In 1976 it was suggested that the British Spotted Horse would have a better export market if it were known as an Appaloosa, as the name carries with it a romantic and exciting history. The Appaloosa Horse Club in Moscow, Idaho, was consulted and agreed to the name's being used provided that ponies and pony breeding were excluded. The British Spotted Horse and Pony Society split at that time and the British Spotted Pony Society was

then formed. At the time there were already a few American Appaloosas, some Knabstrups, and many horses of unknown or partially unknown breeding registered. With sensible outcrossing, mainly to the Thoroughbred, the British Appaloosa has developed into a quality riding type of increasing popularity. In recent years a few additional American and Canadian horses have been imported.

For additional information regarding Appaloosa colored horses in history and color patterns, see Appaloosa.

British Riding Pony

ORIGIN: *Great Britain*
APTITUDES: *Riding, show pony*
AVERAGE HEIGHT: *12.2 to 14.2 h.h.*
POPULATION STATUS: *Common*

The British Riding Pony, News Extra. Photo: Carol Gilson

The British Riding Pony is more a type than an actual breed. This pony is of fairly recent development, bred for the show ring by crossing small Thoroughbred or Arab stallions to native British pony mares, including Welsh, Dartmoor, Exmoor, and most often, New Forest. The ideal cross is a Welsh crossed to Thoroughbred with a bit of Arab blood. The Riding pony is an animal of quality with superb conformation and style. The action is free and straight (from the shoulder and not the knee). The head shows quality and intelligence and is fine with large, well-spaced eyes and small ears. The neck is long and well formed; the withers are well defined and high; the back is straight and medium in length; the croup is muscular and rounded; the tail is set on fairly high; the chest is deep and wide. The legs are hard and clean with broad joints, short cannons, and good, hard feet.

British Riding Ponies are registered in the National Pony Society and British Show Pony Society, and many shows and classes are provided in various divisions. The Working Hunter pony is derived from the Riding pony, which it resembles.

British Spotted ponies owned by Mary Bassett, capturing the *Horse & Hound* Driving Team of the Year award at Wembly Arena in 1987. The right wheeler is a rare Eriskay pony. Photo: Toby Hunt

British Spotted Pony

ORIGIN: *Great Britain*
APTITUDES: *Riding, driving pony*
AVERAGE HEIGHT: *Up to 14.2 h.h.*
POPULATION STATUS: *Rare*

Spotted ponies have existed in the British Isles for many centuries. It is unknown whether they were feral or appeared about the time of the First Crusade, but they are found in many illustrated manuscripts, old paint-

British Spotted Pony foals, from the Broomells Stud. Photo: Mary Bassett

The British Spotted Pony stallion, Wantsley Ariel, at Parnham House. Photo: Mr. and Mrs. Makepeace

ings, and drawings and were obviously widely used both in peace and war in early times. The most common size was a sturdy cob type about 14 hands.

Harsh laws against ponies in the Tudor period seem to coincide with the disappearance of the smaller spotted animals from fashionable equestrian paintings of early times, although the larger Spanish horse was very popular. It was from the Spanish horse, taken to the Americas by European explorers, that the Appaloosa horse descended. It is incorrect, however, to refer to the British Spotted Pony as an Appaloosa. In Britain, despite the emphasis on heavy horses, some people quietly continued to breed many strains of the beautiful native ponies for which the United Kingdom is still so well known, and the spotted ponies were preserved in certain bloodlines, especially in Wales and the southwest. Spotted ponies again became fashionable in the mid-nineteenth century, and there is a print of a lady claimed to be Queen Victoria driving a blanket-spotted pony, but more evidence of its authenticity is needed.

The British Spotted Horse and Pony Society was formed in 1946 to keep a register of the animals and attempt to preserve them. Many good animals were exported to Canada, the United States, Germany, Holland, and Australia. In 1976 the society was split and separate registers were formed. The British Spotted Pony is rather scarce today due to many being sold overseas, but presently there is an awakening interest in the breed and its hardiness.

To demonstrate the versatility of the ponies there is a performance award scheme competition each year, which gives owners a chance to show what they and their ponies can do, whether it be showing, eventing, driving, jumping, long-distance riding, or Pony Club activities.

The society maintains a register for stallions, mares, and geldings of known breeding. Stallions are accepted only after obtaining a veterinary certificate. Presently there is a temporary registration plan for ponies of unknown or unregistered breeding who conform to the required breed characteristics and are of good quality. The society aims for preservation and improvement of the British Spotted Pony with emphasis placed on good conformation, quality, and soundness. Ponies may be of riding or driving type, including cobs, up to and including 14.2 hands with no lower limit.

Coloring and characteristics such as white sclera around the eye, hooves with white and black vertical stripes, and mottled skin are typical of the British Spotted Pony.

Brumby

ORIGIN: *Australia*
APTITUDES: *Riding horse*
AVERAGE HEIGHT: *14 to 15 h.h. (varies)*
POPULATION STATUS: *Common*

Feral brumbies in Australia. Photo: Peter Fischer

The Brumby is difficult to separate from the Waler—and the Waler has a history identical to that of the Australian Stock Horse, at least to a point. Still, the three are distinctly different and no doubt there has been many a lively debate regarding the subject. Understandably, the matter for liveliest debate in recent years has been the practice of government authorities' shooting Brumbies, or feral horses, from helicopters—with the result often being cruelly wounded but not killed animals that are left to suffer.

The question of what a Brumby is, when the Waler descended from exactly the same mixed stock, becomes confusing. Peter Fischer of the Waler Society sums it up this way: "This depends on who is using the term. In some areas any horse in the bush is a Brumby, no matter how 'well bred.' Generally, the term is applied to wild (feral) horses of poor stature, conformation and temperament. That is, a horse of little use for human requirements. A horse well adapted however, to survival." What complicates the matter even more is the fact that there are apparently both feral Brumbies and feral Walers.

The horse is not native to Australia. When the *First Fleet* arrived on Australian shores in 1788, seven horses were aboard, and shipments following that usually included equine stock, which was needed to explore and settle the new land. The Brumbies are descendants of imported domestic stock that have, over a period of many years, escaped captivity to roam freely in the vast Australian wilderness, or outback. Many different breeds were imported to Australia, ranging from heavy draft to small ponies.

Herds of feral horses are found today in several areas of Australia, with major concentrations in the northern part of the Northern Territory, the central part of the Gulf of Carpentaria, around Alice Springs in Central Australia, east of Birdsville, and in parts of Western Australia and Queensland. While some report the Brumby to be rare today (blaming the shooters), the Department of Foreign Affairs and Trade has estimated over 80,000 head in an aerial survey conducted in the area of Alice Springs. In total, there may be as many as 200,000 in the Northern Territory. About 100,000 are thought to inhabit Queensland.

Most of the feral horses in Australia are found in the arid zone of Central Australia (70 percent of Australia is arid, an area equivalent in size to western Europe). The horses are a problem. In the Australian arid zone, as in arid zones everywhere, droughts occur regularly. In the past, a rich native fauna of small mammals coped with drought by retreating to refuges near small springs and seepages, where some of them always managed to survive.

With European settlement and the arrival in Australia of new animals, some of which eventually ran wild, this all changed. When droughts occur the feral animals, and particularly the horses, compete with native fauna for the decreasing supply of food and water. Horses have great mobility and can range further from water, grazing out the native pastures. In the process, they destroy the chances of survival of an entire range of unique Australian fauna. Several native species have become extinct and others are in danger.

The threat to Australia's native wildlife is now so great that culling of feral animals is vital. The Australian authorities are concerned to see that this is done in the most humane manner possible. Under the Australian Constitution, the Commonwealth is responsible for control of native fauna. However, the day-to-day management of feral animals lies with the individual state and territorial governments.

Feral horses are culled in outback Australia when the land is overgrazed and the environment is being destroyed. Similar government-supervised culling programs are carried out on, for example, donkeys in Western Australia, buffalos, pigs, and cattle in the Northern Territory, and cattle in Queensland. Feral pigs are culled in most states.

Although a mixture of many breeds (see Waler), the Australian Brumby has developed through natural selection in its demanding environment into a general type for the majority, though there are exceptions and several types do exist. The Brumby bears little resemblance today to the related Waler and Australian Stock Horse. Usually, the head is somewhat heavy. The neck is short and muscular; the back is short and straight; the croup is sloping with a low tail-set; the shoulder is often quite straight; the legs are sturdy, strong, and have broad joints. However, due to the fact that some Brumbies carry a good deal of Thoroughbred blood, often specimens are found to be quite refined and well conformed.

Most of the Brumbies are descended from great numbers of horses that were turned loose at the time of the great gold rush in Australia in the middle of the nineteenth century. A report from the International League For The Protection of Horses (Australia) indicated another derivation for some feral horses: "I can assure you that the Brumby is not a descendant of the Waler. I can't give you a great deal of history on the Brumby other than to tell you that it was named after a soldier, that there are very few Brumbies left in Australia, and that our wild horses which are being killed from helicopters are not Brumbies, but pure bred horses which were let loose in the 1970s after a particularly serious beef crisis."

As earlier noted, the "Brumby" was well-known for many years prior to the 1970s. This horse remains somewhat of a mystery if it be rare and if the term "Brumby" describes any feral Australian horse—and the Australian government claims more than 300,000 feral horses reside in the bush. As Peter Fischer stated above, "this depends on who is using the term."

Some claim the Brumby to be rebellious and intractable, but it should be noted that any feral horse, whether in Australia or the United States or anywhere else, is naturally intractable due to an inborn fear of people. Offspring of completely gentle animals are "rebellious and intractable" if never handled. In general, the horse, with the exception of Przewalski's Horse, will always respond to kindness and gentle handling and become as tractable as any "domesticated" horse. Some Brumbies are captured and broken in for riding, but this is not a widespread practice. Australia today is a country rich in good horseflesh obtainable by easier means.

There is little doubt that this feral horse would be a good candidate for endurance competitions, as nature is the best breeder of tough, enduring horses.

In the United States, the Adopt-A-Horse Program initiated by the Bureau of Land Management seems to be an excellent solution to management of feral horses (see Mustang). One can't help but wonder if the Australians have considered such a program. Even if there were no market for the hardy Brumby as a riding horse, there are many countries in the world that would no doubt be interested in purchasing horsemeat. To some that is an unappealing solution—but it is equally distasteful to know that thousands of useful horses are being shot and left to rot.

For additional information on history of the Brumby, see Waler.

Poco Frosty Blue, grulla Quarter Horse stallion, also registered in the buckskin association. Photo: Marge Davitt

Leo Reno, IBHA P-1 (deceased), first horse registered in International Buckskin Horse Association. Photo courtesy Rich Kurzeja

Buckskin

(Dun)

ORIGIN: *United States*
APTITUDES: *Riding horse*
AVERAGE HEIGHT: *15 h.h.*
POPULATION STATUS: *Common*

The dun horse is not a breed, but a dominant color found in horses. The first registry for dun and buckskin-colored horses in the United States was the Dun and Buckskin Registry, founded in Sacramento, California. Not much is known about this association and it

Buckskin performing at pole bending, at IBHA World Championship show. Photo: Jeff Kirkbride

did not flourish. I founded the American Buckskin Registry Association in 1962. Research into the ancient strains found in Europe, Asia, and the British Isles revealed some interesting features about the dun-colored horse which seemed to justify a registry association in the United States.

Growing up on the Colville Indian Reservation in northeastern Washington, I had always heard cowboys and ranchers refer to dun horses in the belief they had more stamina and harder hooves and made generally better ranch horses. Research showed that this feeling was prevalent throughout the ranching areas of the United States.

While the dun color (with dorsal stripe and zebra stripes on the legs) is found within many breeds, it is also true that these breeds descended from original or primitive dun types which retained the dun color through the centuries, even with outcrossing to breeds of other colors. With the color, the hard feet and good stamina are retained.

There is more than one type of "dun," and not all breed true for color and tough qualities. It is possible to produce a buckskin-colored horse by mating a palomino to bay, for example, and the offspring will not have the dun factor markings (dorsal stripe, shoulder stripe, zebra stripes on the legs). This type of buckskin or dun horse can in turn produce buckskin, palomino, white (due to the dilu-

tion gene), or any other color. Horses with the dun factor will produce dun offspring at least 50 percent of the time regardless of the color of their mates. Some individuals with strong breeding of dun in the pedigree will produce dun offspring exclusively.

It has been found that even the hair of the true dun is different from any other when viewed under a microscope, with pigmentation concentrated down one side of the hair shaft like a stripe, while the hair of horses not having the dun factor will show pigmentation spread throughout the hair shaft.

The majority of dun-colored horses in America are descended from the horses of the Spanish conquistadores. When European settlers arrived in America, they brought other breeds of predominant dun color such as the Norwegian Fjord, or Norwegian Dun. There has been some speculation among historians that the native, ancient Sorraia of Spain may have obtained its dun color from the old northern dun of Europe. The dun and buckskin color is found most often in the United States within the Quarter Horse and the mustangs, both descending from Spanish and early colonial stock.

In 1970, I resigned from the American Buckskin Registry Association and formed the International Buckskin Horse Registry due to internal problems within the original association. Two associations for the budding buckskin and dun movement in the United States seemed likely to be beneficial to the public preferring this color in their horses. After about a year, when the new association was well started, it was turned over to competent people in Indiana, where the name was changed to the International Buckskin Horse Association. Today, both associations are flourishing and many horses are registered in both the American Quarter Horse Association and one or both of the registries for buckskin and dun horses. Many dun Quarter Horses can be traced in an unbroken line of duns to a foundation sire or dam dating to the 1930s in the United States even when no other dun horses appear in the pedigree, a testimony to the dominance of the dun factor.

Buckskin is described as a body coat of some shade of tan, from very light to very dark bronze, with black mane, tail, and legs. The dorsal stripe is usually absent. Dun is described as a body coat of some shade of tan with black points (mane, tail, and legs), dorsal stripe, and other dun factor markings such as zebra stripes on the legs and shoulders. Red Dun is a horse having a body coat of reddish tan, with red or reddish brown mane, tail, and legs, dorsal stripe, and other dun factor markings. The grullo (pronounced "grew-yo") is a slate-colored horse, not grey. *Grullo* in Spanish refers to a blue crane. The body coat may range from bluish grey to a nearly brownish shade, with black mane and tail, dorsal stripe, shoulder stripe, and barring on the legs.

There are variations within the above colors. Technically, the palomino is a dun horse, as the term dun refers to a horse with a body coat of some shade of yellow. However, palominos are not eligible for registration in either of the associations for dun horses. Both associations still have an open stud book and horses of unknown pedigree are eligible for registration if they meet color and conformation requirements.

Found within the dun factor family of horses is the unusual brindle-colored horse. Essentially a dun, this horse has stripes over the entire body, reminiscent in many ways of the zebra.

The associations for buckskin and dun horses provide many shows for members to compete in and a wide range of youth events. Horses from other countries are welcomed into the stud books.

For additional information see the Spanish-American Horse.

Budyonny. Photo: Nikiphorov Veniamin Maksimovich

Budyonny. Photo: Nikiphorov Veniamin Maksimovich

Budyonny

(Budennovskaya)

ORIGIN: *Former Soviet Union*
APTITUDES: *Riding horse, light draft*
AVERAGE HEIGHT: *15.1 to 16.1 h.h.*
POPULATION STATUS: *Common*

The Budyonny breed was developed in the Rostov region of the Soviet Union during the period from 1921 to 1949 at the S. M. Bud-yonny and Pervaya Konnaya Armia studs. Breeding of this horse paralleled restoration of the Don breed. The main objective in creating the Budyonny was to produce a high grade cavalry horse which could be used in the general improvement of saddle horses in the steppe regions. The breed was developed by crossing the best Don and Chernomor (Black Sea horse) mares with English Thoroughbred stallions. Of the 657 mares used to produce the Budyonny, 359 were Anglo-Don; 261 Anglo-Don × Chernomor; and 37 Anglo-

Chernomor. These mares, which combined the most desirable qualities of the original breeds, were served by the best Anglo-Don stallions, who may rightly be called the founders of the Budyonny breed. Although 100 Thoroughbred stallions were used in formation of the new breed, the descendants of only four were admitted into the breeding nucleus: Sympatiaga, Svetets, Inferno, and Kokas.

The modern Budyonny is a horse of good height with a clean, solid build and heavily muscled body. This breed is good-tempered and energetic, an animal easily broken. It is one of the best racehorses and an able jumper.

The strong influence of the Don breed resulted in the preservation of corresponding interbreed types in the Budyonny—eastern, heavy, and basic. The Don also contributed the golden chestnut color which prevails in the Budyonny breed, although bay and black are also seen.

At first it was necessary to reestablish the Don horse as a basic saddle breed. This goal was realized within a short time, but the Don was unable to meet all of the requirements although it had many good qualities. A horse of more blood was required, larger, able to work under saddle, and suitable for improving the local small-sized horses of the steppe regions. The Don crossed to Kazakh and Kirgiz mares produced offspring of larger size but with coarser heads and lower set necks. The Thoroughbred, used with the Don, produced offspring of well-expressed saddle type but insufficiently tough for *taboon* raising (see glossary).

Breeders were given the task of developing a breed improver, able to produce offspring of a well-conformed saddle type and adapted to *taboon* conditions, from the local mares. For the breeding of such an animal the decision was made to cross Don mares with Thoroughbred stallions. The crosses were made to the best elite Don mares. Selection was made for massiveness, correct conformation, and speed.

In *taboon* raising, conditions are changeable. In spring the steppe is rich in plant life and there are no bloodsucking insects. The weather is mild and the horses feel good,

developing quickly. During summer the plants dry up, flies and mosquitoes appear, and the horses are subjected to difficult conditions, having to walk some distance to water. In the fall, after the rainfall, the period of fall fattening begins. Winter brings severe frosts and winds, wet blizzards, and soil covered with ice without snow, presenting very difficult circumstances for the horse. Under such conditions development of horses is quite inequitable. During the spring they grow quickly and gain weight; in summer and winter they lose flesh rapidly. Youngsters of the native breeds are able to retain their condition much better than crossbreds.

Performance of this breed is improved by correct crossbreeding with Thoroughbred stallions up to the level of five-eighths Thoroughbred blood, but a higher percentage tends to lead to certain weaknesses such as poor bone and decline in fertility and body weight.

Budyonny horses have shown remarkable endurance both in racing and in classic sport events. The stallion Reis held a USSR record and had international success in the obstacle race; Pintset in the three-day event; Gasan in dressage; Priboi in the steeplechase.

The Budyonny is described as being well built, with a comparatively massive body; the blood vessels and tendons show clearly through the thin, supple skin; the head has a straight profile and is well proportioned; the withers are high and the shoulder is of good length and slope; the back is short and straight; the ribs are well sprung and the loins muscular; the croup is long and rounded; the legs are fine, strong, and straight.

This breed is capable of adapting to extremely severe conditions. In an experiment during the 1950s a number of Budyonny horses were turned loose on a large island in Manych Lake in the Rostov district, creating "Budyonny mustangs." In some years their population rose appreciably, but in 1985 only forty head were surviving: twelve stallions, fourteen mares, and fourteen youngsters. The horses have retained the chestnut and sorrel color and live in three groups governed by stallions; three other stallions live alone. The horses live on the island without human assistance and have typical traits of wild horses. The number of horses originally freed on the island is unknown to the author, but the experiment has proven to authorities that a breed such as the Budyonny is able to adapt and survive for an extended period without supplemental feeding or shelter provided by humans.

Buohai

ORIGIN: *China (Northeastern Shandong Province)*
APTITUDES: *Riding, draft*
AVERAGE HEIGHT: *14.3 to 15 h.h.*
POPULATION STATUS: *Common*

Buohai stallion of China. Photo: Weigi Feng and Youchun Chen

Buohai mare of China. Photo: Weigi Feng and Youchun Chen

In the northeastern part of Shandong Province close to the coast of the Buohai Sea there is a large plain where the well-known Yellow River flows. Agriculture, fishing, and fruit raising are all well developed in the area.

Near the city of Qindao and provincial capital Jinan, archaeologists excavated burial grounds containing more than 600 horses. It was determined that the horses had been buried about 2,700 years ago. Examination revealed them to be of Mongolian origin and the same size as present-day Mongolian-type horses in the area.

In 1952, the local government initiated a crossbreeding program designed to improve the local stock. Soviet purebred and high-bred horses (ten stallions) were introduced and artificial insemination was used. In 1956, 2,000 animals of different ages were accounted for. As the first type produced did not meet the requirements needed for heavy work, two stallions and seventeen Ardennais mares and three stallions and thirteen mares of the Soviet Heavy Draft breed were brought in and a stud was set up to cross the improved horses. Thus, the Buohai is classified as a developed breed in China.

In 1963, Shangdong Agricultural College established a good breeding program for state-owned farms. In 1974, a breed association was organized and the first animals were registered; 3,257 animals were accepted as purebreds with the new name, Buohai.

The morphology of the Buohai breed is compact and strong. The head is clean-cut and has a straight profile; the neck is of medium length and well connected to the shoulder, which has good slope; the withers are high. This horse is deep in the chest, flat at the back and loin; the rump is wide and long with some slope. The legs are dry with clearly defined joints and tendons, and the hooves are strong. Predominant colors of the breed are bay and chestnut. A small white star on the forehead is common.

Buohai horses are well adapted to local conditions and do not lose much weight under poor management. They are very fertile. As this breed was developed from both very light and heavy horses, a variety is found in the type. Plans have been made to correct this with more careful selection of breeding stock.

Buryat

(Buryatskaya, Zabaïkal'skaya)

ORIGIN: *Buryatia (former Soviet Union)*
APTITUDES: *Riding, light draft, pack*
AVERAGE HEIGHT: *13.1 to 13.3 h.h.*
POPULATION STATUS: *Rare*

Also commonly called Zabaikal or Trans-baikal, the Buryat breed was once widespread in the Buryat-Mongol ASSR region. A versatile little horse, the Buryat was used under saddle, in agricultural work, and for transport.

The climate of the region is strongly conti-nental and only about one hundred days a year are without freezing temperatures. Rainfall is scarce, and in winter only a few inches of snow may fall. The forest-steppe and moun-tain-taiga regions have been tilled only mini-mally to obtain hay and are otherwise used for grazing. The western area is semidesert with salty, sandy soil.

Bred the year round in *taboons* (see glossary) without supplemental feeding, the Zabaikal or Buryat horse is a hardy individual. This is one of the smallest horses of Siberia, close in origin to the Mongolian breed but for the most part slightly taller. Within the breed, however, members taller than the average are found, and smaller individuals standing only 12 hands also exist. All of these are of the same general type, the primary difference being size.

The Buryat is bony and massive in build with short legs. It is well adapted to local conditions and makes good use of poor pas-turage, putting on weight in summer and a heavy coat of hair for the winter. The head is large and rather heavy, with massive jaws and a straight or slightly Roman profile; the neck is of medium length or often short and muscu-lar. The withers are low; the back straight and strong; the croup sloping; the chest narrow and rather deep. The legs are short, lean, and bony with well-developed joints and small, firm hooves. There is a minimal amount of feather on the legs. The predominant colors are grey, bay, or sorrel, usually with zebra stripes on the upper legs.

The Buryat has the gaits walk, *tropota* (a sort of running walk), and gallop. Especially prized are pacers, although really good ones are rare.

The Buryat has great endurance and is able to travel long distances. It is very healthy and resistant. Local people used to tether the sweating horses after work in the wind and in frost at temperatures of −40°F for three to four hours to harden them. When trained for racing, the horses were often subjected to even more severe treatment. The sweating horses were washed with cold water and then tethered in the wind for some hours in freez-ing weather, and then later supported with food and water. This practice was called *vystojka*, and no other breeds but the Buryat and its close relatives are able to withstand such "strengthening of health."

The Buryat is extremely hardy and fertile, even though wintering on pasture only. Many horses of this breed are still fertile and able to work at over twenty years old. As a larger horse was needed for agricultural work, the Buryat was "improved" in the seventeenth century when Russians entered the Zabaikal territory, bringing heavy draft horses with them. The famous stud farm Chilkovski de-veloped the Chilkov breed, a group of about 1,000 head averaging about 15.1 hands. The Chilkov, however, was more demanding re-garding feed and keeping, and so had no influence on the Buryat breed. Since 1897 Thoroughbred stallions have been brought into Zabaikal territory to improve the breed, and in some regions trotters have been used.

Mares from these improved horses retained the type of the native horse in general, while the stallions lost these traits and were more like the trotters and Thoroughbreds. The fer-tility of the improved horses is lower than that of the original Buryat horses.

The Buryat breed is without official breed status and survives today only in small numbers.

Byelorussian Harness

(Belorusskaya uprazhnaya)

ORIGIN: *Byelorussia (former Soviet Union)*
APTITUDES: *Harness, draft*
AVERAGE HEIGHT: *14.1 to 14.3 h.h.*
POPULATION STATUS: *Common*

Byelorussian Harness. Photo: Nikiphorov Veniamin Maksimovich

Byelorussian Harness. Photo: Nikiphorov Veniamin Maksimovich

The Norwegian Døle, Ardennes, and Brabançon breeds crossed to the native northern forest horses formed the Byelorussian Harness horse. Influence of the Døle was the strongest.

Crossbreeding of the above breeds over a long period produced a horse most suited to the current requirements of Byelorussian agriculture. This horse is well adapted to work in wooded areas with sandy or swampy soils.

Byelorussian horses are medium in size with the typical conformation of a harness horse. The head is not large, the forehead wide; the neck is well muscled and average in length; the withers are average in height and length; the back is long, flat, but often slightly dipped; the loin is flat and short; the croup is wide, nicely rounded, and well muscled; the chest is wide and deep. The legs are clean and solid. Hair of the mane and tail are thick and heavy, but the fetlocks are relatively clean. Usual colors in this breed are dun, bay, chestnut, and light bay.

The mares are often used for milk and the breed is also used for meat production. The Byelorussian matures late and has high fertility and longevity. Mares have often remained fertile to the age of twenty-six years.

Two types, large and medium, are distinguished within the breed. There are six lines of stallions and four mare families. The leading breeding ceners are Zarechye stud and the stud on Pobeda state farm in Byelorussia.

Improvement of the breed is by pure selective breeding.

Calabrian
(Calabrese)

ORIGIN: *Italy (Calabria)*
APTITUDES: *Riding horse*
AVERAGE HEIGHT: *16 to 16.2 h.h.*
POPULATION STATUS: *Rare*

The Calabrian, usually listed by its Italian name, Calabrese, originated in southern Italy in the region of Calabria. Information for this breed is contained in *Il Cavallo Calabrese*, by Pasquale Pisani, a book I received from the Associazione Provinciale Allevatori in Cosenza, Italy.

The Aryans were among the first people to domesticate the horse, and when they commenced their first emigration into Asia they overcame the Mongols in battle and established themselves in Assyria. Later they left this area to invade Arabia (where no horses existed), then Greece, and finally much of the rest of Europe. Due to the extent of Aryan civilization and by means of the Arabs, the Aryan horse predominated.

The precise period in which the Aryans moved into what is now Italy is unknown. According to Pietrement, they arrived a few thousand years B.C. Their arrival into France corresponds with this time frame. Latin authors such as Diodorus Siculus, Dionigi of Alicarnasso, Titus Livius, and Virgil wrote of excellent horses that were raised in various areas in Italy to be used in war against the Trojans.

In a study of horse breeding in Italy prior to the founding of Rome, Bassi noted that there was no single breed, even though all of the horses came from one source. Diverse topographical influences within the Italian peninsula and more importantly the at times unwitting selection of breeding animals led to distinct modifications. These modifications were quickly overrun by foreign breeds.

When the Celtic Gauls invaded Italy, the northern part of the country was populated by energetic, robust nordic-type horses, while in the southern part of the country during the Greek colonial period, smaller, more elegant, and faster horses predominated. Calabria in particular, which was part of Magna Crecia, was the birthplace of a horse with Oriental-African characteristics. Evidence of this is that during the fourth century B.C., horses that competed drawing two-wheeled chariots and quadrilles were imported from Greece into important cities like Thuri, Crotone, and Locri for the Isthmian games.

The mixture of the small, proud Oriental horses with horses from Africa imported by the Carthaginians produced one of the best horses of the time, according to Gennero and Calcagni. Confirmation of the quality of the Calabrian horses came first from Hannibal, who, according to Micali, in preparation for his offensive against Rome in 217 B.C. stole about 4,000 horses from Calabria and Puglia for his cavalry. Julius Caesar also confirmed the caliber of the horses. He tended to favor use of the Bruzia legion's cavalry because of the agility and speed of its horses, notwithstanding the small size of the animals. These factors make up the evidence that the Calabrian horse had Oriental ancestors.

Historical events in Calabria through the following centuries influenced the horses' breeding in various ways. The Calabrian horses were crossed with those of the Roman-occupied provinces, especially those from Gaul and Germany. These added robustness to the Calabrian horse in exchange for less speed and spark. During the rule of the Roman Empire the breeding system crumbled, including that in Calabria. The cause of the decay was the negligence of agriculture by the Romans, for they had by that time risen to the more important stage of military power and world domination. At the same time, they

found it easy to provide and increase the number of horses throughout the various regions of their immense dominion to meet both public and private needs.

After the fall of the western Roman Empire (476 A.D.), horse breeding in Italy went through a period of reawakening and prosperity. With the invasion of the Lombards, Germanic-type horses came into Italy and found their way into Calabria, adapting well to the horse population already in existence there. Certain Lombard (Longobardi) laws protected the horse and its breeding.

The arrival of the Franks in 776 resulted in the formation of many small, independent feudal states. During feudalism, horse production blossomed. The existence of power factions among the various feudal states and continuous discord among the factions stimulated the improvement of horse breeds for use in armies.

Invasion by the Saracens around 900 resulted in horse production in Calabria having priority just a little lower than that in Sicily. Not only had Calabrian horses been revitalized by the influx of Arab horses, but in addition the Saracens provided protection for all things having to do with horses.

With the decline of feudalism, communes were established. Horse breeding continued to prosper due to the importance of military cavalries. Around 1200 A.D., during the same period as the Crusades, there was an influx of new Oriental horses into Calabria and southern Italy. Meanwhile, in the north, the importation of Andalusian horses began. The Andalusians were consequently taken to Calabria to produce larger, stronger horses because of the new, heavier armor worn by knights of the period. Importation of Andalusian stallions into Italy continued under the Aragon (Spanish) rule in the late thirteenth to mid-fifteenth centuries. Of this period, Vallada wrote:

For centuries, the Calabrian horse occupied a shining place among the most distinguished horses of Italy. A descendant of the Arab and crossed with the Andalusian, it is strong, agile, svelte, and long-lived. It has a medium height (about 14 hands) and is rarely any higher. It has a wide forehead, large eyes, arched, well-formed neck, wide chest and croup, the tail is attached high, cylindrical body, small abdomen, very strong, solid hooves, and a coat that was usually bay, and sometimes brown or grey.

Until the early 1700s, there was a decline of horse breeding in Calabria and an increase in the breeding of mules, which were better suited to the mountainous region and difficult travel conditions. In 1742 a place for breeding horses in the province of Salerno, between the Sele and Calore Rivers, was created. Again, the horses were produced through the cross of Andalusians and Arabs.

Breeding of the Calabrian horse hit its lowest point in 1874 when breeding of the Persano horse was suppressed by a decree and the animals that made up the breed were sold. The members of the breed were divided among the Salernian and Calabrian breeders. In 1900 the breed was revitalized, consisting of two groups: the Menton group and the Luati group. The former had English blood, and the latter had Oriental blood.

In Calabria in the early 1900s, of the provinces of Reggio, Cosenza, and Catanzaro, Reggio Calabria was the least and Catanzaro the most productive. While some of the breeding farms had purebred Oriental stallions, others had English and some Hackney stock. Currently, Calabrian breeders are intent on the production of halfbred riding horses for racing.

The Calabrian horse may be bay, brown, black, grey, or chestnut. The head is well formed and proportional, with a straight or slightly convex profile. The neck is well shaped and muscular; the withers are pronounced; the back is straight and short; the loins are short; the croup is sloping and well muscled. The chest is broad and deep; the shoulder sloping and long. Tendons are strong and well defined; the hoof is well formed with strong horn; the legs are most often straight with a good natural stance.

Camargue
(Camarguais)

ORIGIN: *France*
APTITUDES: *Riding horse*
AVERAGE HEIGHT: *13.1 to 14.1 h.h.*
POPULATION STATUS: *Uncommon*

Camargue horses. Photo: André Mauget

Camargue mare and foal. Foals are born dark and lighten with maturity. Photo: André Mauget

Ancient white horse of the Camargue. Photo: André Mauget

The Camargue is located in the south of France in the Rhône River delta, between the Grand and Petite Rhône channels. It is a marshy, swampy region of about 780 square miles. In the northern portion of the delta the alluvium has emerged as dry land, but in the south the highest ground is along the embankments of present and former water courses. The intervening basins are marshes and shallow lagoons. During cool months the Camargue is swept by the powerful mistral, the episodic, cold northern wind that blows down the Rhône Valley.

Sparsely populated with barely 10,000 inhabitants, this region was formerly entirely wild with free-roaming herds of bulls and Camargue horses. The bulls are raised for provincial bullfights and are related to the fighting bulls of Spain and Portugal.

The conquest of the northern Camargue began at the end of the nineteenth century when vineyards appeared, followed by grains and crops. The great herds of cattle and horses for which the region is famous are still found, especially around Vacarrès, in what became the Camargue Regional Park in 1928. A unique area, its swamps and salt marsh smell of the nearby Mediterranean; its flamingoes speak of Africa. It is a place where if one is not to swim or wade, one must ride horseback. The climate of the Camargue is harsh, being either scorchingly hot beneath a fierce sun or whipped by icy winds blowing from the Alps.

The Camargue horse, often called "the horse of the sea," has existed in this region since prehistoric times. Thought to be a descendant of the ancient (extinct) Solutré horse, bones of which were found in southeast France, the Camargue has been influenced by other breeds. Through the centuries many armies have passed by the Camargue, including the Greeks, the Romans (who took Camargue horses into Spain), and the Arabs. The breed was greatly admired by Julius Caesar. Possibly this horse had some influence on the early breeds of Spain. The Camargue horse is born black or brown and turns grey with maturity.

The horses of the Camargue run wild in the marshlands in small herds consisting of one stallion, his mares, and progeny. The fillies are usually caught and branded as yearlings and colts thought to be unsuitable for breeding are gelded at three years old. The breed is strictly protected today. Breeding is semiwild but under supervision by the Tour du Valat experimental station in herds of forty to fifty animals, and some wild breeding areas have also been preserved.

The Camargue is a rugged individual and breeds very true to type—to the extent that it can be difficult to tell one from another. This horse is lively but has a good nature and when trained to ride is used to manage the bull herds in the delta. The Camargue horse is also used for tourism. A robust horse, it has a large and square head with a straight or slightly convex profile; the eyes are large and expressive; ears are broad and short with a broad base; and the neck is short and muscular, deep at the base. Withers are pronounced; the back straight and short; the croup short and narrow; the chest wide and deep. The shoulder is rather straight and quite short; the mane and tail are long and thick. The legs are extremely hardy with clean joints, a long forearm, and very good hooves. They are not shod.

Camargue horses mature slowly and are long-lived. They are able to exist on poor feed and are perfectly adapted to the harsh environment of their habitat. Fast and agile and possessing wonderful endurance, the Camargue is a good riding horse.

Campolina

ORIGIN: *Brazil*
APTITUDES: *Riding horse*
AVERAGE HEIGHT: *14.1 to 15 h.h.*
POPULATION STATUS: *Common*

Breeders of the Campolina like to tell the story that Emperor Dom Pedro II went to Queluz de Minas, Brazil, now called Conselheiro Lafaiete, to inaugurate the railroad. In his honor, great festivities were organized, including horse shows and the traditional contest between the Moors and Christians. Cassiano Campolina, owner of Fazenda do Tanque (Tanque Farm) in the municipality of Entre Rios do Minas, was placed in command. He acquired good stallions and dedicated himself in earnest to the production of strong, courageous, and agile horses so that with time, the "infidels" could be conquered.

The emperor himself, having become the friend of the farm's charming owner, presented Campolina with a stallion from Cachoeira do Campo, a farm in Brazil where there were excellent stallions that had come from Portugal (Alter Real or Spanish Jennet). There were also some descendants of Austrian horses, which had been donated to the cavalry of the army by the emperor of Austria at the request of the Empress Dona Leopolidina. This was the alleged origin of the Campolina horse. It is a story that cannot be proved and is considered by some as nothing more than an innocent legend—perhaps a legend that contains an ounce of truth presented in symbolic form. Cassiano Campolina could represent all of the breeders who for more than a century have worked to produce a type of horse that has beauty, vigor, agility, and comfort. The Moors were not conquered in the return match—but a beautiful breed of horse was created.

There is not much written about the Campolina horse, but there is enough to give information about the various strains that contributed to the breed. At the very least it seems clear that Cassiano Antonio da Silva Campolina had a great interest in producing good horses, since it is apparent in the letters he left that he was a man who greatly appreciated that challenge.

The most complete information regarding the Campolina is found in the thesis of Professor Luiz Rodriquea Fontes. Much of his thesis affirms the story behind the Campolina breed. Origin of formation of the breed is established as 1870, the year Monarca was foaled, a son of Medeia, a native mare presented to Cassiano Campolina by a friend from Juiz de Fora. Medeia was covered by an Andalusian stallion which belonged to Mariano Procopio, a gift from Emperor Dom Pedro II.

Some say that Emperor Dom João VI was the initiator of the Campolina breed. He had imported breeding stock which started the Royal Breed of Cachoeira do Campo. In the meantime, selection work by Campolina began when the Coudelaria strain had become extinct and its animals had been sold at public auction. Cassiano Campolina selected animals that had characteristics he preferred, looking primarily for physical appearance and gait and attempting to produce animals for journeys and riding, for pulling carriages,

and for use by the dragoons of the royal militia. As indicated, his farm Fazenda do Tanque was situated in Entre Rios do Minas, near Cachoeira do Campo.

Characteristically the Campolina has an intelligent, expressive look. The rounded rump, sloping and muscular, and the solid extremities and well-formed hooves are reminiscent of the Andalusian; the full and silky mane and tail are characteristics of the Barb. Experts maintain that during formation of the breed stallions used were of pure English blood (Thoroughbred), Anglo-Arab, and Percheron.

Monarca was the main stallion at the Cassiano Campolina Farm for many years and left numerous offspring. The tireless Campolina later crossed his mares to Menelicke, an Anglo-Norman that had tremendous bearing and beautiful lines. This stallion had come from the Cachoeira Muraux of Rio de Janeiro.

Beyond the Tanque Farm in the municipality of Entre Rios, the cradle of the Campolina breed, other breeding centers emerged. The most outstanding was at the Campo Grande Farm in the neighboring municipality of Passa Tempo.

Upon the death of Cassiano Campolina in 1904, this work of more than thirty years was not interrupted. The Tanque Farm was willed to his special friend, Joaquim Pacheco de Resende. Joaquim Pacheco felt that the developing breed had all the requirements of bearing, robustness, and vivacity necessary but lacked a really comfortable gait—which was indispensable for rides and trips in that era when so much travel was horse powered.

In search of walking horses of more delicate bearing, he brought the stallion Golias to the farm. This bay horse, a good walker, was acquired from Colonel Gabriel de Andrade from the Campo Grande Farm. Pacheco also purchased two Thoroughbred stallions named São Lourenco and Carlito. Golias exerted great beneficial influence on the Campolina breed. His owners and veterinarians who knew and rode him affirm that he was a horse of one-quarter Clydesdale blood, out of a circus horse known as Horse of Cobo.

A Holstein stallion named Trefer, imported in 1908 by Herm-Stoltz & Company, had influence in the breed, primarily to the west of Minas. In the same period, Gabriel de Andrade at Campo Grande Farm imported two American Saddlebred stallions from the United States, named Yankee Prince and Niagara. These left a considerable number of descendants used in breeding the Campolina horse.

Excessive convexity of the head was usually considered a fault, along with sloping of the rump. Additional Thoroughbred blood was infused and the numerous crossbreds were interbred in search of a refined animal with a good walk. Finally the blood of Mangalarga horses was introduced. The Mangalarga Marchador was used until 1934 to give better proportions and refinement to the breed and, in addition, the benefit of its smooth marching gait. After that date there is no record of the introduction of exotic blood to the Campolina breed.

In 1938 breeding of the Campolina horse was formally organized with formation of the Professional Consortium of Campolina Horse Breeders, headquartered in Barbacena. Considerable improvement of the Campolina horse resulted in the organization of breeders and standards. In 1951 the Campolina breeders founded the Campolina Breeders Association in Belo Horizonte, and a breed standard was adopted.

The Campolina may be of any color. The constitution of this breed is strong and sound with good bone. The temperament is docile but active and proud. The head is proportional in size, harmonious, with a flat forehead and well-separated lower jaw. The profile is straight to a subconvex line in the bridge of the nose; the eyes are medium to large in size; the nostrils are large and flexible; the neck is arched, muscular, proportional and well inserted into the shoulder. The withers are well defined and long; the chest is deep, wide, and not too prominent; the ribs are long and arched; the back is medium in length; the loin is short and well muscled; the croup large and long, smoothly inclined; the tail set medium-high; mane and tail are silky in texture. The shoulders are long, oblique, and muscular. The legs are strong with well-defined tendons and strong joints; the hoofs are round, solid, dark, and very hard.

The gaits of the Campolina are a smooth, regular *batida* or a *picada*, the first characterized by a longer time of movement of the bipeds in diagonal and the second only in lateral, both with four-beat movements. The Campolina neither trots nor paces.

Today the Campolina is found throughout Brazil and is gaining in popularity. In 1988, 5,400 animals were registered.

Canadian

ORIGIN: *Canada*
APTITUDES: *Riding, driving, light draft*
AVERAGE HEIGHT: *14.3 to 16.2 h.h.*
POPULATION STATUS: *Rare*

The Canadian stallion, Brandy Creek Fablo Felix. Notice the striking resemblance to the Morgan breed. Photo courtesy Paul E. Belauger

The Canadian is an excellent breed for use in driving competitions. Photo courtesy Paul E. Belauger

The Canadian breed may well be one of the best kept secrets of the twentieth century—not because breeders of the Canadian have tried to hide the animals, but because writers of books on horse breeds apparently have not contacted Canadian authorities to inquire. The Canadian horse is an unsung hero in formation of United States (and Central American) horse breeds. The Canadian horse was well known to the American colonists. Today, however, even many Canadians are unaware of the breed, probably because of its present rarity.

Americans journeyed into Canada during the early years of this country and purchased *thousands* of French-Canadian horses to take home and cross with their mixed stock. The Canadians were talented trotters and pacers, and as roads were developed good roadsters were in great demand. Indeed, the New England states were literally saturated with this blood. Yet when writing about the formation of American breeds the Canadian horse has been overlooked by most writers, who refer vaguely to "horses from Canada that trotted, or paced."

Many purebred French-Canadian horses were entered into the early stud books of the Morgan, Standardbred, and American Saddlebred. Foundation sires of these breeds were often pure Canadian or were mated to pure Canadian mares.

Upon learning of the Canadian breed and gazing at the photographs, the truth about the little bay stallion Justin Morgan became forever settled in the mind of at least this writer. Justin Morgan, famous foundation sire of the Morgan breed, has always been reported to have been a mixture of Thoroughbred and Arab—though his description was like neither of those breeds, nor like that of an Anglo-Arab. In my opinion there is no other possible theory coming close to the obvious, clear ring of truth in regard to his ancestry but that he was a horse of predominant, if not pure, French-Canadian blood.

A description of the old Canadian horse, written many years ago by a man who knew the breed well, could easily describe a typical Morgan: a horse with great powers of transmission (prepotent), on the average 14 to 15 hands, body very solid and compactly built, a well-formed head with small ears and widely spaced eyes with a bold expression, a broad chest, strong shoulder, strong muscling, and extremely hardy feet and legs. Not even bad handling, awkward shoeing, hard travel—nothing, it was said—seemed to affect the feet of the Canadian horse. The mane and tail were described as "peculiar," being heavy and, in almost all cases, wavy. The majority of Canadians were then (and still are) bay or black, although other colors were also seen.

The old Canadian horse was described as being long-lived and hardy, easily kept, and capable of great endurance—heavy enough for the purposes of the farmer or as a roadster while also being a good riding horse. The breed produced both trotters and pacers. Like several other horse breeds esteemed for their high quality and usefulness (such as the Dutch Friesian), the Canadian was drawn on heavily for outcrossing to improve and create other breeds—to the extent that the pure Canadian nearly suffered extinction.

The Canadian horse traces its ancestry to the foundation stock brought to Acadia and New France in the seventeenth century. The first horses were ultimately caught and carried off in 1616 by Samuel Argall's marauding expedition from Virginia. This was the first introduction of French-Canadian blood to America's eastern shores.

The effective introduction of French horses into New France came in 1665 when Louis XIV sent two stallions and twenty mares from the royal stables to the colony. On the voyage eight of the mares were lost, but the king sent additional shipments; in 1667 fourteen or fifteen horses, and in 1670 a stallion and eleven mares. Thereafter, the king sent no more horses, as the colonial governor, Intendant Talon, considered there were now enough in the colony to furnish a dependable supply of colts to all in need of them.

The horses remained the property of the king for three years, but were let out to the leading farmers for an annual rental per horse of 100 livres or for one foal. At the end of three years the horses and any colts not turned over to the intendant in rental became the property of the farmer. Colts that came into the possession of the intendant were raised at government expense until three years old, then parceled out to the farmers on the same terms as the original horses. The breeding program was so successful that in 1679 there were 145 horses in the colony; in 1688, 218; and by 1698 there were 684. This stock was the progenitor of horses on Sable Island.

The horses from the royal stud, with perhaps a few others imported by the seigneurs during later years and at their own expense, formed the basis of the French-Canadian horse of the Old Régime. Whether imported under royal or private auspices, they came from Normandy and Brittany, at that time the two most renowned horse breeding provinces of France. The Breton horse, although small, was noted for its soundness and vigor. The Norman horse closely resembled the Breton, but gave more evidence of infusion of Oriental blood. This strain came from Andalusian sires brought into Normandy and La Perch (habitat of the Percheron breed) for breeding purposes, some direct from Spain and others, between the latter part of the sixteenth century and the end of the War of the Spanish Succession, from the Spanish Netherlands. Influence of the Dutch Friesian is apparent in the notable trotting ability of the Canadian, the feathered legs, abundance of mane and tail, and general appearance.

In the seventeenth and eighteenth centuries there was no standard type in either the Norman or Breton breeds but rather several types in each, being bred with one another in their home province according to the features popular at the moment. Among the horses brought from France into Canada there were various types; some were distinctively draft in type; others were just as distinctively trotters, a type of horse for which France had enjoyed a reputation for generations. A gross error is made by those who attribute all of the credit for the American trotters to the horses of England. Still others were pacers, not descending from the Narragansett Pacers as is often implied but coming from France with that talent.

Although the French-Canadian horse exhibited several "types" due to varied breeding practices of the times, there was no other blood infused into the breed. Many owners bred for the lighter, more refined type, and it is said that the pure breed as it existed in 1850 was scarcely altered from its prototype of a hundred years before. Indeed, in spite of more than a century of separate breeding, Canadian horses still very closely resembled the small horses of Normandy and La Perch.

The Canadian, although usually small, was known to have the quality of breeding up in size when crossed with loftier mares, also giving the foals their pluck, vigor, and iron constitution.

In color, the Canadian was never of any firm, established color, though it was most often solid bay or black, weighing from 900 to 1,000 pounds. Of the Canadian, R. N. Saunders says that "as a farm horse and ordinary farmer's roadster, there is no honester or better animal; and, as one to cross with other breeds, whether upward by the mares to Thoroughbred stallions, or downward by the stallions to common country mares of other breeds, he had hardly an equal."

Very little care was given to the early Canadian breed, and it is testimony to their hardiness that they persisted. In summer, when the horses were little used, they ran loose in the woods, where they were tormented by flies against which they had no defense due to the French practice of docking the tails. In winter they were usually given no shelter at all, especially the young stock which were not in use. The inhabitants cured no hay for their cattle and sheep and so little for their horses that they too existed almost entirely on straw. They seldom received enough grain but were required to work hard. When the owners took to the road they thought nothing of driving the horses as fast as they would go for a dozen miles or more, then leaving them to stand uncovered for hours in blizzard conditions. It was their opinion that harsh exposure was an excellent way to toughen an animal.

New France furnished the horses taken to the western settlements at Detroit and in the Illinois area. Many of these horses were allowed to run loose in large herds and were only brought in when needed for work. Great numbers are known to have escaped to run with the mustangs of the American plains—

an ancestor never mentioned in writings of the American Mustang.

Canadian horses found a ready market in the United States and were also shipped in great quantities to the West Indies, a fact overlooked by enthusiasts of the *paso*-gaited horses. After the close of the American Revolution and until the War of 1812, many Canadian horses were taken into the United States.

One outcome of the horse trade to New England after the American Revolution was the introduction into the St. Lawrence Valley of the once famous Narragansett Pacers. There has previously been a vague tradition that a considerable trade northward in these horses once existed, but historians have found little evidence to support it. There is one report, however, a precise statement made by George Barnard, that in 1846 his father had bought Narragansett Pacers from Rhode Island and took them in droves to the French country around Quebec, where they were readily traded for the stout native horses. The Narragansetts, being mainly saddle horses, were losing favor in the United States due to the sudden popularity of roadsters with the building of good roads. Thus, Canadians readily traded their fine little horses for Pacers, and the Canadian breed came by the hundreds into New England, *the perfect roadster*.

After the War of 1812, the trade in French-Canadian horses grew rapidly. Droves were collected by American dealers each year, mostly at Montreal and Quebec. In 1830 it was reported that most of the trotters then in the northern United States were of French-Canadian origin. The beneficial result of crossing the Canadian on the ordinary stock of the adjacent states was universally admitted.

The popularity of the crossbred horses of northern New England among the stagecoach drivers of Boston is legend. The stallions brought from lower Canada were not entirely responsible, however, for the infusion of Canadian blood into the horses of the United States. Part of it came from both purebred and partbred Canadian mares in Vermont which were mated to American horses of some blood or, at least, some figure.

The Canadian Pacer was a horse bred from the Narragansett Pacer and the old strains of French-Canadian. This breed then returned to the United States and contributed greatly to development of the famous American Standardbred.

So great was the drain into the United States of the pure Canadian horse, particularly during the Civil War, that numbers at home were reduced alarmingly. Another factor involved in the demise of the breed was importation of heavy draft horses for farm work. The Canadian was never considered a work horse although it was worked hard, and it also never qualified as a light breed, being a more medium type (a description also given of the Morgan). By the end of the nineteenth century the breed was in extreme danger of extinction.

In 1886 a stud book was established to record the best of the remaining animals, and the Canadian Horse Breeders Association was formed in 1895. In 1913 a breeding center was opened on the Federal Experimental Farm at Cap Rouge in Quebec, and later at St.-Joachim.

When the federal government, occupied with the war, closed down the operation in 1940 and sold off the breeding stock, the Quebec government reestablished the stud under the auspices of the provincial Department of Agriculture at Deschambault, Quebec. There, the Canadian was bred into a taller, more refined animal, suitable as a hunter or jumper. The more traditional breeders resisted this move and concentrated their efforts on carefully preserving the original Cap Rouge type of animal.

In 1979 the forty-four-head Deschambault herd was suddenly sold at auction and the Canadian was once again threatened with extinction. Thanks to the efforts of a handful of committed breeders, mostly in Quebec, the breed struggles on, its population hovering today at about 1,500.

In recent years this breed has received considerable acclaim as a carriage driving horse. A pair of Canadians, owned by two of the breed's most dedicated patrons, Donald Prosperine and Alex Hayward, won the 1987 North American Pleasure Driving Pairs Championship, in addition to competing throughout Ontario and the northern United

States in combined driving and coaching events.

The Canadian today is a multipurpose animal, suitable for riding, driving, and light draft work. These horses are very hardy, easy to keep, and long-lived. They are described today as being small but solidly built and well proportioned; having strong bones with good muscle development, especially in the forearms and gaskins; having a head that is short with a fine muzzle and great width between the eyes. The neck is strong and arched, mounted high on a long, well-sloped shoulder. The body is long and deep, the barrel rounded. The tail is set high into a heavily muscled rump. Both mane and tail are long and usually wavy.

Virtually unknown even in its native country today, the breed is making a strong comeback. A tough working horse, a dependable riding horse, a jumper, and a trotter with fine action, the Canadian is an honest little horse with a big heart and a great deal to offer.

Canadian Cutting Horse

ORIGIN: *Canada*
APTITUDES: *Riding horse*
AVERAGE HEIGHT: *15 to 16 h.h.*
POPULATION STATUS: *Common*

The Canadian Cutting Horse, Docs Question Mark, ridden by Dave Batty. Photo courtesy Denese Chapman

While not a specific breed, the Canadian Cutting Horse is represented by a very active and popular association. Cutting horses may be of any breed, but requirements for a good cutting horse narrow the range considerably, and the Quarter Horse has been the most popular breed used for cutting cattle. Not all breeds of horses exhibit "cow sense" or are quiet-natured enough to work effectively with cattle. Few breeds have the balance and agility needed for the job.

The main criterion for a good cutting horse is the ability to outthink and outguess the cow it is working. Early cutting horses were small Spanish Mustangs used in the United States. Later, the Spanish Mustang was a main ingredient in formation of the American Quarter Horse, and this is where the breed obtained its great cattle working ability. The horses of Spain have long been famous for their inherent ability to work cattle.

Once the cutting horse was introduced to the rodeo arena the event grew rapidly in popularity. The National Cutting Horse Association and the Canadian Cutting Horse Association were formed to establish adequate rules to judge the performance of the cutting horse. Rules have been designed to result in a pleasing performance in the arena and to show the horse at work comparable to ranch conditions.

A calm, cool performance in ranch work is a great asset to any stock owner, and the same type of smooth performance from the spectator's point of view shows the amazing ability of this valuable animal. The turn-back men and other factors totally unlike ranch work have been added to increase the action and entertainment for the spectator.

A horse that carries its head low and travels cautiously toward the herd, alert and intent on the job, will usually turn in an interesting and action-packed performance. The horse that carries out its duties well, keeps full control of the cow after it has been separated from the herd, and displays continuous cow sense is deserving of a great deal of credit and will score well.

Contestants try to pick from the herd an animal that will give them maximum action to test and show the ability of the cutting horse.

Along with action goes the risk of penalties for being out of position, reining the horse, and many others.

Cows are worked for two and a half minutes with the objective that the horse will outthink the cow with terrific short bursts of speed, the ability to turn in mid-air, fantastic foot work, and coordination between the horse and rider.

The conformation and colors of Canadian Cutting Horses are essentially those of American Quarter Horses, with a vast majority being of the Quarter Horse breed. However, there are no criteria specifying any particular breed.

Canadian Rustic Pony broodmare, second generation. Photo: Dr. P. L. Neufeld

Canadian Rustic Pony

ORIGIN: *Canada*
APTITUDES: *Riding pony*
AVERAGE HEIGHT: *12.2 to 13.2 h.h.*
POPULATION STATUS: *Rare*

The Canadian Rustic Pony is the first equine breed developed in western Canada and has proven even more popular than its founders anticipated.

"By far the smoothest riding and most easily trained horse I've ever had," is a typical comment. "We exercise him behind a truck. He can hard-trot two miles and barely turn a hair; I've never seen a colt with such endurance," is another. Or, "She's just a darling. You can ask her to stand anywhere and she won't move until she's told to. At the same time, she's fast and has endurance; both attributes more pronounced than any other horse I've worked with."

Genetic background and appearance of this unusual breed are unique. Combining to complement similar physical and temperamental traits in the related breeds of Arab, Welsh Mountain Pony, and Tarpan (genetic input of the Tarpan came from the Atlanta Zoo in Georgia), this 50- to 54-inch-high, arresting, primitive looking line-backed grey, buckskin, or bay, with flowing tail and partly raised mane, is the result of two decades of research breeding. These small horses are sturdy, gentle, extremely hardy, smooth-gaited, and

Canadian Rustic Pony stallion, two years old. Photo: Dr. P. L. Neufeld

high-jumping mounts for older children and adults. They are also ideal for driving, chariot racing, show, and family pets. Stallions are slightly more protective of mares than in many breeds, as are mares of their foals.

One enthusiastic breeder wrote: "One thing we have to watch is feeding. They are extremely easy to keep. An ordinary horse will eat a bale of hay in two days, but the same bale will last the Rustic five days. They will thrive on the poorest grass, and we find that grain feeding is just too rich for them."

On January 23, 1989, after lengthy negotiations between breeders and Canadian government officials, the Canadian Rustic Pony Association was incorporated under Bill C-67, the new federal Animal Pedigree Act enacted the previous summer. First officers and directors appointed were Dr. Peter and Elsie Neufeld of Minnedosa, Manitoba; and from Saskatchewan, Holger and Noreen Petersen of Rose Valley, Ken and Donna Longman of Kelvington, Stan and Dorlene Longman of Nut Mountain, and Donalda and Owen Douglas of McTaggart.

As the act is designed especially to promote development of new domestic animal breeds by individuals and groups (as opposed to development primarily by government, as under the old law), and because Canadian Rustics had already been developed more than halfway toward becoming a breed recognized under C-67, many details had to be ironed out regarding where they fitted in. Developed under Agriculture Canada guidelines, this remarkable new breed has been registered through an American registry since 1978. Although initial breed development occurred on the Neufeld forty-acre (Glendosa Research Centre) farm, other than the eight directors above, John Barrett of Moose Jaw, Saskatchewan, Lloyd Gibson of Waskada, and Gen Kurtenbach of Dropmore, Manitoba, also played significant roles.

As of September 1989, seventy-two Canadian Rustic Ponies had been U.S. registered. Of these, detailed documents of sixty excellent breeding-type animals had been submitted to Agriculture Canada in Ottawa to begin the lengthy and rather complex entrance into the Canadian system of registration as a fully recognized breed under the new legislation. That at least fifty will be fully registered under Canadian law by the end of the century is highly probable. U.S. registration and Canadian identification-registration of these ponies will proceed simultaneously for owners who want both, the latter certificates to be issued through Canadian Livestock Records Corporation in Ottawa.

By-laws covering those aspects of operation not already in the incorporation articles of the association were developed in February 1989.

The Canadian Rustic has an overall gene pool of 25 percent Tarpan (reconstructed—it is not known which reconstructed Tarpan was used—that from the Munich Zoo is quite different in type from the Popielno Forest herd in Poland); 37.5 percent Arab (including the Caspian); and 37.5 percent Welsh Mountain (Section A) bloodlines. Phenotype is described as between 50 and 54 inches at the withers; weighing about 550 to 800 pounds. Its color includes various shades of grey or brown but never broken (such as Pinto or Appaloosa). Dark legs, usually with zebra striping and a dorsal stripe, are common. In general appearance the Canadian Rustic is a sturdy line-backed Arab-type pony. The profile is either dished or straight (never convex), tapering to a small muzzle with large and flexible nostrils; the jaw is prominent and wide; the eyes are large, prominent, and widely set; the ears are small and pricked. The mane is partly raised. The neck is thick, medium in length and arched, with shoulders and girth thick as well. The back is strong, short, and almost level and the croup is gently sloped. The legs are strong with hard feet and the action is free and fast at all gaits, especially at the trot, which is very fluid. The Canadian Rustic has smooth jumping ability.

As of September 1989, no Canadian Rustic Ponies were to be found outside the provinces of Manitoba and Saskatchewan. However, this will undoubtedly change quickly as there is strong interest across North America since this unusual new breed recently caught the attention of international horse magazine editors.

Canadian Rustic Ponies are available to serious interested breeders who wish to be involved in increasing this new breed.

Canadian Sport Horse

ORIGIN: *Canada*
APTITUDES: *Riding, Sport horse*
AVERAGE HEIGHT: *16 h.h.*
POPULATION STATUS: *Common*

Canada has long been known for its good horses. As early as 1893, an Englishman, Lieutenant Dan Lysons, stationed in London, Ontario, wrote in his diary that "horses which they procured from the farmers thereabout were remarkable for their ability to jump timber fences."

By the turn of the century many Canadian-bred horses were sold to the United States. Canadian horses were exhibited as hunters and jumpers in both the United States and Europe with gratifying success.

Vast numbers of Canadian horses were shipped by the army to Europe during the First World War, never to return, and a great deal of suitable blood for hunter breeding was lost. However, the Canadian Racing Association imported sixteen Thoroughbred stallions from England and these were made available to farmers in Ontario for halfbred production. The blood of these horses is in the pedigrees of most hunter foundation stock lines.

In 1926 the Canadian Hunter, Saddle and Light Horse Improvement Society was officially incorporated. In 1928 the Foundation Brood Mares Registry was started. The Canadian Hunter Society was incorporated in 1933, under the revised livestock pedigree act, and arrangements were made with the Canadian National Live Stock Records to maintain the Canadian Hunter Stud Book. In 1970 the two were amalgamated into the Canadian Hunter Improvement Society.

In 1984 it was decided to rename the society to include the term "Sport Horse." This decision was made because Canadian horses were competing in all multiple disciplines. Hence in 1987, at the annual general meeting, the membership voted to change the name to the Canadian Sport Horse Association and to work avidly toward developing the sport horses as an evolving breed which will be recognized by the Canadian government and all worldwide equestrian competitions.

The breeding objective is to produce sound sport horses of correct conformation, capable of carrying a rider comfortably and safely over rough country, stiff obstacles, and long distances at suitable speed. High standards of breeding and registration have produced a sport horse so versatile that not only is it found in the hunt fields but it also holds international championships in dressage, hunting, jumping, and eventing.

The association has an open stud book for stallions and mares. In creating and fixing a breed, there must be provisions for eliminating unsuitable animals for each generation.

Through selective breeding and inspections, the association aims to develop a Canadian Sport Horse that is recognizable through performance, athletic ability, and conformation.

Stallions of approved breeds are inspected for registration. Mares of unknown background can, after passing inspection, be accepted in the registry as foundation stock. All foals sired by a registered and licensed Sport Horse stallion and out of a registered Canadian Sport Horse mare are eligible for final registration as Canadian Sport Horses.

The pedigree records of the Canadian Sport Horse are kept by the Canadian National Live Stock Records, Ottawa, Ontario, an organization of breed associations whose records have official standing both nationally and internationally.

Canik

ORIGIN: *Turkey*
APTITUDES: *Riding horse*
AVERAGE HEIGHT: *13.2 to 14.2 h.h.*
POPULATION STATUS: *Uncommon*

The Canik is an ancient breed in Turkey. About 150 years ago some Caucasus light cavalry horses were mixed into the Canik, which had previously been bred pure.

Canik horse of Turkey. Photo: Ertuğrul Güleç

Canik horse. Photo: Ertuğrul Güleç

Champion Cape Boerperd stallion, Mister Quin. Photo: *Landbouweekblad*, Cape Town

This horse is described as alert, speedy, and surefooted. This horse travels like the Rahvan horse—at very fast lateral, four-beat gait. The breed is said to be full of energy and very beautiful.

The head of the Canik is medium in size with great width between the eyes and a straight profile; the croup is gently sloped; the legs and pasterns are very strong and hooves are hard and do not wear down easily. The Canik is found in every common horse color. It is bred in the regions of Samsun, Çarşamba, Tokat, and Amasya.

There are approximately 5,000 of the Canik breed.

Cape Horse
(Boerperd, Kaapse Boerperd)

ORIGIN: *South Africa*
APTITUDES: *Riding horse, light draft*
AVERAGE HEIGHT: *14.2 to 15 h.h.*
POPULATION STATUS: *Rare*

During the first century of European (Dutch) settlement at the Cape in South Africa, a small type of horse known for its stamina was developed. Presently there are two groups attempting to restore and preserve the remaining blood of the famous Cape Horse; the Historiese (Historic) Boerperd Breeders' Society and the Kaapse (Cape) Boerperd Breeders' Society. *Boer* means "farmer" and *perd* means "horse" in Afrikaans.

Historical Boerperd. Photo: *Landbouweekblad*, Cape Town

The horse bred by the South African farmer became known around the world by various names including Cape Horse, Boer Pony, and Boerperd. The Cape Horse was certainly not a pony. There were no indigenous horses in South Africa, and the first arrivals to this area were imported from the island of Java. Various accounts differ; one is that the first equine stock were Java ponies, while another contention is that the horses from Java were Arabs. Despite some accounts that the Javanese pony was a mixture of Barb and Arab, authorities in

Java report that the Java pony descended directly from Mongolian stock taken there by the Chinese (see Java) and was later crossed to Oriental horses from Persia.

The first horses arrived at the Cape in 1652. The first sale of horses to farmers was held in 1665, and this marked the beginning of private enterprise in the horse breeding industry of South Africa. Sixteen horses were sold, the price of each being equivalent to that of five prime oxen.

Further imports were made in 1689 by the governor, Simon van der Stel, introducing Persian Arabs for the improvement of the Javanese stock. The first private importation of horses was allegedly by a Mr. Van Reenen. Beginning about 1769, a lucrative export trade in horses for military use in India started which lasted for nearly a century, bearing witness to the fact that the Cape horse must have remained of good quality and fairly standard type.

In 1778 the arrival is on record of a shipment of South American horses, highly esteemed for their gentleness, beauty, and good service. The South American horses at that period were surely Criollos. Another introduction of Spanish horses followed in March of 1807, during the Napoleonic wars, when two French vessels carrying Andalusian horses were captured en route to Buenos Aires. It is thought that from these were derived the blue and red roans which became famous for their great powers of endurance (personal communication, Dr. F. J. Van der Merwe, director of the Department of Agriculture and Water Supply, Pretoria, South Africa).

In 1782 the first English National Horses were imported. During the first British occupation of the Cape (1795–1803), more English horses as well as horses from Boston were imported, and these and subsequent importations brought about great improvement in the horse stock of the districts where they were standing at stud.

Lord Charles Somerset, the English governor at the Cape from 1814 to 1827, did more than any other person to promote the horse industry in South Africa. He personally imported thirty-four Thoroughbred stallions at great cost, and through his endeavors the export trade to India was greatly increased. It was then that the Cape horse gained world renown as a cavalry horse, being reported by the Indian army officers to have no equal. It was able to stand the extremities of climate and terrain from the arid heat of India to the bitter cold of the Crimea, proving surefooted, hardy, docile, and willing, needing no special care, and living on the scantiest of food.

The Great Trek—an emigration of some 12,000 to 14,000 Afrikaners from Cape Colony between 1835 and the early 1840s, in rebellion against the policies of the British government and in search of fresh pasture—resulted in natural selection among the horses due to the rigors of the journey. Farmers across the Orange River drew their fresh blood from the Hantam studs of Calvinia and Clanwilliam, and some of the finest Thoroughbred stallions imported eventually found their way to the Transvaal, Orange Free State, and Natal.

During the first thirty years after the Great Trek the horses remained unchanged, but during the latter half of the nineteenth century many horses of various breeds were imported to the Cape and the republics of Transvaal and Orange Free State. The firm Hoogendoorn imported black Friesian stallions from the Netherlands to be used for drawing hearses. Many Hackneys, Norfolk Trotters, and Cleveland Bays were imported for use as coach and draft horses. Ironically, as Australia had obtained some of its first equine stock from the Cape, between 1899 at the beginning of the Boer War and 1902, over 16,000 Walers were sent to the Cape from Australia for military use.

The trade with India eventually declined, partly due to lethargic Indian army authorities who were commissioned to buy horses at the Cape and who traded with middlemen and speculators, who provided a poor type of animal, doing irreparable harm to the good name of the Cape horse. Fine stocks were also depleted between the years from 1854 and 1893, when African horse sickness destroyed more than 235,000 horses. After 1868 the desire for fashionable pedigrees swept through the country and numbers of Thoroughbreds were imported by farmers who thought (as many still do today) that the tag "imported" had special meaning. Constitution and substance were overlooked, weedy stallions that looked good only on paper began to head the studs, and the general high standards were irrevocably lost.

The Anglo-Boer War was the final factor in destruction of the old Cape Horse from its last strongholds in the Free State and Transvaal. At the outbreak of the Boer War, there were very few pure specimens of the Cape Horse in the Cape Colony as crossbreeding was the order of the day. In the republics, crossbred horses were common in the cities, but the majority of horses in the farming districts were still the old type of Cape horse. The crossbreds and imported horses were not popular with the farmers, who continued to breed their Cape Horse. The outbreak of the First World War once again threatened the existence of the Boer horse, as thousands of horses were bought and commandeered by the government for its campaign in South West Africa and later in East Africa.

The few Cape Horses that survived the two wars were badly neglected. Shortly after the Second World War the need was again felt for a South African utility horse. Breeders sought horses throughout the area for the true Cape Horse type. The objectives for the breeding of a "national riding horse" were explained in the constitution of the National Riding Horse Breeders' Society of South Africa and Rhodesia. In 1957 this name was changed to the Boerperd Breeder's Society of South Africa. The association's aim has been to breed an indigenous South African riding horse by means of crossbreeding with overseas breeds, especially the American Saddlebred, Arab, Thoroughbred, and Hackney.

The Historic Boerperd Breeders' Society was founded by enthusiasts who disagreed with the practices of the first association, desiring instead to locate animals with the characteristics of the old Cape Horse and to breed these animals selectively without the addition of other bloods.

Dr. Frans Van Der Merwe told me in a letter:

After one group of Cape Boerperd breeders began improving their local stock with American Saddler, from the 1950s onward it was still possible to find the occasional breeder, especially in the Transvaal and Orange Free State provinces who shied away from the Saddlebred. They later began to identify themselves as breeders of the original historical Boerperd and many romantic tales were presented of horses that miraculously escaped the ravages of the Anglo-Boer War. They did manage to locate a few hundred horses of the old type. Not surprisingly, although they are generally larger animals, they bear much resemblance to the Nooitgedacht because the original Basuto stock were an offshoot of the original Boerperd, i.e., farmer's horses of the Cape, Transvaal and Orange Free State. With other interested persons I have been working toward getting the two breeder's societies together because I am convinced they will get much further by pulling together than trying to go their separate ways.

Modern breeders of the Cape Horse are aiming at producing a comfortable, easy-gaited, stylish animal of quality with good action, one that will compare favorably with the existing breeds seen in the show ring today yet have the added asset of being adapted to the environment. The Cape Horse is hardy, tough, and full of substance, with good legs and a sound constitution. The breed is presently more refined than its predecessor, but with no less heart. The ideal Cape Horse has a finely chiseled, dry head with a straight profile and a good deal of character. The eyes are large and placed wide apart; the ears are medium in length; the throat-latch is fine; the neck is long, supple, and finely arched, fitting neatly into the shoulder and withers; the shoulder is sloping and muscular. The back is short; the hips are rounded; the body is deep. The legs are strong with good bone and broad joints. The breed has always been known for its hard, sound feet.

Due to the great amount of American Saddlebred that has been infused into many strains of the Cape Horse, many are shown at three and five gaits.

Carpathian Pony
(Hutsul, Hucul, Huzul)

ORIGIN: *Poland, Romania (Carpathian Mountains)*
APTITUDES: *Riding pony, light draft*
AVERAGE HEIGHT: *12.2 to 14 h.h.*
POPULATION STATUS: *Common*

Carpathian or Hucul, in Romania. Photo: Emil Petrache

Hucul mare and foal in Romania. Photo: Emil Petrache

Hucul or Carpathian, in Poland. Photo courtesy Dr. Jiří Machek, CSc.

The Carpathian, a highly esteemed small horse or pony, originated in the Carpathian Mountain range. Many frontiers run together along this range of mountains, and borders have changed with the passing of centuries. Both Romania and Poland claim origin of the Carpathian, or Hucul, and Czechoslovakia and Hungary also breed this hardy horse.

The Carpathian is a breed dating to the thirteenth century and is believed descended from the crossing of the wild Tarpan with Mongolian horses, both Oriental in type. Horses were brought to the area by migratory tribes, particularly the Mongols.

A stud farm for the Hucul, or Carpathian, was first established in 1856 at Radauti (Romania), collecting some of the aboriginal animals which for centuries had bred in half-wild conditions. Several bloodlines were established and the horses were bred only in purity. From these horses, thirty-three animals were sent to Czechoslovakia in 1922 to establish a herd. The new line of Gurgul was developed there.

The Second World War caused severe damage to the breed and numbers dwindled. An experiment was done in Muran to develop a larger type of horse for work in the forests by crossing Noriker and Haflinger blood to the Carpathian. The new type was called Slovakian Mountain Horse. Information was not received for this book on progress of this new type.

Horse lovers in Czechoslovakia were concerned over the lessening number of Carpathians in their country and established the Hucul Club in 1972, with four elderly mares and one stallion of the Gurgul line. The members have worked diligently to ensure survival of the horses. In 1982 the club had fifty head of purebred animals and established a stud book with the goal of including all living Huculs in Czechoslovakia.

The natural range for this little horse covers a great distance. For hundreds of years Huculs have been bred in the mountains under very difficult climatic conditions. Their habitat is the eastern Carpathians, called Huzulland. Isolated from other horses, the breed developed into a resistant and robust horse. Huculs are bred on bloodlines, the most important being Goral, Hroby, Ousor, Pietrousu, and Prislop, established by the foundation stallions. Goral was a golden sorrel foaled in 1915 at Radauti. Used in the Transcaucasian region from 1920 to 1925, he sired many offspring. All of his sons were named Goral, with numbers. Most were sorrel in color. Hroby (also spelled Groby) was black and sired small but very hardy, tough offspring.

Body type of the Hucul is rectangular with the length being over the height at the withers by two to three inches. The croup is well developed; the chest is deep and broad; the head is expressive with large eyes and small, lively ears. The neck is strong and muscular and the back is very strong. The hooves are small and hard. Usually, the Hucul has no need of shoes.

In color, bay, black, grullo, and chestnut are common. A dorsal stripe and zebra stripes on the legs are characteristic of the breed.

This is a very high quality animal of easy maintenance, capable of living outside throughout the year, finding its own food. Illness is nearly *unknown* in this breed. The legs are extremely strong and resistant, without known ailments. The breed is very sure-footed and able to maintain good speed while trotting on sloping ground or over tight and dangerous trails in the mountains. Due to its docile temperament, the Hucul is a very good riding horse for children, and it displays a natural aptitude for jumping. In addition, it has shown superb endurance.

In Romania Huculs are bred in the Bucovina region at about 4,500 to 5,000 feet in altitude but are found in other regions of the country as well. The Hucul is also bred in the former Soviet Union.

Although small, the Hucul is considered to be a small horse and not a pony. In the Soviet Union, it was developed by Ukrainian people on the eastern slopes of the Carpathian Mountains. As this small indigenous horse was used and bred by various peoples along this mountain range, differing types emerged, and often the Hucul of Romania appears quite different from its cousins. There are three main types found in the Soviet Union: saddle-pack, saddle-harness, and draft. The saddle-pack type contains the lightest horses, very lean and harmo-

nious in conformation. They are found primarily in the foothills. The saddle-harness type shows less "improvement" by other breeds and is numerous, concentrated in the high mountains. The draft type shows characteristics of the Noric horse and Haflinger. Compared to other types of Hucul, the draft-type animals are wide, coarse, and often lymphatic (without energy). These are rare and are not popular among the local people.

Popularity of the Hucul has spread to England in recent years.

Carthusian
(Carthusian-Andalusian, Carthujano)

ORIGIN: *Spain*
APTITUDES: *Riding horse*
AVERAGE HEIGHT: *15.2 h.h.*
POPULATION STATUS: *Rare*

Carthusian stallion. Photo: Evelyn Simak

The Carthusian is not a separate breed from the Andalusian, but is a distinct side branch of that breed and usually considered the purest strain remaining. This is one of Spain's most prestigious lines of the Spanish horse and has one of the oldest stud books in the world. The Catholic kings acquired a group of Spanish mares for breeding and placed them in Aranjuez, near Madrid, which is considered one of the oldest horse breeding farms.

The Zamora brothers, who had mares of this breeding, purchased an old horse named El Soldado. The brothers were blacksmiths, Andrés and Diego. One day Andrés saw a decrepit old horse overburdened with a load of firewood and recognized it as his former mount in the cavalry. He bought the stallion, El Soldado, and bred him to two of their mares before the stallion died. The resultant offspring were a colt and a filly; the former was Esclavo, the foundation sire of the Carthusian strain. Esclavo was a dark grey, considered to be a perfect horse. He produced many outstanding offspring which were purchased by the breeders of Jerez. One day when Andrés was away, Diego sold the twelve-year-old stallion for a high price to Portugal. Upon his return, Andrés was stricken with grief and died a short time later.

Esclavo produced a group of mares that about the year 1736 were sold to Don Pedro Picado, who gave some excellent specimens to the Carthusian monks to settle an outstanding debt he had incurred. The rest of the stock belonging to Don Pedro Picado went to Antonio Abad Romero and were eventually absorbed into the Andalusian breed. The Esclavo stock at the monastery was integrated into a special line and came to be known as Zamoranos.

The Carthusian Order had been founded in 1084 by San Bruno, a master of the cathedral school located at Reims. He and six companions led a reclusive life in the valley of the Isère located in southeast France. The first Carthusian monastery in Spain was established in 1163, and by 1478 the monastery of Jerez was under construction.

The stallion Esclavo is said to have had warts under his tail, and this characteristic was passed on to his offspring. Some breeders felt that without the warts, a horse could not be of the Esclavo bloodline. Another characteristic sometimes seen in the Carthusian is the evidence of "horns," actually frontal bosses (see Moyle), thought to be inherited from Asian ancestors. The descriptions of the "horns" vary from calcium-like deposits on the temple to small horns behind or near the ear. Unlike the warts beneath the tail, the horns were not considered a perquisite for proof of Esclavo descent.

Throughout the centuries that followed, the Carthusian monks jealously guarded their bloodlines, even defying a royal order to introduce Neapolitan and central European blood (see Andalusian).

Don Pedro and Juan José Zapata bought a good number of mares from the Carthusians. In 1854 Don Vincent Romero y Garcia, a Jerez landlord, purchased what he could of that excellent group of horses. Don Vincent lived to be ninety-two years old and because of his knowledge of breeding, greatly improved the quality of the horses without using any outside blood. When he died, most of his horses passed on to Don Roberto Osborne, who later sold a portion of the stock to Fernando C. de Terry just after the Spanish Civil War. Others of the horses went to the Marquis of Salvatierra, Don Juan Manuel Urquijo, Count of Odiel, and others.

Without the dedication of the Carthusian monks, the Zapata family, and a few other breeders who refused to cross their horses with other breeds, the purest line of Andalusian blood would have been lost to the world.

Presently Carthusian horses are raised in state-owned studs around Cordoba, Jerez de la Frontera, and Badajoz. The predominant color is grey, attributed to the important influence of two stallions of this color early in the twentieth century. Some Carthusians are chestnut or black. Nearly all of the modern Carthusians are descended from the stallion Esclavo.

The Carthusian head is light and elegant with a slightly convex profile, broad forehead, small ears, and large, lively eyes. The neck is well proportioned and arched; the chest is broad and deep; the shoulder sloping; the back short and broad; the croup sloped; and the legs are sturdy with broad, clean joints. Nearly all members of this breed have good conformation.

Caspian
(Caspian Miniature Horse)

ORIGIN: *Iran (Persia)*
APTITUDES: *Riding, light draft*
AVERAGE HEIGHT: *9.2 to 11.2 h.h.*
POPULATION STATUS: *Rare*

Perhaps the most exciting equine discovery of this century is the Caspian—not a pony but an ancient breed of miniature horse previously believed to have been extinct for over one thousand years. This tiny horse was probably one direct ancestor of the Oriental breeds and subsequently of all light horse breeds.

Extremely rare and barely pulled back from the edge of extinction in 1965, the Caspian is now being studied by leading archaeozoologists to prove the link beween the modern Caspian and the tiny prehistoric horse of Persia (see Oriental Horse).

Louise Firouz, an American living in Iran, is credited with recognition of this ancient breed. In 1965 while she was searching for small ponies to use in her riding school, an expedition to the Caspian Sea revealed three very fine, small animals. The beauty and grace of these small horses prompted questions as to their origin, for they did not exhibit the usual "pony" characteristics but rather were perfectly proportioned small horses. A study was initiated in the spring of 1965 to determine the range and nature of and historical precedent for a horse of this size in Iran. Over a five-year period, twenty more of the tiny horses were located, mostly overworked and often ill-used. They stood between 10.2 and 12.2 h.h. Their temperament proved such that mares and stallions were ridden together by young children with no problems, and often stud stallions are pastured together peacefully. The Caspians revealed amazing jumping ability and the smooth, elegant paces of a blood horse.

Caspian stallion, Hopstone Jamshyd. Photo: Brenda Dalton

Caspian mare, owned by Louise Firouz, Teheran, Iran. Photo: Caren Firouz

Caspian stallion, Zeeland, owned by Louise Firouz. Photo: Louise Firouz

The Caspian stallion, Zeeland, showing the classic beauty of the Caspian head. Photo: Louise Firouz

The Caspian head is short and fine with large eyes, a small muzzle, and large nostrils placed low. There is a pronounced development of the forehead, the ears are very short; the neck is slim and graceful, well attached to sloping shoulders; withers are pronounced; the back straight; and the tail set high on a rather level croup. The legs are slim with dense, strong bone and no feathering at the fetlock. The hooves are extremely strong and oval-shaped, more like those of the ass than the horse. The overall impression of the Caspian is that of a very small, well-proportioned horse. Subsequent studies confirmed the visual picture osteologically; the Caspian is a miniature horse, not a pony.

A survey was conducted from July of 1965 through August 1968 to determine the range and approximate numbers of the re-maining Caspians. On the basis of this survey it was estimated that there were approximately fifty small horses with definite Caspian characteristics along the entire littoral of the Caspian Sea. It was concluded that due to the fact the individuals were widely scattered, it was virtually impossible for any of them to be considered completely pure.

During the period from 1965 through 1970 seven mares and six stallions were used for breeding at Norouzabad Stud, producing fourteen foals. Only one of the foals did not exhibit Caspian characteristics. The remaining thirteen all had the small ears, characteristic bulging of the forehead, slim, dense bone, and oval-shaped hooves. The growth rate of the Caspian is distinctive in that most of the height is attained within the first six months of life and subsequent growth is minimal, mostly being in width and secondary sexual characteristics. Sexual maturity is reached in both colts and fillies at about eighteen months.

The mares have a strong tendency not to ovulate until about a year after foaling, making a continuous breeding program difficult. In spite of improved conditions and feed, the mature height of offspring born at Norouzabad is smaller than the average height of sire and dam, possibly indicating that the original size of the Caspian is closer to 9 hands. This also indicates that the present stock is not completely pure and that breeding to type will further emphasize the true conformation of the Caspian and lead to a return of the natural size.

In the summer of 1969 a mature Caspian skeleton was studied in comparison with bones of *Equus hemionus onager*. The onager had been excavated from first-century graves of Shahr-e-Qumis near Damghan in northwestern Iran. This was a predominantly Parthian site dug by David Stronach and John Hansman of the British Institute of Persian Studies in Teheran. The study concluded that since the slenderness index of the metacarpal and metatarsal bones of the Caspian fell within the range of what is considered normal for *Equus asinus* and *Equus hemionus onager*, particular care should be taken in evaluating equid remains from archaeological sites in the future as identification on the basis of the slenderness index of the long bones was no longer conclusive evidence of the particular species of Equidae.

Comparative anatomy studies of Caspian, Turkoman, Plateau Persian (see Oriental Horse), and one Percheron skeleton, together with live comparisons, revealed five basic skeletal differences between the Caspian and all other breeds of horses studied: (1) The Caspian skull shows a pronounced elevation or bulging of the forehead (bulging of the interparietal and parietal bones) and, most distinctive, the Caspian possesses no parietal crest, the interparietal continuing unbroken to the nuchal crest of the occiput. (2) The scapula of the Caspian is wider than in other breeds, resembling more that of a ruminant than of a horse. (3) The metacarpal and metatarsal bones of the Caspian are much longer and slimmer in comparison with the height of the horse than those of other breeds. (4) The Caspian shows in the spinous processes of the first six thoracic vertebrae (T.1 to T.6) a pronounced elongation as compared with other breeds, resulting in the withers being higher than the croup. (5) The hoof of the Caspian horse is narrow and oval-shaped, resembling the hoof of *Equus asinus* more than that of *Equus caballus*.

For many years archaeologists have been aware of the existence of a tiny horse found on stone carvings in Iran. One example is the trilingual seal of Persian king Darius the Great (ca. 500 B.C.), now in the British Museum, showing a pair of tiny horses with very slim legs, small ears, and slightly convex faces pulling the royal chariot on a lion hunt. Positively linking the present Caspian miniature horse to the ancient miniature will depend on careful evaluation of equid material from archaeological sites in the future. Establishing continuity for a breed for more than 3,000 years will be difficult, but failure to take an exhaustive look into the past of this unique little horse may result in the loss of the last vestige of the wild horse of the Middle East. The greatest importance of the Caspian lies in the fact that if continuity can be proven, the breed is

likely the ancestor of all modern breeds of hotblooded horses.

Construction of the Caspian head renders it unlike any other modern breed in the world. Stone carvings, especially at Persepolis, show that the horses of the Achaemdenian period (Iranian dynasty 559–330 B.C.) all possessed massively developed foreheads. The powerful head of the Nisaeans of the great king showed a pronounced swelling beginning at the occiput and extending through the parietals, frontal, and nasal bones. This formation has occasionally been referred to as "Roman-nosed." However, the typical Roman nose begins with a bulging in the frontal bones which extends through the nasal bones, but on no account does the convex profile extend to the occiput. The Lydian ponies of early stone carvings show the vaulted forehead development, giving them a typical "Arab" look. This head formation, so prized among lovers of the Arabian horse, is present today in the Caspian to a degree unknown in any other breed, including the Arab.

During the summer of 1976, the Caspian herd of Louise Firouz came under attack by wolves and three mares and a foal were lost. In an effort to ensure the safety of at least some of the herd, an emergency flight was arranged to bring seven mares and one stallion to a stud already established in Shropshire, England. The flight proved to be a lifeline for the entire breed. Export had been banned by the Royal Horse Society in Iran in 1967 to preserve and maintain numbers in the country, and this spelled tragedy for the Caspians, which were nearly eradicated during the Revolution. Few presently remain in their native country.

The Caspian is no longer in danger of extinction, although the breed is still extremely rare. Several studs now exist in Britain and a few individuals have been exported to Australia and New Zealand.

A versatile little horse, the Caspian is becoming known in the show ring with its exceptional jumping ability. The driving prowess that endeared it to Darius the Great still makes this horse a favorite in harness.

Cerbat

(Cerbat Mountain Spanish Mustang)

ORIGIN: *United States (Cerbat Mountains)*
APTITUDES: *Riding horse*
AVERAGE HEIGHT: *14 to 15 h.h.*
POPULATION STATUS: *Rare*

Cerbat mare, Dama Roja, owned and bred by M. A. Thompson. Photo: M. A. Thompson

Two-year-old Cerbat colt. Photo: M. A. Thompson

Though the actual date of influx into the Cerbat Mountains northeast of Kingman, Arizona, will never be known, it is authenticated that when the first white settlers arrived in the area in the 1860s, the Cerbat horses were well established. A local family history suggests that this may have been one of the few authentic mustang bands at large in the United States to continue to exist unchanged into modern times.

The Neal family was among the earliest settlers and has ranched in the area since 1860. In an interview in 1966, rancher John Neal said that to the best of his knowledge no other breed of horse was ever introduced into the Cerbat herd in past decades and it was doubtful that any domestic horse could survive such a rugged life. The local Indians have never claimed the horses and say "they have always been here." Though that statement cannot possibly be true, it does indicate that the herd has been in the area as far back as tribal memory now reaches. Spanish influence in what is now Arizona was extremely widespread in the past with expeditions of Spanish explorers, topographers, priests, and soldiers extending into all areas. The Apaches were experts at stealing Spanish horses which they used for barter with other tribes, and by these means the Spanish horses spread throughout Arizona.

The Cerbat Mountains range from 5,000 to 7,000 feet in elevation with some of the roughest terrain in Arizona. Consequently, the Cerbat horses have evolved into extremely sturdy, well-boned animals with the ability to forage for themselves and dig for water in drought conditions. However, in 1971 a drought caused problems with the ranchers in the area and the horses became a liability, drinking water needed for range cattle. The horses were being eliminated by cowboys, although they were digging holes into the dry creekbeds and thus providing water for the cattle as well as themselves. A group of approximately eighteen animals were water-trapped by private mustang fanciers and dispersed elsewhere in Arizona and to Washington State. The Washington herd subsequently was further dispersed and disappeared, but part of the Arizona herd was placed in private hands and maintained as a special breeding herd existing into the present and registered with the Spanish Mustang Registry.

In early 1990 the Bureau of Land Management, which now controls the Cerbat area, discovered a small group of animals believed to be descended from the remnants of the original herd. To ascertain whether this was so, blood samples were drawn and compared to samples taken from the privately owned herd. Though the feral group proved more inbred than the originals, the genetic markers did show the relationship, proving that the few individuals that escaped capture in 1971 had provided the genetic material for the 1990 group. One unusual feature of Cerbat horses is that inbreeding has not been a destructive force. Though the original Cerbats of 1971 were shown genetically to be highly inbred and the later Cerbats are more inbred yet, no defects have ever shown up either in the wild or in the now domesticated horses. Due perhaps to inbreeding over the centuries, they are extremely prepotent (see Namib).

In the mid 1970s, after the original capture, there were only twenty-two horses inventoried in the Cerbat Mountains. As in all feral populations in remote areas of the West, mountain lion predation takes a high toll on the very young and the old, sick, or disabled; therefore, the Cerbat herd has never consisted of a large number of individuals.

Showing typical Spanish conformation and appearance, the Cerbat horses are always bay roan, chestnut roan, bay, or chestnut. Though size of the original horses captured was small, ranging from 13.2 to a maximum of 14 hands, second generation Cerbats consistently attain a height of 14.2 to 15 hands. Heads are typically subconvex with high-set eyes and small, curved ears. The backs are short; croups are rounded with very low-set tails. The chest circumference is large with well-sprung ribs and medium to narrow chests. The shoulders are well laid back with medium to heavy arched necks in both stallions and mares. Manes are somewhat unique, being very soft, full, and falling on both sides of the neck in many individuals. The tails are long and full. Cerbat horses are well balanced with legs set well under the body. Legs are consistently well boned, short in the cannon, with well-muscled upper forearms. The hooves are extremely thick walled with a high dome. Fore chestnuts are small and flat, rear chestnuts extremely small or nonexistent, as are ergots. The fetlock hair is sparse. Both solid-colored and roan individuals may possess dark-pigmented areas approximately three to four inches in diameter on one or both hips. The disposition is calm, quiet, and intelligent.

Genetic blood studies show Spanish "markers" as well as a gene in the biochemical markers carried also by the Peruvian Paso breed. This may explain to some extent the production of laterally or *paso*-gaited offspring from non-laterally gaited parents.

Though Cerbats are few in number, efforts are underway to preserve these unique horses. To that end, members of the Spanish Mustang Registry acquired eight more in March of 1990 from the Marble Canyon area of the Cerbat Mountains, bringing the total of purebred horses up to twenty-one, with breeder M. A. Thompson of Willcox, Arizona, possessing sixteen and being the primary source of purebreds. Crosses of registered Cerbats to other registered Spanish Mustangs have given excellent results; however, the ultimate goal is to preserve the Cerbat horse in its pure form while at the same time providing genetic outcrosses through the stallions to the overall gene pool of the Spanish Mustang Registry.

For additional information see Spanish-American Horse.

Chakouyi

ORIGIN: *China*
APTITUDES: *Riding horse*
AVERAGE HEIGHT: *12.2 to 13 h.h.*
POPULATION STATUS: *Common*

Chakouyi broodmare of China. Photo: Weigi Feng and Youchun Chen

and well muscled. Joints and tendons are well developed in this breed and there is little hair on the legs. The legs and feet of this horse are strong with few problems.

Chakouyi stallion of China. Photo: Weigi Feng and Youchun Chen

Chakouyi horses are distributed in the vast pasturelands along the ancient and well-known silk route. "By-route transmission station" is the meaning of the word *chakouyi*. In ancient times (206 B.C. to 907 A.D.) there was a great need for fast horses with good endurance qualities for use in carrying information, military orders, and the instructions of the emperor. Horse ranches were established along the silk route. Horse shows were common events, both private and imperial. At that time the Buddhist temples owned the best stallions. Transmission of messages today is not as important as in times past, but the need for good, powerful riding horses ensures the continued breeding of the Chakouyi horse. Horses are still necessary for everyday life in China. This breed is a natural pacer, inheriting the gait, so it is called the "placental pace" by the Chinese. The Chakouyi is classified as a native breed.

Common colors found in the Chakouyi are bay, with few greys or blacks. An elongated star is seen on many members of this breed.

The Chakouyi is strongly built with a square shape. The head is straight and of medium size with large, bright eyes and small ears, set up straight. The face is dry. The neck is of medium length; withers are long but not high; the chest is wide and deep; the back moderate in length; the loin short and wide; the abdomen rounded; and the rump is sloped

Chara Horse

(Altai Productive)

ORIGIN: *Altai District (former Soviet Union)*
APTITUDES: *Harness, saddle, meat production*
AVERAGE HEIGHT: *14.3 to 15.2 h.h.*
POPULATION STATUS: *Rare, new breed*

The Russians are proven experts in horse breeding and presently several new breeds are in the process of development in the former Soviet Union. One new breed in formation is the Chara horse, or Altai Productive. This breed is aimed primarily at meat production but with an eye also toward utility as a working and riding animal. It has been proposed to name the new breed Chara, after the river of that name. The following is a look into the scientific formation of a new breed.

The Altai is an area of *taboon* horse breeding (see glossary), and groundwork for development of the new breed was done in the 1920s with the improvement of local horses. Use of saddle and trotter stallions in previous times and draft stallions in recent years has resulted in a considerable increase in size of the local horses. The import of stud farm stallions results in additional expenditures, lowering the effectiveness of productive horse breeding—and the excessive use of stud farm horses also lowers fertility of mares. Adaptability of crosses to severe conditions of *taboon-tebenevka* raising (*tebenevka* means year-round pasture raising without supplements) is also reduced. It is necessary to obtain horses that meet the needs of productive horse breeding under local climatic conditions. Such horses are crossbreds of local mares and stud farm stallions that have themselves been raised locally under *taboon* conditions.

The basic goal for development of the new breed is production of meat with horses adapted to year-round natural pastures. Horses of the new breed could also be used for improvement of Siberian horses and for work in harness and under saddle. Recently the work has been concentrated on six farms of the Upper Altai Autonomous District and the Chara region of Altai District.

In 1978, 845 mares were selected out of 3,100 that were evaluated. These were crosses of heavy draft, trotter, and saddle breeds. The most valuable were draft crosses, rather massive with a long, well-muscled croup; a wide, deep chest; and, compared to saddle and trotter breeds, a better ability to retain their weight during winter. They also regain lost weight more quickly in spring.

The major fault lies in soft hooves and lymphatic joints. A great part of the selected group were trotter crosses. Crossbreds of trotters are comparatively large, of medium massiveness with a long trunk, a deep but not especially wide chest, and fairly good muscling. Their fault is also soft hooves. Don and Budyonny crosses made up only a small number in the selection group and their rate is declining. Their value is that they have, besides a massive trunk, a light skeleton and very hard hooves.

In 1978, from 140 stallions in ownership of stud farms, only fifteen purebred draft horses were selected: ten Russian, two Soviet, and three Lithuanian. Of the draft crosses, ten were selected, seven of Russian breeding and three of Eston-Arden (Estonian Draft). They were mated to 665 mares, and the remaining 180 mares were mated to Zhmud and trotter cross stallions. In 1986 the stock of selected mares was doubled to 1,716 head and the number of selected stallions increased to thirty-three purebred horses—twenty-nine Russian, three Lithuanian, and one Soviet. From these, the most valuable proved to be the Lithuanian sires.

For development of the new breed, creative crossing of crossbreds is used and special attention is paid to saving adaptability to year-round *taboon* raising and quick reestablishment of physical condition after wintering.

Presently there are three types, called desired, universal, and aboriginal. Horses displaying the best adaptive qualities are crosses of the desired type, and, fortunately, in the selection of breeding stock this type increased while the numbers of the other two decreased. Mares of the universal type are mated to pure draft stallions or massive stallions of the desired type. Aboriginal-type mares are bred to big draft crosses or massive Orlov Trotters to increase their height.

The crossbreds of Lithuanian Draft are larger than crosses of Russian Draft. Breeding sires having at least ten foals from one to five years of age are evaluated according to live weight, size, type, and appearance of their offspring. So far, the Lithuanian stallions have proven to be superior sires.

Development of the new breed is now at the stage of concentrating the best crosses by selection and further crossing within that group. At the beginning of 1989 there were 1,620 crossbred mares in the selection group, and 1,030 of these were used for interbreeding with fifty of the crossbred sires. As a result of well-planned work, the live weight of both mares and stallions has increased appreciably.

Although the work is only in the consolidation stage of horses of the desired type, the youngsters have already received high evaluation from specialists. Breeding farms in the eastern regions of Russia and Kazakhstan are anxious to buy great numbers of them.

Cheju

ORIGIN: *Korea (Cheju Province)*
APTITUDES: *Riding, light draft*
AVERAGE HEIGHT: *11 h.h*
POPULATION STATUS: *Rare*

Cheju Island is off the southern coast of South Korea. The island is the Province of Cheju, an independent kingdom in ancient times until 938, called Tamra-guk. It is located in the East China Sea and comprises 705 square miles, including twenty-six small associated islands. From the year 935 until 1910 it was used as a place of political exile and for grazing horses.

According to Professor Dominicus C. Choung of the Cheju National University in Cheju City, it has not been firmly established

Cheju pony of Korea. Photo: Prof. D. C. Choung

Cheju pony. Photo: Prof. D. C. Choung

A group of Cheju ponies in Korea. Photo courtesy Kim, Duk-rak

Cheju pony foal. Photo courtesy Kim, Duk-rak

when and what kind of equines were first introduced to the Korean Peninsula, but it is assumed that they came to Korea from China. Ancient records reveal that horses were among the most important animals used for agricultural and military purposes from the period of tribal states (before the first century B.C.) through the Choson dynasty (fourteenth to nineteenth centuries).

The Cheju native pony has been known to exist for about seven hundred years on the island, and was used mainly for draft work. The island has rough, narrow roads and this pony is well adapted for use in the area. Mild weather and sufficient precipitation on Cheju Island assure abundant forage for horse production.

The Cheju native pony may have existed since prehistoric times, although no clear record confirms this. During the Koryo dynasty (1276–1376), Mongolians governed Cheju and introduced their horses to the island. One record shows that 160 breeding horses were brought from Mongolia to Cheju and used for improving the native ponies. Since that time, horses raised on Cheju have been exported to the mainland of Korea and to

China. Native ponies were also used for cross-breeding with Mongolian and some other exotic breeds.

During the Koryo and Choson dynasties, Cheju was a major horse producing area and 25 percent of the island's farm households were engaged in horse production. At one time there were as many as 20,000 native ponies in Cheju, but this number decreased with mechanization of farming and transportation to only 2,500 by 1989. The breed is in serious danger of extinction. For their preservation, in 1987 the Korean government designated the Cheju native pony as National Treasure No. 347.

The hardiness and draft ability of the Cheju native pony is outstanding, especially considering its small size. The ponies survive the most severe winters without artificial shelter and are highly resistant to both disease and ticks. Mares foal regularly up to twenty or more years of age. Cheju ponies are able to carry heavy loads up to 230 pounds.

Predominant colors of the ponies are chestnut, bay, and black and occasionally grey, black, white, *cremello* (cream colored), or pinto. This pony has a nicely shaped head with a straight profile, large eyes, and small ears. The jaw is deep, tapering to a small muzzle. The neck is short and well muscled; the back is short and straight; the croup is gently sloped but the tail is set fairly high; the shoulder is often quite straight; the legs are strong and well muscled with clearly defined joints and tendons. The Cheju shows influence of both the Arab and Mongolian breeds.

To help promote the Cheju pony industry, the Korean Equestrian Association organized pony racing in the spring of 1990. Horse racing is an old sport in Korea. The ancient Koreans were horse riding people, living in an agricultural society. During the Three Kingdoms period (from the fourth to seventh centuries A.D.), horses were a major trade item for Korea. During the Koryo dynasty (tenth through fourteenth centuries), studs were set up to breed horses, and according to ancient records approximately 40,000 horses were bred in some 160 studs across the nation following the Choson dynasty. Horse breeding continued to expand, and by the end of the

1930s there were some 80,000 horses in Korea. Most of these were used for either labor or military purposes.

Chickasaw

ORIGIN: *United States*
APTITUDES: *Riding horse*
AVERAGE HEIGHT: *14 h.h.*
POPULATION STATUS: *Rare*

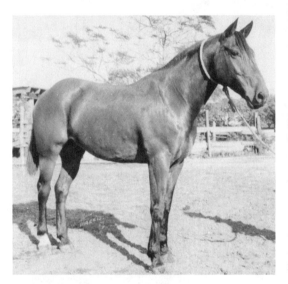

Chickasaw. Photo courtesy Ellenore Barker

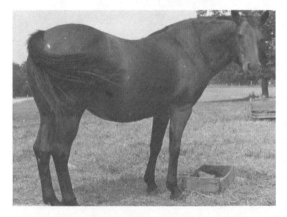

Chickasaw. Photo courtesy Ellenore Barker

The Chickasaw horse was one of the early descendants of the Spanish horses brought to the Americas in the sixteenth century. They were called Chickasaw as they were brought north from Florida by the Chickasaw Indians, and many were traded to the colonists. Settlers in Virginia and North and South Carolina were among the first to become interested in the horses of the Chickasaw Indians. Noted for speed at short distances, this plucky Spanish horse was partly the foundation of the early Quarter Horse.

Chickasaw horses are easily recognized by a short head, short ears, short back, stocky hips, short neck, wide chest, and unusual width between the eyes. They may be black, bay, brown, chestnut, grey, roan, sorrel, or palomino.

The Chickasaw Horse Association is small and not very active, but the strain is worthy of preservation. For additional information see Spanish-American Horse.

Chilean Criollo. Photo courtesy Carlos Fresno L. De L.

Chilean
(Chileño)

ORIGIN: *Chile*
APTITUDES: *Riding horse*
AVERAGE HEIGHT: *13.2 to 14.2 h.h.*
POPULATION STATUS: *Common*

The Chilean is the *criollo* (native) horse of Chile. The breed does not differ greatly from the Criollo of Argentina, descending from the same Spanish stock as other South and Central American breeds.

Chilean horse. Photo courtesy Carlos Fresno L. De L.

The Republic of Chile, situated on the Pacific coast of South America, is an extremely long and narrow country extending about 2,700 miles from its boundary with Peru in the tropical zone to the tip of South America at Cape Horn. This country has an average mainland width of slightly more than 100 miles. It is bounded in the north by Peru and Bolivia, on its long eastern border by Argentina, and to the west by the Pacific Ocean. Chile's relief is mostly mountainous and dominated by the Chilean Andes range, which extends almost the length of the country. The Chilean Andes include the highest segment of the Andes mountain chain, acting as both a physical and human divide.

The Spanish conquest of Chile began in 1536 to 1537, when Diego de Almagro invaded the region as far south as the Maule River in search of "another Peru." He was met with disappointment, finding neither gold nor a high civilization, and returned immediately to Peru. His reports forestalled further attempts at conquest until 1540 to 1541, when Pizarro, after the death of Almagro, granted Pedro de Valdivia license to colonize the area. Valdivia, with about 150 companions, entered Chile in late 1540 and founded Santiago (February 12, 1541). The horses taken into Chile by Valdivia and his party were the first to arrive and stay in Chile.

The Chilean is slightly heavier than the Criollo of Argentina. This breed has the same qualities of endurance as other Criollo horses of South America (see Argentine Criollo for additional history) and is primarily used for work with cattle and in rodeos.

The Chilean is a muscular horse, powerfully built but very agile. Selective breeding did not begin until 1913. All colors are accepted in the Chilean except "albino," which is eliminated. The skin is thick; the mane and tail are wavy and thick, but there is little hair at the fetlock. The head is of medium length with a wide, flat forehead and slightly convex or straight profile. Any horses with a concave profile are eliminated from the breed. The eyes are expressive and ears are small and mobile. The nostrils are flared. The physical stamp of the Chilean is preferably expressed in the profile and characteristics of the head.

The horse's neck is of medium length, wide at the base and inserted strongly into the shoulder and finely attached to the head. The withers are lean, not very prominent, and blend smoothly into the back; the shoulder is of medium length and sloping; the back is short, strong, and wide; the loins are short, wide, and muscular. The croup is long, with pronounced musculature, wide, and lightly sloping; the tail is set low; the trunk is well developed with arched ribs. The forearm is tied strongly to the arm, with abundant musculature; the knee is low, short, and well shaped; the legs have strong, lean bone with straight tendons which are strong, pronounced, and as short as possible. The thighs are well-muscled, deep, with exterior lines straight, set well apart; the pasterns are short but sufficiently sloped for elasticity; the hooves are preferably black, of proportionate size, with a concave sole and tough horn.

The breeding aim for the Chilean is to produce a horse that is suitable and functional for saddle and stock work.

Chilote

ORIGIN: *Chile (Isle of Chiloé)*
APTITUDES: *Riding, light draft*
AVERAGE HEIGHT: *Uncertain, about 12 h.h.*
POPULATION STATUS: *Rare*

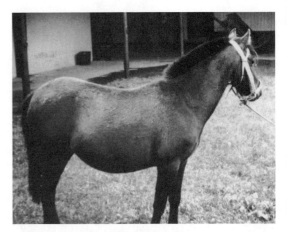

A Chilote mare in Chile, pure descendant of the horses of the Conquest. Photo courtesy Prof. E. Gus Cothran

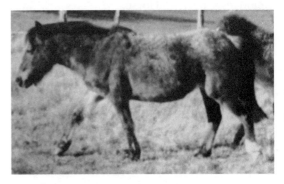

Chilote horse. Photo: Dr. Rafael Mancilla

Chiloé, the largest island of Chiloé Province (3,241 square miles) in the Los Lagos region of southern Chile, is the habitat for a genetically unique equine breed. Chiloé is an extension of the Chilean coastal mountain range, from which it is separated by straits called the Paso del Chicago. Nearest of the myriad islands and archipelagoes to the south are the Islas Guaitecas, which lie across the Boca del Guafo. To the east, thirty miles across the Golfo Corcovado, is mainland Chiloé province; to the west is the Pacific Ocean. Chiloé is situated in an area of heavy rainfall, and the island's inland is heavily forested with evergreens. The island is mostly undeveloped.

The Spaniards captured this island from the Indians in 1567 and controlled it until 1826, making it the last foothold in Chile of royalist forces during the wars for independence. Most of the inhabitants of the province live on the island, especially in Ancud and in the oldest city and provincial capital, Castro, founded in 1567. They are engaged in agriculture, raising potatoes and grains; livestock, raising sheep and cattle; fishing; and lumbering.

The only horses of Chiloé are the small Chilotes, pure descendants of the horses of the conquistadores. The horses have never been crossed with other breeds and, although very small, bear many characteristics of the present Spanish horses. Whether their small size is due to island habitat and environment or inbreeding is unknown. Possibly both played a role, and the horses of the Conquest were small to begin with. Uniquely, the Chilote is the only confirmed pure descendant of the horses brought to the Americas by the conquistadores. Horses of other breeds were never allowed on the island. Most Spanish horses in other areas of the world have been crossed with other breeds at least to some extent through the years. Not much is known about this breed and my correspondent, Dr. Rafael Mancilla in Santiago, reported that he was unable to find any literature regarding them.

Although Chilote horses are rarely seen on the mainland of Chile, the accompanying photos are of specimens in Santiago. The most predominant colors in the Chilote are white, grey, bay, black, and "muddy" *(barroso)*. Most of them are domestic and owned by local peasants who possess a few individuals each, but feral Chilote horses are also found on some small islands of the Archipelago of Chiloé. Apparently selective breeding is not practiced.

Genetic blood testing of the Chilote breed done by Professor Gus Cothran at the University of Kentucky shows the Chilote to be unique among Spanish horse breeds.

CHINESE HORSES

Some general information regarding the horses of China seems to be in order, as little information has been made available regarding them. Most volumes on horses present the "Chinese pony" and merely state that there are several types. But there is no breed in China known by the name Chinese pony, and the horses of China are as distinct from one another as are the Thoroughbred, Trakehner, Quarter Horse, or other commonly known breeds.

The original Chinese horses were of Mongolian type. Other than these, a certain number of exotic breeds were introduced over a long period prior to the birth of Christ and blended with some Chinese breeds. Some strains, resulting from crossbreeding of European and Russian breeds on Chinese horses and then interbred for many generations, are considered as new breeds. Thus, in China, horse breeds are classified as *native, introduced,* or *developed breeds*, and they are classified as such here.

Among Chinese horses today there are five distinct types, which may be classified as: Mongolian, Southwest, Hequ, Tibet, and Kazakh types. Among these basic types there are over thirty distinctly different breeds which are continually being improved (see China, appendix A). Only one Chinese equine breed is a pony; all others are small horses, the progeny of which show an increase in size when raised in warmer climates and under better feeding conditions.

There are approximately 11 million horses in China, accounting for about one-sixth of the world's horse population. To understand the distribution of horses in China one should locate on a map the city of Lanzhou in the province of Gansu. Using this as a point of orientation, one may then refer to the northwest, northeast, north, southwest, and the area of Tibet.

Northwest China is a large steppe separated by desert from most other Chinese territory; the northeast is a vast plain with developed agriculture under moderate rainfall the year around; the north has quite a complicated range of topography and climate; Tibet is characterized by a very high plateau with an average altitude of more than 13,000 feet above sea level. Southwest China is situated between the Tibetan plateau and the rice plains of southern China, broken country of mountains reaching almost 11,500 feet and deep canyons with an extremely humid climate.

Chinese horses vary in size, conformation, and temperament corresponding to their location. For example, horses in southern China are small, short, and slender in build. In northern regions they tend to be larger, of heavier bone, and more massive in body construction.

Most horses found in China have a long recorded history. They are called "short horses" and are *not* ponies. When moved to warmer areas with better feed, especially in their first year, there is an increase in size. The only "pony" of China is called the Guoxia, and even this old breed may well be a miniature horse and not a pony. In any case, the Guoxia, in a warmer climate and on better pasture, only contentedly grows fatter—there is no increase in size.

While some Chinese horses are of recent, purposeful development, other breeds date back 4,000 years in a pure state. China was among the first cultures to use the horse other than as a food animal. Some of the earliest chariots have been found in ancient tombs there, and with them, the hame collar that was unknown in western Europe until 1000 A.D. Due to the vastness and varied topography of China, different horse breeds have remained pure, isolated and unchanged for centuries.

Horses in China have not suffered the decline to recreational use only, as have most other horses in the world; they are still of vital, daily importance to the Chinese people. Much of the terrain is rugged and best traveled by horseback, and the endurance and strength of the horses of China are phenomenal. Weakness or problems in the legs are extremely rare, and these horses are remarkably disease-free and long-lived. Their feet are exceptionally good, of hard, black horn, and usually well shaped.

In addition to the present Chinese breeds, research is under way to present eventually the entire story regarding the ancient "Dragon Horse," which appeared first about 4,000 years ago and was written about in the biography of King Mu Wang, who reigned during the years 1023 to 982 B.C. This horse is again mentioned in a bamboo carved book of the Sui dynasty. As information regarding horses with horns is mostly absent from historical books or mentioned only as a myth—or contained only in ancient Chinese writings, any

amount of information regarding this famous strain would have been impossible without the aid of Professor Youchun Chen of Beijing, to whom this writer is extremely grateful (see Moyle).

All of the breeds of China are presented in this volume under their respective names.

A most exciting and important archaeological find was unearthed in 1974 at Xian, China, by people digging a well. It turned out to be the beginning of one of the most spectacular excavations of the century. The Wei Valley tombs have long attracted the interest of scholars, but no archaeological digs of the tombs had been authorized until the amazing discovery was made by the well diggers. East of the tumulus where China's first emperor, Qin Shi Huang (259–210 B.C.), is buried, life-sized terra cotta soldiers and horses were unearthed. Evidence that such figures might be concealed in this area had appeared sporadically between 1932 and 1970. The continuing excavation of the tomb, which startled and thrilled the archaeological world in 1974, has since yielded an army of 6,000 terra cotta warriors and horses, arranged systematically around the tomb to guard the deceased emperor.

Despite the staggering number of figures required to fill the ranks of this pottery army, the warriors and horses were not stamped from molds but were individually modeled. Each body was made up from coils of coarse grey clay with the hollow torso fully supported by solid columnar legs; once the desired shape was achieved, the surface was finished with a slip of fine clay. Small details—such as the elements of armor—were modeled individually and applied to the surface while the clay was still pliant. Then the entire assemblage was fired at high temperature and mounted on a prefired base. The hands and heads of the warriors and the tails and forelocks of the horses were fashioned and fired separately and attached with clay strips. Finally, each figure was painted in bright colors and fitted with genuine weapons or chariot gear. It is believed that the actual soldiers were happy to model for the statues, knowing that these would be buried in the tomb in lieu of themselves.

In December of 1980, after five years of work at the site, two full-scale chariots, complete with life-size horses and charioteers and all in bronze, were discovered seventeen meters west of the grave. These are the oldest bronze chariots and horses ever found in China. Chinese archaeologists estimate that the figures are only the vanguard of a chariot procession, which may contain hundreds if not thousands of replicas of ceremonial chariots carrying the images of the crown prince, the empresses, royal concubines, noble families, and court officials. They were created 2,000 years ago to accompany the dead emperor on his celestial journey and were buried in an underground vault near a subterranean palace containing the magnificent sepulcher of Emperor Qin Shi Huang.

The horses are harnessed, four abreast, before their royal chariots. Unlike their pottery counterparts, which had been broken by marauders, the bronze statues have remained intact. In twenty-two centuries their original coat of white paint had faded to grey but their bodies, including their muzzle halters decorated with gold and silver ornaments, were all found in remarkable condition. Each chariot contained a bronze box about 39 inches wide, 47 inches long, and 17 inches deep for the passenger to ride in. It was shaded by an awning made of thin sheet bronze, painted elaborately in color with symbolic cloud formations and geometric designs. The ornate, spoked wheels are 23 inches in diameter. Each chariot has a single shaft 98 inches long, with a 30-inch crossbar fixed at a right angle near the end. The box measures 39 inches wide, 47 inches from front to back, and 16 inches deep and has an awning, an armrest bar in front, and a door at the back. The box, door, and awning are made of thin bronze sheets. Each chariot's axle measures 59 inches.

One of the charioteers of painted bronze is depicted kneeling, while the other is standing. Their garments show them to be *dafu*, or ninth grade officials under the Qin dynasty.

The horses were also modeled after local horses, according to Professor Youchun Chen of Beijing. The area in which the tomb is located is known as Quanzhong Plain. In the western part of this plain there were good pastures, and horses of that time were of draft type, medium heavy. Their head was light. Usually they were used for pulling carriages and plows.

A certain amount of Dawan (Akhal-Teke) blood was introduced about 900 B.C. during the reign of King Xiao in the Zhou (or Chow) dynasty. In a book, *Story of Emperor Jing, Dynasty Han*, there are records of the body size of horses used for palace ceremonies and other purposes. It was given as 5.9 chi, or 135.7 centimeters at the withers (about 13.2 hands).

The pastoral horses from the northwest and horses from Tibet also had an influence on the local horses from the time of the Chow dynasty to the Song dynasty in the Guangzhong area. The horses were known as Qin horses. During the latter Song dynasty this horse completely disappeared due to extended and debilitating wars. Since 1949 a new Guangzhong breed has been developed using exotic breeds.

The location of the tomb site and the form of the chariots suggest that they were made between 221 and 211 B.C., within the decade after Qin Shi Huang unified the country. This was the emperor who built the Great Wall, burned the books valued by Confucius, and declared himself China's first sovereign emperor; prior to this, Chinese leaders had been called kings. He asserted that his rule would last for 10,000 years, but in reality it turned out to be the shortest dynasty in the history of China. He ruled for only fourteen years. Nevertheless, he accomplished sweeping changes that made his dynasty a turning point in the history of China. His tomb is said to have taken 700,000 conscripts thirty-six years to complete.

The design of the chariots and the tack for the horses illustrates China's early knowledge and use of the horse.

Chinese Kazakh

ORIGIN: *China (Western Xianjiang Uygur Autonomous Region)*
APTITUDES: *Riding, packing*
AVERAGE HEIGHT: *13 to 13.2 h.h.*
POPULATION STATUS: *Common*

Chinese Kazakh horse. Photo: Weigi Feng and Youchun Chen

Chinese Mongolian

ORIGIN: *Northern China and Mongolia*
APTITUDES: *Riding, pack horse, farm work*
AVERAGE HEIGHT: *12 to 14 h.h.*
POPULATION STATUS: *Common*

A distinction is made here between the Mongolian horse of China and that of the Mongolia People's Republic because, while having the same roots, the horses have been bred differently (see Mongolian).

The Mongolian is one of the oldest known breeds and has had a much greater influence on other breeds than is generally acknowl-

A Kazakh horse of China, ready for work. Photo: Youchun Chen and Tiequan Wang

Chinese Mongolian mare. Photo: Weigi Feng and Youchun Chen

The Chinese Kazakh originated in the western part of the Xianjiang Autonomous Region. There were good horses in ancient China about 200 B.C. when the Kazakh tribe was settled in this area. Chinese Kazakh horses had common ancestors with the Kazakh of Russia and some crosses with Mongolian horses. The majority of Kazakh horses are bay with some chestnut, black, and occasionally grey. The Chinese Kazakh is classified as a native breed.

This is a versatile horse with good tolerance for hard work and harsh environment. It is considered more durable and is larger than the Mongolian horse.

Mares are milked, and this breed is also used for meat. The Kazakh is a very fertile breed; often ten to fifteen foals are produced in a lifetime.

Chinese Mongolian stallion. Photo: Weigi Feng and Youchun Chen

A young child on his truthworthy Mongolian horse. Photo: Youchun Chen

Mongolian horse with his "pastor," holding a "horse-catching stick." Photo: Tiequan Wang and Youchun Chen

edged. The Mongols used this breed on raids into neighboring territories, spreading the Mongolian horse to many parts of the known world. The horses of Indonesia are descended primarily from the Mongolian with the addition of Oriental blood, as horses of this breed were presented as gifts to Indonesia. Also, Chinese cavalry landed on the Indonesian islands in early times (see Java).

Mongolian horses are widely distributed in China and number about 1.7 million head—nearly one third of the Chinese horse population. Mainly, however, they are found in the Inner Mongolian Autonomous Region, western parts of the province of Helongjiang, Jilin, and Liaoning, and partly in Xinjiang Uygur Autonomous Region.

In an ancient Han book a chapter regarding the Xiongnu nation or tribe says that "as early as in King Yao's time [4,000 to 5,000 years ago] this horse lived in the river basin depending on the natural pastures along with cattle and sheep." This is the earliest written record of the Mongolian horse and was probably recorded before people learned to ride. The breed was domesticated about 2,000 B.C. in China in the northern region near the Mongolian border. It is believed to be related from ancient times to Przewalski's Horse of Mongolia. (However, see Przewalski's Horse.)

The area of domestication of this horse in China was the Mongolian Plateau. Pure breeding has been carried out for approximately 4,000 years, with polygamous stocking on pasture.

While the Mongolian has been described as treacherous by some writers, Professor Youchun Chen in Beijing, the source for information on Chinese horses for this volume, completely refutes this claim. The Mongolian is raised in *taboons* and before training is extremely wild. Once captured, this horse becomes gentle and tractable very quickly and responds readily to commands. In the wilderness a child is safe riding bareback on a Mongolian horse, even in the presence of wolves.

The Mongolian is a very adaptable and useful breed with a tolerant and willing disposition. It is active, alert, and possesses a remarkable working ability. The breed is primarily used for riding (including traditional Mongolian racing) and carting. In carting, four horses can easily pull a load of over 4,000 pounds for a distance of thirty to forty miles in a day. Mongolians are excellent mounts for long-distance riding, as they are very tough, willing, and tireless.

The Mongolian is also used for milk and meat, which is necessary for the pastoral life of the people. Mares are milked four to five times a day, producing about two pounds per milking.

The Mongolian horse shares with the Kazakh the valuable characteristic of remarkable adaptation to unfavorable environment, grazing year round without additional or supplemental feed even when the ground is covered with snow and in temperatures below −40°F in severe winters.

Bay, grey, and black are the most common colors. White is extremely rare in this breed.

The head is heavy with a straight profile and usually almond-shaped eyes; the ears are small and active; the neck is short, thick, and muscular; the withers are rather low; the back is very strong, short, and straight; the loins are short; the croup is long and sloping; the chest is deep and usually of good width; the shoulder is fairly well sloped and quite muscular. The legs are sturdy with solid bone and good muscling. The joints are broad and strong and the hoof is well rounded and extremely hard. The Mongolian horse rarely needs to be shod.

The Mongolian is a small horse and not a pony; in those raised on better pastures and in a less severe environment, the size of offspring is slightly increased. Various types have developed within the Mongolian breed in both China and Mongolia. In China there are three main types—Wuzhumuqin (also spelled Wuchumutsin), Baicha, and Wushen—as well as the Ujumqin and others.

The Wuzhumuqin, averaging about 12.3 hands, numbers approximately 100,000 head. This is a horse of the steppe type and is mainly located on the Wuzhumuqin steppe in the east-central part of Inner Mongolia.

The Baicha horse numbers about 4,000 head and is referred to as the mountain type, also called the "iron hoof horse" for obvious reasons. Baicha horses are found in the southeastern Mongolian area.

The Wushen type, averaging about 12 hands and numbering about 18,000, is found on the west-central dry steppe area. This horse is famous for "walk racing" and is born with an amazing natural gait: a lateral four-beat gait that is easily done extremely fast. The Wushen, racing at its "placental walk," bears an astonishing likeness to the Icelandic Horse at the tølt and the Rahvan of Turkey, which is also raced at the running-walk. All of these horses travel with their heads held very high. Other types of horses would be forced to travel at an extended trot or even a lope to keep up with these exceptionally gaited horses. (See Icelandic Horse and Rahvan.)

This is the gait at which the Mongolian horses carried the great hordes of warriors during the Mongol invasions. The gait is very smooth and comfortable for the rider, and an interesting contrast is created with the horse moving over the ground at great speed and animation while the rider seems to sit motionless.

Chumbivilcas

ORIGIN: *Peru (Cuzco Province)*
APTITUDES: *Riding horse*
AVERAGE HEIGHT: *14.1 h.h.*
POPULATION STATUS: *Common*

The Chumbivilcas is a type of the Peruvian Andean, identified with the area of Cuzco and Apurimac. This horse is more highly developed than the Morocucho.

Like all high altitude horses, the Chumbivilcas has short, straight pasterns and is considered a magnificent horse with great endurance. The predominant colors are bay and grey in differing shades.

This animal has no equal in the Puno region, as it thrives on the sparse natural grasses. The Chumbivilcas is used by the army because of its strength, agility, and ability in the high altitude regions.

It would be useful to improve the living conditions of this hardy horse, as that would produce great benefits to the people in the region. It is their only means of transportation.

For additional information on the horses of Peru, see Andean, Peruvian Paso.

Chumysh

(Chumyshkaya porodnaya gruppa)

ORIGIN: *Altai region (former Soviet Union)*
APTITUDES: *Riding, light draft*
AVERAGE HEIGHT: *15 h.h.*
POPULATION STATUS: *Rare*

Chumysh horses are bred in the Altai region and were developed by peasants near the Chumysh River. Attempts have been made since the nineteenth century to improve the small-sized local horse by using various breeds of draft and saddle animals. As a result of crossbreeding and then interbreeding of the best offspring, a valuable type of agricultural horse has been produced. Of great importance in developing the Chumysh breed were the Kuznet and Orlov Trotter breeds.

Chumysh horses have outstanding working ability in harness pulling freight and as cow horses on farms. They have been highly prized for pulling artillery in wartime. Well adapted to the harsh Siberian conditions, this breed is long-lived and healthy and the mares are "milky," producing strong foals.

The typical Chumysh horse has a large head, usually with a Roman nose; the neck is straight and medium in length; the withers are short; the back straight and wide; the croup wide and sloped; the ribs rounded; the legs long, bony, and strong. Occasionally the hind legs are cow-hocked. The pasterns are short; the hooves of medium size and strong. The skin is thick with short hair, and the legs have well-developed feather. Colors vary, with bay, black, and grey being predominant.

Breeding of the Chumysh horse was done at the Sorokin stud farm but stopped when this farm ceased operations. Various breeds were imported into the area without any system, and this had a negative effect on both the number and quality of Chumysh horses. Recently, in the former Soviet Union as in other countries, a great deal of work is under way to save the gene pools of farm animals, and the Chumysh should be included in the breeds to be preserved, possessing great value as both a working and a meat horse.

The purebred Chumysh survives today in small numbers—some 450 head—and there are many crossbreds with trotter blood. Two stud farms have been established in the Zarin and Zalesovsk Districts, and plans have been made to breed and improve the horses. Horses for the stud farms were chosen with the help of archival material to assure having animals of the proper type.

Cirit

ORIGIN: *Turkey*
APTITUDES: *Riding, sport horse*
AVERAGE HEIGHT: *13.1 to 14.1 h.h.*
POPULATION STATUS: *Common*

A group of Cirit players and horses, ready for a game. Photo: Ertuğrul Güleç

A Cirit player throwing the cirit. Photo: Ertuğrul Güleç

A typical Cirit horse. Photo: Ertuğrul Güleç

Cirit horse of Turkey. Photo: Ertuğrul Güleç

Cirit, also called *jereed* (pronounced jereet), is an ancient game played on horseback. While cirit may be played on any horse, the Cirit horse has been carefully bred for this sport for hundreds of years.

Cirit horses have been produced from crosses of Anadolu and Turkish Arabian race horses, then crossed to the East Anadolu breed. More recently, Karabair horses have been crossed with Arabs to produce some Cirit horses.

The game of cirit originated in Central Asia, and twice this dangerous sport was banned by the Ottoman sultans. Jereed was to the Ottomans what jousting was to the Christian nobility of Europe. On Friday afternoons after the noon prayers, courtiers, pashas, and local dignitaries gathered to watch, or, if they were young and energetic enough, to play the game themselves. The sultan watched from the imperial box. Serious injuries and deaths were not infrequent, and the sport was prohibited by Sultan Mahmud II in 1826 and by the government in the 1890s. The sport is currently allowed. Riding native Turkomans descended from the steeds of Genghis Khan and Attila the Hun, the Turks enthusiastically play the game to the delight of onlookers.

Cirit horses are beautiful animals with great agility, endurance, and intelligence. They are rather short with strong shoulders and legs, short backs, and strong, thick necks. Those with Arab-type heads and bay or grey color are most favored.

The Turks are credited with first domestication of the horse, and there is still a special closeness with horses among tribal people in Anatolia. The horse is so important to some nomadic tribesmen that it is said they and their children will go hungry, they will steal and even kill, but they will feed the horse, the mare, and the foal.

Rhodes Murphy, a Columbia University expert on the Ottoman Empire, stated that the Turkoman breed, ideally suited for jereed because of its wildness and agility, was most likely nonexistent as we now know it when the sport was first introduced to the Ottoman court. The Huns arrived on horses in the fifth century and their herds interbred and intermixed indiscriminately with local horses.

By the time of the Ottoman Empire, the Turkoman breed was well established and the horse was a muscular animal with remarkable endurance. The Ottomans got their horses in large numbers from the Tartars in Crimea, the Voynuks in present-day Bulgaria, and from the grasslands of the Danube. The old Turkoman breed likely was the result of Mongolian horses crossed to the ancient Turkmenian horse.

Cirit, or jereed, is played on horseback by two teams; each rider is armed with a meter-long heavy stick of oak or persimmon. The teams line up on opposite sides of the field. The game starts by one player challenging another in the rival team by calling out his name or number. The challenger then sets off in pursuit of the challenged, attempting to strike him with the stick, which must be thrown. If the stick is caught in the air by the defender, the latter wins a point; if the player throwing the stick strikes his quarry, the attacker gets a point; but if the stick strikes the horse ridden by the player on defense, the attacker is disqualified from the match. Players take the field as team strategy dictates, depending on their distance from the action and on the speed, agility, and other characteristics of each horse and rider. The sport of cirit is widespread throughout Turkey, and other Middle Eastern countries have their own version of the game.

Currently there is no registry association for the Cirit horse, but breeders are hopeful the Turkish government will see the need and start a stud book for this ancient strain.

Cleveland Bay

ORIGIN: *England (Cleveland)*
APTITUDES: *Riding, light draft*
AVERAGE HEIGHT: *16 to 16.2 h.h.*
POPULATION STATUS: *Common*

Oldest of the established British breeds, the Cleveland Bay is a wonderful utility horse, meeting a variety of needs. When the Cleveland Bay Society was formed in 1884, the

Bantry Bere, a Cleveland Bay premium stallion with an impressive show record. Photo: Trevor Meeks

Forest Saga, Supreme Champion Cleveland Bay of G.B., 1984 and 1985. Photo: Charles Medforth

Clevelands, known at the time in their native Yorkshire as Chapmans, already had a history dating back 150 years. Even then they were very fixed in type. They were strong and short-legged and capable of carrying great weights. At that time the Clevelands were used extensively in hilly areas on farms and as pack animals. They were reputed to have been capable of carrying 600 pounds of iron ore out of the mines in the North York Moors. It is thought the early Cleveland Bay horses carried some blood of the racing Galloway, now extinct. Also, Andalusian and Barb infusions occurred in the seventeenth century.

The old name of Chapman was given to the early Cleveland Bays, as they were used by traveling salesmen known as chapmen prior to the advent of wheeled carriages. By the mid-nineteenth century the Cleveland's ability to produce an unsurpassed carriage horse when crossed with the English Thoroughbred nearly led to the breed's extinction. The Cleveland Bay Society was formed to protect the pure breed.

The Yorkshire Coach horse of the late nineteenth century resulted from the cross of Cleveland Bays to the English Thoroughbred. The Yorkshire was an extremely classy carriage horse, and this excellent cross represented the main threat to the Cleveland Bay as a pure breed. With the advent of motorization and the end of the carriage era, the Yorkshire became extinct in the 1930s.

The Cleveland Bay horse is always bay in color. A small star and a bit of grey are allowed in the mane and tail. It is as a carriage horse that the Cleveland Bay excels, and the consistent color makes matching pairs and teams a simple matter. Cleveland Bay horses are also excellent hunters, powerful and eager but with a calm, steady nature.

This breed has been bred pure since the middle of the eighteenth century, when all infusion of Thoroughbred blood ceased. Cleveland Bay horses have been used throughout Europe to improve other breeds, including the Holstein, Hanoverian, and Oldenburg during their formation. Today the breed is perhaps most valuable as a cross to the Thoroughbred, producing excellent hunters and competition horses for all of the disciplines. The society maintains a part-bred register so that horses can be traced and followed during their careers. Lieutenant Colonel Sir John Miller, when Crown Equerry to the Queen, was very successful in crossing Oldenburg mares with Cleveland stallions to produce Prince Philip's well-remembered team of international carriage driving horses. The Queen has been patron to the society since 1977 and was its president during the centenary year in 1984.

The strength of the Cleveland Bay lies in its nucleus of knowledgeable pure breeders who continue to nurture the old bloodlines and ensure that there are enough good mares to allow for some to be used for crossbreeding. It is interesting to note that several of the continental breeds, having lost their own foundation stock by too much crossbreeding, are now using the Cleveland Bay to reintroduce bone and substance. This horse is widely known as an improver of stock, and there are thriving studs in Japan, Pakistan, the United States, Canada, Australia, and New Zealand. All have reported excellent results when crossing Cleveland horses to their native stock.

Cleveland Bays must be bay with black points—black legs, mane, and tail. Grey hair in the mane and tail have long been recognized as a characteristic in certain strains of pure blood.

The body is deep and wide. The back should not be too long and should be strong with muscular loins. The shoulders are sloping, deep, and muscular. The quarters are usually quite level, powerful, long and oval, the tail springing well from the quarters. The head is bold and not too small, well carried on a long, lean neck. Often the profile has a pronounced convex shape. The eyes are large, well set, and kindly in expression. The legs and thighs are muscular with large, strong joints, sloping pasterns, and very good feet.

Clydesdale

ORIGIN: *Scotland*
APTITUDES: *Heavy draft*
AVERAGE HEIGHT: *16.1 to 18 h.h.*
POPULATION STATUS: *Common*

Indigenous to Scotland and deriving its name from the district in which it was developed, the Clydesdale is a heavy draft horse. Modern history of the breed began about the middle of the eighteenth century. The native, hardy breed founded in Lanarkshire, through which the River Clyde flows, was being upgraded to greater weight and substance by the use of

Bouncing baby Clydesdales at the Clydesdale breeding farm at Grant's Farm (former home of Ulysses S. Grant) in South St. Louis. Photo: Fleishman-Hillard

The famous Budweiser Clydesdales of Anheuser-Busch. Photo: Fleishman-Hillard, Inc.

The Clydesdale mare, Blueton Blossom, and her foal. Photo courtesy Clydesdale Horse Society of Great Britain and Ireland

The Clydesdale stallion, Muirton Superior. Photo courtesy Clydesdale Horse Society of Great Britain and Ireland

Flemish (Belgian Draft) and Friesian stallions and at least one outstanding native stallion named Blaze. Blaze was brought from Ayrshire into the Upper Ward of Lanarkshire. (The old name for Lanarkshire was Clydesdale, just as the Mearns is the old name for Kincardineshire, Angus for Forfarshire, and the Merse for Berwickshire.)

Sometime in the eighteenth century one of the dukes of Hamilton, a wealthy local landowner, imported six Flemish Great Horses, a breed already regularly shipped to Scotland for use as war horses and in farm work. The duke made his six prize horses available for breeding to local mares. People from outside Lanarkshire began to refer to the big, powerful horses as "the Clydesman's horses." The early Clydesdales quickly garnered attention as a breed more powerful than any available before, said to be capable of pulling loads of more than a ton at a walking speed of five miles an hour.

The characteristics of the Clydesdale breed were molded by farmers of the Upper Ward of Lanarkshire to meet the demands of commerce when the coalfields of Lanarkshire began to be developed, roads to be improved, and haulage by the shoulder to be substituted for carriage on the back of the native breed of horses. The official debut under the present name of Clydesdale was at the Glasgow Exhibition of 1826.

Soon, the breed developed more than a local reputation and during the last quarter of the nineteenth century many young colts and fillies were annually drafted from fairs held at Lanark, Biggar, Rutherglen, and elsewhere in the country into the Midland counties of England.

The system of district hiring of Clydesdale stallions was an early part of Scottish agriculture. There is evidence of its operation as early as 1830, and through the medium of hiring as well as by purchase, Clydesdale stallions were drafted for service into Galloway, Kintyre, and Aberdeenshire in Scotland and into Cumberland and the Glendale district of Northumberland in the north of England. Mated in these areas with the native breed, the Clydesdale stallions stamped a similarity of type upon the draft horses of Scotland. By the year 1840 the name Clydesdale had become a synonym for the Scottish draft horse breed.

A stallion show was in existence in the Glasgow Cattle Market for years prior to 1844. In that year the stallion Clyde (155) won the first prize. This horse did much to mold and develop the modern breed. Classes for agricultural horses and mares exhibited were of Lanarkshire breeding or descent. In the west of Scotland, the modern type owed much to the horse Broomfield Champion (95); he also traveled in Aberdeenshire, where he was known as Aberdeen Champion. He had a son called Clyde (alias Glancer 153), a virile and most impressive sire. Clyde had seven sons that traveled in districts as far apart as Wigtownshire, Midlothian, Bute, Kintyre, Renfrewshire, and Ayrshire.

When the Muir family moved in 1840 from Sornfallo on the slopes of Tinto in Lanarkshire to farms in Galloway, a notable transfer of Clydesdales of the best type took place. The results can be traced to this day through the influence of sires such as Lochfergus Champion (449). Agnew's Farmer (292) won first at a major horse show at Dumfries in 1830. He was sired by a horse called Clydesdale, his name bespeaking his origin.

Clydesdale horses were used on the farms and for coal hauling and, in Scotland, eventually completely replaced the Shire as a heavy carriage and draft horse.

The Clydesdale Horse Society was founded in June 1877 and the first volume of the stud book was published in December 1878. The Clydesdale was the first of the British draft horses to have its own society.

The outstanding characteristics of the Clydesdale horse are the combination of weight, size, and activity with the exceptional wearing qualities of the feet and legs. These qualities have been consistently aimed at by breeders throughout the history of the breed. Clydesdales are long-lived and hardy. This breed has been exported to many other countries since about 1850.

The Clydesdale suffered a decline in numbers along with other draft breeds when mechanization replaced the horse, and by 1980 this breed could have been listed among the rare breeds. Today, however, interest and consequently population have once again increased, and Clydesdales are often seen in driving and pulling competitions.

The Clydesdale Breeders of the United States was incorporated in December of 1879 with members from both the United States and Canada. This breed has enjoyed popularity in the United States, and possibly the best promotion for the breed has been the Budweiser Clydesdales.

The Budweiser Clydesdales were formally introduced to August A. Busch, Sr., and the Anheuser-Busch brewing company on April 7, 1933. Prohibition (of sale of alcohol) had just been repealed and August A. Busch, Jr., wanted to commemorate the special day. To his father's delight, the hitch thundered down Pestalozzi Street carrying with it the first case of post-Prohibition beer from the St. Louis brewery. Today, the tradition and quality the Budweiser Clydesdales have come to represent for more than fifty-five years continues. Today's Budweiser Clydesdales are larger than their Scottish ancestors. To qualify for the world-famous eight-horse hitch, a Clydesdale must meet certain requirements, size and quality being the most important considerations.

Standing at 18 hands when fully mature, Budweiser Clydesdales weigh about 2,000 pounds. They must be geldings, bay in color, and have four white stockings and a blaze of white on the face as well as a black mane and tail. A gentle temperament is also important, as hitch horses often meet millions of people each year.

The Budweiser Clydesdales make more than three hundred appearances annually, spending most of the year on the road. Pulling the red Budweiser beer wagon with their beautiful harness, perfectly groomed and exhibiting their stylish gait, these horses have made friends all over the world.

Clydesdales are active and lively but also have a gentle and willing nature. The head must be strong, intelligent, and carried high. The forehead is open and broad between the eyes; the horse has a wide muzzle, large nostrils, bright, clear eyes, big ears, and a well-arched, long neck. Usually the profile is strongly convex. The shoulder is sloping and the withers are high. The body is deep, the back short, and the ribs well sprung. The loins and croup are wide and the croup is slightly sloping. The legs are robust and feathered, with strong joints; the pasterns are long, and the foot rounded and tough.

Colombian Criollo

(Colombian Walking Horse, Colombian Paso Fino)

ORIGIN: *Colombia*
APTITUDES: *Riding horse*
AVERAGE HEIGHT: *13.1 to 14.2 h.h.*
POPULATION STATUS: *Common*

It has been firmly established that the Paso Fino horse was derived from a similarly gaited horse with a Spanish and Barb origin (obviously the Spanish Jennet), which disappeared in Spain as other crosses were made to horses having a trotting gait.

As a result, the modern Colombian horse bears little resemblance to the modern Andalusian horse in phenotype, stature, or movements. Unfortunately, there are no books dealing with any similarity in movements between the progenitors of the *paso*-gaited horses and the present-day Paso Finos. The term *paso fino* means "fine step," and the gait is praised for its comfort to the rider.

It is known that Canadian pacing and trotting horses were shipped to the West Indies in early times for work on the sugar plantations, especially to Cuba. Possibly, some of this stock was introduced to Colombia.

In 1881 Santiago de la Villa y Martin published a book in Madrid in which he analyzed and studied the horse, including the imperfect ambling gait and the rack gait. In an 1886 book published in Barcelona, Dr. Carlos Fernando de Controverde discussed the natural gaits of the horse, defining the paso gait as unusual.

Some writers have claimed that the horse taken to the Americas by the Spaniards was native to Andalusia, but with Arab connections—that is, Hispanic-Arab. However, Colombian authorities have searched diligently and have found no documents that indicate the arrival of authenticated Arab horses to the Iberian Peninsula during that period.

The American races, largely derived from the horses brought by the Spaniards, developed into different varieties in each country and region depending on selection, feeding habits, infusion of other blood, and environment. The horses of Colombia became superior in appearance, elegance, nobility, and especially in their peculiar way of moving. It cannot be determined when this gait occurred in the Colombian horse, since no historian has recorded it.

Today the Colombian Paso Fino horse is a focus of national pride, the Colombians considering this horse the best and purest of the "Paso" breeds existing today. All Colombian Paso Fino horses move in the same way, indicating that they descended from a single breed. Undoubtedly, other types of horses with different gaits were also brought to the Americas, and they gave origin to the trotting and galloping Colombian native horses.

In height, Colombian Criollo mares average 13.39 to 14.37 hands and males stand on an average 13.58 to 14.57 hands. Horses are measured when over five years old. Individuals failing to reach the minimum required height are disqualified from competition. Those surpassing the maximum are not disqualified but lose points when judged.

In temperament, the Colombian Criollo or Paso Fino is sprightly, gay, regal, and gentle. This breed is distinguished by its extraordinary *brio*, a word referring to vigor, spontaneous spirit, and fire. There are various manifestations of brio in the purebred horse: to the rein, to the ear, and to the leg. Brio to the rein is the willingness or nimbleness a horse shows at any movement of the rein, without needing demands from the heel. Brio to the ear denotes a horse that responds to any sound produced by its rider. Brio to the heel or leg is response to the slightest pressure. It is not uncommon for a Colombian horse to exhibit all three types of brio.

There are several colors in the Colombian Paso Fino horse. The rules and regulations accept any defined color (but only with pigmented skin). The most common colors are *bayo* and *zaino* (see below) or grey. Less common are chestnut, dun, "deer-colored," strawberry roan, and swan. White markings are allowed but must not touch the knee or hock joints, and markings must be continuous. White markings that do not detract from the beauty of the face are preferred.

A description of its coat colors is as follows. Chestnut is described as yellowish red, very similar to cinnamon, with varieties called roast chestnut (similar to roasted coffee); light chestnut (the tone is light and resembles *café au lait* in color); and gold chestnut (the tone is reddish and yellowish with highlights, and generally the legs, nose, tail, and mane are lighter).

Colombian Criollos may be dapple grey or white, but "albino" (white hair but pink skin) is usually a cause for disqualification. Blue-grey is the term for a mixture of white and black hair; "porcelain" is white with black skin visible through the hair; greyish means white and black hair mixed together; and slate grey is used when the mixture of white and black hair gives a blue sheen. "Straw-colored" is a mixture of the hair resulting in a light tone resembling dry straw, and *"verbeno grey"* is a mixture of white and reddish hair that yields a dark tone with the head very dark.

Black means the majority of hair is black, but various tones are recognized: jet black (bright and shiny); reddish black (when the color resembles that of a ripe blackberry); and pitchy black (when the color resembles a fish color—usually, this black coat fades when not well kept). When a black horse does not show a single white hair in its coat, it is called *hito*, and these animals are very difficult to find.

Hazel is the term used for bay, where the body is reddish but tail, mane, and legs are black or very dark. *Zaino* is a dark bay with a mixture of black and reddish hair but having a great predominance of the black coat over the reddish one. *Bayo* is a buckskin color that shows a yellowish red, with black mane, tail, and legs; usually there are stripes on the legs and a dorsal stripe. The bayo color may be found in many tones. Roan is similar to bayo but with white, red, and black hair mixed together and with darker extremities. "Deer-like" is similar to the color of deer; the base of the hair is yellowish, and the forehead is black.

The gait of the Colombian Paso Fino is natural, exhibited by foals the day they are born. Some horses are more *fino* than others. A trotting horse cannot carry out the *paso* (lateral, four-beat) gait properly. Projection of the gait is carried out by the horse's changing its center of gravity. It first rests the rear legs and pushes the spine forward, placing its center of gravity to the fore part of the body, then raises its forelegs, with which it carries out four beats. First, it raises the leg from the ground; second, it maintains the leg in the air and advances; third, it brings the leg down to the ground; and fourth, it supports and pushes the weight forward. The beat of the leg com-mences when, first, it supports and gives the push; second, it raises the leg; third, it maintains the leg in the air and advances; and fourth, it places the leg on the ground. Each fore or rear leg moves the center of gravity; afterwards, the animal receives it again. It completes the action taking the weight in order to later abandon it. The gait is lateral and diagonal biped in four strikes, carrying out four diagonal movements and four lateral moves in a cycle of sixteen times. When making these changes, the animal rests three legs on the ground, four times, with alternate rear legs. These movements are carried out rhythmically and harmoniously between rear and fore legs, allowing four strikes carried out with marvelous agility, spring, and flexion of the horse's joints. The rider experiences a sensation of smoothness that is characteristic of this breed.

There are several distinct varieties of the paso fino gait found among Paso Fino horses, but the Colombian Walking Horse is considered unique in the world.

Colorado Ranger
(Rangerbred)

ORIGIN: *United States*
APTITUDES: *Riding horse*
AVERAGE HEIGHT: *14.2 to 16 h.h.*
POPULATION STATUS: *Common*

The cornerstone for breeding of the Ranger-bred horse was laid in 1878 when General Ulysses S. Grant, during a world tour, visited Sultan Abdul Hamid of Turkey. As a token of friendship the old Sultan presented Grant with two desert stallions, an Arab named Leopard and a Barb named Linden Tree.

These stallions reached Virginia early in 1879, where they attracted the attention of Randolph Huntington of New York and Virginia. One of America's great horsemen, Huntington had spent nearly fifty years breeding trotters or roadsters. He saw the potential in the two stallions to improve his new breed of light harness trotters, which he proposed to name the Americo-Arab. General Grant gave his friend permission to use the stallions as he

Colorado Ranger. Photo courtesy Laurel Kosior

thought best, and the next fourteen years saw the Americo-Arab brought to a high degree of perfection. Financial problems in 1906 caused the complete dispersal of the fine Huntington horses, numbering about one hundred head.

This great horseman was often asked why he was so enthusiastic about using General Grant's stallions in his breeding program. In response, he said his objective was to establish a truly American breed of national value in blood qualities for use not only in the United States but for export. He felt the best results obtained by crosses were not through abrupt but by affinity crosses. Like produced like only when the blood was "like." Crosses must be sustained by crosses. They could not be successfully interbred for any length of time without degeneration. Purebreds on the other hand, Huntington held, could be and had been interbred from the beginning.

While this may seem a strange idea to many, it is noteworthy that in many instances around the world where ancient, primitive, or very pure breeds are inbred due to low numbers, inbreeding depression does not seem to occur. A most interesting case is the feral

horse of the Namib desert in the south of Africa, naturally inbred for many years without care or any infusion of new blood (see Namib).

Often, circumstances which at the time seem of little matter later prove to be events of lasting historical importance. In 1894 General Colby, who had established large ranch holdings at Beatrice, Nebraska, received permission to bring Leopard and Linden Tree (now aged stallions) west for a single breeding season. Here, in one short summer, the two desert stallions left an indelible impression upon the offspring of native mares on the Colby holdings. A new breed of cow horses emerged, later to be named the Colorado Rangers.

Colorado horsemen were destined to write the next chapter of the Colorado Ranger story. By the late 1890s many good reports had spread across the cow country concerning the excellent horses being bred on the Colby holdings in Nebraska. Several of the large ranchers on the eastern Colorado plains grouped together and purchased a herd of mares and a stallion named Tony, a white horse with black ears. Tony was a double-bred grandson of the stallion Leopard.

Early breeders of the Rangerbred were not interested in creating special color in their horses. Their objective was to produce horses to work the cattle effectively on their ranches. A wealth of odd, barbaric color patterns did emerge from the intensive line breeding program. Horsemen on the high plains had seldom seen such leopard-spotted, rain-dropped horses. Many were blanket-hipped or with a snowflake pattern. The W. R. Thompson Cattle Company of Yuma County, Colorado, added a new infusion of Barb blood to this plains breed in 1918 by purchasing a pure Barb stallion, Spotte, for their daughter as a wedding gift. This stallion is found in the pedigree of many Rangerbreds.

In the same year, a stud colt named Max was born at the headquarters of Governor Oliver Shoupe at Colorado Springs. He, too, was to leave a lasting impression on the Rangerbred horses of the high plains. This stallion was white and covered with black leopard spots.

Colorado Rangers are registered according to strict requirements and must meet standards of conformation and pedigree. Several outcrosses are allowed, including Thoroughbred, Quarter Horse, Appaloosa, Arab, or AraAppaloosa. From 1980 through 1987 some Lusitano crosses were allowed. Pony and draft blood are excluded from registration.

Comtois

ORIGIN: *France (Franche-Comté)*
APTITUDES: *Heavy draft and farm work*
AVERAGE HEIGHT: *14.1 to 15.1 h.h.*
POPULATION STATUS: *Common*

A breed of very early origins, the Comtois is thought to be descended from horses brought into France by the Burgundians, a people coming from what is now northern Germany around the fourth century.

The Comtois was used in 1544 to improve the horses of Burgundy but became most famous as an army horse. Louis XIV used this breed for his cavalry and artillery, and it was taken to Russia during Napoleon's campaign.

During the nineteenth century the Comtois was bred to Norman, Boulonnais, and Percheron horses. Since 1905 the use of small Ardennais sires has developed a stronger horse with improved legs.

The Comtois is a stocky horse with a square head, lively eye, and well-set, mobile ears. The neck is straight and muscular; the withers

Comtois. Photo courtesy of UNIC, Paris, France

French Comtois. Photo: Y. de la Fosse-David, courtesy UNIC, Paris, France

are well set; the chest is wide and deep; the rib cage rounded; the loins short and strong; the croup wide; the thigh well developed. The legs of this horse are strong with good joints, clean tendons, and nicely shaped feet. Coat colors of the Comtois are generally chestnut or bay.

The Franche-Comté and, more particularly, the High Jura around the Besançon stud are the original breeding ground for the Comtois breed. However, breeding of this horse has spread to the Massif Central, the Pyrenees, and the Alps—in all the mountain areas for which they are perfectly suited. The Comtois has good qualities of endurance, hardiness, and balance in rugged landscapes.

A spectacular show is held each fall at Maiche in the Doubs. Breeding classes as well as riding and harness classes enable one to judge the characteristics of liveliness, energy, and endurance of the breed. Only the Breton outnumbers the Comtois in France as a draft horse.

Well-known as a surefooted breed, the Comtois is still widely used for hauling wood in the high pine forests of the Jura and for work on the hilly vineyards of the Arbois area. The breed is also used extensively in France for meat production.

Well adapted to different living conditions, the Comtois is currently used for free breeding in the semiwild breeding farms with excellent reproduction results.

Connemara

ORIGIN: *Ireland (Connacht)*
APTITUDES: *Riding pony, light draft*
AVERAGE HEIGHT: *13 to 14.2 h.h.*
POPULATION STATUS: *Common*

Supreme Champion Connemara, Village Laura. Photo: Michael Connaughton

Connemara, on the western seaboard of Ireland, is the area of origin for the Connemara pony and the source of its name. It is an exposed, mountainous region, a land of stone, bog, and wave-lashed coastline. This wild beauty is deceptive, however, for the Gulf Stream provides the area with an average temperature of 42°F. Nevertheless, it is the ruggedness of this environment that brought about development of many of the Connemara's prized qualities; hardiness, intelligence, agility, and extraordinary jumping ability.

Origin of the Connemara breed is somewhat obscure. There is evidence of the existence of horses in this area of Ireland from very early times. The Celts, who came to the region from the Alps and the valley of the Danube through Spain and Gaul in the fourth

century B.C., are reputed to have been skilled in horsemanship. Most likely, they brought horses with them. Interestingly, the Gaelic tongue is still the living language of much of Connemara.

Many centuries later there is evidence that Spanish horses were imported by the rich merchants of Galway City at the height of its commercial prosperity. It is known that they carried on extensive trade with Spain. The Spanish Barb and Andalusian horses imported then were, it is said, the best that money could buy. Some of them are thought to have found their way to the adjacent district of Connemara and interbred with the native ponies. In more recent times, as late as the middle of the nineteenth century, there are accounts of Arab horses being imported by certain estate owners in Connemara. All of the stories of early importations of horses for mating with the native ponies state common sources— Spain, Morocco, or Arabia—and the type of horse was said to be Andalusian, Spanish Barb, or Arab. More recently the Congested Districts Board, which was established under British rule in 1891, introduced a number of Welsh stallions for mating with the native Connemara mares, and certain strains gave satisfactory results, notably that of Cannon Ball.

In 1900 the old dun type of Connemara pony was described as being capable of living where all but wild ponies would starve; the ponies were strong and healthy as mules. In 1923 the Connemara Pony Breeder's Society was formed in Ireland and a stud book established for the preservation and improvement of this versatile and valuable breed.

Connemaras are a product of their original environment. Only animals of hardy constitution could survive the storms that blow in from the Atlantic Ocean. Native Connemaras live outdoors from the day they are foaled and are extremely tough. Anthony A. Dent, well-known author and historian, stated in a personal communication to the writer, "It is of course the waterproof quality, the ability to survive weeks and weeks of rain by day and by night that singles them out as environment proof. The winter coat is not at all long, but it is intensely dense and plushy in texture; rain doesn't penetrate any more than it does on a duck's feathers."

Connemaras are one of the most popular breeds for use as mounts for children. They are capable of all forms of equitation from dressage to hunting. Also excellent in harness, they are becoming increasingly popular for use in competition driving. They are agile, fast, and athletic and have great stamina. This pony is also noted for being sensible and cool-headed, a quality quite common in primitive type breeds that have survived by taking care of themselves.

The typical Connemara has a well-shaped head with a straight profile, small ears, large eyes, and flared nostrils. The neck is long and well formed with a full mane; the withers are pronounced; the back is long and straight; the croup is muscular and slightly sloped; the chest is wide and deep; the shoulder is long and sloping. The legs of this pony are sturdy and well muscled, with clean joints, long cannons, clearly defined tendons, and a well-formed hoof. The action of the Connemara is free and true. Coat colors are grey, black, brown, bay, and dun, with an occasional chestnut or roan. Dun shades were once predominant but are becoming increasingly rare. Many are dun when foaled and turn grey by seven years.

The Connemara is popular in France, where the breed is recognized by the State Stud Service, and is also highly regarded in Germany.

Corsican

ORIGIN: *France (Island of Corsica)*
APTITUDES: *Riding, packing*
AVERAGE HEIGHT: *13 to 13.3 h.h.*
POPULATION STATUS: *Rare*

No equine fossils have been found on the island of Corsica, indicating that the island was not connected to the mainland at any known time. Horses and asses found on the island were taken there by people, as were the dog, pig, and black rat.

The aboriginal Phocaeans, harassed by the Etruscans and Carthaginians, finally submit-

Corsican horse. Photo courtesy François Pietri

A group of Corsican horses. Photo courtesy François Pietri

ted to the latter before the Romans took over the whole island in 259 B.C. Much later, horses must have furnished the means of transportation for the Vandals, Lombards, Saracens, Genoese, Pisans, Spaniards, English, and finally the French. Known as the Isle of Beauty, Corsica must have become fairly well battered as these different peoples, either together or separately, disputed possession of all or part of it.

The inhabitants of Corsica made their living through the use of horses. Although the centuries witnessed the arrival of many different types of Equidae, the climate and rugged terrain gave rise to a breed of horses and one of asses particularly adapted to survive there. Natural selection had a great deal to do with the emergence of a hardy, strong breed.

The Haras National (Remounts Department) began operations on Corsica in 1861. With the goal of producing light cavalry horses, the state studs used stallions of Anglo-Arab type. Mule breeding was also pursued.

When the army remount production ceased, control of horse breeding on Corsica once again passed to the inhabitants. Natural genetic selection and the conditions of life in the open have ensured a certain homogeneity in the Corsican horse.

The horses of Corsica are small and termed ponies by most horse enthusiasts, although they bear no pony characteristics. They are predominantly bay with few white markings. The head is somewhat heavy with a straight profile. They have a compact, square frame; pronounced withers; the croup is sloped; the hooves are narrow and very hard. There are approximately 1,000 Corsican horses, and an increasing number of feral animals in the scrub-covered hills of the island.

Traditionally Corsicans are not horse eaters, so there are no horse butchers on Corsica, and those who keep horses for pleasure would not deal with them anyway. Of the horses that work, most are used for trail riding, either privately or as part of the tourist industry.

Recently a society has been established for the Corsican horse, with organized shows. Plans are to establish competitive endurance rides to demonstrate the endurance of this small, tough horse.

The Costa Rican Saddle Horse, Legendario, National Champion 1987 and 1990. Photo: Manuel Carazo

Costa Rican Saddle Horse
(Caballo Costarricense de Paso)

ORIGIN: *Costa Rica*
APTITUDES: *Riding, stock horse*
AVERAGE HEIGHT: *14.1 to 15.1 h.h.*
POPULATION STATUS: *Common*

The Costa Rican horse originated from Barb and Spanish bloodlines. At the time of the European discovery of the New World, the Arab occupation of Spain was coming to an end. Barb horses were easier to obtain than pure Arab horses. Arab people had much esteem for their horses, and certain religious prejudices toward the Roman Catholics made them reluctant to sell their horses to them. In fact, it has always been considered dishonorable for an Arab to sell his horses, a belief never understood by non-Arabs.

Costa Rican Saddle Horse. Photo: Manuel Carazo

Also important when considering possible Arab influence in the Spanish horse is the fact that Spanish horses do not exhibit Arab influence—the tail is set low, the croup is rounded and sloping, the typical arching neck of the Arab is absent, and the profile is nearly always slightly convex, never concave.

The Costa Rican Saddle Horse, or Caballo Costarricense de Paso (meaning Costa Rican Paso Fino), is a medium-sized horse; a handy size for riding, pleasure, or any type of work. This horse is capable of carrying heavy weight for long distances and is surefooted in mountainous terrain.

Since 1850 breeders of the Costa Rican horse have paid more careful attention to their selection of breeding stock and have begun to use only outstanding stallions. A few stallions were imported from Spain to introduce new blood, as the population in Costa Rica was small and it was feared that too much inbreeding would spoil the breed. In importing stallions from Spain, however, breeders encountered the problem of finding the right type of saddle horse for their needs. The ideal Costa Rican saddle horse possessed a smooth ambling gait, while the Andalusian was a trotter.

A few stallions were imported from Peru and used on native mares of Costa Rica with excellent results. This improved the breed enormously and established the Costa Rican Saddle Horse.

This breed is distinguished by the beauty of its lines and excellent, natural action of the forelegs. The leg action must be high and smooth, with good shoulder action and sharpness. To a lesser degree, action of the hind legs must also be good. One thing that makes this horse great is the graceful natural gait, which is executed with determined "piston action," as it is called by native breeders.

Almost all colors are accepted into the breed association with the exception of pinto. White markings on the legs and a blaze are preferred by most breeders.

Costa Ricans are avid horse lovers and have worked diligently to perfect their own type of Paso Fino horse. Parades are held every Sunday in one town or another around the country. The Christmas Tope (parade), traditionally held on December 26 every year, is the largest parade of horses in the world. Some 800 to 1,000 horses dance through the main avenues of the capital, San Jose, horses and riders outfitted in everything from elegant Spanish costumes to traditional Costa Rican dress and to cowboy garb.

As Costa Rica is essentially an agricultural and cattle-raising country, the horse still plays an important role in everyday life. In the north of the country there are great cattle ranches, where horses are used daily. Many of the ranches are only accessible by horseback.

Costeño

(El caballo costeño de paso aclimatado a la altura)

ORIGIN: *Peru*
APTITUDES: *Riding horse*
AVERAGE HEIGHT: *14.1 to 14.2 h.h.*
POPULATION STATUS: *Common*

Horses in Peru, descendants of the horses of the Spanish Conquest and later crossed with imports from the United States and other countries, developed into three main types; two distinct horses called Costeño, and the Andean. The *Costeño* (coastal) *de Paso*, which translates as "Coastal gaited horse," is commonly known as the Peruvian Paso; the *Costeño de paso aclimatado a la altura* translates as "Coastal gaited horse acclimated to high altitude." The Andean has the subtypes known as Morochuco and Chumbivilcas.

The Costeño is dun, grey, sorrel, black, bay, or other common colors. The head is small with a broad face, large, expressive eyes, and a fine muzzle. The neck is short, arched, and muscular; the shoulder is sloped and long with powerful muscling; the back is short and strong; the croup rounded and sloping; the chest wide and deep; the legs are clean with short cannons.

The Costeño developed at high altitude is a strong, elegant saddle horse. It is widespread in Peru, particularly in the regions of Cajamarca, Callejon de Huaylas, and Huancayo. At more than 9,000 feet above sea level, this small horse works magnificently as a saddle horse. Unfortunately, it shows signs of degeneration due to the low mineral content in its diet because of leaching by the continual rains.

The Costeño represents a race that is highly esteemed. Possessing the smooth, lateral gaits of the old Spanish Jennet, the Costeño horse was noted in early times for its endurance over long distances and the comfort given to its rider. It has been used extensively to work cattle in the mountains and is a surefooted horse with renowned "cow sense."

For additional information see Peruvian Paso and Spanish-American Horse.

Cuban Paso
(Cubano de Paso)

ORIGIN: *Cuba*
APTITUDES: *Riding horse*
AVERAGE HEIGHT: *13.3 to 15 h.h.*
POPULATION STATUS: *Common*

Cuban Paso. Photo courtesy Dr. David William Cantera, director, CIDA, Havana, Cuba

At the time of European discovery, Cuba had no indigenous horses, and all Cuban horse breeds are descended from horses of the second voyage of Christopher Columbus, and later, of the conquistadores.

After the Indians were nearly exterminated, many of the Spanish horses became feral in Cuba, breeding and roaming freely throughout the countryside. Due to natural selection, the climate and terrain developed a new type of horse unlike its famous ancestors. The Cuban Paso, or Cuban gaited horse, emerged with the gait of the renowned Jennet. This is a lateral gait, called *marcha* or *andaduras*.

As in the case of all "Paso" horses found in the West Indies and throughout Central and South America, the Cuban Paso is characterized by its specialized four-beat lateral gait, which is very smooth to ride. The horse is able to maintain the paso gait for long distances with no discomfort to the *jinete* (rider).

Cuban Paso horses have regal bearing and are energetic and animated animals of great *brio* (vigor). The conformation is harmonious, and the breed has maintained many of the characteristics of ancient Iberian ancestors.

While the Paso breeds all resemble one another, due to differing environments and breeding practices each country has developed a special type of horse. Little crossbreeding has been done between countries, the breeders preferring to maintain their national lineages in pure form.

The Cuban Paso has a small, refined head with small ears and large, expressive eyes. The profile is straight with a wide forehead. The neck is of medium length, arched and muscular; the back is of moderate length with a short loin and sloping croup. The croup is wide and muscular; the chest is wide and well muscled; the ribs are arched and long. The legs are well muscled and short with large, strong knees, short cannons, and well-defined tendons.

Cuban Pinto
(Pinto Cubano)

ORIGIN: *Cuba*
APTITUDES: *Riding horse*
AVERAGE HEIGHT: *14 to 14.3 h.h.*
POPULATION STATUS: *Common*

The Cuban Pinto has existed in Cuba within the native (Criollo) variety since horses were first taken to the island in the fifteenth century, but it was not until 1974 that serious work was done on a project of genetic improvement by crossing three breeds; the Pinto Criollo, Quarter Horse Pinto, and pure English Blood Pinto (Pinto-Thoroughbred).

The Cuban Pinto is a medium-sized horse, compact, with well-defined musculature. This is an improved Criollo type, retaining its resistance to disease and fatigue while combining height and beauty from the Thoroughbred and musculature from the Quarter Horse, along with superb stock working ability. The character of this breed is docile and obedient.

Cuban Pinto. Photo courtesy Dr. David William Cantera, director, CIDA, Havana, Cuba

Cuban Trotter. Photo courtesy Dr. David William Cantera, director, CIDA, Havana, Cuba

The Cuban Pinto is an excellent trotter; the gait is straight, suave, and elastic. Both the *tobiano* and *overo* varieties are found in the coat color of the Cuban Pinto. This horse is used primarily as an excellent cow horse.

Cuban Trotter
(Criollo de Trote)

ORIGIN: *Cuba*
APTITUDES: *Riding horse*
AVERAGE HEIGHT: *13.3 to 15 h.h.*
POPULATION STATUS: *Common*

The Cuban Trotter is a descendant of Spanish horses brought to the Americas during the Conquest. The constitution and morphology of this breed is not greatly different from those of horses of Argentina, Uruguay, Chile, Peru, and Brazil.

Prior to the American Revolution great numbers of Canadian horses were shipped to Cuba for work on sugar plantations, and these no doubt had a strong influence in producing this Cuban trotting breed when crossed to the Spanish horses. The horses of Canada were always noted for their excellence at the trot.

The temperament of the Cuban Trotter is tractable and friendly, but this horse is animated, energetic, and has abundant vigor. It is used for agricultural work and has superb endurance. Cuban Trotters are agile and very responsive and sensitive to the rein. They are esteemed for their intelligence and willingness to obey. In addition, they are able to make efficient use of meager pastures.

The Cuban Trotter has well-balanced conformation and musculature. The head is of small to medium size with a straight to subconvex profile, the jaw muscular; the ears are small and mobile, set wide at the base; the forehead is wide with large, expressive eyes. The neck is muscular and of medium length, well inserted into the body; the back is strong and short, the croup rounded and muscular with good width; the tail set rather low and carried close to the body. The chest is wide, deep, and muscular; the ribs are long and rounded; the forearms are short and well muscled; the knees are large, strong, and clean; cannons are short and strong with well-defined tendons; hooves are wide and strong. The trot is the best gait of this breed, traveling straight and smooth. The dominant colors are dark and solid, bay or black of various shades.

Çukurova

ORIGIN: *Turkey*
APTITUDES: *Riding, harness*
AVERAGE HEIGHT: *14.3 to 15.3 h.h.*
POPULATION STATUS: *Rare*

Çukurova of Turkey. Photo: Ertuğrul Güleç

Czechoslovakian Coldblood. Photo courtesy Ing. Otakar Dočkal, CSc.

Çukurova horses were developed through crosses of Uzunyayla and Turkish Arab horses, then crossed to the native Anadolu breed. Larger than the Anatolian horse, the Çukurova is a fast, spirited animal. Two types are found in the Çukurova breed—harness and saddle. The harness type has a greater percentage of Uzunyayla blood, while the riding type has more Arab lineage.

There are approximately 3,000 Çukurova horses, and presently there is no breed association or stud book for them. Breeders hope that the Turkish government will see the need to record and preserve this breed.

The Czech Coldblood is a heavy type of draft horse, robust and with a longer body frame than some other draft breeds. The musculature is good and this horse has a strong bone structure. The hooves are also well formed and hard.

The Czech Coldblood is usually chestnut, less often bay. Other colors are exceptions. Besides forestry work, the breed is also used to a lesser extent as a draft horse in agriculture.

Czech Coldblood
(Český chladnokrevník)

ORIGIN: *Czechoslovakia*
APTITUDES: *Draft work*
AVERAGE HEIGHT: *16 h.h.*
POPULATION STATUS: *Common*

Development of the coldblood horse in Czechoslovakia dates from the beginning of the twentieth century. After 1900 many imports were brought from the Netherlands and Belgium, and later, Noric coldblood, originating from the Alpine lands, was also introduced.

Presently the Czech Coldblood horse is bred primarily for use in forestry. In Czechoslovakia several thousand Coldblood broodmares are bred and about 200 Coldblood stallions are used. The breed is based now on Belgian and Noric blood.

Czechoslovakian Small Riding Horse

ORIGIN: *Czechoslovakia*
APTITUDES: *Riding horse*
AVERAGE HEIGHT: *13.2 to 13.3 h.h.*
POPULATION STATUS: *Rare, new breed*

Development of this small riding horse began in 1980 at the Agricultural University in Nitra. The initial breeding stock, about seventy mares, were kept at the Nová Baňa farm.

Broodmares were selected mostly of Arab stock and also included Hanoverian, Slovak warmblood, and Hucul (Carpathian). A herd of twenty-seven mares averaging 14.2 hands was selected for the first crosses to a Welsh pony stallion, Branco. The first offspring, foaled in 1981, were kept outdoors on the rough terrain. Additional crosses were made to another Welsh stallion, Shal, imported from West Germany.

The first foals were trained under saddle and in harness in 1984 and nearly all successfully passed the working trials. Animals of the new breed are docile, with an alert but calm temperament, modest in feeding requirements and effective in utilization of food. They have good gaits and jumping ability.

In 1989 a conference was held at the Agricultural University of Nitra regarding the breeding program. It was agreed that methods being used to breed a small sport horse for children from the ages of eight to sixteen years were correct and that efforts should be continued.

The number of broodmares will be increased to one hundred in 1990, and plans are being made to establish a club for private breeders of small horses in Nitra.

Czechoslovakian Warmblood. Photo: MRDr. Karel Dvorak

Czech Warmblood
(Český teplokrevnik)

ORIGIN: *Czechoslovakia*
APTITUDES: *Riding, sport horse*
AVERAGE HEIGHT: *16 h.h.*
POPULATION STATUS: *Common*

Original horses in Czechoslovakia were largely warmblood in type. Initially, they were improved through use of Spanish and Oriental blood; in the twentieth century improvement has mostly been through English blood. The first laws directing horse breeding were issued by the Empress Maria Theresa in 1763.

Czechoslovakian Warmblood. Photo: Marie Goldbergerevó

After World War I, Czech Warmblood breeding was influenced by Thoroughbreds and robust Oldenburg stallions. After World War II the Warmblood breeding went through a complicated stage characterized by an extreme drop in the number of horses because of mechanization in agriculture and motorization of transport and arms (requisition by the German army, 1939–45). After the war horses were no longer used on farms as in the past due to the use of tractors.

The modern Czech Warmblood is the most widespread breed in Czechoslovakia. An elegant, universal Warmblood horse is required with predominance of riding type and sporting qualities. The aim for improvement is a systematic increase of performance potential for the disciplines of riding sport. The original line, Przedswit (a Thoroughbred halfbred), asserts very well, with improving influence of Hanoverian, Trakehner, and Thoroughbred. The main breeding facilities are the studs Albertovec, Kladruby, and Netolice.

The Czech Warmblood is an aristocratic, robust horse of riding type. The character is good and this horse is animated, a good jumping animal. The movements are elastic and easy.

Predominant colors in Czech Warmbloods are bay and chestnut, less often grey and black. Some individuals are *isabella* (see Palomino) or dun. The breed is used mainly as a sport horse for all the disciplines of riding and as a leisure horse. It is also very suitable for driving competitions.

Many Czech Warmbloods are exported as riding horses into western Europe, especially Italy and Germany and, recently, England.

Dales Pony

ORIGIN: *England*
APTITUDES: *Riding, driving, light draft*
AVERAGE HEIGHT: *13.2 to 14.2 h.h.*
POPULATION STATUS: *Common*

Dales pony, showing action and style in harness. Photo courtesy Dales Pony Society

Champion Dales pony mare, Brymor Mimi. Photo by Stuart Newsham, courtesy of Dales Pony Society

Dales pony working sheep in the Pennines. Photo by Kit Houghton, courtesy of Dales Pony Society

Native to the upper dales of Tyne, Allen, Wear, Swale, and Tees, the Dales pony stems from the Pennine pony, with infusions of Scotch Galloway (which carried Friesian blood), Norfolk Trotter, and Wilson pony blood. Bred especially for the Pennine lead industry as pack animals, Dales ponies were famous for their ability to carry heavy weight over rough country at good speed, traveling distances of up to 200 miles each week. They were called Dale Galloway ponies.

The Scotch Galloway was a fast, sturdy pony. This was Britain's original race horse, and was also part of the foundation for the English Thoroughbred. The breed became extinct by the end of the nineteenth century.

With the advent of railways and improved roads, the ponies found use on small farms of the inhospitable upper dales. Due to their strength and surefootedness, they were capable of doing all the work on small hill farms where larger horses were at a disadvantage. Dales ponies carried shepherds and burdens of hay for great distances on the fells. Stylish and fast in harness, they were also successful in trotting races when harvest was over. The ponies were used by the army as pack and Mountain Artillery ponies.

During the eighteenth and nineteenth centuries there was some influence on the Dales pony from Clydesdales, Norfolk and Yorkshire Roadsters, and Welsh Cobs. Throughout such infusions of new blood, the Dales pony has retained its similarity to the Dutch Friesian, including its excellent trotting ability.

In 1916 the Dales Pony Improvement Society was formed to preserve this ancient pony, but World War II brought near destruction to the breed. By 1955 only four ponies were registered. Formation of the Dales Pony Society in 1963 led to distinct improvement in both quality and quantity.

The coat color of Dales ponies is usually black, bay, brown, or grey. White markings are rare. The head is small with a straight profile, broad forehead, small ears and eyes, and flared nostrils; the neck is rather short and well muscled; the withers are prominent; the back is short and sometimes hollow; loins are broad and muscular; the croup wide and sloped; the chest deep and wide; the shoulder rather straight. The legs are long, muscular, and feathered from the knee down; the hoof is bluish in color and very tough. The mane and tail are thick and full.

The action of this pony is distinctive, again reminiscent of the Friesian, with great flexion of the joints giving high knee and hock action.

The combination of good conformation with energy and agility makes this pony an excellent mount. Dales ponies have a fast walk and the stamina and determination to go long distances, making them favorites for trekking and long rides. Being willing and clever jumpers, they are successful competitors in cross-country trials, dressage, and performance competitions. They are strong enough to carry adults easily and, while gentle enough to carry children, they are best suited to competent adult and teenage riders.

Danish Oldenborg

ORIGIN: *Denmark*
APTITUDES: *Riding, sport horse*
AVERAGE HEIGHT: *16 h.h.*
POPULATION STATUS: *Rare*

Danish Oldenborg. Photo: Jørgen Bak Rasmussen, courtesy Dansk Oldenborgavl

Danish Oldenborg. Photo: Jørgen Bak Rasmussen, courtesy Dansk Oldenborgavl

The Danish Oldenborg is not the same as the Oldenburg found today in Germany. The Oldenburg in Germany has been crossed with various breeds to produce a warmblood sport horse of a different type. The Oldenborg of Denmark originated in West Germany prior to the time that crossbreeding occurred, and today Denmark is the only country where one may find pure breeding stock of the original Oldenburg. Crossbreeding to Thoroughbred is allowed, providing no more than 50 percent of Thoroughbred blood is present.

Systematic breeding of Oldenborg horses in Denmark has been going on for more than a century, and tradition combined with small-scale breeding has produced some of the world's finest horses. The Danish breeding association has about 450 members. The goal in breeding is partly to preserve the pure

Oldenborg horse for breeding and partly to make top class performance horses by crossbreeding to Thoroughbreds. The crossbred is very popular throughout the world. Its positive attitude and enormous capacity have made this horse well suited for international equestrian sports.

The main colors of Danish Oldenborg horses are black and brown.

For history and early development of the Oldenborg breed see Oldenburg (Old type).

Danish Sport Pony

ORIGIN: *Denmark*
APTITUDES: *Riding pony*
AVERAGE HEIGHT: *Maximum of 14.2 h.h.*
POPULATION STATUS: *Common*

The Champion Danish Sport Pony, Carina. Photo courtesy Danish Sports Ponyavl

In Denmark serious work has been concentrated on the pure breeding of the British pony for about thirty years. Previously, either the Icelandic horse or Norwegian Fjord was used if one wanted a small horse for children or the easier work in the fields.

As pony riding became more popular at many riding competitions and pony riding schools, demand for a good riding pony of standard type increased. The Danish Sport Pony Breeding Association (DSP) was founded in 1976, and under the auspices of this organi-

zation ponies were crossbred with other races to develop a good, uniform riding pony.

The most important goal in producing riding ponies for children is good temperament, and breeders strive toward this end. Conformation is carefully monitored to maintain pony type and avoid producing a small horse.

Ponies for competition may not be over 14.2 hands and there is great demand for animals of that height. There is also use for good riding ponies that are smaller. The riding classes for Danish Sport Ponies are separated into three categories: category 1, ponies up to 14.2 hands; category 2, ponies up to 13.2 hands; and category 3, ponies up to 12.2 hands.

It is important that the pony is well built in both body and limbs so that it has good action with speed and agility, is easy to ride, and is well adapted to dressage or competitive jumping.

A great deal of emphasis is directed to exterior judging of the Danish Sport Pony, with the desire for a beautiful, harmonious animal. The head should be beautiful with an alert, intelligent expression and well set on a good, well-placed neck. The shoulder is sloping; the withers well defined; the back muscular and straight; the thighs are strong and muscular; and the limbs must be absolutely sound.

Previously dappled grey was the dominant color in this breed, but today chestnut, bay, and especially black are very popular. Walleyed ponies are not desired for breeding, but all colors are allowed.

Registration of the Danish Sport Pony is through the national society called Landsudvalget, which registers all special races. In the purebred horse and pony breeds, one must prove ancestors for at least three generations for breeding. At the annual DSP show, a mare may be brought before the Breeding Committee. If her conformation satisfies the judges, the mare is entered into the pre-register. Perhaps the next generation may be registered in the register proper, denoting a very good pony seen as an asset for the breeding program.

To have a stallion accepted for breeding by the society and to have his progeny in the stud book, again one must show him before the committee of judges. Upon passing inspection the stallion may then receive breeding authorization. Later, he must be presented for a test, being ridden and driven in harness. Having passed this test, he can be shown a year later for conformation judging and may then be accepted and authorized as a breeding stallion. A number of his progeny must be judged four years later if he is to continue as an authorized breeding sire. This practice assures the continued production of useful, good quality ponies.

Stallions used for crossbreeding are usually Arab horses, which add elegance and good size to the breed. Connemara, New Forest, and Welsh B stallions are also used.

Danish Warmblood

(Dansk Varmblod)

ORIGIN: *Denmark*
APTITUDES: *Riding, sport horse*
AVERAGE HEIGHT: *15.3 to 17 h.h.*
POPULATION STATUS: *Common*

The Danish Warmblood is relatively new, as the breeding of riding horses in Denmark was formally organized in 1962. Since that time this outstanding horse has risen to the ranks of

The Danish Warmblood stallion, Rastell. Photo courtesy Danish Warmblood Society

The Danish Warmblood, Marzog. Photo courtesy Danish Warmblood Society

world leaders. Many claim the Danish Warmblood Society to be the most successful breeding organization in modern times.

Success of the Danish Warmblood is based on several factors. Most of the breeding takes place on Danish farms, in a farming community known for a hundred years for its high international standard. Stallions are selected under possibly the most extreme critical standards. Danish breeders have shown a willingness to invest in already proven bloodlines from neighboring countries and have skillfully combined those lines to achieve their goal—a champion riding and performance horse.

In the beginning there was no set tradition for breeding riding horses in Denmark. The resulting openmindedness on the part of breeders and their willingness to learn and take advice has impressed judges and consultants.

Danish Warmblood horses are exported by the thousands, to lands as far away as the United States and Australia.

Following the old saying, "A good horse has no race," Danish riding horse breeders have been able to utilize their knowledge of sport requirements of the horse. The result is that Danish horses are now in demand by riders from all over the world. There are Danish-born Warmblood horses on the national dressage teams of Italy, Germany, Finland, Norway, and Australia in addition to the Danish team.

Danish Warmblood breeding is unique in that a large number of approved stallions participate actively in riding competitions, enabling the riders and future buyers to follow the work of the breeders.

The committee in charge of breeding follows progress of the young horses from the beginning. As two-year-olds the best mares are presented at local shows in each of the seven regions under the organization.

Young Danish stallions are first judged at two and a half years, when preliminary selection is carried out in each of the seven regions. In March, as three-year-olds, approximately fifty young stallions are allowed to enter the stallion selection. Only a mere ten to fifteen receive a one-year covering approval. Having received this, they are automatically obligated to participate in the hundred-day test, which begins in the fall of the year. If a stallion passes the test with good results, he is again presented before the judges as a four-year-old and may then be finally approved.

The hundred-day test is carried out at the Queen's Regiment, but all costs are paid by the stallion owners and the organization. There is no state subsidy in Denmark for breeding horses.

Amazingly, Denmark has changed in the last thirty years from a country forced to import good horses to one of the major exporters of fine sport horses.

Danubian

(Dunavska)

ORIGIN: *Bulgaria*
APTITUDES: *Draft work*
AVERAGE HEIGHT: *15.3 to 16 h.h.*
POPULATION STATUS: *Common*

The Danubian stallion, Hissar. Photo: G. Dimitrov

A Danubian mare working on the farm. Photo courtesy State Corporation of Cattle and Sheep Breeding, Bulgaria

A Bulgarian halfbred, the Danubian or Dunav horse was developed by crossing halfbred mares with stallions of the Nonius breed and selection toward a type suitable for the country. Production of this breed began in 1924 in the G. Dimitrov stud near Pleven.

Since that period the Dunav horse has been bred for a massive and compact constitution, well-developed and healthy skeletal system, good conformation, light gaits, a good temperament, and low requirements regarding nutrition and management.

Prevalent colors are black and bay and, rarely, chestnut. Horses of this breed have very good draft qualities and with their mild temperament are extremely well suited for all kinds of transport and farm work.

The main habitation for this breed is the Dunav (Danube) valley and part of the plains region of southern Bulgaria. During the last two or three decades some Thoroughbred blood has been used, resulting in horses with excellent competitive abilities.

Dartmoor Pony

ORIGIN: *England (Dartmoor, in Devon)*
APTITUDES: *Riding pony*
AVERAGE HEIGHT: *11.2 to 12.2 h.h.*
POPULATION STATUS: *Uncommon*

The Supreme Champion Dartmoor stallion, Allendale Vampire. Photo courtesy Dartmoor Pony Society

Ponies have existed in Britain for over 10,000 years and remains have been found in many parts of the country. As human settlement expanded and enclosure and cultivation reduced the habitat of the ponies, they were driven back into bleaker, more desolate areas. Different breeds evolved in their own districts.

The earliest reference to the Dartmoor pony appeared in 1012 in the will of a Saxon Bishop, Aelfwold of Crediton. Much later, during the heyday of the tin mines on Dartmoor, the ponies were used extensively for carrying tin. Later they were left to roam free, apart from those required for work around the farms.

In 1898 the Polo Pony Society (now the National Pony Society) set up local committees to produce descriptions of each native breed. Apart from the heights, the original description is nearly identical to the present breed standard.

The Dartmoor pony breed was hit very hard by the First World War. About the same time, the Duchy Stud, owned by the Prince of Wales, began buying many Dartmoor ponies to use in a breeding program aimed at producing an all-around saddle horse. One stallion used with success was a desert-bred Arab, Dwarka, a bay with a typical pony head. Two outstanding ponies came from this breeding program, The Leat and Heatherbelle VI.

The Second World War took its toll, and only a few registered ponies were left. Registration by inspection was begun, and prize winners at various selected shows were automatically eligible for registration.

The Dartmoor is an excellent first mount for children. It has a quiet, gentle disposition and is economical to keep. It is a versatile breed, having talent for jumping, and is excellent when used for driving. Bred wild on the moors, the Dartmoor is hardy and strong and well adapted to harsh weather conditions. The rough, rocky terrain and sparse grazing have produced a surefooted pony easily capable of carrying an adult. First and foremost a riding type, the pony has a sloping shoulder and long, low stride that give the rider a feeling of riding a much larger mount.

In appearance the Dartmoor is sturdily built yet full of quality. The head is small with a broad forehead and small, alert ears. The neck is of average length, strong but not heavy, and stallions have a moderate crest; the shoulders are well sloped; the back is of medium length; the quarters are well muscled. The tail is very full and set high. The legs have dense, flat bone, and the hooves are tough and well shaped. The action is straight, free, low, and fluent.

The most common colors in the Dartmoor breed are bay, brown, black, and grey. A few chestnuts are seen as well as an occasional roan. Piebald or skewbald are not permitted for registration, and white markings on the face and legs are allowed but discouraged.

In the United States the Dartmoor was first introduced in the 1930s with additional shipments following in the 1940s and 1950s. Many of those first ponies were used for producing polo ponies and for improving local stock, with the effect that most of the pure blood was lost. Later imports were more successful in maintaining the pure Dartmoor in the United States. Today there is renewed interest in this beautiful and versatile pony.

Datong

ORIGIN: *China (Qinghai Province)*
APTITUDES: *Riding, farm work*
AVERAGE HEIGHT: *12.2 to 13.3 h.h.*
POPULATION STATUS: *Common*

Datong horse of China. Photo: Weigi Feng and Youchun Chen

The Datong horse is improved by better range management. Photo: Tiequan Wang and Youchun Chen

The Datong breed of China dates back some 4,000 years. The Datong River basin, 8,000 feet above sea level in northern Qinghai Province, is the range of this ancient breed, which is classified as a native breed of China.

This breed was famous in China for centuries because of the "Dragon Horse" and is mentioned in the biography *Son of Heaven Mu*. Mu Wang, also called Xiao Wang, was the king of the Chow or Zhou dynasty during the years 1023 to 982 B.C. The Dragon Horse was an exceptional strain with great powers of endurance; it appeared suddenly during the reign of Mu. Many references are made in ancient Chinese art and history to the eight celebrated horses of King Mu, which were said to be able to gallop "1,000 li in a day." One *li* is equal to about a quarter-mile, but in any event, to gallop 1,000 li in a day is a figure of speech in China and denotes horses of great endurance, able to travel long distances without tiring. The Dragon Horse had small, bony projections on the forehead directly above the eyes that resembled small horns (these were not true horns but frontal bosses, found in several mammals but not common in the horse).

As in the case of the famous "blood sweaters" of China (see Akhal-Teke), things pertaining to the emperors were called "heavenly," as emperors were considered the sons of Heaven.

The ancient Dragon Horse is also mentioned in writings in a bamboo carved book of the Sui dynasty. According to Professor Youchun Chen of Beijing, this book reported that wild mares were caught on the plateau near the Qinghai Sea, the place of origin of the ancient Dragon Horse, and this strain reappeared in foals born to those mares. Whether or not the mares that had been caught exhibited characteristics of the strain is unknown.

The history of the Dragon Horse is now so remote that modern breeders, when seeing bony projections that still occasionally occur above the eyes, have thought this an abnormality and have eliminated such horses from breed selection. Many equine historians have believed the stories of horned horses to be nothing more than myth (see Moyle).

The best individuals of the Datong breed are over 13.3 hands at the withers. Two types exist in the breed today—heavy and light.

The Datong tends to be coarse and compactly built with a gentle temperament. The structure of this horse is very strong and well muscled. The body tends to be long but still in harmony. The head is heavy with a straight or "rabbit-like" profile and of medium length; the ears are of medium length, the nostrils well opened. The head is wide at the jaw. The neck is short and thin, not very well connected to the withers; the withers are low; the chest is deep and ribs are well rounded; the back and loin are level. The abdomen is long; the croup is sloped and short; the legs are of medium length with thin cannon bones that are quite hairy; the joints and hooves are very strong. The hind legs tend to be slightly sickled. The hair of the mane and tail is thick and long.

The riding type of Datong has a dry head, longer limbs, and more refined cannon with less hair and is described as being strong and speedy. Predominant colors in the breed are bay, black, chestnut, and sometimes grey. Some white markings may be found on the face or lower legs. Appaloosa coloring is rarely found. The size of the Datong horse was increased during the last century by using Hequ blood.

Both good trotters and pacers are found in the Datong, and this is a horse of exceptional endurance qualities, well-known for excellent performance on long distance rides at very high altitude. The breed is also used for agricultural purposes.

Deliboz

(Delibozskaya)

ORIGIN: *Azerbaijan*
APTITUDES: *Riding, light draft*
AVERAGE HEIGHT: *14.2 h.h.*
POPULATION STATUS: *Common*

An Oriental riding and pack horse that has been bred in the Kazakh, Akstafa, and Tauz regions of Azerbaijan and adjacent areas of Georgia and Armenia for hundreds of years, the Deliboz was described in earlier times as the Kazakh horse of Azerbaijan or as the Azerbaijan horse. Actually the modern Deli-

The Deliboz of Azerbaijan. Photo: Nikiphorov Veniamin Maksimovich

boz is a descendent of the ancient Azerbaijan breed, now rare and persisting only in small numbers.

The breed was influenced in its evolution by the Persian and Turkmene breeds. Widely used within its own restricted area, the Deliboz was virtually unknown elsewhere for many years. Horse specialists singled out a specific Deliboz type among large horse populations studied in the late 1930s and late 1940s. These horses were improved by Arab and Karabakh saddle-type stallions. Deliboz horses were widely spread throughout western Azerbaijan. In 1943 a state breeding cooperative named Gosplemrassadnik was established to improve horse populations of the republic.

Since 1950 the Azerbaijan horses have been improved by Arab and Tersk sires, while the Karabakh was bred separately. Crossbreeding involved horses of the Deliboz type.

The breed differs from others in having a clean, short head with a broad forehead and narrow nose; a compact, heavy neck; a massive body with good top line and an even, long back and loin. The legs are clean and well proportioned and the cannon bone girth is larger than that of many other horses of the same size.

The Deliboz is characteristically an animal of uncertain temperament, easily aroused, and the majority have a natural pacing gait. One interesting feature of the breed is a peculiar lengthwise fold on the tongue giving the impression of a forked tongue.

The breed possesses good working ability, easily covering twenty-seven to thirty-four miles in a day at high altitude and carrying a pack weighing from 250 to 280 pounds. The horses routinely carry a rider forty-five miles in a day.

Presently the pedigree nucleus of the breed is represented only by descendants of the Tersk stallions Tselostat and Pygmalion, so careful steps are being taken to restore the breeding nucleus of local horses. The Dashyuz stud farm has 140 of the most typical Deliboz horses, including mares with a high percentage of Tersk blood. Plans call for breeding the Deliboz pure, with single backcrossing to one or two stallions with a little Tersk blood in order to preserve the valuable traits of the crosses.

The predominant colors found in the Deliboz are grey and bay.

Døle Gudbrandsdal

ORIGIN: *Norway*
APTITUDES: *Draft work, trotting racer*
AVERAGE HEIGHT: *14.1 to 16 h.h.*
POPULATION STATUS: *Common*

Døle Gudbrandsdal of Norway. Photo: Knut Davidsen

Known as Dølehest (Døle horse), Døle Gudbrandsdal, Ostlandhester (Eastland horse), Døle Trotter, and Norwegian Trotter, this breed is the most widespread in Norway. Organized breeding dates from the middle of the nineteenth century. The area of origin is the Gudbrandsdal Valley, connecting the Oslo region with the North Sea coast.

It is likely that the Døle horse descended in part from the Dutch Friesian, as there is similarity between the two. During the period 400 to 800 A.D. the Friesians traded extensively with Britain, Norway, and the Rhine Delta. Undoubtedly, they took their black Friesian horses with them, and it is interesting to note that the Fell and Dale ponies of Great Britain also resemble the famous Friesian.

The Døle horse of today is found in two types—the heavy horse and the light draft horse (the coldblooded trotter). The difference between the two was formerly much greater than it is today. Interbreeding of the types is now a usual practice, and the breed is becoming increasingly uniform.

Show classes in Norway are not just a showcase for the public but a means of grading animals to be used for breeding programs. The two types are judged in separate classes, but both groups are entered into the same state-authorized stud book.

Horses achieving the accepted minimum trotting speed are entered with a "T" after the stud book number. Mares of trotting blood that have not gained the minimum trotting speed can be judged and graded as heavy horses if their conformation is satisfactory.

In 1928 trotting races were organized at the Bjerke trotting track and later at nine local tracks. Interest in trotting races has enjoyed a continued increase in Norway. Some of the profit from the races is put back into promotion of the horses in both sport and breeding. Of the trotting races, 50 percent are for warmblooded trotters (Standardbred, and so forth) and 50 percent are for coldblooded trotters.

The National Dølehorse Association was established in 1967 with the aim of promoting and preserving the heavier type of draft horse. Inbreeding has been practiced to a certain degree, resulting in a homogenous, good type. Negative effects at one stage were low fertility and weakness of the legs, but these weaknesses have been corrected.

Mechanization within agriculture and forestry has left little use for the heavy horse, but there is growing interest in the trotter type and other light breeds.

The predominant colors (60 percent) found in the Døle horse are bay or brown; 25 percent are black and 10 percent are chestnut. There are a few greys and duns. Some white markings on the head and limbs occur, especially among animals of the trotter type.

The heavy type of Døle, when presented for grading, must pass a test to denote its pulling power as well as a simple trotting test. X-rays of the knee and lower leg are also taken, and animals with defects are not used for breeding.

Stallions of the lighter trotting type must have achieved satisfactory results on the track before being put to stud. Three-year-olds may be issued a temporary stallion license if their conformation, breeding, and trotting ability are acceptable. Quality of the Døle has improved during the last several years due to the organized selection of breeding stock.

Horse breeding in Norway is state directed, led by the government consultant in cooperation with the various breed societies. A law defining selection of stallions for breeding purposes and rules for entry into the respective stud books provide a framework for the work. Only selected stallions may be used for breeding. Crossbreeding is allowed only in an organized form. A crossbreeding program must be approved.

Despite the fact that the first cross of warmblood and Døle produced animals of generally good riding type, the practice of crossbreeding with the Døle breed has more or less ceased. The warmblooded stallions used were Swedish Warmbloods. Use of the English Thoroughbred has been very restricted. One might suppose that a first cross could produce a nice hunter type, but this has not been proven.

The Døle horse is one of the smallest coldbloods. This horse has an excellent trot and great pulling power. It is agile and active and shows great stamina. The head is small and somewhat heavy and square with a straight profile, long ears, and small eyes; the neck is rather short and well muscled; the withers are broad and pronounced; the back is long and straight; the loins are strong; the croup is broad, muscular, and gently sloped. The legs are short and sturdy, well muscled with broad,

strong joints, and show heavy feathering from the cannon down in the heavy type; the hooves are broad with very hard horn.

The Døle horse contributed greatly to formation of the North Swedish Horse, another excellent trotter.

Don
(Donskaya)

ORIGIN: *Don River steppes (former Soviet Union)*
APTITUDES: *Riding, light draft*
AVERAGE HEIGHT: *15.2 to 16.2 h.h.*
POPULATION STATUS: *Common*

Don horses. Photo: Nikiphorov Veniamin Maksimovich

Don horse. Photo: Nikiphorov Veniamin Maksimovich

The modern Don horse developed on the steppes of the Don River and its tributaries. It descended primarily from the south Russian steppe horses crossed with stallions of Oriental breeds such as the Persian, Karabakh, and Turkmenian. The most ancient ancestor of the Don breed was the horse of the south Russian Tartars, which was later influenced by the horses of the Polovtsy people, nomads of Turkish origin. Their small, tough horse, alert and full of endurance, was called the "old Don horse." The first recorded mention of the name Don Kazakhs appears in documents in Moscow dated 1549.

Traditionally associated with the Don Cossacks, the modern breed evolved during the eighteenth and nineteenth centuries. It became famous in the campaigns of 1812 to 1814 when 60,000 Cossacks fought Napoleon's army. In the severe winter of 1812, Napoleon's army lost many thousands of French horses while being relentlessly pursued by Cossacks riding their hardy Don horses. After driving the French back to their homeland, these horses carried their riders all the way back to Moscow, an almost unbelievable feat and one unparalleled in the history of cavalry. Since that time, however, the breed has undergone many changes.

While intensive horse breeding in the Don area began in the late eighteenth century, selective breeding of the Don did not begin until the 1830s. The modern Don had emerged by the early twentieth century. In the west Don region, the population was improved primarily by horses of the Orlov-Rostopchin (Russian Saddle Horse) and Thoroughbreds. Horses on the steppes east of the Don River retained many features of the native steppe horses and Oriental breeds. Good pasture in the virgin lands contributed to establishment of the breed with its strong constitution and remarkable adaptability.

This led to extensive use of the Don in all of the steppe regions—west and south Kazakhstan, Kirgizia, southern Siberia, and the Transbaikal area—both as a purebred to improve the local horses and to form new breeds. The Don quickly adapted to the mountainous regions of Kazakhstan and Kirgizia and was used in development of the New Kirgiz (Novokirgiz) and Kushum breeds.

The main objective in breeding Don horses was the production of a solid army remount requiring a minimum of care and attention. Dons were raised in herds on the natural pastures of the steppe, foraging for themselves and pawing away the snow in winter to reach the grass beneath. This is one of the few European breeds that adapts easily to climatic hardships when living on pasture throughout the year, and for this reason it is much valued to improve local stock in areas where horses are raised in herds.

By the twentieth century the Don population had risen to the largest in Russia. In the Transdon steppes it reached 70,000 head. Then, during World War I and the Civil War, the breed nearly vanished. In 1949 there remained only 89 purebred stallions and 466 mares. Reestablishment of the breed was carried out at S. M. Budyonny and Zimovnikovski studs in the Rostov region and at Issyk-Kul stud in Kirgizia. Numbers rose and fell, but by 1980 the total population of purebreds numbered 17,000, once again a healthy population.

The modern Don has a distinctive appearance. The head is of medium size and light, with a straight or slightly dished profile. The forehead is wide, the jaw fleshy; the neck is of medium length, the crest prominent with an insignificant throat-latch; the withers are of medium size; the back and loin are straight, flat, and wide; the croup is sloping or sometimes straight; the chest is well developed; and the trunk is wide.

The forelegs of this breed have been improved considerably from former times. During the period from 1950 to 1960, 20 percent of Don horses exhibited underdeveloped knee joints and knees set too far back. With careful selective breeding this problem has been nearly eliminated. The most common coat color in the Don breed is chestnut, usually with a golden, metallic sheen.

Today there are three types found in the breed; eastern, heavy, and saddle. An extensive effort is being made to use horses combining the qualities of the heavy and eastern types.

Pure breeding and infusion of Thoroughbred blood showed that while an increase of 8 to 10 percent did not decrease height, body size, or performance, an increase of Thoroughbred blood over 25 percent sharply reduced fitness for *taboon* conditions (pasture raising with a herdsman) and eventually resulted in a decrease in body size, accompanied by an increase in abortions.

Dons are versatile working animals. While not exceptionally fast, they have remarkable endurance and excellent jumping ability. They are popular for mounted tourism.

Dongola
(Dongolawi)

> ORIGIN: *Cameroon (West Africa)*
> APTITUDES: *Riding horse*
> AVERAGE HEIGHT: *15 to 15.2 h.h.*
> POPULATION STATUS: *Common*

The Dongola or Dongolawi horse is thought to have originated from the Dongola Province in Sudan, but information was not received from that country concerning the breed. Although it is thought to be a degenerate type of Barb, there is good reason to believe that, in part, the Dongola descended from Iberian horses taken to Egypt from the area of Numidia in the thirteenth century. Attachment of the breed to the Cameroonian Foulbés is historical, and these people originated in the valley of the Upper Nile. Dongola horses predominate in northern Cameroon, which is heavily inhabited by the Foulbés people.

While much has been said by horse book writers of the poor quality of the Dongola breed, this should not be attributed to lack of genetic potential but rather to lack of good management. Dongola mares are rare, and fertility is insignificant. Mostly owned by Muslims, the Dongola is exploited of possible quality by the fact that owners prefer to ride stallions. Stallions are considered more prestigious and are used for competitions and recreation. Rather than breed their own horses, they prefer to buy animals from neighboring countries like Chad and Nigeria. Inferior stallions are left to mate with the Dongola mares, and quality in the breed continues to degenerate.

In 1772 the King of Sennar (in Sudan) obtained horses of the Dongola breed described as being very fine in quality, as strong as coach horses but light and free in their actions. Less than a century later, a German traveler recorded that the original Dongola horse was nearly extinct. The Dongola has been interbred with Arab, Barb, and Arab-Barb crosses.

The Dongola of Cameroon is characterized by a convex profile which in many cases is very pronounced. Many individuals have large, unattractive heads, long backs, poorly attached loins, flat croups, and thin legs. The chest is flat and high. Dongola horses are usually a deep reddish bay, sometimes chestnut or black. White markings on the face and lower legs are common.

In the north of Cameroon, despite many breeding shortcomings, the Dongola is often characterized by beauty, vitality, and energy. The Foulbés people spend a great deal on the care of their horses.

The Institute of Animal Research in Garoua is doing much in the way of selection of the Dongola, and improvement of this neglected breed is on the rise.

The horses collectively termed West African Dongola are descended from the Dongola breed and the name applies to several strains found in West African countries and known as Fulani (across North Africa), Bahr-el-Ghazal (in Chad), Hausa and Bornu (in Nigeria), and Bandiagara, Djerma, Mossi, Songhai, and Yagha (all in Niger Bend). These horses are also sometimes referred to as the West African Barb. Many are small, resembling the Pony Mousseye of Cameroon. See Nigerian.

Dülmen Pony
(Dülmener)

> ORIGIN: *Germany (Westphalia)*
> APTITUDES: *Riding pony*
> AVERAGE HEIGHT: *12 to 13 h.h.*
> POPULATION STATUS: *Rare*

Dülmen pony in Germany. Photo: Erbprinz von Croy

A herd of Dülmen ponies. Photo: Erbprinz von Croy

West of the town of Dülmen is the Merfelder Bruch. The only natural breeding ground for wild ponies on the European continent is located in this extensive region of wood, moor, and heath. According to a pamphlet sent by Erbprinz von Croy, the ponies can be traced back for more than 600 years and are the only remaining ponies of German origin living in completely natural conditions "since always" with never an interruption in natural selection.

Wild ponies are mentioned in a document dated 1316. At that time wild horses were being persecuted, and the Lord of Dülmen obtained rights to the wild ponies in the area, providing them a refuge. The Merfelder Bruch originally covered a large area, the ownership and running of which were shared by the lord of the manor and the farmers. At the beginning of the nineteenth century there were still several natural breeding grounds in Westphalia. With the common lands being divided up and settlements growing, reservations for the wild ponies gradually disappeared. As division of the common land of Merfeld between 1840 and 1850 devoured more and more of the area for the wild ponies,

the Dukes of Croy had them rounded up and gave them a place of refuge on their own lands.

The enclosed *wildbahn* (area for wild animals) of the Merfelder Bruch covers an area of about 865 acres (350 hectares). In this area, which is marked by the colorful contrast of meadow, moor, heath, briar thickets, conifers, and deciduous forest partly in its primeval state, the wild ponies find all kinds of food and the shelter they need in bad weather. The herd, including foals, now numbers an average of 300.

The ponies spend the entire year in the open without supplemental food. Only when deep snow covers every blade of grass and every bush are they offered hay at a few feeding places. They are otherwise left completely to themselves and have to cope with sickness and birth without human help. The weak are eliminated by nature. These hard, natural living conditions have made the Dülmen a pony of self-reliance and hardiness.

From a zoological standpoint the Dülmen can no longer be classified as "wild." It shows the marks of domestication and breeding, obvious in the hanging manes and various colors. The existence of animals of mouse-grey (grullo) or dun gives evidence that there is still blood from the original wild ponies in their veins. Probably the Dülmen is in part a remnant of the wild Tarpan. Other than the predominant dun and mouse coloring, the Dülmen may also be brown, black, or chestnut.

Due to its hardiness, self-reliance, endurance, and long life, the Dülmen used to be in great demand as a working horse in nurseries and on farms. Today it is highly esteemed, especially for children, for riding, acrobatics on horseback, and pulling carts, and can in general be harnessed from the age of two or three years. These little horses are intelligent, good natured, and easy to tame. They often reach an age of twenty-five to thirty years.

Each year on the last Saturday in May the traditional roundup is held. The male yearlings are captured and then auctioned in the presence of thousands of spectators. Mares, however, do not come up for sale. Born in the *wildbahn*, they also die there. After the roundup of the male yearlings, one or two stallions are introduced for breeding purposes. They stay with the herd until September.

In the beautiful countryside of the Münsterland Bruch, among moor and heath, the wild ponies have withstood the storms of time. It is due to the generosity and sense of tradition of the Dukes of Croy that in our age of technology this unique natural monument from times long past has been preserved.

Good weather permitting, the *wildpferdebahn* (area for wild horses) is open to the public from March 1 until October 31 on weekends and holidays.

Dutch Draft

(Nederlandsche Trekpaard)

ORIGIN: *Holland*
APTITUDES: *Heavy draft and farm work*
AVERAGE HEIGHT: *16 h.h.*
POPULATION STATUS: *Uncommon*

Dutch Draft. Photo: Persburo Jacob Melissen, courtesy Koniklijke Vereniging Het Nederlandsche Trekpaard

The Royal Society, "the Dutch Draft Horse" (Koninklijke Vereniging "Het Nederlandsche Trekpaard") was formed on December 22, 1914. This society is also the stud book organization for the Haflinger horse in Holland.

The purpose of draft breeding in Holland is to produce a large, massive horse, strong and solidly built. It should have a head with a straight profile, short, lively ears, a strongly muscled front, good withers, a heavily muscled loin and hind quarters, with good muscle in the legs and good feet. The movements should be spacious and easy. The Dutch Draft must be able to do hard work for extended periods of time. Coat color is usually chestnut, bay, or grey, and sometimes black.

Closely resembling the Belgian Heavy Draft (Cheval de Trait), the Dutch Draft was developed from the Zeeland horse with crosses to the Belgian and Belgian Ardennes. This breed was found in great numbers in the provinces of Zeeland and North Brabant, where it did excellent work on large arable farms in the heavy marine clay. It was also represented in fairly considerable numbes on large farms in the marine clay areas of the province of Groningen, although it did not succeed in ousting the Groningen horse, whose qualities were more in keeping with the character of the farming population. In the province of South Holland, and more particularly on the islands, on large farms where practices were influenced by those of Zeeland and western North Brabant, the Dutch Draft horse was also used.

The most valuable characteristic of this draft horse is its exceptionally quiet disposition and the unhurriedness of its movements. However, for all its quietness, this horse can "turn a lively foot" when required. This makes it a good animal for many small farmers on mixed farms in sandy regions and largely explains why the Dutch Draft is so often found in Gelderland and to some extent in North Brabant and Limburg. The popularity of this horse in the provinces of Overijssel and Drenthe, adjoining Gelderland to the north, may well have been due to these same qualities. All this, however, is now largely a thing of the past.

This is the heaviest of all Dutch breeds and is particularly suitable for prolonged and heavy haulage or traction. Increasing mechanization and motorization have led to a decline in demand for heavy horses, but a number of good, sound breeding animals still exist.

Dutch Tuigpaard
(Dutch Carriage Horse)

ORIGIN: *Holland*
APTITUDES: *Carriage horse*
AVERAGE HEIGHT: *16.2 h.h.*
POPULATION STATUS: *Common*

Dutch Tuigpaard of Holland. Photo: Ellen Van Leeuwen

Dutch Tuigpaard in driving competition. Photo: Persburo Jacob Melissen

Among horses registered by the Royal Warmblood Stud Book of the Netherlands (KWPN) there exists a separate type called the Dutch Tuigpaard, which can be roughly translated as "carriage horse." This exceptional animal is unique in the world.

Used on the farm for field work in days gone by, this was the horse selected to lend distinction to the family or call attention to his status when the farmer went to church or to town. While on these outings a certain degree of competition naturally arose among the farmers surrounding the movement and presence of their Tuigpaarden. This special pride in and involvement with their harness horses have never left the Dutch breeder.

When mechanization made the working farm horse superfluous, the pleasure of the Tuigpaard (spoken in English as "toagpaard") was transferred to the show arena. While many breeders began to concentrate on riding horses, the exceptional carriage type generated enthusiasm among a select group of farmers who chose to continue their efforts with and commitment to the breeding of their show horses.

The Tuigpaard breeders were quick to organize and formed their own program for stallion and mare selection based on presence and movement. They began to introduce some Hackney blood, which increased the class and hard trotting action of the horses.

Today, Tuigpaarden are in evidence at many Dutch horse shows, adding excitement to the events. From the in-gate bursts a line of Tuigpaarden, snapped up to full speed by their drivers. The flashing blur of black, chestnut, and bay is accompanied by the pounding rhythm of hooves and squeak and rattle of carriage and harness as the horses sail around the ring bringing cheers and applause from the spectators. There is a unique grace and power to these horses, tightening the heart of any horse lover.

For competition the Tuigpaard is usually harnessed to a show trap. The light weight of the vehicle allows the horse full use of its active and powerful stride. While bred for this special type of movement, correct training and special competition shoes enhance the action of the horse. The judges look for a horse with a proud presence and bearing. The haunches must lower as the hind quarters engage well under the body, creating propulsion. From this tremendous thrust, the front of the horse elevates, the shoulder swings fully and freely, allowing for the forward snap and extension of the forelegs, and the horse then settles into a rhythm that brings applause from the crowd and a smile to the driver.

The underlying qualities of all Dutch

horses are especially present in the Tuig-paard—good temperament, willingness to work, and soundness. Even outside the show ring many Tuigpaarden are used by horse lovers. They may be seen along a country road on a beautiful day with only their driver sharing the sound of their hooves, or driving clubs may fill the air with laughter and calls as people congregate to enjoy an afternoon with their horses and carriages.

Dutch Warmblood

(Nederlandsche Warmbloed)

ORIGIN: *Holland*
APTITUDES: *Riding, sport horse*
AVERAGE HEIGHT: *16.2 h.h.*
POPULATION STATUS: *Common*

The Dutch stallion, Sultan. Photo: Ellen Van Leeuwen

Dutch Warmblood broodmare. Photo: Ellen Van Leeuwen

Dutch horses are famous the world over. The brand displaying the proud Dutch lion is the trademark of the Dutch Stud Book, called the Koninklijke Vereniging Warmbloed Paarden-stamboek Nederland (Royal Warmblood Stud Book of the Netherlands). This organization, with over 25,000 members, breeds riding and harness horses with only one philosophy: to breed the best warmblood horse in the world.

Dutch Warmbloods are appealing, modern horses with good character and strikingly good health. They are bred to perform for years in all disciplines, and their willingness, reliability, intelligence, and conformation have made them loved by riders everywhere. Careful breeding, selection, and use have molded a horse that is a durable and loyal companion for either recreation or champion equestrian events.

Horse breeders in Holland have been united for more than a century. King William III recognized the first stud book in 1887, thereby stimulating the breeding of registered horses in the Netherlands. The national KWPN stud book originated in 1969 from a combination of various regional books and was granted the royal title (K) in 1988 by Queen Beatrix.

Dutch breeders have shown uncanny feeling for husbandry and breeding, adjusting to market needs and demands and becoming famous for their cows, pigs, chickens—and flower bulbs. That instinctive feeling along with extensive knowledge and experience gathered in many decades of breeding have combined to result in horses that rank among the world's finest.

Roots of the Dutch Warmblood are deep in two areas of Netherlands soil—Gelderland and Groningen. In Gelderland, located in the central Netherlands, the soil is sandy, and here the lighter type of horse developed. In Groningen on the other hand, where the soil is heavy clay, a much heavier and denser horse was produced. The two remained genetically compatible and to refine their horse Groningen breeders used Gelderlanders, while in order to add more mass to the Gelderlander, Groningen blood was used (see Gelderland and Groningen).

The farmers earned their living with

horses, so strict breeding practices have long been used. Faults in soundness and character were quickly eliminated, along with horses that lacked intelligence. This severe culling made a great contribution to the Dutch Warmblood or sport horse of today. As mechanization of equipment gradually caused the horse to become superfluous on farms, riding clubs became more popular and interest in the sport horse increased steadily.

Using the fine old established bloodlines, the Dutch changed direction in their horse breeding to develop a true riding horse. Thoroughbred stallions were used, as well as riding-type stallions from France, Holstein, Hanover, and elsewhere, with breeders taking care to retain the time-tested farm horse qualities of the old Dutch breeds.

Over the years the modern KWPN has divided into two streams; 80 percent riding horses and 20 percent carriage horses (see Dutch Tuigpaard). A small group of breeders has also remained dedicated to maintaining the original farm type of light draft horse, to which the Dutch Warmblood owes much of its strength and enduring quality.

KWPN horses are high achievers, but achievements are not permitted to overshadow the good conformation, beauty, or charm that have made Dutch horses famous. Modern breeding selection methods, new insight into genetics, and combining sport results and breeding through the use of a central computer have all contributed to the quality of the Dutch Warmblood horse. The feet and legs of all stallions used for breeding have been examined by X-ray to detect possible heritable defects. Plans are under way also to X-ray the Dutch mares.

About 12,500 mares are serviced each year and 8,500 foals are born. Of the 4,500 colts, between 600 and 800 are selected at the age of three for evaluation as approved KWPN stallions. After a primary selection the best of them are taken to Utrecht in February where, during an internationally renowned stallion show, only a small, select group stands the test of two appraisals. These elite horses then undergo semen and X-ray tests before being sent to a central training facility where they are tested for one hundred days for character,

dressage, jumping talent, and overall health. Only then are the best stallions admitted to the KWPN stud book for breeding purposes.

After the first year a selected group and random sample of the first foals from each stallion are inspected. Quality of the foals must indicate that a stallion exerts a positive influence on the breed. After another two years, the progeny are reinspected. When they have reached six or seven years of age, their performance in competition is an important factor in the ongoing approval of stallions. Dutch Warmblood stallions are reviewed each year to assess whether each one fits the picture of the continually developing breed. The Dutch Warmblood stallion must also prove himself in competition, showing his ability in dressage and jumping at the stallion selection show.

Older stallions are often competitors in national and international shows; Gondolier, Nimmerdor, Jasper (Little One), Vincent, Orpheus, and Olympic Treffer are only a few examples of Dutch stallions participating at the top.

Only those stallions demonstrating a decidedly positive effect on the breeding of the KWPN horse come under consideration for the classification *keur* (choice). The highest classification is *preferent* (preferred), an honor awarded to famous stallions such as Amor, Doruto, Joost, Lucky Boy, and Nimmerdor.

In mare selection, broodmares are judged in regional shows throughout Holland in June and July. Three-year-old mares can be registered in the stud book and branded with the proud lion of the KWPN. The best mares from regional shows are entered in one of the twelve central shows in July and August, events drawing thousands of spectators. The highest honor for a breeder is to be allowed to present a mare at the national mare competition in Utrecht, a two-day show in September for the cream of the crop.

Mares that achieve the KWPN jury's demanding standards at the central show are granted the classification of *star*. The higher classification, *keur*, can be awarded to mares that have produced a foal and that have demonstrated their ability in jumping and dressage in a performance test. If a mare has produced three foals considered superior in both conformation and gaits, she may earn the *preferent* classification. If three of a mare's progeny are also top performers, she is awarded the *prestatiemerrie* (achievement mare) classification.

With such careful selection of breeding stock the Dutch breeders have produced one of the very best sport horses anywhere in the world; a horse that is beautiful and talented with a constitution as strong as steel and a willingness to work hard at its job.

Dutch Warmblood horses average about 16.2 hands with some reaching 17 hands. Coat colors are chestnut, bay, black, or grey, and white markings are often present. The head is well shaped, usually with a straight profile, and the neck is arched and well muscled, merging neatly into the withers, which are fairly prominent. The back is straight and fairly long, with the croup short, broad, and flat. The tail is set high. The chest of this horse is deep and full and the shoulder is well sloped. The legs are strong with a long forearm. Hind quarters are powerful and highly muscled, a characteristic inherited from the original and powerful farm horses of the Netherlands and a feature necessary for strong movements. Over fences, this breed is outstanding.

Although they have bred a high quality performance horse which combines beauty and symmetry with ability, the Dutch are still striving to improve their breed to reach the optimum product.

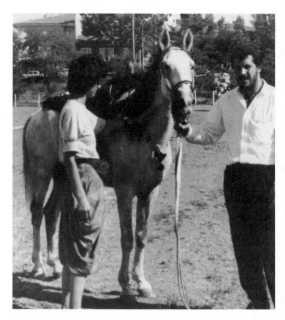

East and Southeast Anadolu horse. Photo: Ertuğrul Güleç

East and Southeast Anadolu (Anatolian). Photo: Ertuğrul Güleç

East and Southeast Anadolu

(Doğu Güneydoğu Anadolu Ati)

ORIGIN: *Turkey (Anatolia)*
APTITUDES: *Riding horse*
AVERAGE HEIGHT: *13 to 14 h.h.*
POPULATION STATUS: *Rare*

This breed dates back 1,000 years and is bred in the mountainous area of the east and southeast of Turkey. Some authorities claim the ancestors of the Kurds (Kurmanj) were the Turkmen who came from the Caucasus, so horses from this area were termed Turkmene.

However, the notable horseman, the late Professor M. Nurettin Aral, author of *The Types of Animals Bred in Turkey and Breeding History*, stated that in early times these horses were called Kurds as they were mainly used by the Kurdish people but that it is more correct to call them East and Southeast Anadolu (or Anatolian).

Currently there are few of this breed in the southeast of Turkey as the region now breeds more Arabs and Turkish Arab race horses, so the breed would perhaps be better called sim-ply East Anadolu horse. There are only about 3,000 of this breed today.

According to the late M. Nurettin Aral, this breed was developed by crossing the Turkish Arab and Persian breeds. They are very beau-tiful horses with good bone and plenty of spirit and stamina.

There is no breed association or stud book kept for this ancient breed, but it is hoped by breeders in Turkey that the Turkish Ministry of Agriculture will see the need for such a stud book.

The East Bulgarian stallion, Gordeliv. Photo courtesy Dr. P. Chakarov, State Corporation of Cattle and Sheep Breeding

East Bulgarian

(Istochno-b''lgarska)

ORIGIN: *Bulgaria*
APTITUDES: *Riding, light draft*
AVERAGE HEIGHT: *15.3 to 16 h.h.*
POPULATION STATUS: *Common*

The East Bulgarian is a breed of recent origin, created in the studs Kabiuk, near Shumen, and Bojurishte, near Sofia. Development of the breed began at the end of the nineteenth century using local horses, improved local horses, Arab, Anglo-Arab, and half Thor-oughbred mares crossed to Thoroughbred,

East Bulgarian horse in jumping competition. Photo courtesy Dr. P. Chakarov, State Corporation of Cattle and Sheep Breeding

halfbred, and Arab stallions. By reproductive crossing and line breeding, a halfbred horse was produced which is suitable for riding, racing, and work. The East Bulgarian was officially recognized in 1951.

Looking back through the ages at Bulgar-ian history, it is clear that the horse has long played an important role in the life of the people, who first came to the Balkan penin-sula on horseback 1,300 years ago. Their lives have always been related to use of the horse, especially as a means of transportation and in battles during the almost endless wars and migrations. Horses were a source of meat and milk and later of draft power for farming. Bulgarians have traditionally loved and re-spected the horse, and no man was treated as a true Bulgarian if he was unable to ride. As a token of their special attitude, the Bulgarians carried a horse tail as their flag for ages.

Horse breeding was especially important in the period of stabilization and development of the Bulgarian state before it fell under Turkish rule in 1396. For a period of five centuries, horse breeding in Bulgaria suffered a serious decline because the people were not allowed to keep good horses.

After liberation following the Russo-Turk-ish War (1877–78), systematic work for im-provement of horse breeding was resumed. In a rather short time the first horse studs and stallion farms were organized at the end of the nineteenth and beginning of the twentieth centuries, using imported stallions and mares of various breeds. These became the foundation for development of modern horse breeding in the country. Accompanying the state horse studs, a special importance should be attributed to the Union of Horse Breeding Societies (1935–47) and the organizational structures implemented after that period.

Improvement of horse breeding gradually extended to all parts of the country, with mares of local origin being crossed with stallions of different breeds. Due to the creative efforts of several generations of breeders and specialists and the active participation of Bulgarian scientists, breeds with extremely valuable characteristics have been produced, such as the East Bulgarian, Pleven, and the Dunav (Danubian) horse. At the same time Arab horses were bred, and since about 1955 the Thoroughbred, Trakehner, Hanoverian, and Haflinger have been introduced. Small populations of draft and mountain horses have also been produced. The most valuable breeding animals are kept at the state studs, stallion farms, and a number of breeding farms around the country where breeding and improvement are carried out under the control of the State Corporation of Cattle and Sheep.

Horse breeding in Bulgaria improved further after World War II, reflected in development of the economic and competitive qualities of the basic breeds, expansion and enrichment of the gene pool, and some positive results in international competition. Presently there are 120,000 horses in Bulgaria to satisfy the draft, competition, and export needs of the country.

The modern East Bulgarian horse is characterized by good conformation, robust constitution, a nice exterior, and free movements. Predominant colors are chestnut, bay, or black. The head is light and elegant with a straight profile and large, lively eyes. The neck is quite long and well formed; withers are prominent; the back long and straight; the croup slightly sloping. The chest is full and deep; the shoulder is sloped and muscular; tendons are well defined and clean; legs slender and solidly built; and the hooves are well shaped and of tough horn. The breed has a well developed genealogical structure of family lines which are subject to constant improvement.

Due to the variable qualities of this breed it became widely popular throughout the country, but mainly in East Bulgaria. Statistical data for 1987 showed the East Bulgarian representing 39 percent of the total purebred horses and 42.6 percent of improved horses in the country.

During the last three decades a significant improvement of the economic properties of this horse has been achieved, mainly in competition. This has been seen especially in the success of the East Bulgarian in the difficult steeplechase competitions such as at Pardubice, Czechoslovakia, with twenty-eight first, thirty-three second, and twenty-nine third places.

East Bulgarian horses have demonstrated significant success in the classical horse disciplines in recent years. Especially successful have been the horses Bakara, Blinhuli, Likvidator, Talk, Euphoria, Gloria, Eleonora, Ferry, and Fall.

East Friesian (Old type)
(Ostfriesisches)

ORIGIN: *Germany (East Friesland)*
APTITUDES: *Carriage horse, light draft*
AVERAGE HEIGHT: *15.2 to 16.2 h.h.*
POPULATION STATUS: *Rare*

The old type of East Friesian horse is nearly extinct, with approximately fifty pure specimens alive in western Germany today. The number of living individuals of this breed in what was East Germany is unknown to the writer, but even there the breed is rare.

Although some conflicting information has come in, documents received of the old Ostfriesisches Stutbook (East Friesian Studbook) in Norden dated 1962 indicate that the East Friesian breed *was* based in part on the black Friesian horse of Holland.

East Friesland is bounded on the north by the North Sea: westward by the Kingdom of the Netherlands; to the south by the bailiwick Osnabruck; and to the east by the Grandduchy of Oldenburg. Forming a part of the north German plain, this land is protected against daily tidal flow by embankments surrounding the coast and by the dunes of the Friesian chain of islands, which first break the north waves of the sea. The soil of the East Friesian mainland consists of marshlands, sand, and moors.

Horse breeding in 1900 formed an essential branch of East Friesian agriculture. The object was to produce a strong, elegant, and easily managed carriage horse which, when suitably fed, would develop early. Such horses were ready for light work at the age of three.

The document *Deutsche Pferde*, published in Paris in 1900, records that the East Friesian breed descended from "that Friesian race, already well known in the Middle Ages," but that in ensuing ages the breed had been developed according to demands of the time. The blood of highly bred horses was introduced and changed the original heavy East Friesian to suit it better for carriage work. Breeding was conducted exclusively by farmers, who also furnished the necessary stallions. Selection of stallions for breeding was established in 1715, and then only approved stallions were allowed to be used for breeding purposes.

In 1898 six royal stallions and sixty-three sires owned by private breeders were kept for breeding purposes. Mares used for breeding numbered 5,624. In 1900 the extent of horse breeding in eastern Friesland was proportionately more extensive than in any other part of the empire.

The East Friesian stud book was founded in 1893, and first entries included the stallions of the royal country stud at Celle. Horses from Celle had been used for breeding since 1850.

In a letter, I. A. Fegter, director in 1962 of breeding for the Ostfriesisches Stutbuch, said that up to the beginning of the twentieth century the receipts from selling of horses made up the main source of income from agriculture in East Friesland. Breeding methods were determined by wishes of the buyers, representing every country in western Europe. For a number of years before and after the turn of the century, the United States government purchased from 100 to 150 East Friesian stallions each year from the stallion market in Aurich, the only place in western Europe where a central exchange was conducted. From 300 to 400 young stallions were exhibited each year at the peak years of the breed. This connection was broken off in 1911 due to a change of import regulations in the United States.

A large number of East Friesian horses were also exported to England during the nineteenth century. They were the preferred breed of undertakers, harnessed to hearses and crowned with black ostrich plumes to make them appear taller.

With the development of farming technology and mechanization, the need for draft horses decreased and East Friesland suffered a decline in demand along with other areas. Necessary transition of the breed from a pure draft animal to an all-purpose horse began in 1945. Pure Arab horses were among those used for crossing to the East Friesian for refinement.

An excellent carriage horse in its time, the East Friesian was an outstanding representative of the German Coach horse for many years, with global appeal. Most of the population of old type East Friesian horses was absorbed into the modern Oldenburg breed.

East Friesian Warmblood (Modern type)
(Ostfriesisches Warmblut)

ORIGIN: *Germany (East Friesland)*
APTITUDES: *Riding, sport horse*
AVERAGE HEIGHT: *15.2 to 16.2 h.h.*
POPULATION STATUS: *Common*

While the East Friesian originated in East Friesland in western Germany, with the division of Germany its breeding was carried out mostly in the German Democratic Republic.

The modern East Friesian bears little resemblance to its immediate ancestor, the old East Friesian breed, so popular around the turn of the century as a coach horse. The market for this old breed was spread internationally. In 1962 the white horses for the state carriage of the Queen of England originated from the East Friesian breed.

The modern East Friesian is closely related to the Oldenburg (as was the old type). The

The East Friesian horse (modern type). Photo: Sally Anne Thompson

Mecklenburg stallion. Photo: Prof. H. C. Schwark

newer breed was founded upon the most cosmopolitan stock imaginable—English, Irish, Polish, Hungarian, Spanish, Turkish, and other blood all being thrown, as it were, into the melting pot in the effort to produce a modern horse to meet the growing demands for sport. Cleveland Bay and Norman stallions were also introduced with great success. After World War II Arab stallions were used with satisfactory results, and most recently the Hanoverian has exerted the greatest influence on the breed.

The East Friesian is bay, brown, black, grey, or chestnut. The head is well shaped with a straight profile, lively eyes, and flared nostrils; the neck is long and slightly arched; the withers are usually prominent; the back is straight and long; the croup gently sloped; the chest deep and wide; the shoulder long and well sloped; and the legs have clean, broad joints with good bone. The thighs are muscular, as are the forearms.

Elegant Warmblood

(Edles Warmblut)

ORIGIN: *Germany (Brandenburg)*
APTITUDES: *Riding, sport horse*
AVERAGE HEIGHT: *15.3 to 16 h.h.*
POPULATION STATUS: *Common*

Horse breeding in Mark Brandenburg has a long tradition. Earliest activities concerning the breeding of horses can be traced to 1694, when the Duke of Hessen-Homburg founded a stud in the area. For economic reasons, the business of horse breeding changed to that of mule breeding in 1718.

The Prussian king, Friedrich Wilhelm II, ordered the construction of the Nuestadt stud, which was completed in 1788. Still remaining is a bronze plaque acknowledging foundation of the stud: "Friedrich Wilhelm II built this stud for the good of the country in the year 1788."

The first director of the stud was Graf von Lindenau. His mandate was to produce horses for military use, cavalry horses for the Prussian army, horses for various uses for the king's stables, and stallions for use on native mares belonging to the farmers. From 1788 to 1833, primarily Thoroughbred and Anglo-Arab mares were bred to Arab and Turkmenian stallions, of which Turkmen Atti (see Akhal-Teke) was an important one, influencing the breeding to a great extent. From this time the stud brand was introduced—an arrow (for high speed) with a wiggling snake (for agility). After 1833 a strong Thoroughbred type was bred and from 1843, when training tracks were built, races were held under difficult conditions to test the horses.

In the following years the farmers wanted a strong horse for farm work. As this wish could not be met immediately, the stud was dissolved and another nearby was devoted to producing heavier horses with the use of stallions from the Saxon stud at Döhlen. With the import of coldblood stallions from Denmark a period of crossbreeding began, but it resulted in a poor horse of little use to anyone. From 1895 this crossbreeding was stopped and once again the aim was to produce good, hardy horses for military use. However, the farmers still needed a strong warmblood horse for work in the fields. The use of Trakehner stallions and some from the Hanover stud soon produced what the farmers needed. Until 1945 the Neustadt stud bred both military and warmblood horses.

The horse breeding branch of Neustadt stud became specialized. In 1970 the agricultural branch, which previously was primarily for milk production and cattle raising, was separated. Although the elegant horses were bred to a high degree, selection was mainly for the work horse and for producing stallions for its breeding. The 1960s saw a change in the goals of breeding, with the stud trying to meet demand for a modern sport horse. Due to the high percentage of blood horses in the mare herd and use of modern-type stallions, the new goal was soon achieved. The main aim today is to keep a good herd of broodmares.

There are four warmblood breeding studs in eastern Germany, responsible for breeding and keeping excellent mare stock: Neustadt/Dosse, in the region of Potsdam; Ganschow, near Schwerin; Radegast, near Halle; and Zöthen, near Gera. Breeding objectives for the Elegant Warmblood are to produce a versatile, handy sporting horse. In all, a population of 450 breeding mares is kept.

Besides the warmbloods, these studs also keep coldbloods that are used for crossbreeding. Warmblood breeding is constantly upgraded through the use of Thoroughbred, Trakehner, and Hanoverian blood.

After World War II there were four warmblood breeds in East Germany: the Mecklenburg, bred on Hanoverian bloodlines; the Brandenburg, bred on Hanoverian and Trakehner bloodlines; and the Sächsische and Thüringische, bred on Oldenburg and East Friesian bloodlines. All of these breeds were modernized in order to produce the desired sporting horse.

The Mecklenburg and Brandenburg warmbloods already represented the modern-type riding and sporting horse, while the Sächsische and Thüringische (Oldenburg–East Friesian) represented the heavy warmblood of old type. The Thoroughbred and Trakehner warmbloods in the German Democratic Republic studs were used systematically to upgrade the old-type horses.

The Brandenburg, from a historical viewpoint, was a subpopulation of the saddle horse in East Germany, a blend of warmbloods rather than a particular breed.

Presently there are about 100 broodmares at Neustadt stud. Through stud breeding as well as purchases, about forty young stallions are raised and trained each year. From 1961 through 1986, 134 young stallions—99 of them from the production of Neustadt mares—were officially approved and sold to the stallion depots. In addition, breeding stallions have been sold to Russia, Czechoslovakia, Bulgaria, and Hungary.

The stud is constantly working to improve its produce. The mare families are genetically stable, and only the best stallions are used for breeding. Selection is based on producing horses with nationally and internationally desired riding horse characteristics such as elegant heads, long necks, roomy gaits, elastic movements, and high performance.

In 1987 about 44,000 horses (40 percent of the total GDR horse population) were registered at Neustadt. There are approximately 6,200 registered mares, of which 3,163 are of the Elegant Warmblood type. Of these, 1,148 are kept at eighty-nine different government-approved horse breeding studs. In five other studs, coldbloods and Haflinger horses are bred.

The Elegant Warmblood has been developed from the best warmblood stock available in East Germany, a horse of elegance, useful for sporting or hobby riding. The gaits are balanced and smooth. Horses offered for sale at Neustadt are tested for their usefulness for show jumping, dressage, cross-country, hunting, or hobby.

Carriage drivers are able to find trained carriage horses at Neustadt which are either Elegant Warmbloods or heavy warmbloods bred on Oldenburg–East Friesian bloodlines. The heavy warmblood is kept primarily in the southern areas, where it is used for work in the forests and on farms as well as for carriage driving. The pure population is about 400 broodmares and fifteen approved stallions. The aim is to breed horses with no more than 12.5 percent of Thoroughbred blood. During recent years there has been an increasing demand for this type of horse for work, primarily in the forests.

Eleia

(Andravida breed)

ORIGIN: *Greece (Andravida)*
APTITUDES: *Riding horse, draft and farm work*
AVERAGE HEIGHT: *15 to 16 h.h.*
POPULATION STATUS: *Uncommon*

The Eleia (Andravida) horse. Photo courtesy A. M. Zafracas, Aristotle University, Thessaloniki

Developed in the plain of Eleia, especially in the regions of Andravida, West Peloponnesus, this breed, according to information obtained from the University at Thessaloniki, was formed by crossing native mares of the plain type to Anglo-Norman stallions at the beginning of the century. After 1920, crossbred mares were mated to Nonius stallions.

The main colors of the Eleia breed are brown, bay, chestnut, red roan, and, rarely, grey.

The head has a rectangular shape with a straight profile and rather long ears. The chest is broad; the back slightly dipped; the shoulder well sloped; the croup gently sloped. The legs of this horse are free of excess hair and very strong with good bone.

English Cob

ORIGIN: *England*
APTITUDES: *Riding horse*
AVERAGE HEIGHT: *14.2 to 15.1 h.h.*
POPULATION STATUS: *Common*

English Cob. Photo: Bob Langrish

The Cob is not a breed but a well established type. Generally a Cob can be described as compact, muscular ("large bodied, short legged and stuffy" was the way one person put it), with plenty of bone. The term came into use about the eighteenth century, although these horses were used long ago in feudal times by squires in battle. Estate stewards rode them in peace time. They were commonly known as "rouncies."

Aside from the well-known Welsh and French Cobs, they are not a breed and are often produced by chance. Breeding stock producing the Cob is widely varied, but a good example would be a Highland and a small Thoroughbred or perhaps a Suffolk and an Arab.

The Cob has pony-like qualities. The head is usually small, refined, and well set on an elegant neck which is arched; the back is short with great girth; the tail is carried high; the horse is balanced and bold but with a gentle temperament. The shoulder is sloped and the quarters rounded. A requirement of a good Cob is that the temperament be tractable and well mannered. Cobs make good mounts for older people or novices and are also used to pony race horses to the track.

Classes are held through the British Show Hack, Cob and Riding Horse Association in England, and Cobs must be registered as either heavyweight or lightweight. They may be shown only in the class for which they are registered.

English Hack

ORIGIN: *England*
APTITUDES: *Riding horse*
AVERAGE HEIGHT: *Not to exceed 15.3 h.h.*
POPULATION STATUS: *Common*

English Hack. Photo: Bob Langrish

A Hack is not an established breed, but it is a recognized type. "Hack" is a term that has been used to describe a variety of horses over the years—a horse for hire; a horse ridden to the meets before changing to the hunter; and also a refined, elegant horse ridden by those considered fashionable.

The ideal breeding for a show Hack is Thoroughbred crossed to Arab or pony blood to keep the height down. The Hack must be a pleasure to ride, having excellent, impeccable manners and good balance and being light in response to the rein. The ideal Hack of today is a well-trained animal and very easy to ride. Hacks are supposed to be horses that give pleasure to their riders, who enjoy "hacking" on a good-looking, well-mannered horse.

Hacks are judged on conformation, action, response to "aids," manners, and overall presence. The head should have great quality and elegance, the latter stemming from a well set on head and neck combined with a good length of shoulder. The movement should be smooth and graceful with a true pointing of the toe. To achieve this, a Hack must be extremely well schooled.

Classes for Hacks are held by the British Show Hack, Cob and Riding Horse Association.

English Hunter

ORIGIN: *England*
APTITUDES: *Sport, riding horse*
AVERAGE HEIGHT: *14.2 to 17 h.h.*
POPULATION STATUS: *Common*

English Hunter. Photo: Carol Gilson

The English Hunter is a type and not a breed and therefore varies considerably depending on its use. Basically it should be a horse capable of carrying a considerable amount of weight across country, jumping natural obstacles such as hedges and ditches while proceeding at a good pace. Old-fashioned hunters were usually bred from mares that worked on farms, using Premium Stallions selected by the National Light Horse Breeding Society.

At a meeting of the Hackney Horse Society's council on November 5, 1884, it was decided to include a class for Thoroughbred stallions suitable for use as hunter sires. A committee of management was formed to administer the new venture.

On March 4, 1885, at a general meeting of the Hackney Horse Society, it was proposed that a new organization be formed for improving the breed of hunters. The new body was called the Hunter's Improvement Society; some years later its name was changed to National Light Horse Breeding Society.

At the turn of the century the government encouraged the breeding of horses in England, mainly due to the vast numbers that had to be imported, primarily for use by the army. During the years 1873 to 1882, 197,000 horses were imported. In 1894 premiums were offered by the Ministry of Agriculture and the War Office and twenty-eight stallions received these awards. The army, through its remount officers, continued to play a major part in the society's activities, and financial backing came from the same source until the outbreak of World War II. Backing for the society was then taken up by the Totalisator Board. Today it comes from the Horserace Betting Levy Board, without whose generous support the society would not be able to operate its stallion and mare premium schemes at their present level.

With the army now almost completely mechanized, emphasis has changed. Breeders are required to be much more selective. With no market for the "misfit," it is a question of quality and not quantity. The middleweight horse with good quality and movement, now referred to as the Competition horse, is what breeders are striving for, and the market for such horses is very strong. While continuing with the established policy of making strong, sound Thoroughbred stallions available for breeding, the society has also registered a number of stallions of other breeds such as Cleveland Bay, Irish Draft, and Welsh Cob to maintain bone and substance in the British horse. The result is production of some horses for uses other than hunting or eventing. With more leisure time available people are using horses increasingly for sport, and the dressage animal and show jumper enjoy a steady demand.

The National Light Horse Breeding Society concentrates great attention on registration and grading of mares. Inspections are held annually at centers throughout the country where mares are graded and examined for hereditary diseases and defects. Those attaining the highest grade are awarded an acceptance premium and details of their progeny are recorded. Eventually, specific lines will be established for breeding horses suitable for various or specific types of competition.

English Thoroughbred

ORIGIN: *England*
APTITUDES: *Riding, flat and jump racing*
AVERAGE HEIGHT: *15 to 17 h.h.*
POPULATION STATUS: *Common*

English Thoroughbred. Photo courtesy UNIC, Paris, France

Thoroughbreds in cross-country racing in France. Photo courtesy UNIC, Paris, France

Of the many equine breeds in existence, only a few have spread to the far corners of the earth in pure form and have dazzled and quickened hearts as has the Thoroughbred. While racing has existed since the horse was first ridden, the sport became a science when the Thoroughbred was developed in England early in the eighteenth century.

It is customary to trace origin of the Thoroughbred to the importation into Britain of the three famous foundation sires; the Darley Arabian, Byerley Turk, and Godolphin Barb. However, development of this great running horse actually began much earlier with the importation from Spain, Italy, Africa, and Turkey of great numbers of Iberian, Barb, and Turkmen or Turkmenian (commonly called Turk) horses over a period of 665 years. The "royal mares," usually thought to be the forty Spanish mares bred to the three foundation stallions to develop the Thoroughbred, were but a small number of the Iberian and Oriental horses on British soil at the time.

Beginning with King Stephen, who reigned from 1135 to 1154 and had imported "royal mares and stallions" in his stables, until the Thoroughbred was well developed, horses used in the royal stables of England were always referred to in manuscripts, records, and books as "hot blooded" or "imported." The records show heavy importations from Spain of the famous Jennets (a type of Andalusian), Andalusians, Barbs from the north coast of Africa, and "Turks" from Turkey and Syria; Arab horses were difficult to obtain due to political problems. Often overlooked by writers is the fact that Arabs have always been reluctant to sell their good horses, especially their good mares. Not only were these

Oriental horses bred at the royal stables, but common people had access to the blood as well, so that England became a rich breeding ground of imported hotbloods. As a result, several excellent breeds evolved there.

As is often the case when a great breed emerges, controversy has existed since formation of the breed regarding the exact bloodlines of the foundation animals. It is hotly contended by many passionate admirers that the three famous foundation sires were meticulously pure Arabs. However, due to the extreme difficulty of obtaining good Arab horses, one should be realistic. In fact, the term "Arab" was greatly overused or used loosely much of the time. Syria, Turkey, Persia, Iraq, and parts of North Africa were all considered "Arab" countries in those days, as great numbers of Arab peoples inhabited those areas.

The Darley Arabian, foaled about 1700, appears to have been accurately named. He was brought to England directly from the Syrian horse market at Aleppo in 1704 and was certified to be of the best Maneghi blood, a strain used for racing. However, Maneghi blood is known to have always carried many Turkmenian crosses. He was given by Thomas Darley to his brother. Although the Darley Arabian's services were restricted with outside mares, he stood at service in Yorkshire, and one particular mare was to have great significance in development of the Thoroughbred. That mare was Betty Leedes, whose dam was of the Leedes Arabian line, progeny of another important stallion of the day. One of the foals resulting from the mating of Betty Leedes and the Darley Arabian called Flying Childers became the first outstanding race horse. His full brother, Bartlett's Childers, unable to race himself because of a tendency to break blood vessels, stood at stud where he sired among others, Squirt, who in turn sired Marske. Marske produced one of the greatest Thoroughbreds of all time, Eclipse.

The Byerley Turk, imported to England in 1689, is believed by many historians indeed to have been a "Turk" or Turkmene horse (hence his name), likely an Akhal-Teke, directly descended from the ancient Turkmenian breed, and not an Arab. He is said to have been captured from the Turks at the Siege of Vienna in 1689 by a Dutch officer who then sold him to Byerley. Another account claims the horse was captured by Captain Byerley at Buda in the 1680s. Some say the captain rode him at the Battle of the Boyne. In any event he was later sent to England to stand at stud, first in County Durham and later in Yorkshire. Although the number of well-bred mares sent to him was low, he got one horse, Jigg, good enough to establish a male line that can be traced down to important twentieth-century stallions such as the Tetrarch, a stallion never defeated on the race track. The Byerley Turk's great-grandson, Tartar, sired the top horse Herod, whose progeny won over 1,000 races and established him as one of the most important sires in Thoroughbred history.

It is generally agreed that the Godolphin Barb, often referred to as Godolphin Arabian, was in fact a Barb horse from Morocco. Brought into the stud of Lord Godolphin about 1730, he was first used only as a teaser. His real worth was not realized until, mated to the mare Roxana, he produced Cade, who founded the Matchem line. He lived to thirty years and although he did not cover vast numbers of mares, he too had a profound influence on the Thoroughbred breed through his grandson, Matchem.

The controversy regarding Arab, Barb, or Turk will rage on, and perhaps I have only thrown more fuel on the fire—but there is no doubt the Thoroughbred carries the blood of all three, and probably a good deal of other "native" British and *other* blood as well. The Galloway, always mentioned as a foundation of the Thoroughbred, was not a native to Britain. The English Thoroughbred owes no explanations or excuses for its greatness. The breed stands completely alone in the world on its own achievements as a race horse, unmatched by any other breed at any other time in history.

Three of the modern Thoroughbreds' four great tail-male lines—Herod, Eclipse, and Matchem—became thus established. The fourth was that of a horse named Highflyer, a son of Herod, bred by one of the most influential racing men of the day, Sir Charles Bunbury.

As important as the main foundation sires are in the pedigrees of registered Thoroughbreds, one should be cautious in imagining that no other stallions had a profound influence on the breed. Horses such as the Unknown Arabian, Darcy's Chestnut Arabian, the Helmsley Turk, and the Lister Turk all played important parts in the early days. Lady Wentworth, author and breeder of Arab horses, calculated that the pedigree of Bahram, winner of the English Triple Crown in 1935, contained an incredible number of crosses to each of these horses. It is largely because we can trace *direct* tail-male lines back to the "big three" that they are considered traditionally the Thoroughbred's foundation sires.

No breed of horse can run quite as fast, quite as far as the Thoroughbred. Some can run faster for a short distance, and some can gallop further, but none can match the English blood horse in both speed and distance. The breed is famous for its great "heart," and many gallant Thoroughbreds have attempted to finish a race even with a fractured leg. The Thoroughbred is well known as a horse of great fire, endurance, and agility.

There was a time when Thoroughbred horses bred in the United States were not allowed admission "in the book" unless they could be traced without flaw on both sire's and dam's side to stallions and mares already accepted in earlier volumes of the General Stud Book. This legislation came about after closure of the U.S. race tracks at the beginning of the twentieth century due to widespread corruption in the "Sport of Kings." Many breeders and trainers in the United States set their sights on Great Britain, and the English racing authorities became concerned that too many American horses of what they considered doubtful pedigree would be introduced. Of course, there were many American-bred horses that failed to meet the requirements of the Jersey Act and were deemed "halfbreds." The Jersey Act was not repealed until 1949, by which time a large number of the top races had been won by the so-called halfbreds, to the great chagrin of British breeders. The horse Tourbillon, whose American antecedents had been accepted into the French Stud Book before that country's version of the Jersey Act was put into effect, was just such a "halfbred" and became a very influential stallion.

While the British wrote of "purity of blood," respected American journalists wrote of the fact that a bit of peasant blood in royal families had been highly effective now and then and further contended that the ruling might even be a good thing; Americans could import Sir Gallahad III, Blenheim II, St. Germans, and Pharamond, but the British could not come and get Equipoise, War Admiral, Seabiscuit, or Black Toney. The best American blood could not get away, yet Americans could avail themselves of the best English strains. One journalist, addressing what he considered English arrogance in regard to purity, wrote that when the English Thoroughbred was being developed, the "purebreds" were not there but plenty of cart horses *were*. The Jersey Act was rescinded in June of 1949, and from that time admission requirements into the General Stud Book read: "Any animal claiming admission from now on must be able to prove satisfactorily some eight or nine crosses of pure blood, to trace back for at least a century, and to show such performances of its immediate family on the turf as to warrant the belief in the purity of its blood."

The English Thoroughbred does not exhibit a standard morphology because it is a composite breed, and in fact some say that without heedful selection and new blood, the breed tends to degenerate. Three basic types are found in the breed: the sprinter, tall with a long body, very fast; the stayer, a smaller horse with shorter body length and good stamina; and the middle distance horse, with well-sloped shoulder, shorter back, and sloping croup, well suited for cross-country events.

Coat colors in Thoroughbreds may be bay, dark bay, chestnut, black, or grey; roans are seen only rarely. White markings are frequently seen on both the face and legs. The head is small and elegant with a straight profile, with well-proportioned ears which are mobile and active; the eye is large and lively; nostrils are flared. The neck is usually quite long and straight but may be slightly arched; the withers are prominent; the back is usually long; the loins are well attached to the croup, which may be quite sloping; the tail is set on high; the chest is high and wide in the sprinter and tends to be deep in the stayer; the shoulder is very well sloped and muscular; the legs are long and have large, clean joints; the forearm is long and muscular; the cannon bone is usually quite slim; pasterns are long and sloped; the hooves are small. The skin is very thin. In general, everything about the appearance of the Thoroughbred is indicative of speed.

The English Thoroughbred has been used to improve or to help develop many other breeds of the world, adding refinement, endurance, speed, and heart.

Eriskay Pony

ORIGIN: *Scotland (Eriskay Island)*
APTITUDES: *Riding, light draft*
AVERAGE HEIGHT: *12 to 13 h.h.*
POPULATION STATUS: *Rare*

The Eriskay pony is one of the last pure surviving native ponies of the Western Isles of Scotland known as the Hebrides. This pony is of ancient origin with Celtic and Norse connections. Related to other northern breeds such as those on the Faeroes and Iceland, the

A rare Eriskay pony. Photo: Alex McIntosh

Aged Eriskay mares with their foals. Photo: Alex McIntosh

ponies of the Hebrides may have been as much the prototype of these ponies as the other way around.

In previous centuries this pony was quite numerous, but numbers were reduced rapidly around the nineteenth century due to increased crossbreeding. Breeding was directed at producing larger, heavier ponies for field work. Heavy horses from the mainland were used to produce what became known as "the cross," while a variety of horses including the Norwegian Fjord were used to produce the present-day Highland pony from smaller island ponies.

Eriskay is the most southerly island of the Outer Hebrides now inhabited. South of it is the now-deserted Vatersay. The nearest island to the west is Newfoundland, in Canada. Eriskay measures about two and a half miles long by one and a half miles wide. The human population is about two hundred strong but, almost uniquely in the Western Isles, there are thirty children in the island school and more boarded out across the water. All of them have ponies to ride if they want to. Almost every family is Gaelic speaking and Roman Catholic. In these two respects, Eriskay resembles Connemara on a tiny scale. Additionally, just as the seaboard district of County Galway afforded the last refuge for what was once the unique and original Irish horse that we now call a Connemara pony, so Eriskay was the last place where people knowingly, deliberately, and lovingly went on breeding what we

have good reason to believe was once the unique and original Scottish horse.

From its inception in the late nineteenth century, the Highland Pony Society recognized two different types of pony native to the whole Highlands/Western Isles region, a mainland type and an island type. From the outset it refused any sort of stud book classification such as the Welsh Pony and Cob Society set up. This controversy raged from the beginning, in terms of stud book and show class, until the late 1950s and thereafter with diminishing fury until 1972. Some Highland fans maintained, not entirely in jest, that there were as many breeds of pony in the Hebrides as there were islands. Some deplore this "fact," while others rejoice in it. Of course it is an exaggeration, but certainly at one time or another many of these islands have had a distinctive breed; for instance Barra, slightly larger than Eriskay and lying between it and South Uist, was famed in the eighteenth century for its ponies, which were smaller than Shetlands. They vanished without a trace sometime in the reign of Queen Victoria.

In February of 1972 an association was formed to snatch the Eriskay ponies from the brink of the abyss. Its nucleus included the priest on Barra, whose parish included Eriskay; the local physician, also resident on Barra; four veterinary surgeons, one of them then resident on Tiree and a member of the Crofter's Commission; the principal of Glasgow Veterinary School and his wife, who also taught in the school; and the official Department of Agriculture veterinarian for the area. The immediate crisis that faced them was the death of the only purebred stallion on the island. This was resolved by the fortunate discovery of a purebred stallion on South Uist, two islands to the north of Eriskay. More serious was the shortage of known purebred broodmares; there were twenty, including five on Barra, two on Tiree. There have ensued twenty years of almost constant lobbying—of the Department of Agriculture, of the Crofter's Commission, of the National Pony Society—on the one hand, and on the other of careful search through the islands and beyond for authentic foundation stock to be added to the registry.

The association defiantly took as its official title in Gaelic Comman Each Nan Eilean (Island Horse Society) and only secondarily Eriskay Pony Society, as a reluctant concession to foreigners who will want to know *which* island.

The commonest color among these ponies is variable grey, never chestnut, piebald, skewbald, or with white marks on head or feet. The average height is 12.2 hands high. None have been registered above 13 hands high or below 11.3 hands high. The conformation in general is very similar to that of the Exmoor except that the Eriskay is shorter in the back. It is noted for its equable temperament and suitability for children's mounts and also for its usefulness on small agricultural holdings such as the Highland crofts.

Eriskay ponies are extremely capable and willing to work and make good riding animals, moving more like a horse than a pony. Foals are born dark, brown or black, and turn in degrees as they mature to grey. In winter the ponies have a dense coat that is notably waterproof, and the forelock, mane, and tail are substantial, indicating their importance as a survival strategy in a fierce, hostile climate.

The action of Eriskays is comfortable, making them particularly suitable as mounts for children. Their temperament is very good and they are easy to train. Recently this pony has had spectacular success in driving competitions, demonstrating the great rewards it offers to people with suitable experience. Typical comments are that the Eriskay will train in as short a time as or less time than any other breed.

An Erlunchun stallion of China. Photo: Weigi Feng and Youchun Chen

Erlunchun broodmare. Photo: Weigi Feng and Youchun Chen

Erlunchun

ORIGIN: *China (Xingan mountain area)*
APTITUDES: *Pack horse*
AVERAGE HEIGHT: *12 to 12.3 h.h.*
POPULATION STATUS: *Common*

Dating to the middle of the seventeenth century, the Erlunchun is a special type of Mongolian horse. The breed was developed by crossing different Mongolian strains such as the Helongjiang and Soulun. The area of development is in northeastern China. In earlier times this horse was used for military action and later for hunting. The Erlunchun is a native breed of China.

This is an excellent hunting horse, sure-footed and enduring in the mountains, sensible and steady in temperament. This horse can travel in rough terrain about four and a half to five miles per hour and about forty-seven miles a day with a pack weighing upward of 400 pounds. This is amazing considering the size of the animal. The Erlunchun also has remarkable staying power. It seems almost unaffected by hard work and is able to cover close to twenty-five miles every day for a month before needing a short period of rest.

When snowbound in its severe habitat, the

Erlunchun can stand for days without feed, often fighting off wolves or other wild animals in temperatures of from −40° to −58°F. This incredible little horse survives the most severe conditions and usually seems little affected by its hardships. Predators inhabiting the range of the Erlunchun include bears and tigers, in addition to large wolves.

Most Erlunchun horses are grey or bay. The body is very compact and strong, built close to the ground. The head is heavy but tapering, wide between the eyes with small, erect ears; the neck is short, thick, and heavily muscled; the withers are short but pronounced; the back is long, straight, and strong; the croup is sloping. The shoulders are sloped and muscular; hind quarters are strong and well muscled. The legs are short with good bone and joints and very strong, and the hooves are well shaped and extremely hard.

In winter this horse grows a heavy, protective body coat and the mane may grow to below the point of the shoulder.

The Erlunchun is raised in polygamous stud groups. The herd instinct is strong, and stallions are excellent protectors of their group. The mares are very defensive of their foals. During severe winters hunters may use wild boar meat to feed their horses.

Estonian Draft

(Estonskiĭ tyazhelovoz)

ORIGIN: *Estonia (former Soviet Union)*
APTITUDES: *Heavy draft work*
AVERAGE HEIGHT: *15.2 to 15.3 h.h.*
POPULATION STATUS: *Rare*

The Estonian Draft horse (also called Eston-Arden) was developed by suppressive crossing of Estonian Native horses with pure bred Swedish Ardennes stallions. The goal was to develop a tough working horse, and greatest attention was paid in selection toward leanness, good conversion of food, calm temperament, and quick movement. Recently bred Estonian Draft horses are bony, of harness type. The breeding work is done with eight lines.

Eston-Arden horses are widely used in various agricultural work and are very much in demand although numbers are low. They are bred at forty farms, but the basal nucleus of the breed is concentrated at twelve of them.

Estonian Native

(Mestnaya estonskaya)

ORIGIN: *Estonia (former Soviet Union)*
APTITUDES: *Riding, draft*
AVERAGE HEIGHT: *13.1 to 14.3 h.h.*
POPULATION STATUS: *Rare*

Estonian Native. Photo: Nikiphorov Veniamin Maksimovich

A group of Estonian Native horses. Photo: Nikiphorov Veniamin Maksimovich

An ancient breed, the Estonian Native is one of the few breeds in the former Soviet Union that has resisted all effects of crossbreeding, retaining the characteristic features of the native northern horse. This breed played an important role in formation of the (now extinct) Obva as well as the Vyatka and Tori or Toric breeds.

This horse first penetrated Russia via Novgorod as early as the fourteenth and fifteenth centuries, due to its working abilities and high adaptability. The Estonian was widespread throughout Estonia until the nineteenth century. As agriculture was developed and demand for large working horses increased, many of this native breed were crossed with larger breeds, though they were also bred pure. Pure members of the Estonian breed were increasingly displaced to the Baltic seashore and neighboring islands. Reliable information on the improvement stages of the Estonian Native dates back to the origins of the Tori stud in 1856. This stud was engaged in breeding native horses and crossing with light harness and saddle breeds. The best crossbred mares were subsequently used to develop the Tori. The first pure breeding stage yielded good results.

Horses from the mainland were slightly larger than those from the islands. As agriculture became more intensive and roads for transportation needs were developed the Estonian Native lost ground to the new breed (Tori), and it is no longer used in mainland Estonia, surviving only on the islands of Saarema Hiyumaa and Mukhu. The total purebred herd numbers about 1,000 head.

The modern Estonian Native is not large in size. The head is well shaped and has good proportions with a wide forehead but is occasionally coarse. The neck is on the short side or medium in length and fleshy; the withers are low and wide. The loin is well muscled; the croup is of average length and has normal slope; the chest is very wide and deep. The legs are short, properly set, and distinguished by firmness and cleanness. The hooves of the Estonian Native are extremely hard.

This horse is undemanding and well able to forage for itself and thrive. It has extraordinary endurance and the action is very good. It is slow to mature, but young horses are often able to work at the age of three years. In addition, the breed has a willing disposition. Predominant colors are chestnut, bay, light bay, dun, and grey. Grullo and roan coloration are rare. Commonly, the Estonian Native has a dark dorsal stripe.

Inbreeding of the Estonian Native has become widespread due to the limited numbers of breeding stock. However, almost no in-

breeding depression has been noted regarding work performance, measurements, or conformation, although closely inbred individuals do seem to take longer to mature. The Estonian Native is long-lived and healthy. The mare Tenki, foaled in 1946, was still alive in 1983 at Syrve state farm in Kingisepp region.

Presently the local Estonian is used for light agricultural work, as a saddle horse for children, and in tourism. It represents good breeding material for the production of ponies of various crosses. It is also being used to cross with the disappearing Ob (Priob) breed.

Exmoor

ORIGIN: *Great Britain (Devon and Somerset)*
APTITUDES: *Riding pony, light draft*
AVERAGE HEIGHT: *11.1 to 12.3 h.h.*
POPULATION STATUS: *Rare*

Exmoor pony. Photo: Sally Anne Thompson

Exmoor pony. Photo courtesy Exmoor Pony Society

A breed of great antiquity and purity, the Exmoor of the distant past was essentially identical to the Exmoor of today. As the oldest of the pure native breeds of Great Britain, the Exmoor has attracted the interest of many scientists, resulting in a great deal of valuable research into development of the breed as well as into evolution of British horses in general. Early in the twentieth century a good deal of research was carried out by Professor and Mrs. Speed at the Royal "Dick" Veterinary College in Edinburgh.

The Exmoor is believed to be the oldest pure descendent of ponies that inhabited Britain and many areas of the world between the latitudes of forty-five and fifty degrees north approximately 100,000 years ago.

The exact area of origin is still uncertain, but fossil remains point to North America. Scientists do not agree on exact details of early distribution, but fossils show that a pony closely resembling the Exmoor in structure, coat color, dentition, and grazing habits was widespread one million years ago in many areas of the world. (See equine prehistory under Arab.)

What has been established is that the Exmoor-type pony, which may have been trapped by ice barriers in Alaska for generations, thus adapting to the hostile conditions, was only one of a number of Equidae evolving separately in different geographic areas during this time. A number of indigenous types became established and continued to reproduce their own distinctive type. The Exmoor is the only living breed to show jaw development similar to that found in fossilized bones in North America, also showing the beginning development of a seventh molar, found in no other living breed of horse or pony.

These ponies roamed the British Isles in pre-Celtic times, and later, stallions often mated with domesticated Celtic mares. Such crossbreds, however, did not survive. Unfortunately there are no early stud books of the breed, although there are records of the ponies in the Doomsday Book of 1086. Nothing further is recorded until 1818, when the last warden of the Exmoor Royal Forest, Sir Richard Acland, drove 400 of the ponies to his own land on Winsford Hill. Eventually this herd, known as the Anchor Herd, came into the ownership of Frank Green and has passed down through his family to Mrs. R. Wallace, the present owner. Descendants of Sir Richard's ponies, now numbering about forty-five, still run on Winsford Hill and Ashway Side and are branded with an anchor, which is the Acland brand.

In 1818 the remainder of ponies on the moor were sold at a forest dispersal sale. Fortunately some were purchased by the ancestors of today's moorland breeders and remained on their homeland. These included Samuel Milton, whose ponies continue as Herd 23 on Withypool Common, and a Mr. Crockford's Herd 12, running on Codsend and now belonging to J. Western. Herds 44 and 10 have also remained in the same families to the present day.

In the 1820s and 1860s successful crossbreds were produced, but again they could not survive a hard winter on the moor without extra concentrates and attention. Sidney's *Book of the Horse*, published in 1893, describes the Exmoor as having an average height of 12 hands and being generally bay with a wide forehead and nostril, mealy nose, black points, small, sharp ears, good shoulder and back, short legs, and good spine—much as they are today. Recorded history of the breed started when the N.P.S. (National Pony Society) began registering purebred Exmoors and crossbreds in its stud books during the late 1800s.

The Exmoor Pony Society was formed in 1921 by Reginald Le Bas and others with the aim, as it is today, of improving and encouraging the breeding of Exmoor ponies of moorland type. The society produced its first stud book in 1963. The First World War made little difference to the breed, but the difficult times of World War II left only a few purebred mares on the moor. When peace came, the moorland farmers began once more to rebuild the herds.

Although the Exmoor had withstood the rigors of nature for hundreds of thousands of years, the breed was put on the road to destruction by humans, who have made continuing attempts since earliest times to "improve" the stock by crossbreeding with imported exotic horses. The imports may have had many attributes of beauty, but they were not adapted to live unsupported in the harsh climatic conditions of the moor. Nature quickly eliminated them.

Very fortunately, Exmoor being an isolated area of severe conditions where the earliest settlers were resistant to change, the ponies largely escaped "improvement." The wise conservatism of some moorland farmers has succeeded in preserving for the world ponies that can claim direct descent from the indigenous horse stocks. Therein lies their immense value both as subjects for research and as a pure breed for reintroducing pony blood and attributes to other stock.

Publicity in 1981 brought about a welcome interest from "up country" breeders, but although upward of eighty foals are now registered each year, the Exmoor is still numerically a rare breed with about 290 mares and forty stallions in the United Kingdom. Future difficulties may lie in maintaining the true type away from the moor. The importance of moor-bred ponies cannot be overemphasized. Fortunately, judges, especially mountain and moorland judges, are now looking more favorably on the pony's intrinsic performance, characteristics, and qualities as an Exmoor.

The color, along with uniformity of type, are the most distinctive features of this breed. Color ranges from a special kind of dun unique to Exmoors to bay or brown with black points. There is a mealy color (light buff) around the eyes, on the nose, inside the flanks, and under the belly. No patches or collections of white hairs or white markings are permitted (they are rarely seen), as these are thought to indicate foreign blood.

The Exmoor is a sturdy pony with good bone, capable of carrying great weight in relation to its size. The action is that of a lighter type of animal due to a well-proportioned frame and good shoulder. The stride is long, low, and smooth, giving a comfortable ride. This pony is well balanced and extremely surefooted.

The coat of the Exmoor is interesting, having evolved to withstand prevailing winter conditions. An undercoat of short, wool-type hair is topped by a longer, greasy coat that repels rain. Whorls of hair strategically placed help to direct water away from sensitive parts of the body. So important are the quality of the coat and positioning of the whorls that research has shown that ponies without this "primitive" coat structure often do not survive. The special two-layered coat retains body heat so well that snow can lie on the pony's back for days. So little body heat is lost that the pony remains warm and dry while the snow stays frozen.

The Exmoor also has what is known as an "ice tail," being very thick with a fan-like growth at the top. In many breeds a small, pretty head is desired. Having evolved for survival and not fashion, the Exmoor has a large, well-shaped head allowing plenty of space for the essential functions of breathing, thinking, and eating. The nostrils are wide to allow inhalation of a liberal supply of air, and this is warmed while passing over the generous area of mucous membrane before entering the lungs. The eyes of the Exmoor are dark and large, appearing to stand out due to the surrounding light hair. They are protected by a heavy ridge of bone above the eye.

Accustomed to looking after themselves on the moor, Exmoors are independent—their survival depends on the ability to think and act for themselves. It takes those born on the moor a little time to grow accustomed to human ways. Once confidence has been gained, however, they become active and dogged performers.

Each year in October an event occurs on Exmoor which few people see—something of a cross between the Running of the Bulls at Pamplona and the Calgary Stampede. It is the rounding up of Exmoor ponies to bring them down from the moor to the Home Farm for the annual inspection. The ponies of the moor live in isolation and never see vehicles or people apart from this one yearly intrusion. Consequently the driving and shepherding of these wild, unbroken creatures on their five-mile journey from moor to Home Farm is an exciting adventure.

Foals are examined by two inspectors appointed by the Exmoor Pony Society. Colts not meeting exacting standards of type are gelded. Those that pass examination are branded on the near shoulder with a star. Beneath the star is the number of the herd and on the left hind quarter the number of the pony within the herd.

The day after the inspection there is little

problem returning the ponies to their moorland home, even though part of the way is over public roads. Whether by instinct or intelligence, they know the way. Once they are back the gates are closed, more to keep intruders out than to keep the ponies in.

While there are only a few herds running on the moor today, there are a number of Exmoors bred in studs throughout Great Britain. Bred and born away from the moor, they tend to lose type and grow slightly larger.

Despite their increasing popularity, Exmoor ponies are wild in the sense that they remain on the moor throughout the year, find their own food, care for their young, and wander free over huge areas of the moor surviving bad winters without additional shelter or food (although their owners do keep a watchful eye on them). Each pony belongs to someone and is branded to denote ownership. The Exmoor Pony Society is determined to ensure the continuance of the moorland herds to enable "up country" breeders to return to the foundation stock in future generations.

This pony is quite literally the "child of the moor." The food, climate, and living conditions produce the pony as it is today. Exmoor ponies do need support, but the emphasis must be on conserving of those herds that are on the moor now and have run there for generations. The unique qualities of the Exmoor, enabling it to be self-supporting in a harsh environment, are a legacy of its direct descent from its Ice Age ancestors. From them it has retained its coat, adapted to withstand extremes of cold and wet, and its digestive system, which enables it to survive on roots and fodder other breeds cannot tolerate. If the hardiness and vigor of the stock is to be maintained, the preservation of its natural habitat—the extensive, unimproved, solitary moorland—is a conservation must of the highest order for the continuance of these proud, untamed masters of the open moor.

Faeroe Island Horse

ORIGIN: *Faeroe Islands (Denmark)*
APTITUDES: *Riding horse*
AVERAGE HEIGHT: *11 to 12.1 h.h.*
POPULATION STATUS: *Rare*

The horses of the Faeroe Islands (in Danish, Faeroese or Føroyar) represent one of the oldest and purest horse breeds on earth today and one of the few that has not been altered through crossbreeding. The only breed comparable is the Icelandic horse, also of ancient origin and very pure due to total isolation.

The Faeroe Islands are located in the North Atlantic between Iceland and the Shetland Islands. They form a self-governing community within the kingdom of Denmark. There are seventeen inhabited islands and several islets and reefs (a total area of 540 square miles). The principal islands are Streymoy (Strømø), Eysturoy (Østerø), Vàgar (Văgø), Sudhuroy, (Suderø), Sandoy (Sandø), and Bordhoy (Bordø). The capital is Tórshavn (Thorshavn) on Streymoy. In 1984 the population was estimated at 44,800.

The climate on the Faeroes is oceanic and mild, with little variation in temperature and frequent fog and rain (approximately sixty inches per year). There are no reptiles, toads, or indigenous land mammals. Rats and mice have gone to the islands with ships over the centuries. Puffins and other seabirds are numerous. The islands are naturally treeless due to strong winds from the west and frequent gales, but a few species of hardy trees have been planted in sheltered places. The warm Atlantic drift keeps the harbors relatively free of ice.

The islands were first settled by Irish monks around 700 A.D. and were colonized by the Vikings about 800. They were Christianized by the King of Norway some 200 years later.

The Faeroe Island horse resembles horses which came to Europe from Asia about 200 B.C. The Scandinavians and Celts brought horses with them to the islands, where they readily adapted to the climate.

The horses are small. Prior to formation of the association for the Faeroes horse there were only five individuals remaining. By 1988 numbers had risen to twenty-seven as dedicated breeders worked to preserve this hardy horse. Most Faeroes horses are red in color (bay), some black, seldom brown, and *never* grey or skewbald. Occasionally a palomino or pale dun appears in the breed. The hair is thick and grows very heavy in the winter. The belly has the thinnest protective cover, but the horses keep it dry by standing up all winter, even in their sleep.

All of the Faeroes horses have been entered into the stud book, and their blood types have been identified. In addition to this, to avoid inbreeding, Peter Hahr, the leader of breed experiments with this breed, has taken five or six horses to be used for breeding for two or three years and then returned them to the owner.

The Danish consultant Henning Rasmussen has judged the Faeroes horses, and twenty-four of the present stock have been approved as breeding animals. The horses have almost no faults in conformation and are beautiful in appearance. I did not receive information regarding their gaits. They are said to be clever, sure on their feet, and better than humans at avoiding swamps.

Falabella

ORIGIN: *Argentina*
APTITUDES: *Light harness*
AVERAGE HEIGHT: *6.1 to 7 h.h.*
POPULATION STATUS: *Rare*

The Falabella is a special and unique breed which, after many generations of selection, has acquired stable characteristics of build and height. The name comes from the family

Maria L. B. de Falabella with Elisa, a two-year-old, 22″ tall filly. Photo: Maria L. B. de Falabella

The Falabella stallion, Falabella Comodin. Photo: Maria L. B. de Falabella

who developed the breed in the middle of the nineteenth century.

In 1845, south of Buenos Aires on the Argentine meadowlands, tribes of Pampas Indians had some unusually small individuals among their horses. These attracted the attention of an Irishman named Patrick Newtall, who managed to obtain some. After many years of experimenting and selective breeding, in 1853 he was successful in creating a herd of perfectly built, tiny horses, characterized by a nature of their own and lower than one meter (39.37″) in height (less than 9 h.h.). In 1879 he transferred all of his findings and knowledge to his son-in-law, Juan Falabella.

Juan Falabella added some specimens from other breeds to the developing "minihorse," including small English Thoroughbred, Shetland, and Criollo. Through successive crossing and careful selection, by 1893 the Falabella horses had achieved harmonious conformation and an average height of under thirty-three inches. In 1905 the herd was turned over to Juan's son, Emillio, who proceeded with breeding efforts. Through strict selection, he managed to fix the established conformation.

In 1927 Julio César Falabella inherited the establishment along with knowledge gained through the years by the Falabella family. He kept careful genealogical records and began to spread the breed internationally. It was Julio César Falabella who gave the Falabella breed the title "minihorse." The herd was expanded to 700 breeding mares, all characteristic of the breed. In 1937 the stallion

Napoleon 1 was born; he became one of the best foundation sires of the breed. Many registered Falabella horses trace to this sire. Napoleon 1 was a chestnut piebald. Another important sire was the chestnut Jauncito 168 Falabella, foaled in 1968. With international interest surging, the Falabella attracted study from Europe, the United States, and the Far East.

Maria L. B. de Falabella took over the Falabella establishment in 1980, successor to generations of breeders of this unique miniature horse.

The Falabella horse is intimately linked to the earliest modern horses in the New World. Andalusian horses brought to the Americas and later left to survive on their own underwent a series of structural changes to meet the needs of the variable climate of the Pampas region. The Argentine Criollo horse, so distinct today from its ancestors, illustrates what can occur in nearly 400 years of adaptation and natural selection. While retaining many of the characteristics of its ancestors, the Criollo has been changed through adaptation to survive and flourish in the harsh environment of the Pampas. Strong sun, cold winds, and fierce storms are common to the area. The land is arid, and horses were obliged to travel great distances to water and pasture. Often they fled from both Indians and pumas. Some strains of the Criollo underwent changes to smaller size, resulting in the horses that attracted Patrick Newtall in 1845.

The Falabella is a horse and not a pony, possessing every feature and attribute of its taller relative. The fixing of the genetic characteristics enables natural reproduction to yield offspring true to type and size. Other kinds of miniature horses do not always retain small size without careful selection.

Falabella horses are gentle and docile and possess strength far beyond their small proportions; they have been tested as both draft and saddle horses. In the 1970s a team of Falabella horses standing only twenty-five and a half inches tall pulled a cart loaded with liqueur drums from Usuahia, at the southern end of the American continent, to La Quiaca in northern Argentina. The distance covered by the Falabellas was more than 1,860 miles, across plains, valleys, and mountains.

The standard height is 75 centimeters (29.25 inches) for an animal of controlled production. For breeding and nurse animals of certified reproductive capacity, the size varies for each category from 70 cm or less to 80 cm. Many Falabellas are only twenty-three inches in height.

Due to their natural rearing, the Falabellas show a degree of rusticity that enables them to survive under the most severe weather conditions without special care. This breed is extremely long-lived, often reaching forty to forty-five years.

Veterinarians in France examining horses of the Falabella breed discovered the absence of a lumbar vertebra and two pair of ribs in all horses examined. Japanese veterinarians who also examined Falabella miniatures noted that two lumbar vertebrae and three ribs may be missing.

The Falabella has fine, silky hair and skin that is thin and supple. The hooves are narrow with an oval shape. There is diversity in the manes in this breed, some having manes that are short and straight while in others it is long, falling on both sides of the neck. The gait is spontaneous and very energetic.

The first established stud for Falabella horses in Europe was the Kilverstone Miniature Horse Stud in England. Lord and Lady Fisher traveled to Argentina to visit Sr. Falabella and see the amazing miniature horses, which were in different herds, all in varying stages of reduction to the small size. The Fishers were able to purchase four stallions and some mares, all of the smallest sizes and including some of the rarest colors. Today the Falabellas are well established and breeding at Kilverstone.

Fell Pony

ORIGIN: *England*
APTITUDES: *Riding, driving, light draft*
AVERAGE HEIGHT: *13 to 14 h.h.*
POPULATION STATUS: *Common*

Ponies of the British Isles (see appendix A) descended from the wild European pony and are not indigenous to the islands. They either migrated or were driven across the English

Fell pony. Photo: Clive Richardson

Channel, which was once a marshy stretch of water, by nomadic hunters as long ago as 1500 B.C. Archaeological discoveries enable us to envision what the original ponies looked like. They resembled a cross between their ancestor, the Tarpan, and the present-day Exmoor, oldest of the British native breeds and little changed over the centuries. Distribution throughout the islands brought the ponies into various climatic and regional environments which molded them into distinct types.

The Romans first landed on the shores of Britain around 55 B.C. and made many subsequent landings before settling in various parts of the country. Their supremacy in the north was constantly challenged by the warlike Picts, who came from further north in what is now Scotland. In 120 A.D., Emperor Hadrian decreed that a wall should be built to a height of fifteen feet, complete with gates, sentry stands, and fortifications at every mile point, stretching across Britain from coast to coast. Although the native ponies were strong, their small size limited their working capability and the Romans looked elsewhere for more suitable horsepower.

Prior to their invasion of Britain, the Romans had traveled through Friesland in the Netherlands. There, they had seen and were greatly impressed by the Friesian horse, one of the oldest breeds in Europe. To supplement the labor force engaged in building the wall, the Romans employed the services of some

six hundred men from Friesland. The black Friesian horses came with them. When finally the Romans withdrew from Britain to go to the aid of their besieged city of Rome, they left approximately 1,000 Friesian stallions in northern England. From these, not only the Fell pony was bred but also the Old English Black, now merged into the Shire breed, and the extinct Galloway and Fen ponies.

The Fell pony remained free of any further introduction of foreign blood for many years and breeds very true to type. Fells are very good trotters, and when fast trotters came into vogue in the late eighteenth century, some Norfolk and Yorkshire Roadster stallions were used. The Welsh stallion Comet also stood at stud in the Orton district and was put to both Fell and Dales mares.

Fell ponies are probably most famous as pack ponies. They were also used in the thirteenth century in pack pony droves, with one pony being ridden and used to guide groups of free ponies, which were heavily laden.

The Fell Pony Society was formed in 1912. Fell ponies are used today for riding, trekking, and driving competitions. In that they breed uniformly true to type, it is easy to find a matched pair.

This pony has a small head with a straight profile, small, pricked ears, lively eyes, and flared nostrils. The neck is well proportioned, muscular, and well set into the shoulders; the withers are pronounced. The back is long and straight; the loins are broad and strong; the croup is short and sloping; the chest is deep and broad; and the shoulder is long and sloping. The legs are sturdy and well muscled; the cannons are large and the fetlocks are feathered. The mane and tail are heavy and full.

Most Fell ponies are bay, brown, black, or grey. A small white star and small amount of white on the legs is tolerated but not desired.

Finnhorse
(Finnish Universal)

ORIGIN:	*Finland*
APTITUDES:	*Light draft, trotting races, riding horse*
AVERAGE HEIGHT:	*14.3 to 15.2 h.h.*
POPULATION STATUS:	*Common*

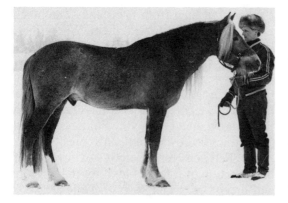

The Finnhorse stallion, Vekseli, a leading sire in Finland, and "Trotting King" 1979–82 and 1984. Photo: Juhani Länsiluoto

The Finnhorse mare Vekkuliina, "Trotting Queen" 1984–86. Photo: Juhani Länsiluoto

The Finnhorse is a descendant of the northern European domestic horse. Having both warmblood and heavier draft blood influence, it belongs to general horse breeds. This breed is called Finnish Universal as it is able to fulfill all needs for horses in Finland, from draft to speedy trotting races and riding. The principle of pure breeding has been followed since 1907, the year the stud book was founded.

The Finnhorse is easily handled, with a compliant nature. Considerable speed, liveliness, endurance, and longevity are typical characteristics of the breed. Finnhorses are dry and strongly muscled, with hard legs and good hooves. The coloring is mainly chestnut, and white markings are often found on the face and legs. Bay, grey, and, more rarely, brown or black are also seen.

In 1924 it was ordered that breeding was to be divided into two branches. Breeding of the heavy working horse for draft and forestry use went on as in the past, and in addition breeders began to concentrate on developing the lighter type suitable for riding and racing and especially good as a trotter.

The importance of the working horse has lessened as agriculture has become mechanized. However, horses are still necessary for light transportation in agriculture and for moving timber in the woods. The Finnhorse is good in rough terrain. A horse is particularly useful for thinning out a young forest, as horses inflict less damage on young trees than does heavy equipment.

Trotting is popular in Finland. In races, about 40 percent of the starts are for Finnhorses. Speed, well-balanced trot, endurance, and eagerness for victory characterize a good trotting horse.

The number of Finnhorses decreased after 1950, but in the 1970s there was a revival of interest in horse sports. The Finnhorse has its own special position in Finland as a sport horse.

In 1971 the riding type of Finnhorse was given a separate branch of the stud book, and since that time it has had its own classes in all kinds of riding competition. With the increasing popularity of riding sport in Finland, there is a growing demand for good riding horses. The Finnhorse is an excellent mount in riding schools, for both young and adult riders. It is calm, patient, and lively. In riding competitions the Finnhorse has had its best success in show jumping, but the breed also shows talent in dressage. The Finnhorse is truly a universal horse.

Flores

ORIGIN: *Indonesia (Flores Isle)*
APTITUDES: *Riding, light draft*
AVERAGE HEIGHT: *12.1 h.h.*
POPULATION STATUS: *Rare*

Little information was furnished for the Flores pony. It has been linked to the Timor in some references, but the Indonesian authorities gave different statistical information for this pony.

Dominant color in the Flores breed is given as "red," which must refer to bay or chestnut. The Flores pony is said to be known for its patient attitude.

For additional general information regarding Indonesian ponies, see Bali.

Florida Cracker Horse

(Seminole)

ORIGIN: *United States*
APTITUDES: *Riding horse*
AVERAGE HEIGHT: *14.2 h.h.*
POPULATION STATUS: *Rare*

The Florida Cracker horse, on the Paynes Prairie State Preserve. Photo: Jack Gillen

Florida Cracker horses. Photo: Jack Gillen

One of the newest horse registries in the United States has been formed to preserve remnants of the oldest breed in this country— the Spanish horse brought to the Americas by the conquistadores. The Florida Cracker Horse Association was formed on August 17, 1989.

There is some dissension among the various groups breeding authentic descendants of Spanish horses in the United States, with most contending that horses in other groups are less pure or not pure at all. However, genetic blood group testing being done by Professor Gus Cothran at the University of Kentucky has proven that nearly all of those various groups are dealing with horses that positively do carry the blood of the Spanish horses. Blood from these groups is currently being compared to that of the Andalusian, Sorraia, Lusitano, and Alter Real breeds as well as to other breeds, and a clear distinction is emerging between horses of Spanish descent and those with no Spanish ancestors.

The Floridian horse was popular and well known in the early days of development of eastern America. A wealth of information received from authorities in Florida has documented the authenticity of the small horse commonly known as the "Cracker" horse. A frontier Florida cow camp at Lake Kissimmee State Park in central Florida is brought to life through interpretation by a park ranger, dressed as a cow hunter using a cow whip and McClellan saddle. Virtually everything a visitor sees and hears is authentic for the year 1876, the peak of Florida's nineteenth-century cattle industry. The cow pen near the thatch-roofed shelter holds half a dozen scrub cows, descended from the Spanish stock introduced to Florida during the 1600s and 1700s. Another twenty graze in view of the camp on the nearby prairie. The cattle are small, lean, multicolored, and wild. The large steers tend to have long horns, while the bulls have shorter horns and are stocky, showing characteristics of Spanish fighting bulls.

Scrub cattle were once common throughout Florida, but the breed was nearly extinct when the Florida Park Service began the cow camp project in 1962. Two bulls and two cows were purchased from a cattleman at Fort Basinger near Lake Okeechobee. Eight cows were acquired near Old Town in Dixie County on the Suwannee River. From these original twelve head, a herd of about twenty-five is maintained for exhibition at the cow camp and about one hundred head are maintained at Paynes Prairie State Preserve near Gainesville, Florida. The preserve also shelters the few remaining feral Cracker horses, as seventeen head graze with the cattle.

Believed descended from Spanish horses brought by early explorers, the animal later came to be called the Florida Cracker Pony, its name derived from the riders who had become known as "crackers" because of the crack of their cow whips. Research has shown, however, that the Cracker horse most likely descended from Spanish horses brought from west of the Mississippi at a slightly later time than that of the conquistadores.

In describing the importance of the cracker pony, Ed Smith noted in *Them Good Ole Days* that some hands would ride for over sixty hours on the same horse, riding all day and most of the night, and both would be ready to go the next morning. He doubted if any horse had ever had more endurance.

Seed stock for the preservation of the cracker pony was given to the Department of Agriculture and Consumer Services by pioneer Hernando County cattleman and former State Representative John Law Ayers.

When William Bartram visited Paynes Prairie (the Great Alachua Savannah) in 1774, the Seminole Indians were grazing herds of wild Spanish cattle on the natural prairie. He noted that they rode the sprightly Seminole horse.

In 1838 Comte de Castelnau wrote: "The Floridian horse, such as it is today, and which is called generally 'Indian pony,' is small, long haired and bright eyed, lively, stubborn, and as wild as the Indians themselves; it has a wonderful endurance of fatigue and hardship; it has a singular instinct in finding its way in dense woods. Its food consists only of the high grass that covers the prairies and it does not require any care."

From the beginning of the cattle industry in Florida, cattlemen favored the small, wiry, active, and speedy cow ponies with extraordinary stamina and recuperative powers. Many of the range riders were large men, yet their tough and tireless cow ponies weighed only from 750 to 800 pounds. They carried their riders here, there, and everywhere at polo pony speeds, over ranges festooned with palmetto scrub and pockets, gopher holes, and other pitfalls without developing serious leg or ankle ailments.

The Choctaw and Chickasaw horses emerged from the same early stock, and from the same area as the Seminole or Cracker horse. It has been a matter of conjecture whether or not the different soil and situation of the country may have contributed in some measure to the differences in size and other qualities of the horses. Horses in the hilly country of Carolina, Georgia, Virginia, and all along the shores were of a larger, stronger variety than those bred in the flat country near the coast. A buckskin of the Upper Creek and Cherokee Indians often weighed twice as much as those of the Seminole or Lower Creek Indians.

Although the Cracker horse is nearly extinct today, with only about 150 remaining, the newly formed Florida Cracker Horse Association hopes to promote and preserve this strain of Spanish horses. While only a handful roam the prairie at the state park, a few ranching families have retained their historical Cracker horses through the years, refusing to outcross them.

For additional information see Spanish-American Horse.

Frederiksborg

ORIGIN: *Denmark*
APTITUDES: *Riding horse, light draft*
AVERAGE HEIGHT: *15.1 to 16.1 h.h.*
POPULATION STATUS: *Rare*

The Frederiksborg is the oldest horse breed in Denmark, traceable to the horses of the royal stud at Frederiksborg. In its prime, this was one of the finest studs in Europe. Reformation came in 1536 and King Frederik II (1543–88) gained possession of all the abbey studs, including some of the finest horses in the country. The best breeding material from these studs was collected in a breeding center for the royal horse breed situated at the manors Hillerødsholm and Favrgaard in North Zealand. Frederik's son, Christian IV, later added several good Spanish stallions, and in the eighteenth century the Frederiksborg was one of the most famous horse breeds in the world. The breed was highly valued as a good school horse, reliable and elegant, and it also made a high-class carriage horse and military charger.

Frederiksborg. Photo: Sally Anne Thompson

Frederiksborg in harness. Photo: Sally Anne Thompson

Frederiksborg horses were famous through the seventeenth, eighteenth, and nineteenth centuries, when they were exported to many countries and used to form or improve other breeds.

Breeding stock at this stud was generally separated, one stallion to ten to twenty mares. Typically they were chosen by color, and the studs were named accordingly; a red stud, a black stud, a grey stud, and so forth. Therefore, breeding was essentially coincidental and aimless.

In 1690 a stud regulation, reducing the breeding to a system and concentrating particularly on pure breeding, was set forth. Pure breeding was the only method used from 1700 to 1770. During that period there were approximately 200 mares and 20 stallions plus fillies and colts, about 800 horses in all. The stallion Pluto was the progenitor of one of the lines in the breeding of Lipizzans.

The breeding aims were a horse for riding, with beautiful and supple action, and another horse, larger, for pulling the royal carriages. For the carriage horses, the color was of utmost importance in order to present a complete team of six to eight horses of exactly the same color.

Due to its popularity this breed was exported in great numbers to many areas of the world. It was exported, however, to the detriment of the breed, leaving a low number of good animals in its homeland; the breed began to decline, until by 1839 the stud was closed. A small number of pure individuals were maintained over the years by private breeders devoted to the breed.

In 1939 efforts were started to reestablish the breed, using Friesian and Oldenburg blood, and later Thoroughbreds and Arabs were used. Currently numbers are low. Connections between the Frederiksborg of today and the old stock are very slight, but the breed still exists. After the royal stud was closed there were still enough Frederiksborg horses to continue the breed. Until about 1920, the aim was to produce a light carriage horse that could also be of use in farming. As motorization developed, the demand changed toward riding horses.

The Frederiksborg is a strong, attractive horse, always chestnut in color. Usually the mane and tail are a lighter, flaxen color. White markings on the face and/or legs are common. The head is well proportioned, with a straight or sometimes convex profile, pointed, mobile ears, and large, expressive eyes; the neck is of medium length, well proportioned and slightly arched and muscular; the withers are pronounced, broad and muscular. The back is short and straight; the croup rounded and broad; the chest high, full, and deep; and the shoulder sloping and muscular. The legs are well muscled with strong, broad joints, good bone, and clearly defined tendons. The feet are small and tough.

The Knabstrup is essentially a lighter version of the Frederiksborg, with the distinguishing feature of a spotted (Appaloosa) coat.

Freiberg

(Freiberger, Franches-Montagnes, Jura)

ORIGIN: *Switzerland*
APTITUDES: *Riding, light draft*
AVERAGE HEIGHT: *14.3 to 15.2 h.h.*
POPULATION STATUS: *Common*

A light to middleweight warmblood, the Freiberg horse was developed around the end of the nineteenth century by crossing native coldblooded Jura mares (Franches-Montagnes) with the English Thoroughbred and Anglo-Norman. Some Breton and Belgian blood was also used and, more recently, Arab. The Freiberg is one of the last light draft breeds bred in western Europe. Purebred selection began in 1910 after a long period of systematic crossbreeding. During the difficult times of the two World Wars, agricultural work and military transport could not have been accomplished without the vital services of the Freiberg horses.

At the Avenches stud strict control is maintained over quality and type of the Freiberg, which has two types, one more broadly built with greater muscle development and the other lighter in build and better suited for riding. Due to the mountainous terrain of Switzerland, horses are still considered vital to the army for patrols or possible transport of artillery. The Freiberg horse is also important to farmers who cannot use tractors on the steep hillsides.

The Freiberg is a compact horse with sturdy conformation. Modern Freiberg horses show more Arab influence than earlier horses of this breed. The breed is docile but energetic, easily trained, and agile. This is a good animal for use in the mountains as it is surefooted and calm. The gait is animated and lively and the breed has good stamina.

The Freiberg has a small head showing Arab influence, although sometimes rather heavy, with a straight profile, broad forehead, and small ears; the neck is muscular, arched, and broad at the base. The withers are broad

The old-type Freiberg horse of Switzerland. Photo: Evelyn Simak

The modern-type Freiberg horse. Photo: Evelyn Simak

and pronounced; the back is straight and short; the loins are powerful; the croup is slightly sloping; the chest deep and well developed; the shoulder sloping. The legs are short with good joints, plenty of bone, and tough hooves. The mane and tail are abundant, but the legs are free of much feather.

With the increase of mechanical agricultural equipment since 1950, draft horses are not needed as before. Meanwhile, demand for sport and riding horses has increased. By careful crossbreeding by Freiberg mares with warmblood stallions from Sweden and France, the Freiberg has evolved into a high-quality riding horse.

French Anglo-Arab

ORIGIN: *France*
APTITUDES: *Riding, sport horse*
AVERAGE HEIGHT: *15 to 16 h.h.*
POPULATION STATUS: *Common*

French Anglo-Arab. Photo: Pennington Galleries, courtesy UNIC, Paris, France

The Anglo-Arab is bred throughout the world, a cross between the English Thoroughbred and Arab breeds. However, France has excelled in the breeding of Anglo-Arabs to the degree that the French animals have almost become a new and distinct breed. Many Anglo-Arab progeny in France are the offspring of Anglo-Arabs. All horse breeding in France is monitored by the national organization, Union Nationale Interprofessionnelle du Cheval (UNIC).

The cross of Thoroughbred and Arab horses began officially during the second half of the nineteenth century in the Limousin and southwest of France. Domestic mares also bred from Thoroughbred and Arab horses were added later. The Limousin foundation mares were a light type which had been bred in this area for centuries. They had a very strong Oriental appearance and were known as horses of the Navarre, Bearn, and Gascogne. The French Anglo-Arab was first popular as a mount for the cavalry, then as a general riding horse. It was the first French breed to be created and bred for use in a wide variety of sporting events.

The profile of the French Anglo-Arab can be straight or Roman. The croup is long and horizontal. The general bone structure is rather light with good limbs. This breed was originally just slightly taller than the Arab, but the height has a tendency to increase and today horses over 16 hands are no longer rare. All coat colors are acceptable, but bay, chestnut, and grey are most frequently seen. This is a distinguished horse, very elegant about the

French Anglo-Arab. Photo courtesy of UNIC, Paris, France

French Anglo-Arabs. Photo courtesy UNIC, Paris, France

head and harmonious in the body with supple and dazzling gaits—and tremendous personality, which gives the breed charm.

The traditional breeding areas for the Anglo-Arab are around the Pau, Tarbes, Villeneuve-sur-Lot, Aurillac, Rodez, and Uzès studs and around the Pompadour stud, which keeps a famous national Anglo-Arab brood mare stock. They are also bred around Compiègne (Osie) and in other parts of France. There were 256 Anglo-Arab stallions in 1983 that were bred to over 6,689 mares, of which 1,521 were Anglo-Arabs. Many Thoroughbred and Arab stallions were bred to Thoroughbred, Arab, or Anglo-Arab mares to produce Anglo-Arabs.

Its aptitude for jumping and galloping as well as its elegance and resistance (overall hardiness) make the Anglo-Arab a prized horse for show jumping, eventing, and dressage. Many Anglo-Arabs have very good selection indexes, providing an objective statement on the value of the animal. This breed also finds a good flat racing as well as hurdling program in the southwest of France. However, the major qualities of beauty, height, harmony, resistance, skill, and physical aptitude to bear a rider make this breed above all a wonderful saddle horse. The Anglo-Arab is also recognized and sought after to improve other sport and saddle breeds.

Recently the Service des Haras altered the requirements for registration of Anglo-Arabs in France. Animals will now be admitted having, in the fourth generation back, one ancestor that is neither Arab nor Thoroughbred. This relaxation of registration requirements has previously been adopted for limited periods (1940 to 1946) due to the ravages of the military and, later, German requisitioning, and in 1958 in response to a shortage of numbers (personal communication, Anthony A. Dent).

French Ardennais

ORIGIN: *France*
APTITUDES: *Heavy draft and farm work*
AVERAGE HEIGHT: *15 to 16 h.h.*
POPULATION STATUS: *Common*

French Ardennais. Photo courtesy UNIC, Paris, France

The French Ardennais, doing what it does best. Photo courtesy UNIC, Paris, France

The French Ardennais is the most ancient draft breed in France. (Ardennais is the French spelling for the Ardennes breed.) This breed is considered to be a direct descendant of the ancient Solutré horse, a small breed (13.2 to 15 hands) originally found only in the French and Belgian Ardennes.

All of the Roman emperors beginning with Julius Caesar drew on this equine population, and since that time evolution of the breed has been largely affected by needs for war purposes. This breed pulled Napoleon's cannons and other guns in other wars. During the Middle Ages the morphology of the breed was made heavier for agricultural purposes. Later the dispersion of cultivated lands on various soils and under different climates diversified the original conformation and produced horses fitted for various areas. Oriental stallions were used at one time, and over the years the Ardennais has become a small horse known for its hardiness and endurance qualities.

In the nineteenth century an important influx of Belgian draft blood brought an increase in size of the breed. However, the horses introduced were not always the best stud horses, and they are sometimes blamed for reducing the quality of the limb tissues of the original Ardennais. This continued until World War I. The purpose of breeding was to produce a horse with more bone and power, able to work the heavy soil of the large farms in the east of France.

The Ardennais is a compact type of horse. Mature animals may weigh 1,540 to 2,200 pounds. The coat color is bay or roan, sometimes chestnut, dark grey, or palomino. The head is expressive with a slightly dished or straight profile; small ears which point forward; and the head well set on the neck. The body shows bulk, density, and power, wide and deep at the chest. The breed has a strong back and loins; wide hips; long and muscular hind quarters; a shortened tail; slanted shoulder; muscular forearm and legs; low and wide joints; and wide and sound feet.

The Ardennais horse is found in nearly all of northeastern France from the eastern border and Paris basin to the Rhine and from the Jura and Massif Central to the Belgian and German borders (areas of Montier-en-Der and Rosières-aux-Salines studs). Breeding of the Ardennais is now spread outside the original breeding grounds, especially toward the Massif Central and the southeast of France.

Today there are about 280 stallions and more than 4,700 mares of this breed, putting it in third place behind the Breton and Comtois and at about the same level as the Percheron for the number of heavy mares covered in France.

After use as a draft horse for many centuries, the Ardennais is now directed primarily toward meat production. Nevertheless, this breed is compact and docile and is still used as a draft horse on some farms.

The stud book for the French Ardennais was started in 1929.

French Cob
(Cob Normande)

ORIGIN: *France*
APTITUDES: *Carriage horse*
AVERAGE HEIGHT: *15.3 to 16.3 h.h.*
POPULATION STATUS: *Uncommon*

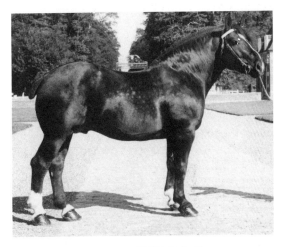

French Cob. Photo courtesy UNIC, Paris, France

The Cob has always existed in Normandy. This is the typical dual-purpose horse, used for carriage work as well as light field work. The origin of the French Cob was the Carrossier Normand (Norman Coach Horse), and the origin certificate bore the term "Half Bred" for many years. There are two types: light and heavy. The light Cob is close to the blood horse, with long tail, built close to the ground, deep in the chest, and with wide hips. This type was absorbed by the saddle horses. The large Cob exhibits a docked tail and is now associated with the draft horse breeds.

The Cob is more often bay or chestnut in color, sometimes roan or grey. The head and general features resemble those of a heavy Norman saddle horse. This is a rather squarish horse with good limbs; it is strong, harmonious, well balanced, and easily distinguishable from the heavy draft horses by its "halfblood" type. The majority of Cob breeding is concentrated in the Manche, around the Saint-Lô stud.

Docile, energetic, and having exceptional gaits, the French Cob is especially adapted to carriage work, a task for which it is in great demand, not only to work the fields but for leisure and harness. Because of its origins, which are very close to those of the blood horses, the Cob is an appreciated meat breed, giving a somewhat light carcass but a meat of refined taste. The Cob remains a dual-purpose horse, able despite small numbers to see a certain breed improvement nowadays.

There is no stud book for the Cob. As a rule, if the sire of a foal is a Cob, the foal is a Cob no matter what the breed of the dam may be. The best colts and fillies participate in breeding competitions reserved for them, and exceptionally good colts are retained as stallions and purchased by the National Stud when they are three years old. At the present time, all Cob stallions are owned by the National Stud.

French Saddle Pony
(Poney Français de Selle)

ORIGIN: *France*
APTITUDES: *Riding pony*
AVERAGE HEIGHT: *12.1 to 14.2 h.h.*
POPULATION STATUS: *Common*

Of recent origin, the French Saddle Pony is the result of a cross between pony mares of French origin and Arab, Connemara, or Welsh sires. The French Pony Stud Book also includes crosses using the blood of New Forest, Selle Français, Mérens, Basque (Pottok), and Landais mares to Arab stallions.

The desire to produce a high-quality riding pony for children has been fulfilled in the French Saddle Pony, which excels in jumping and dressage and is also a first-class harness pony.

All coat colors are permitted. The head is small with a straight or slightly convex profile, the ears of average length, and the eyes large and lively. The neck is long and well shaped; the withers are well defined; the back is straight; the croup sloping; the chest wide and deep; the abdomen tucked up; and the shoulder is long and sloping. The legs are strong with good, clean joints and hooves are well shaped with tough horn.

French Trotter
(Trotteur Français)

ORIGIN: *France*
APTITUDES: *Racing trotter*
AVERAGE HEIGHT: *15.1 to 16.2 h.h.*
POPULATION STATUS: *Common*

French Trotter. Photo courtesy UNIC, Paris, France

French Trotter. Photo courtesy UNIC, Paris, France

The French Trotter is also known as the Norman Trotter, and its origins are closely related to those of the old Norman horse. Breeding selection began in 1836 when the first trotting races took place in Cherbourg. The breed was developed from crosses of English Thoroughbred and mainly Norfolk Trotters from Great Britain to Normandy mares. Some American Trotter (Standardbred) blood was also used.

Interest in trotting races has increased since races were first held in 1836, and the selection for speed at the trotting gait has continued to develop.

More than 662 French Trotter sires were at stud in 1983 and there were 14,206 blood mares, of which 14,032 were of Thoroughbred or Trotter breeding. Trotters are numerically the most important in breeding stock in France.

The French Trotter has no standard morphology. Some characteristics are nonetheless found in most individuals of the breed: a handsome head, usually with a straight profile, prominent breast bone, and long, wide croup. The shoulder, originally quite straight, is becoming more sloped to allow a broader movement of the forearm. The hind quarters are rather short.

The French Trotter retains a broad build from its Norman ancestors but continues to show more and more refinement. The most common colors found are bay and chestnut. Breeding of this horse is concentrated in Lower Normandy and especially in the areas of Le Pin and Saint-Lô studs. The breed is found more generally in the northwest of France around Cluny, the Loire area, and the southwest. Many breeders have no more than two to three mares. They often breed, school, and race their horses themselves.

French Trotter mares registered in the stud book can benefit from "selection premiums" and "conformation premiums" given out at regional shows or an annual national show at Vincennes. These premiums permit a maintenance of conformation and selection of the breeding stock.

Sires of the breed are only allowed for public mating services for pure breeding if they obtain minimum references: having been in the first five of a classical, semiclassical, or international race, or holding a registered record on an officially confirmed track. These requirements vary with the animal's age and the type of race (driven or mounted).

Given its selection, the French Trotter is essentially destined for trotting races, whether driven or mounted. The races are very popular in France. To take part in a trotting race, all trotters must pass qualification tests by running the kilometer in a specified time. These tests eliminate almost half of the presented horses from races and are therefore valuable for improvement of the breed.

French Trotters are also used for breeding saddle horses, giving good results due to their musculature and good temperament. Some riders use French Trotters for leisure mounts.

The stud book for this breed was started in 1922.

Friesian
(Friese)

ORIGIN: *Holland (Friesland)*
APTITUDES: *Riding, carriage horse*
AVERAGE HEIGHT: *15 to 16 h.h.*
POPULATION STATUS: *Rare*

There is little written in English concerning the Friesian horse, bred for centuries in Friesland, the Netherlands. In fact, the breed is so new to the United States and so rare in the world that many people have never heard of this centuries-old breed. But having once seen the Friesian in action, one will never forget it.

When these high-stepping black beauties with their proudly arched necks, small pricked ears, and flowing manes and tails strut into a

Frâns, one of the few qualified Friesian stallions in the United States. Photo by owner, Carolyn Sharp

Frâns, Friesian stallion. Photo courtesy Carolyn Sharp

Friesians in harness. Photo by L. E. Huijing, courtesy of Het Friesch Paarden-Stamboek

A Friesian stallion in Germany, showing the typical regal bearing seen in this breed. Photo: Evelyn Simak

asm. Friesians have an appearance all their own—they are unlike any other horses in the world. The Friesian is one of the oldest fixed breeds in Europe, and its blood was used for centuries to create and improve many other breeds.

Friesland (Fryslân in the Friesian language) is one of the eleven provinces of the Kingdom of the Netherlands, situated in the northwest of Europe. It is an old country dating to 500 B.C., when the Friesians settled along the coast of the North Sea. They were tradesmen, seafarers, farmers, and horse breeders.

Gradual rising of the sea, caused by melting ice at the poles and sinking of the earth, forced the Friesians to build mounds upon which to place their homes for flood protection. Most towns and villages along the coast were built thus. Around the year 1000 when the territory of the Friesians was restricted to the north of the Netherlands and bordering Germany, sea walls kept the land free of the floods. These walls are continually heightened and presently are nearly four times as high as they were four hundred years ago.

Primitive drawings in caves in Spain and southern France and the bones of game found there and elsewhere show that even during the Ice Age there were both large and small types of horses in existence. Labouchère (1927) found bones of both large and small horses in the Friesian mounds. From the types *Equus occidentalis* (western horse) and *Equus germanicus* (German horse), he gave his find the name *Equus robustus* (big horse). As for the smaller bones found, he supposed these belonged to the *Equus przewalski* (Przewalski's Horse). Slipper (1944) believed these to belong to the *Equus gmelini* (Tarpan). It is difficult to determine if crosses were made and, if so, to what extent.

The Friesian breed is believed descended from *Equus robustus*. During the sixteenth and seventeenth centuries and possibly earlier, blood of the ancient Iberian breed, the Andalusian, was introduced. There were also infusions of blood from other horses of western Europe. From these crosses the Friesian developed high knee action and elegance. The natural action at the trot is so high that many

show ring in a carriage class, there is little need for translation. Their presence draws the crowd to its feet roaring approval and enthusi-

Friesians brush their elbows with their front heels as they move. The Friesian has been kept completely free of influence of the English Thoroughbred and during the last two hundred years has been bred completely pure.

Records of the past tell us that the Friesian of old was highly esteemed and famous. There is information from 1251 (Cologne), 1276 (Münster), 1466 (Aduard), 1617 (Markham), and 1771 (Kladrub), and the Friesian is mentioned and praised in books published in 1560 (Blundeville), 1568 (Guiccairdini), 1629 (Pluvinel), 1658 (Duke of Newcastle), 1680 (De Solleysel), 1687 (von Adlersflügel), 1734 (Saunier), 1741 (Guérinière), 1744 (Oebschelwitz), 1779 (Le Franq van Berkheij), 1802 (Huzard), and 1811 (Geisweit van der Netten).

The Friesian breed had great influence on many other breeds. The Hungarian King Louis II used a Friesian stallion when he took to the field against the Turks on June 15, 1526, in a campaign that culminated in the battle of Mohacs on August 29, 1526. Etchings by Stradanus (Jan van der Straat, 1568) show a Friesian stallion from the stables of Don Juan of Austria. Because of their excellent qualities, Friesian stallions were imported by the Electoral Prince George William of Prussia in 1624 and later by the famous Danish stud at Frederiksborg, the stud at Salzburg, and by the Kladrub stud in 1771 and again in 1974.

The well-known English author Anthony Dent and others are of the opinion that the Friesian horse influenced the Old English Black horse and the Fell pony. Dent proposes that the Norwegian Døle (Gudbrandsdal horse), which shows a great likeness to the Friesian, must have gotten there from Friesland either as booty or by regular trade. The North Swedish Horse, in turn, was greatly influenced by the Norwegian Døle. Dent also suggests a Norwegian influence on the English Dales pony. In the Pyrenees in southern France there is a pony, the Mérens, also bearing a marked resemblance to the Friesian breed.

The Friesian was first introduced to the Americas when the Dutch settled on the southern tip of Manhattan Island in 1625. Like other Dutch interests, the Friesian didn't flourish there for long. The Dutch settlement fell into British hands in the second Anglo-Dutch war (1665–67) and New Amsterdam quickly became New York. Newspapers advertised in 1795 and 1796 "trotters of Dutch descent." These must have been Friesians. Jenne Mellin proposes in *The Morgan Horse* (1961) and *The Morgan Horse Handbook* (1973) the possibility that this well-known American horse is of Friesian descent. The fast trot and heavy mane and tail of the original Morgan may be an indication (see Canadian and Morgan).

Trotting breeds seem invariably to trace to the Friesian, both in Europe and North America—the Standardbred, Canadian, Norfolk Trotter (now extinct), French Trotter, Orlov Trotter, Hackney, Døle Trotter, North Swedish Horse, and Finnish Universal. Many other breeds are known to carry Friesian blood.

It is possible that the Shire horse of England is a direct descendant of the Friesian, as Dutch engineers were contracted to drain the Fens in the sixteenth century and it is difficult to believe they did not bring their beloved Friesians with them. The Shire, it is said, is a descendant of various heavy horses native to Lincolnshire, usually referred to as the "Lincolnshire Blacks." The great horses of the eastern areas had been greatly influenced by imported stock from the "low countries." It is interesting to note that from the time of the Dutch engineers working in the Fens, a predominantly black horse emerged in Lincolnshire.

Always black, the Friesian stands between 15 and 16 hands at the withers and weighs from 1,250 to 1,450 pounds. It has a compact build and superb, almost regal bearing. Rarely there is a small white star on the forehead, usually no more than a few hairs, and this is allowed only on geldings and stud book mares. The head is long although small, with a straight profile and a somewhat hollow nasal bone; the eyes are large and shining; the ears are small and point inward at the tips. The neck is graceful and swan-like, carried high and proudly. This horse has an abundant mane and tail and feathering on the lower legs. Under no circumstances is it permissible to dock the tail of a Friesian and, in fact, trimming of any hair from mane, tail, or legs is frowned upon.

Exhibiting the very high, showy action that is natural to the breed, the Friesian is difficult to surpass as a carriage horse. The temperament of this breed is remarkable in friendliness and gentleness, always calm and willing. Friesians create an interesting contrast with their extreme calmness and draft-like characteristics coupled with their high, showy action and elegant carriage when in motion. There is hardly a horse more proud or regal in appearance than this one in harness.

Armored knights of old found this horse very desirable, having the strength to carry great weight into battle and still maneuver quickly. Later, its suppleness and agility made the breed much sought after for use in riding schools in Paris and Spain during the fifteenth and sixteenth centuries. Before an elegant carriage this breed has few rivals, and throughout Europe the royal courts used them as coach horses. The kingdoms of France, Austria, Prussia, and Holland are known to have had Friesians in their royal stables.

An excellent trotter, the Friesian was used for racing short distances in Holland, the winners being awarded silver or golden whips. There is a fine collection of these in the museum at Leeuwarden (Holland). King William I introduced a race at Leeuwarden, held annually for over eighty years, known as the King's Golden Whip Day. This was celebrated in remembrance of Napoleon's defeat at Waterloo.

Today in Friesland there are many carriage events and often the *sjees* are seen. This is the Friesland form of the French word *chaise*, the vehicle after which the Friesian version is patterned. This unique two-wheeled cart may be drawn by either one or two horses, and aboard are a gentleman and a lady dressed in the traditional costumes of the 1880s. The *sjees* is one of the few carriages in which the driver is seated on the left; the right-hand side, the place of honor, is occupied by his lady.

Four-in-hand carriages are common, and as many as ten-in-hand can be seen in front of light carriages. These large, unusual hitches used for demonstration purposes are becoming very popular. The Friesian people take great pride in the natural ability of their black horse in harness.

Reintroduced to the United States in 1974 by Tom Hannon of Canton, Ohio, the Friesian has steadily gained in popularity, causing excitement wherever it is seen. Frank Leyendekker of Visalia, California, himself a native of Friesland, imported a stallion and two mares in 1976, and in April of 1979 he took delivery of three additional mares, one of which was in foal.

By 1983 the popularity of the Friesian in America had grown enough to support a national association and a national show, which is held each September at the Los Angeles County Fair in conjunction with a four-day open carriage competition. And, to show that the Friesian has lost none of its versatility, the national show provides both halter and saddle classes as well as driving competitions. Judges from the Netherlands officiate for several classes. After the show, the Friesland judges make their annual inspection tour of U.S. breeding operations, checking horses for eligibility for registration and breeding.

The Friesian was nearly lost during the First World War, and by 1913 there were only three remaining registered stallions in the stud book. A group of concerned breeders instituted a strict breeding program to preserve the breed. World War II again threatened the breed with total extinction. German troops took horses as they needed them, and more than one Friesland farmer drove a nail into a black hoof to lame his horse to prevent its loss. When the troops were gone the foot was lovingly treated. Today these wonderful horses are making a comeback. At this writing the latest report shows thirty-seven living registered qualified (see below) stallions in the entire world.

Friesian horses have two basic *keur* times: as weanlings and again as three-year-olds. The Dutch judges inspect them for conformation and balanced, rhythmic action. They look for animation in both the hind and front feet. As weanlings are keured at their dam's side, this also gives the judges a chance to view the mother and see how her offspring are turning out. At this time they are given a foal book number and the underside of the tongue is tattooed. The next important inspection is at three. This is the first time that the horse can be keured to move into the adult books. The procedure varies for stallions and mares.

In February, from forty to sixty of the best three- and four-year-old Friesian stallions come together in Leeuwarden, Holland. They spend two days being judged for conformation. At the end of this time the chosen stallions (anywhere from two to fourteen have been chosen in previous years) go to the government-run stallion school in Ermelo. Here they spend fifty days away from their owners for training. They are trained for riding and driving and their manners and temperament are carefully judged. At the end of the training time, the stallions that have passed all the tests become "qualified" to breed and produce registered foals. They are branded with an "F" on the left side of their necks. These stallions will stand to Friesian mares until they are eight years old. Then forty of their offspring are picked randomly and judged. The stallion must show that he is benefiting and improving the breed. He will be inspected in this manner until he is twelve years old. If he is not considered to be improving the breed and producing excellent offspring, he is disqualified. At any of the above judging junctures, if the stallion is turned down for any reason he will not be considered again.

Mares may be judged for entry into the stud book at three. At this time the mare is presented to the judge. She is shown at a walk and trot and basically viewed for the same attributes as the stallion. If she is good, she may be entered as a *Stud book* mare and can produce registered foals when bred to a qualified stallion. If she has very good conformation and action, she may be entered as a *Star* mare. Stud book mares are branded with an "F" on the left side of the neck and Star mares receive an "F" plus an "S". Presently, there are approximately 1,500 Star mares in the world. At seven years of age, when a mare is fully matured, she may be shown again for her *Model* classification.

A Model mare is a mare of fantastic Friesian conformation plus incredible action. There are only approximately forty Model mares in the world. The Model mare is branded again with an "M". Things are a bit more liberal for the mares, and they may be presented for upgrading more often than the stallions. Mares may be shown to the judges every year.

Both stallions and mares can become *Preferred* by points they earn from their offspring. If they gather enough points they receive an additional brand, this time on the right side of the neck in the shape of a crown.

Such careful qualification of the Friesian has assured excellence in the breed and is quite typical of the dedication of the Dutch breeders to producing high-quality animals.

Furioso
(Mezöhegyes Halfbred)

ORIGIN: *Hungary*
APTITUDES: *Riding and light draft*
AVERAGE HEIGHT: *16 to 17 h.h.*
POPULATION STATUS: *Rare*

Furioso stallion. Photo: Dr. Csaba Németh

Breeding stock of the Mezöhegyes Stud Farm, founded in 1785 by the emperor of Austria and king of Hungary, Joseph II, was initially grouped by color. As a consequence of the outstanding breeding impact of the stallions Furioso, who arrived at Mezöhegyes in 1841 at the bay-colored stud, and North Star, imported in 1852 from England, the descendants of these stallions were inbred. The Furioso breed evolved as a result of this purposeful breeding activity.

In 1943 the stock was divided genealogically into the Furioso A and B lines and North Star A and B lines. The latter North Star Line has since become extinct. In 1961 the breeding stock was transferred to the Nagkunság State Stud Farm, where breeding ceased in 1966. The present breeding stock is partly the surviving Nagkunság stock and popular stock gathered from the former breeding region. Descendants of Furioso and North Star have maintained their characteristics.

Continued breeding of this diversely utilized breed is due to the serviceable and economical use of the Furioso for farming as well as saddle use. In its present form the Furioso is one of the most versatile universal warmblood horses, able to satisfy any need of agriculture, and when crossed to English Thoroughbred stallions it produces animals highly qualified for sporting purposes.

A dual-purpose, medium-heavy harness and saddle type horse, the Furioso has a large frame, strong body constitution, and noble appearance. The head is well shaped, proportional to the body, and usually has a straight profile; the neck is straight or only slightly arched and muscular. The withers are medium to high and long; the back and loins are said to be stiff; the barrel is deep and well sprung; and the shoulder is lean but muscular and well sloped. The legs have good muscle and solid, clean joints which are clearly defined, and the hooves are well formed and tough. The Furioso is often pigeon-toed with the hind legs being somewhat cow hocked.

The most common color seen in the Furioso is bay, with black being rare. Chestnut is sometimes seen.

This is a tough, constant farm horse of brawny constitution, and it can be worked or used for long periods. The breed is generally easy to break in and useful for almost any purpose. Jumping ability is usually mediocre, but some individuals are excellent over obstacles. This is a uniform, very valuable breed.

The Furioso exists in small numbers and within the breed only a few lines remain. Thus in breeding of this horse the most important task is protecting the breed against gene erosion. An increase in body mass is desirable. Selection is according to conformation and judging of descendants. Before individuals are used for breeding they are tested for harness or saddle use.

Galiceño

ORIGIN: *Mexico*
APTITUDES: *Riding horse*
AVERAGE HEIGHT: *12.2 to 14.1 h.h.*
POPULATION STATUS: *Rare*

Galiceño. Photo: Jack Frazier

Although relatively new to the United States, having been officially introduced in 1958, the Galiceño has been in the New World since the Spanish conquistadores brought horses to the continent in the early sixteenth century. Descending from the Galician of northern Spain and the Garrano of Portugal, the Galiceño was among the first sixteen horses landed by Hernando Cortés when he invaded Mexico from Cuba in 1519.

Prized for its intelligence, beauty, smooth gaits, and endurance, the Galiceño became the cherished possession of natives in the coastal regions of Mexico. Although they were in part ancestors to the thousands of mustangs that have endured for centuries in the U.S. Southwest, pure Galiceños never migrated north to join other equine breeds in the United States until recent interest prompted their importation.

While usually featured as a child's pony due to its small size, the Galiceño is well able to carry a man all day in rough country. All of the cow horses of Mexico and the southwestern United States were small in the early days of settlement.

The Galiceño Horse Breeders Association was formed in 1959, one year after the first horses were imported to the United States. Its purpose was to record, perpetuate, and preserve the breed. Through association efforts to interest new breeders and owners and through sponsorship of shows and contests, the Galiceño has gained in popularity with children and adults alike. A good riding horse for pleasure and equally outstanding for shows and contests, the Galiceño today still exhibits the versatility and eagerness to please that made it a choice of many conquistadores of long ago.

Galiceños are extremely gentle in nature and easy to handle. Their good disposition makes them an ideal family horse. From the time of the conquistadores this small horse has been asked to perform every conceivable task and has accomplished each with a remarkable native intelligence. Bright, alert, and very quick to learn, the Galiceño demonstrates its worth to the novice and expert alike.

Although small in stature, the Galiceño possesses hardness, courage, and stamina that are difficult for many larger horses to match. Its gaits are smooth and not tiring to the rider, a gift inherited by most horses of Spanish descent and readily apparent in all breeds found in Central and South America. Agile and fast, the Galiceño is a good cutting horse and is excellent in reining classes or timed events.

This breed possesses substance, style and beauty, and a natural, ground-covering running-walk which sets it apart from pony breeds. All solid colors are accepted for registration, while albinos or pintos are not allowed. The head shows refinement with good width between the eyes, pointed ears, a large, lively eye, and a small muzzle; the neck is slightly arched with a clean throat-latch and fits smoothly into the withers, which are prominent. The body is smoothly muscled; the back short and straight; the croup slightly sloped with a moderately high tail-set. The hind quarters are set slightly more under the body than many other breeds; the joints are strong and well shaped; the shoulder is well sloped, giving a long stride; the forelegs have well muscled forearms. The feet are well shaped and hard, open at the heel. For additional information see Spanish-American Horse.

Garrano
(Minho, Garrano do Minho, Tràs os Montes)

ORIGIN: *Portugal (Minho and Tràs os Montes)*
APTITUDES: *Riding and pack pony*
AVERAGE HEIGHT: *10 to 14 h.h.*
POPULATION STATUS: *Common*

The Garrano is usually referred to as a pony, yet many feel this is a small horse. The breed is of ancient origin and resembles horses depicted in cave paintings dating to the Paleolithic era, indicating that the Garrano has not changed in appearance for thousands of years. The Garrano has a concave profile, and while the breed has received infusions of Arab blood in the last century, ancient cave art reveals that the concave profile did not come from the Arab.

Largely used as pack horses, Garranos may be seen moving over steep, narrow paths of the thickly wooded countryside of northern Portugal. Often they are used to pull small carts. The breed is also popular with the Portuguese army as a pack pony for machine guns on military exercises.

The word *garron*, used in Scotland for Highland ponies, may have come from the same Celtic background as the name Garrano. Dr. Ruy d'Andrade, the noted Portuguese hippologist, believed that all of the ponies of Europe came from one common ancestor. Possibly they migrated down into France and Spain from northern Europe to escape the advancing cold of the Ice Age.

The Garrano is a very hardy individual, strong and surefooted. Generally the conformation is good. At one time trotting races were traditionally held for this pony, which is quite fast.

The Garrano is thought to be one of the ancestors of the Asturian and Galician horses of northern Spain (the pure Galician is nearly extinct today). The ponies or small horses introduced to the Americas during the sixteenth and seventeenth centuries no doubt included the Garrano, traditionally used as a pack animal. The Garrano breed also may have been involved in early formation of the Andalusian breed and may be responsible for the occasional concave profile seen in the Andalusian.

Coat color is usually bay, brown, or dark chestnut with an abundant mane and tail. White markings are the exception rather than the rule. The head is small but sometimes tending toward heaviness, with a concave or straight profile, large, lively eyes, and small, mobile ears. The neck is fairly long and well muscled; the withers are rather flat; the back is straight and short; the croup sloping; and the tail set low. The chest is deep and wide; the abdomen well rounded; the shoulder usually straight. The legs are strong and solid, with broad joints, short cannons, and well-shaped feet of tough horn.

Gelderland
(Gelderse paard)

ORIGIN: *Holland*
APTITUDES: *Riding horse, light draft*
AVERAGE HEIGHT: *15.2 to 16 h.h.*
POPULATION STATUS: *Rare*

Gelderland. Photo: Sally Anne Thompson

The Gelderland is a warmblood horse which originated in the sandy, mid-eastern province of the same name in Holland. Although it has not been officially bred since the late 1960s (the Gelderland has been absorbed into the Dutch Warmblood), some breeders still maintain this fine old breed. Gelderland horses trace to native mares of the Gelderland province, crossed to Andalusian, Neapolitan, Norman, Norfolk Roadster, and Holstein stallions. In the nineteenth century East Friesian, Oldenburg, Hackney, and Thoroughbred blood were introduced. The Gelderland is an elegant carriage horse and also possesses talent as a show jumper. Its action is flowing and the breed has an effective high-stepping trot. Gelderland horses have made a major contribution to the quality and style of the Dutch Warmblood and Dutch Tuigpaard, a showy carriage horse. Gelderland horses have always served as riding and working horses for the farms in Holland.

The Gelderland is chestnut, bay, black or grey, often with white markings, with an occasional skewbald. The head is long and somewhat flat with a straight profile, large, expressive eye, and nicely set, mobile ears; the neck is well shaped and muscular, usually with an arch. Withers are usually prominent and broad, merging into the line of the back; the back is straight and long; the croup is short and usually flat; the tail set high; the chest full and deep; the shoulder sloped and long. The legs are muscular with a long forearm and strong, broad joints; the hooves are broad and strong.

Additional information can be found under Dutch Warmblood and Dutch Tuigpaard.

German Riding Pony
(Deutsche Reitpony)

ORIGIN: *West Germany*
APTITUDES: *Riding pony*
AVERAGE HEIGHT: *13.2 to 14.2 h.h.*
POPULATION STATUS: *Common*

The German Riding Pony is a new breed, developed from crossing various pony breeds such as Welsh, Connemara, New Forest, and Dülmen with Arabs and Anglo-Arabs. The ponies are meant to represent small horses

German Riding Pony. Photo courtesy Rheinisches Pferde-stammbuch e.V.

Gidran stallion. Photo: Dr. Csaba Németh

without pony type for children to ride, but they are also excellent for driving. They have good temperament and are generally easy to handle, although spirited.

The German Riding Pony is now consolidated as a breed, and crosses to other breeds are no longer used. This is an elegant pony with very good conformation and action. Selection of breeding stock is strictly maintained. There is no distinct type as a breeding basis, so individuals differ in type and size.

Apart from its use for children, the German Riding Pony is seen in increasing numbers at show jumping and other sporting events.

In East Germany, the counterpart to this breed is the Lewitzer pony (see Lewitzer).

Gidran

ORIGIN: *Hungary*
APTITUDES: *Riding, harness, light draft*
AVERAGE HEIGHT: *16.1 to 17 h.h.*
POPULATION STATUS: *Rare*

The Gidran is descended from Siglavy Gidran, a chestnut stallion imported in 1814 from Arabia, through his son, Gidran II. This horse was brought to the royal stud farm at Mezöhegyes as a breeding stallion. Horses and breeds at this stud farm were separated according to color at that time. By 1850 the "yellow" stud at Mezöhegyes Stud consisted mostly of Gidran-bred horses, although purposeful breeding of these did not commence until some time later. To improve some of the breed faults English Thoroughbred stallions were used, and as a consequence many individuals of bad temper resulted. To eliminate this problem as well as to recover losses of the stock, which had been greatly reduced during World War I, Arab and Kisber stallions were used.

The Gidran breed was arranged genealogically in 1943 to A, B, and C bloodlines, of which only the B strain remains today. The breeding stock, greatly reduced in number, was transferred in 1958 to the Dalmand state farm. Meanwhile, the stock was further reduced. In 1977, through purchasing rustic-bred Gidran mares, the OTAF (National Feeding and Breeding Inspectorate) founded a new stud farm in Borodpuszta. Two stallions were imported from Bulgaria for introduction of fresh blood. The aim is to breed a race of large-framed multipurpose horses with an appealing exterior. They are also used to improve the rustic public stock.

The Gidran has a vigorous constitution. This is a saddle horse with a large frame and a medium to heavy harness horse, for there are two distinct types found within the breed today. The head is small and well set on, elegant and proportional to the body mass; the neck is slightly arched and muscular; the back and loins are tight; the barrel is deep and well sprung; the shoulder is muscled with good slope. The joints are strong, the legs well muscled with short cannons, and the feet of natural good shape with hard horn.

The usual color of the Gidran is chestnut, but all colors common to the Arab horse occur in this breed.

Gidran horses can be classified as a warmblood saddle and harness type, useful for farming, saddle, touring, and equipage sports. Jumping ability is mediocre. Previously some individuals were unmanageable due to bad temper or weak serviceability, but that type has virtually been eliminated today.

As the Gidran now exists only in small numbers, the primary task on the breeding agenda is maintenance of the gene pool to prevent genetic erosion. Since the number of purebred stallions is low, crosses to Kisber Halfbred and Thoroughbred stallions have been used in numerous cases. The breed received fresh blood in 1976 from a Bulgarian stallion. Selection of breeding stock is directed to conformation and type, and candidates for brood stock must pass tests under saddle or in harness.

Gotland

(Skogsruss)

ORIGIN: *Sweden (Island of Gotland)*
APTITUDES: *Riding pony, light draft*
AVERAGE HEIGHT: *11.2 to 13 h.h.*
POPULATION STATUS: *Common*

Prior to 1955 there was practically no pony breeding in Sweden except for the Gotland in the Baltic. This is an old native breed, the subject of much interest due to its antiquity.

During the 1980s, good economic circumstances and increasing holidays and leisure time have caused an increase of interest in the hobby-horse.

The Gotland, which has existed on the island of Gotland since the Stone Age, descended from the ancient northern European

The famous Gotland stallion, Dröm. Photo: Kurt Graaf

forest horse. The breed is very reminiscent of the British Exmoor, and many branches of the breed are difficult to distinguish from Exmoors. Gotland ponies have a strong constitution and are noted as naturally good jumpers and fantastic trotters.

The Goths established their civilization on Gotland by 1800 B.C., and their sculpture shows that they used Gotland-type horses to pull chariots by 1000 B.C. Beginning about 200 B.C., waves of Goths began to migrate from Gotland Island and south-central Sweden into eastern Europe. They traveled with their families, household goods, and livestock, setting up settlements. Their line of travel extended through what are now eastern Germany, Poland, and western Russia to the Black Sea, where they settled for approximately one hundred years. Even today, there is a town in Russia on the Dnieper River known as Gulsvenskbi or Gammalsvenskby, meaning "Old Swedish Town." Around 270 A.D. they continued into the Danube River basin and from there were driven into the Roman Empire by the Huns about 370 A.D. The Goths first cooperated with the Romans but later overthrew them, sacking Rome in 410 A.D. They invaded the Spanish Peninsula in 414 A.D., eventually setting up their capital in Toledo and ruling until the Moslem conquest in 711.

The Goths left no literature, only fragmentary runic inscriptions. From their travels however, it is easy to suppose that many breeds in Europe are in part descended from the ancient Gotland horse. The (restored) Tarpan, Konik, Hutsul, Dülmen, Vyatka and Pechora, Bosnian, Garrano, and possibly even the ancient Sorraia of Spain and Portugal had genetic links to the Gotland breed. Nearly all of the breeds mentioned range in size from 12 to 14 hands and are predominantly of dun coloration. Originally the predominant color of the Gotland was some shade of dun with a dorsal stripe.

The Gotland has changed little since the Stone Age model. The pony is also known as Russ and is sometimes called Skogsruss, meaning "little horse of the woods," or "little goat," due to its surefootedness.

Gotlands are light and elegant. The head has a straight profile, small alert ears, and large, lively eyes; the neck is short and well muscled. Withers are pronounced; the back is long and straight; the croup slightly sloping; the chest deep; and the shoulder muscular and long with a good angle of slope. The legs are strong and well muscled with solid joints, clearly defined tendons, and long cannons. The hoof is strong and hard.

Gotlands are often kept on small farms in Sweden for light work, but today the primary use of this ancient breed is as a child's mount.

The only breeder I was able to locate in the United States is Leslie Bebensee of Corinth, Kentucky. She is working to restore the small number of Gotlands in this country and at the time of this writing has a herd of twelve fine animals.

Groningen

(Groninger, Groningse paard)

ORIGIN: *Holland*
APTITUDES: *Riding, light draft*
AVERAGE HEIGHT: *15.3 to 16.1 h.h.*
POPULATION STATUS: *Rare*

A Dutch farm horse which has been successfully used as a heavyweight saddle horse and an excellent carriage horse, this neighbor to the Dutch Gelderland, like that breed, has also been absorbed for the most part into the world-renowned Dutch Warmblood. The breed

Old-type Groningen of Holland. Photo: Evelyn Simak

Groningens of the old, original type. Photo: Sally Anne Thompson

originated in the northwestern province of Groningen through crossing of the East Friesian and Oldenburg breeds on native stock. An attractive carriage horse, the Groningen has a quick, stylish action and also exhibits great endurance, docility, and obedience.

The Groningen breed nearly became extinct in the 1970s, when only one purebred stallion remained (other than its crossbred descendants in the Dutch Warmblood breed), but breeders have taken steps to assure preservation of this old breed. Oldenburg blood has been used to prevent too much inbreeding. The powerful quarters of the Groningen have contributed greatly to the jumping power of Dutch Warmblood horses, and the Groningen is established as a basic type in the Dutch Warmblood stud book.

Groningen horses have a long head with straight profile and long ears. The neck is of medium length and well muscled, wide at the base; the withers are prominent and long; the back is long; the croup is flat with a high tail-set; and the quarters are powerfully muscled. The chest is wide and deep with great depth to the girth. The legs are short, well muscled, and strong with excellent joints and well-shaped hooves. The primary colors are black, bay, or brown.

Additional information can be found under Dutch Warmblood.

Guanzhong

ORIGIN: *China (Weihe basin)*
APTITUDES: *Riding, draft*
AVERAGE HEIGHT: *14.3 to 15 h.h.*
POPULATION STATUS: *Common*

The Weihe basin is a famous agricultural region in the province of Shannxi. Historically, this area produced good horses for many neighboring regions. The basin ranges from 1,100 to 2,300 feet in elevation. With good irrigation and established food and cash crops, it is an ideal region for developing agriculture and animals. Besides the Guanzhong horse, the famous Qinchuang cattle and Guanzhong donkey were developed in this old political, economic, and cultural center.

The Guanzhong region, irrigated by the Weihe River, nursed the ancient Chinese culture to a high level of agriculture. Horses were indispensable for many purposes. In the early 1950s a great need for food and transport called for a new, stronger type of horse. A breeding goal was set to produce a horse of high performance with greater body weight, speed, and easy trot, and plenty of pulling strength. This program was started in Liuling stud farm in the province of Shannxi. The Guanzhong is a developed breed.

Budyonny, Karabair, and Soviet "high-blood" (improved Thoroughbred) stallions were imported from the Soviet Union. Cross-breeding was graded up until the second generation. Immediate results did not meet the requirements, as the body of the horse was still too high, resulting in unsatisfactory pulling power. After 1958 Ardennes stallions of different types were used to produce heavier body weight, conformation, and pulling pow-

Guanzhong stallion of China. Photo: Weigi Feng and Youchun Chen

Guanzhong mare of China. Photo: Weigi Feng and Youchun Chen

er. In 1965 the ideal animals were mated using the principle of homogenous pairs with a complimentary practice of crossing. In 1970 the population consisted of 12.5 percent local blood, 25 percent light type, and 62.5 percent heavy type. The inbreeding coefficient was 3.38 percent.

Guanzhong horses have a strong constitution and good conformation, dry and compact. They have a good temperament and bold attitude. The head is of medium size, clean-cut with a straight profile and straight set of ear; the neck is of medium length, slightly crested. Back and loins are wide and straight; the rump is square and well connected; the legs are squarely set with good joints and hooves and clean of excess hair. The primary coat color in this breed is chestnut, though some are bay.

Noted for endurance, the Guanzhong horse returns to normal respiration and heart rate rapidly after a hard workout. The breed is well adapted to local conditions, even in high temperatures over 100°F. It is used for improvement of other local breeds.

Guizhou

ORIGIN: *China (Guizhou Province)*
APTITUDES: *Riding, draft*
AVERAGE HEIGHT: *11.2 h.h.*
POPULATION STATUS: *Common*

Guizhou stallion of China. Photo: Weigi Feng and Youchun Chen

Guizhou is a mountainous province. Agriculture was developed as early as 770 B.C. in the basin area of the province, and trade in horses and salt was of great importance. Some breeds from outside the country were introduced in the 1950s but did not bring beneficial results, so the Guizhou is still bred in pure form and is classified as a native breed.

In plantation areas the horses were used mainly for cultivation and transport. In hilly regions this breed has always been a good pack horse, somewhat smaller than the type used for agricultural purposes.

The Guizhou horse is active and vigorous, good for travel in hilly regions, and can trot at a steady speed for long distances. This breed is the main power source for the mountain farmers.

Approximately half of Guizhou horses are bay or chestnut, others being grey, black, and dun. This horse is compact and short bodied with a solid build. The head has a straight profile, ears small and set up. In the riding type the neck tends to be sloped, while the pack type has a level neck. The withers are medium in height and length; the chest is of good width and depth; the rump is short and sloping but well muscled; the shoulder is short and tends to be straight. The forelegs have good posture while the hind legs are often sickled; the hoof is solid and tough and shoes are unnecessary even in the mountains. The skin is thin; mane and tail are thick.

Guizhou horses are willing and obedient and have a kind, patient temperament.

Guoxia

ORIGIN: *China (Southwest China)*
APTITUDES: *Riding, light draft*
AVERAGE HEIGHT: *9 to 10 h.h.*
POPULATION STATUS: *Rare*

A breed of great antiquity, the Guoxia is found in Debao, Jingxi, and Tianyan Counties, where there are many rugged and rocky mountains. Its name means "under fruit tree horse," illustrating the small size and early use of this animal—standing beneath trees so that baskets could be easily filled with fruit. This pony was used as a plaything for the concubines and wives during the Han dynasty. In museum records of the Song dynasty it was said the Guoxia was used as an interesting diversion for leisure time in the life of the emperors. This record is believed to be the oldest reference to amusements in the imperial courtyard.

A set of bronze statues of the Guoxia with its handler (called a pastor) was excavated in Guanxi province, dated at 2,000 years old.

Forgotten and thought to be extinct, the

A stone carving from the Ming dynasty showing the Guoxia. Photo: Tiequan Wang

Guoxia ponies pulling a trailer in China. Photo: Tiequan Wang and Youchun Chen

A tiny Guoxia mare and her foal. Photo: Tiequan Wang and Youchun Chen

Guoxia was rediscovered by officials in 1981. About 1,000 horses standing less than 10 hands were found in a rocky mountain area. A local name for the breed is the Rocky Mountain Pony. The average temperature for this area is 30° to 68°F and the rainfall is heavy.

The mane, tail, and limb hair is thick and heavy in the tiny Guoxia, distinctly different from that of other horses in China. The Guoxia is classified as a native Chinese breed.

A breed association was formed in April of 1981, registering 390 animals. The head of this association is Professor Wang Tiequan, an associate author of *Horse and Ass Breeds in China*, one of the very few books ever published on the subject.

Today the Guoxia is used in children's parks in twenty large Chinese cities. This small horse is lively and pretty with good conformation. The breed is extremely hardy. Small size remains consistent regardless of improved climate or feeding conditions, making the Guoxia a true pony or miniature horse as opposed to other Chinese breeds.

This breed is based in the same area as that where the Baise breed originated and they have similar conformation, although the Guoxia is much smaller. Determination as to whether the Guoxia is a pony or a miniature horse has not yet been scientifically made. The Guoxia does not appear to have typical pony characteristics. If it is a miniature horse, it is an extremely rare find. The only other natural miniature horse in the world is the Caspian; all others have been humanmade by selective breeding.

The head is short and slightly heavy with a straight profile and small ears; the neck is short, thick, and level; the withers, back, loin, and croup are at the same level, but the croup is gently sloped. The abdomen is rounded; limbs are straight with good joints and strong feet. Main colors found in the Guoxia are bay, roan, and grey. For additional information see Chinese Horse.

Hackney

ORIGIN: *England*
APTITUDES: *Harness horse*
AVERAGE HEIGHT: *14 to 15.3 h.h.*
POPULATION STATUS: *Common*

The Supreme Champion Hackney stallion, Hurstwood Gondolier. Photo: J. Moor

Hackney horse in America. Photo: J. E. McNeil

History makes it evident that England had a breed of trotting horses during the Middle Ages and the Tudor era. Trotting horses from Norfolk were valuable in those days and brought high prices.

The Hackney breed originated in England about the same time as the English Thoroughbred. East Anglia and the East Riding of Yorkshire had for many years each bred a distinctive type of animal: the Norfolk Trotters and Yorkshire Hackneys.

Generally the Norfolk type was cob-like and the Yorkshire type showed more quality. These were practical strains of light working horses which traveled the roads of England long before the advent of motor transport and rail.

The Norfolk Trotter, later known as Norfolk Roadster, was a strong, well-built carriage horse with great stamina. The blood of this breed was used to improve and develop several other breeds, including some in the United States.

Both the Norfolk Trotter and the Yorkshire Coach Horse were used in their day much as we use a taxi service today. Both types depended ultimately for their future success on a stallion born in East Anglia, the Original Shales, whose number in the Hackney Stud Book is 699. Shales was foaled around 1755, sired by a Thoroughbred called Blaze, a grandson of the Darley Arabian. His dam is described only as a "Hackney mare."

Much credit for development of the Hackney breed must go to two men who gave the area of East Riding its name as one of the homes of the Hackney breed—father and son Robert and Philip Ramsdales, who lived at Market Weighton. It was they who took to the East Riding area stallions such as the black Wroot's Pretender and the red roan Norfolk Phenomenon, thus enabling Yorkshire trotting mares to be crossed with the blood of the Original Shales. When blood of the Norfolk Trotter and Yorkshire Coach horse were blended, the modern Hackney was the result.

It seems the custom to credit only the foundation stallions with the breed they helped to spawn—but what of the mares to which they were mated? Both the Thoroughbred and Arab are gallopers, but the developing Hackney was a trotter. Some of the accounts of Hackney history, especially in speaking of the Norfolk Trotter and Yorkshire horse, mention Thoroughbred stallions being bred to "native road mares." It is also said that the old Norfolk Trotter resulted from Danish horses introduced into England with the army of King Canute in the eleventh century. They were reportedly mated to "native mares" as well as to Dutch and Danish Harddravers, a term used when speaking of the Dutch Friesian breed—which would certainly explain the talent for the trotting gait found in those early English roadsters. The Hackneys and roadsters thrived from the early part of the nineteenth century until railways reduced the demand for horses. Hackney horses found another niche in the show ring due to their style, while the sturdier, less spectacular breeds of roadster died out. In 1833 the stud book for Hackneys was founded.

Between 1939 and 1945 the Hackney horse and Hackney pony met and overcame the most critical period in their history. At times their existence as a national breed was in danger. Though there were no large shows to attract breeders, foreign trade was at a standstill, and commercial trade in England was limited, a few owners kept their studs operating. By the end of 1942 there were signs that the market for pedigreed Hackneys and Hackney ponies was beginning to reawaken. Once again there was demand for Hackneys for the road and for exhibition at charity shows. By the end of the war, prices were reaching record heights. In 1946 the Hackney Horse Society resumed its national breed show at Crewe, and by 1947 Hackney horses and ponies had once again taken a foremost place among the attractions at show rings throughout the country.

The modern Hackney is a harness horse. It is known also as the "roadster" and the "stepper" and attains its glory in the show ring, where its unique and elegant action makes it an exhibit of unsurpassed attraction. This is a horse of beautiful conformation and style.

The first Hackney imported to America was named Jary's Bellfounder, a beautiful bay with superior bloodlines, conformation, and action. He was foaled in 1816 and had great influence on development of the American Standardbred through his daughter, the Charles Kent mare, dam of Hambletonian 10 (Rysdyk's Hambletonian).

The modern Hackney is black, brown, bay, or chestnut, with or without white markings. It possesses a small head, muzzle, and ears; the body is compactly constructed; the neck should be long and blend into a broad chest; the shoulders are powerful and sloping; the back is level; the ribs well rounded; the loin short; the croup level with a high-set tail. The legs are of medium length with large, strong joints; the thighs and quarters are well muscled. The pasterns are of good length and slope. The Hackney has excellent feet and a reputation for soundness.

The hallmark of this breed is its spectacular and highly distinctive action. The shoulder action is fluid and free with very high, ground-covering knee action. Motion of the hind legs is similar but to a lesser degree. The hocks should be brought under the body and raised high. All joints should show extreme flexion. The action must be straight and true, and the entire effect must be arresting and show great brilliance.

Hackney Pony

ORIGIN: *England*
APTITUDES: *Harness Pony*
AVERAGE HEIGHT: *12.1 to 14.1 h.h.*
POPULATION STATUS: *Common*

The Supreme Champion Hackney Pony, Hurstwood Untouchable. Photo: J. Moor

Hackney pony in America. Photo courtesy American Hackney Horse Society

The beautiful Hackney pony was developed in England in 1872 through scientific selection to establish a fixed pony type within the Hackney breed. In addition to Hackney blood, this pony carries the blood of the Fell and Welsh ponies.

Breeding of the Hackney pony was started by Christopher Wilson of Westmorland. Using a roadster stallion with Yorkshire and Norfolk blood by the name of Sir George, he bred the best Fell mares available. The blood of Sir George contributed speed and elegance, and the Fell mares added high knee action and substance developed on the moors.

The success of these ponies was immediate. They were fast, fancy trotters that, when hitched to light carriages, made a speedy and impressive mode of travel. In 1896 a Hackney filly sold for the equivalent of $1,000, a grand price for that day!

Hackney ponies have been exported to many countries of the world and are popular in driving and carriage classes. The pony, spirited and very full of itself, with bright eyes and exuding a look of fire-under-control, is an impressive sight in any show.

The Hackney pony may be bay, brown, grey, black, or roan. White markings are permitted. The head is light and somewhat long with a straight profile, small, pointed ears, and large, expressive eyes. The neck is arched, muscular, and set well onto the shoulders; the back is short and straight; the croup long and slightly rounded; the tail set high and carried gaily; the shoulders are muscular and sloping. The legs are slender but very strong with good joints and hard feet.

In general this pony presents beautiful conformation combined with a great deal of style. The Hackney pony does not have its own stud book and is registered in the same book as the Hackney horse.

Haflinger. Photo: Elisabeth Brotherton

The Haflinger in driving competition. Courtesy Haflinger Society of Great Britain

Haflinger

ORIGIN: *Austria (southern Tyrol)*
APTITUDES: *Riding, light draft*
AVERAGE HEIGHT: *12.3 to 14 h.h.*
POPULATION STATUS: *Common*

Emerging high in the mountains of the south Tyrol, the Haflinger is a popular small breed. Of ancient origin, this horse was isolated from other breeds by the terrain of its habitat. Harsh environment shaped it into a healthy, robust animal that breeds very true to type—

A group of Haflingers at Chatsworth in 1982. Courtesy Haflinger Society of Great Britain

so true that individuals are universally the same color, a rich, golden chestnut with white mane and tail. Most Haflingers have a full blaze on the face, but some have a star or narrow strip.

Early origin of the Haflinger is uncertain. The main area of breeding was around the village of Hafling in the valley of Sarn. Early inhabitants are believed to have been descendants of the Ostrogoths who settled in the area in the middle of the sixth century. They had been driven there by Byzantine troops after the fall of Conza in 555 A.D. They crossed their small Arab-type horses with the cold-blooded native mountain ponies. Another belief is that the Haflinger traces to a stallion sent as a gift to the Margrave Louis of Brandenburg as a wedding gift in 1342.

Later, Arab blood was infused when the stallion El Bedavi XXII was used beginning in 1868. A son of his, Follie, foaled in 1874, became the foundation sire of the breed. This stallion is said to have been a chestnut with white mane and tail, and inbreeding during the nineteenth century fixed this color in the breed.

Haflingers are sturdy and well built with plenty of bone. The disposition is calm and willing and, although small, this horse is strong and capable. The breed has characteristically been used by farmers and has also served as a pack animal during both peace and war time.

The Haflinger has a small, well-shaped head showing Arab influence; the neck is strong, muscular, and broad at the base; the shoulders are nicely sloped and well muscled. The back is straight, strong, and short; the croup muscled and sloping; the tail set low; the chest deep and wide; and the quarters are powerful and muscular. The legs are short with broad, hard joints, muscular forearms and thighs, straight hocks, and cannons free of much feather. The hooves are hard and healthy. Mane and tail are full and long.

The Haflinger has spread far beyond its original home and is bred today in over twenty countries. In Austria, state studs own the stallions and careful quality control is maintained. Colt foals are carefully inspected, and only a few are chosen for use as future sires.

The first Haflingers to reach Bavaria in Germany were imported by Major Rudolf Erhard and the breeding director, Dr. Franz Gentner, in 1935 for use as pack animals. Bavarian farmers rapidly took a liking to the breed, and currently Bavaria has 2,500 breeding mares and more than 100 registered stallions.

Recently experiments have been carried out in Bavaria breeding the Haflinger to Arabs, and the cross is called Arabo-Haflinger. The first generation is usually sorrel with a red mane and tail. When crossed back to the pure Haflinger the next generation again has the obligatory white mane and tail. Most Arabo-Haflingers are 25 percent Arab and 75 percent Haflinger. This crossbreeding is not practiced in Austria.

The Haflinger is long-lived. While it is not usually put to work prior to four years of age, many claim this breed is still able to work at over forty years.

In Italy descendants of this same small mountain horse are called Avelignese. Haflingers were first imported to England in the 1960s by the Duchess of Devonshire, followed by other enthusiastic breeders. The first imports were on the whole thickset and study, but after more than twenty years the breed has changed somewhat, being taller and more refined.

Half Saddlebred

ORIGIN: *United States*
APTITUDES: *Riding horse*
AVERAGE HEIGHT: *15 h.h.*
POPULATION STATUS: *Common*

The Half Saddlebred Registry of America was formed in 1971 and has enjoyed steady support and growth. The American Saddlebred, known for quality, refinement, substance, endurance, speed, elegance, and wonderful gaits, is a good cross to almost any breed.

Half Saddlebreds have proven to be "horses for all reasons" and have won championships in dressage, open jumping, pleasure riding, and driving. They are horses of beauty, grace, and intelligence regardless of the pedigree of the non-Saddlebred parent and can be trained

La Bomba, Half Saddlebred stallion. Photo: Valerie Kieser

Half Saddlebred stallion, Syrian's Genius. Photo: Sandra Anderson

Hanoverian

(Hannoveraner)

ORIGIN: *Germany (Lower Saxony)*
APTITUDES: *Riding, sport horse*
AVERAGE HEIGHT: *15.3 to 17 h.h.*
POPULATION STATUS: *Common*

Hanoverian stallion. Photo: Werner Ernst

Hanoverian. Photo: Werner Ernst

for any field of horsemanship. The registry now has horses registered in England and Scotland, and four Half Saddlebreds have been exported recently to Japan.

For registration in the Half Saddlebred Registry of America, either the sire or dam must be a registered American Saddlebred. The HSRA supports all youth organizations and is a yearly sponsor for National 4-H Council activities.

Known today as a world-class show jumper, the Hanoverian originated in the seventeenth century when Oriental, Spanish, and Neapolitan stallions were imported and crossed with local mares. In 1735 George II, King of England and Duke of Brunswick and Lüneburg, founded the state stud at Celle, and fifteen black Holstein stallions were included in the breeding program. The stud was established to provide farmers with the service of good stallions for their mares at a nominal fee. For the first thirty years, Holstein blood was dominant. Then Thoroughbred blood was used and a lighter horse was produced. The goal in breeding was to produce a horse suitable for riding while still useful for driving and farm work.

Evolution of the Hanoverian may be broken down into four main periods, even though transition from one to another is not sharply defined. The first period extended from establishment of the Landgestüt Celle in 1735 to the wars for liberation from Napoleonic domination, from 1812 to 1813; the second period, from 1815 through 1870, saw the increased use of privately owned Thoroughbred and English halfbred stallions; from 1870 through 1945, the breed was genetically being consolidated; and 1945 to the present has seen refining of the type with the aim of producing a superior quality riding and competition horse.

From the beginning, mares produced by the state-owned stallions were registered and detailed pedigrees were kept by the end of the eighteenth century.

During the Napoleonic wars breeding stock was severely depleted and only thirty of the one hundred stallions returned to Celle when efforts were made in 1816 to reestablish the stud. To increase numbers, horses were obtained from England and Mecklenburg. At one time, 35 percent of Celle's stock was Thoroughbred. It became evident that too much Thoroughbred blood was producing a horse too light in structure to perform the work that was required. After World War II many Trakehner stallions were sent to Celle.

As agricultural needs for horses diminished, the breeding goal changed to producing a good competition horse. Trakehner and Thoroughbred blood was once again used to refine and upgrade the breed.

Today the Hanoverian enjoys high demand for use in dressage and show jumping. This horse is large, strong, and has good conformation. Hanoverians are used in the production of another warmblood strain in Germany, the Westphalian.

Since its beginning in the early eighteenth century as a farm and coach horse, the Hanoverian has changed dramatically in its physical characteristics. The breed is still found in a range of sizes and types, but they all possess the ability to move in a spectacular manner and have a temperament that enables them to accept discipline and direction from their rider without rebellion. Many Hanoverian horses have a rather plain head with a straight profile; the neck is long, muscular, and well set on; the shoulder is powerful and sloping; the body is deep; the hindquarters are very powerful. This is a strong, well-made horse with good feet. The main colors in the Hanoverian breed are chestnut, bay, brown, black, and grey. White markings are frequently seen.

This is probably the most famous breed of competition horses in the world. Selection of breeding stock is done through testing and careful evaluation of both conformation and pedigree. The auctions at Verden attract buyers from many foreign countries.

The aim in breeding the Hanoverian horse is to produce an elegant warmblood of large proportions, with action full of verve and elasticity and covering a lot of ground; a horse that, on the basis of its temperament, character, and ridability, is suitable for all equine activities. This breeding aim is being achieved through the method of pure breeding and strict selection.

Four hundred top-class riding horses go through the Verden auction each year. At that time there are great comparison shows for young Hanoverian horses in all variations of type, model, and capability.

The Hanoverian horse is well represented by world champions and Olympic winners. Stallions and broodmares are regularly bought by breeders in Europe, the United States, Asia, Africa, and Australia.

Heihe

ORIGIN: *China (Heilongjiang River basin)*
APTITUDES: *Riding, draft and farm work*
AVERAGE HEIGHT: *14.2 to 15.2 h.h.*
POPULATION STATUS: *Common*

Heihe horse of China. Photo: Weigi Feng and Youchun Chen

Along the boundary between China and the former Soviet Union there is a long river basin called Heilongjiang. Heihe city is the center of the Heihe horse breeding region. Beyond the forest/tundra to the north, the terrain is hilly with an altitude of 1,640 to 3,300 feet and a gradual slope. The area is rich in water resources. Average temperature is near 32°F, with the minimum about −58°F and the maximum in summer about 89°F. Horses are absolutely necessary for both agriculture and transport in this area.

The Heihe horses developed on the basis of Soulun and Mongolian stock. In the *Longsha summary* it was recorded that prior to the seventeenth century the Heihe area was occupied by the Soulun nation, which was connected to the Manchu. To enforce defense, the Qing government sent a large regiment of cavalry there that contained many Mongolian horses.

After 1820 lumber and mining for gold again brought Mongolian stock to the area. In 1908 Russian immigrants brought horses of unknown origin, and after 1930 some Orlov Trotter and draft stock was added to the growing mixture. During the years from 1937 to 1939 a Russian strain of Mongolian was brought in and at the same time the Keshan stud farm was organized. Anglo-Norman stallions were mainly used and four insemination centers were established. By 1940 there were sixteen stallions at this stud farm, among them two Anglo-Arabs and four Anglo-Nor-

mans, and others were crosses from Anglo-Norman and Percheron. Systematic selective breeding was not done until 1955, when the North Horse Farm was founded. Finally, in 1963, the new breed was officially recognized. Thus the Heihe is a developed breed.

The Heihe breed has good, uniform conformation and a willing, obedient nature. The head is medium in size with a straight profile, the eye is large and open, and the ears are long. The neck is medium in length and wide at the base; the withers are high; the croup tends to be short; the limbs are strong with long forearms and short cannons. The hock is usually not straight enough in this breed. The Heihe is found in two types, riding-draft and draft-riding. The main coat color is bay or chestnut. A few are grey or black, and other colors are rare.

These are powerful horses with good endurance qualities, easily able to pull heavy loads. They are also used for plowing and other farm work. Well adapted to the cold, the Heihe stays out overnight in temperatures ranging to −40°F with no ill effects.

Heilongkiang

ORIGIN: *China (Song-liao plain)*
APTITUDES: *Riding, draft*
AVERAGE HEIGHT: *14.2 to 15 h.h.*
POPULATION STATUS: *Common*

Heilongkiang stallion of China. Photo: Weigi Feng and Youchun Chen

Heilongkiang mare. Photo: Weigi Feng and Youchun Chen

Song-liao plain in the wide Heilongkiang basin in northern Heilongkiang Province is the region of development of this breed. There are a large number of state-owned farms, and good grasslands played an important part in formation of the Heilongkiang breed. This region has a very complicated and varied topography. Extremely rich humus is found in agricultural and semiagricultural zones; there are large steppe areas, and forest surrounds the entire region.

The Heilongkiang breed was developed through an intricate crossbreeding system. Records beyond 1950 are varied and uncertain. In 1950 there were 510 Ardennes, 425 Soviet Heavy Draft, 30 Vladimir Heavy Draft, 130 Orlov Trotters, 79 Soviet Thoroughbreds, 143 Don, and 390 Kabarda horses introduced to cross to local stock. Heavy and light breeds in turn were used to breed the next generation of animals as one variant. Rotational crossing with light breeds was rare in this system. Artificial insemination was important in founding the new breed. Among the above mentioned breeds, the Ardennes, Orlov, and Soviet Highkravnaya (improved Thoroughbred) had the greatest influence. This program of crossbreeding was carried on throughout a large area and involved many animals. Herds with one stallion were organized in many stud farms, and a large number of separate lines were recognized. One particular stallion started two important lines

which, through selection of offspring, influenced half of the existing pedigrees. Eventually, eight lines were selected for recognition and the breed was considered established by 1975.

There are two main types in the Heilongkiang breed, riding-draft making up 55.5 percent and draft-riding the other 44.5 percent of the total. Horses of this breed are generally of good conformation and posture, well muscled and solid in hoof and joints. Chestnut is found in 56.6 percent of the breed with the rest being bay in color.

The Heilongkiang is a horse of strong endurance and constitution, keeping its condition under hard work and well able to withstand cold and poor forage.

Hequ

ORIGIN: *China (Qinghai Province)*
APTITUDES: *Riding, racing, draft, pack*
AVERAGE HEIGHT: *12.2 to 13.2 h.h.*
POPULATION STATUS: *Common*

The Hequ breed originated in the border area of Qinghai, Sichuan, and Gansu Provinces where the large "zig" of the Yellow River is found. (Hequ in Chinese means "river zig.") The name Hequ has only been used for this breed since 1954; prior to that time it was called Nanfan, a name often misapplied to the modern-day Tibetan horse. During the T'ang dynasty the Hequ was known as Tu-fan and was a horse often plundered in war.

In the nineteenth century the Hequ horse was taken from southern Qinghai to the northern part of the province where the Datong horse is found and used to improve that breed. The Hequ today is one of the most numerous breeds in China, with about 180,000 counted in 1980.

During the T'ang dynasty (618–917), the emperor established a large horse ranch to develop the cavalry, importing considerable numbers of horses from western Asian countries to cross with the local Tibetan stock. The Hequ bred was strongly influenced by the Dawan horse from western Asia, famous for "blood sweating." Dawan was the ancient Chinese name for the kingdom that included

Fergana, home of the ancient Akhal-Teke breed (see Akhal-Teke). In the Yuan dynasty (1279–1368), Mongols of the Xianbei tribe invaded this area and Mongolian stock was mixed with the Hequ and Datong stock.

The Hequ is versatile, being used for riding, racing, and draft and as a pack horse. It is strong and well able to carry 200 pounds a distance of thirty miles a day for several days in succession. It is classified as a native breed of China.

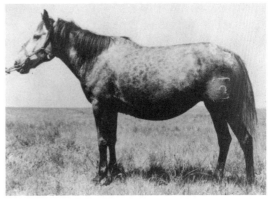

Hequ mare of China. Photo: Weigi Feng and Youchun Chen

Hequ stallion of China. Photo: Weigi Feng and Youchun Chen

Hequ mare. Photo: Youchun Chen

The habitat of the Hequ is at high altitudes, some 11,000 feet, where temperatures are low, but the climate is temperate and moist during the summer. Hequ mares mature at two years of age and are bred at three. The conception rate under year-round pasture conditions is 70 percent, but may be as high as 80 to 90 percent under good feeding and management. Mares may produce twelve to thirteen foals in a lifetime.

The Hequ is described as a draft-riding type, solid but rather coarse in build. The head is of medium length with straight or often "rabbit-like" head (a term used to indicate a profile that is convex above the nostril), long ears, large, open nostrils, and wide jaws tapering to a small muzzle. The neck is of medium length, sloped, and well connected to the shoulder; the withers are medium in height and long; the chest is wide and deep; the shoulder well sloped; the loin level; the croup slightly sloped. The limbs are medium in length; joints and tendons are well developed; the foot is large but not as strong as it should be.

There are three varieties or types of the Hequ breed. The Jiaoke type is located in the southern part of Gansu province. Horses of this variety are rougher in constituion, have a larger, coarser head, and do not have strong hooves. The main color of the Jiaoke is grey, and it stands on average around 13.3 hands. The Suoke type is located in western Sichuan province and is now raised at Baihe farm. This strain has a relatively large head and ears and short loin. The Suoke horse carries its tail very high, as did the T'ang horse of ancient times. The Kesheng type is located in the Kesheng Mongolian Autonomous Region in the province of Qinghai and is mixed with Mongolian horses.

Hequ horses may be black, bay, or grey.

Hessen
(Hessischer)

ORIGIN: *Germany (Hesse)*
APTITUDES: *Riding, sport horse*
AVERAGE HEIGHT: *16.2 h.h.*
POPULATION STATUS: *Common*

Hessen horse in dressage. Photo: Werner Ernst

Hessen horse. Photo: Werner Ernst

The tradition of breeding fine horses has existed for many centuries in all parts of Hesse. In the early days, studs of the reigning princes played the leading role. Some brief accounts of the activity of those princes follows, taken from the information sent by the Hessen breed association (see appendix C).

As early as 1490 the "Wild horses from the Sababurg Stud" *(Wilden aus der Stut Zapfenberg)* were mentioned—horses that were kept in the isolated forest area of the Reinhardswald. They were described as being hard as nails, strong, agile, with plenty of stamina, and were therefore the best barter items of the Hessen-Kassel court.

In 1724 Landgraf Karl elevated the Berberbeck Stud—together with the Bergpark (mountain park), which was enclosed by a high wall—to the status of court stud. Berberbeck's significance increased primarily when it became the principal stud of Prussia in 1876. In the following forty-five years this stud suppled 21 main sires and 418 state sires that gave important blood to almost all German breeding areas (especially Hanover and eastern Prussia).

Also about 1700, the court stud of Ulrichstein, which later became the Grand Duchy of Hessen, was set up in Vogelsberg. Until 1849 Oriental mares were paired there with Prussian, Mecklenburg, and English stallions to produce a strong horse of medium size with lean flanks, a high tail, and plenty of power and stamina. This type of horse was exactly suitable for the particular needs of the court and later served as a sire at stud.

In the area of Nassau (a historic region in the western part of the Land of Hesse), the court stud of the Dillenburg dukes became especially well known. The Dillensburger Ramsnasen (now extinct), bred here by crossing Holstein and Danish mares with Spanish, Neapolitan, and Oriental stallions, were strong, attractive, medium-sized horses with lean, hard limbs and a high gait. They were much sought after at horse markets in the sixteenth century.

Later, noblemen with an eye to the future set up provincial studs (Landgestüte). As early as 1737 Landgraf Wilhelm VIII established the Kassel Stud (which followed the example of Celle Stud founded two years earlier). In 1811 Grossherzog Ludwig I established the state stud of Darmstadt. Also in 1811, Herzog Friedrich August from Nassau established the state stud of Weilburg, and in the same year, Prince Friedrich of Waldeck set up the state stud of Arolsen. In these studs stallions for the area stood at service, at first free of charge and later for a small serving fee, with the obligation that breeders give the state first option on foals or young horses the breeder did not need. In 1870 the Prussian administration brought together the 103 stallions from Kassel as well as the newly established Hessen-Nassau provincial stud at Dillenburg. In 1956 the provincial stud in Darmstadt was closed and these stallions also came to Dillenburg.

For more than two hundred years the Hessian provincial studs have exercised strong influence on developments in local horse breeding and breeders have received advice concerning the keeping and breeding of horses.

Hessen, between the Weser and Neckar, forms a link between northern and southern Germany and is one of the most attractive German states, with beautiful countryside—chains of mountains forming close ranks and abundant woodlands between streams winding northward and westward. The conditions for horse breeding are not ideal, however. The limestone soils do make it possible to grow fodder rich in minerals and of high nutritional value, but only the most industrious farmers can earn a living from these small holdings. They compensate for the lack of surface area with a type of animal breeding far superior to cattle farming, and their success is noted in the quality of their horses.

In the first half of the twentieth century, breeding of the coldblooded horse was of prime importance. This horse was used for work purposes in Hesse. A parallel development was the working warmblooded horse, based on the Holstein and Oldenburg breeds. After World War II, as motor vehicles replaced the horse and needs and uses changed, these willing, friendly draft horses formed the basis for an attractive, large-framed, versatile riding horse.

Today a great majority of Hessian horses are raised on farming establishments with only one or two mares. In this way the relationship with human beings has not been destroyed—the horses are members of the family.

The hard breeding conditions in the Hessian hills guarantee superb health and a long life. Owners of a horse raised in Hessen have a long-lasting possession. The advice from the network of studs concerning breeding was and is taken very seriously by Hessian breeders, and today it is evident at international horse shows that the advice was effective and successful.

With the increase in popularity of riding as a sport, the foundation of breeding has been extended. The number of mares entered in the association's register has increased steadily since 1965. The age of distribution of the breeding population is of special interest: the young generation is gaining ground. Until 1985 an increase in the number of foals born was recorded. The number of one- to ten-year-old mares increased accordingly. A decisive factor for the work of the association is, however, that the proportion of the young generation in the total stock of mares has increased at the expense of mares over ten years old. This also means the desirable elimination of female animals that do not correspond with the type of horses in demand today. Purposeful selection and well-planned use of sires have exerted great influence on the breeding effort.

Stallions used in breeding the Hessen horse are mainly from the Hanoverian, Holstein, Thoroughbred, and Trakehner breeds. As a result the pedigrees of the new generation of Hessen riding horses include the names of internationally famous sires and are thus comprehensible to everyone.

Annual shows for young stallions as well as the evaluation of show and competition results enable a constant check on the performance levels of the stallions and give breeders decisive information to help in choosing partners for their mares.

In Hessen only "elite auctions" are organized. This means that from a large number of riding horses, only the best are chosen by a highly qualified commission, and all horses entered in the auction must bear the Hessian brand and be owned by members of the association.

In recent years many buyers from abroad have found their way to the Hessian auctions, including some from the United States, Belgium, Canada, England, Finland, France, Holland, Italy, Japan, Luxembourg, Mexico, Switzerland, and Taiwan.

The Hessen horse is a big, strong, well-made animal with good conformation and elastic, free-moving gaits. This is an excellent dressage or show jumping horse. Closely resembling the Hanoverian breed, its most common colors are bay, dark bay, and chestnut.

Highland mare. Photo: E. Owen

Highland

ORIGIN: *Scotland*
APTITUDES: *Riding, pack pony, light draft*
AVERAGE HEIGHT: *12.1 to 14.2 h.h.*
POPULATION STATUS: *Common*

A very old breed, the Highland pony inhabited Scotland before the Ice Age. In the past there were two distinctive types found in the Highland pony: the Scottish Mainland and the Western Island, the latter being lighter in build and smaller than the Mainland type. Recently there has been a good deal of inter-breeding between the two types and the distinction is not as marked as in the past.

This is another old pony breed whose hardy constitution was formed by survival under harsh conditions. The breed is very strong for its size with a well-proportioned body. The head is well carried, attractive, and broad between bright and kindly eyes, short between eye and muzzle, with wide nostrils. In color the Highland varies between several shades of primitive dun; yellow, cream, mouse — most have a dorsal stripe and zebra markings on the legs. Greys are numerous also, and some are black or brown.

Docility and a calm nature in a pony are essential, and these are born characteristics of this breed. A good pony will choose its own way over soft ground and cannot be induced to go on ground which it believes to be unfit to

The Highland mare, Tamara. Photo: E. Owen

bear its weight. Young ponies destined for hill work are often kept on the hills so that they gain experience on rough ground.

The Highland pony is used in the hilly country for riding and for carrying deer or panniers of grouse. The height of 13 to 14 hands makes it easy to load a stag, and ponies of this size have the strength and endurance to carry either a heavy man or a large stag.

With the growth of mechanization the Highland is not as popular today on farms as in former times. There is, however, still a place on the farm for a good pony, and still a market for the pony in agriculture. With the introduction of cattle on the hills, farmers began using the Highland on rough and boggy ground for roundups. On many hill farms the Highland is found to be of great service. Being hardy, it is less expensive to keep than any other type of horse or pony.

Molded by the harsh environment of the Highlands into a strong, hardy survivor, this pony has been familiar through the ages to king and commoner alike. Earliest records refer to such a pony's being ridden by southern Scots in the fourteenth century, including no less than King Robert the Bruce himself.

The Highland was favored over the larger war horses then for the same reasons it is liked today—ease of keep, strength, hardiness, and adaptation to hard environment. Highland trade inevitably looked toward the south, and ponies were of necessity included in the droves of animals heading for the markets and towns of Lowland Scotland and England.

When the new 1813 Commission roads were built, giving easier access to the Highlands, sportsmen, tourists, and new landlords, eager for wild scenery, flocked to the area. Pony trekking was founded on the broad backs of Highland ponies.

In 1952, to give summer work for deer ponies, Evan Ormiston opened the first trekking center at Newtonmore, Scotland. Since then the activity has spread throughout Britain, taking the breed with it. Gentle nature and quick learning ability make the ponies easily broken to the daily routine of a trekking center. With a good range of sizes, the Highland can accommodate all abilities and weights of riders, from the lightest novice to the more experienced adult.

No better breed for the hill, a fit Highland will also give a lively hack for trail riding. This breed is a natural choice for commercial uses, as winter keep is economical with the pony's ability to live outdoors. First contact with the Highland has come to many through trekking, resulting in a constant trade of ponies to new private owners. In both build and temperament the Highland makes an ideal family riding pony for all ages, sizes, and weights.

Properly broken and handled, Highland ponies are excellent mounts for riding clubs and pony club activities. Many of this breed are natural jumpers and in the show ring can increasingly hold their own in side-saddle, mountain and moorland, and working hunter pony classes (up to Olympic and Malvern competition standard). As useful, safe hunters and long distance ponies they excel in rough or steep country, where cleverness and steadiness are essential. The Highland has become increasingly popular with riding schools, and some riding centers in France now use these ponies exclusively for tutoring of both riding and driving.

The wide diversity of Highland pony type makes the breed an excellent choice for any harness work, whether pleasure driving, wedding hire, or in eventing. The breed is renowned for its strength and resilience, and these qualities, along with good temperament, are vital to success in driving. Easy to match in size and color and easily broken to harness, these ponies with their smart and striking appearance give pleasure wherever they go.

In recent years success has come to an increasing number of whips driving their Highlands, notably as overall champion at the Royal Highland Show and high placings in pairs in the BHS (British Horse Society) National Driving Trail Points Championship. Not only singles or pairs but also tandems and four-in-hand teams are now in use. Demands of the modern sport are intensive and increasing, but the Highland's scope and history of work in draft is more than keeping pace with developments.

Pedigree records have been kept since 1896, and the Highland Pony Society was founded in 1923. Affiliated societies now flourish in Australia and France, and the export of ponies is worldwide. The royal connection continues today at Balmoral Estate with the Queen's expanding stud, where ponies earn their keep in traditional manner carrying deer, driving, and trekking.

Hinis
(Hinisin Kolu Kisasi Ati)

ORIGIN: *Turkey*
APTITUDES: *Riding horse*
AVERAGE HEIGHT: *13.2 to 14.1 h.h.*
POPULATION STATUS: *Rare*

While thought by some authorities to be extinct, the Hinis horse still exists in small numbers in Turkey. The name Hinisin Kolu Kisasi Ati means "the short front legs of Hinis horse." Ertuğrul Güleç, my correspondent in Ankara, reported that he was alerted by friends that the breed still exists in Erzurum and in Hinis village, and he traveled to view them.

The Hinis horse descended from horses

The Hinis horse, drawing a crowd. Photo: Ertuğrul Güleç

Hinis horse. Photo: Ertuğrul Güleç

Hinis horse of Turkey. Photo: Ertuğrul Güleç

brought to Hinis by Turks during the Ottoman Empire. These were Turkish Arab horses which were crossed with the Anadolu breed. Selection of good animals from the offspring of this cross resulted in emergence of the Hinis horse. The breed is about one hundred years old.

The Hinis horse has an extremely wide and deep chest, allowing great capacity for heart and lungs. Due to the deep chest, the front legs are very short. This breed has exceptionally good bone, and the hooves are hard and always black. Hinis horses are found in all common horse colors.

The head of the Hinis horse is much like that of the Turkish Arab, with great width between the eyes. The breed has been crossed with the Turkish Arabian to produce excellent horses for the game of cirit (see Cirit). The breed is very hardy and enduring.

Only about 500 Hinis horses remain in pure form. It is hoped that the Turkish Ministry of

Agriculture will take steps to preserve this ancient and worthwhile breed that tenaciously survives at the brink of extinction.

Hirzai

ORIGIN: *Pakistan (Baluchistan)*
APTITUDES: *Riding horse*
AVERAGE HEIGHT: *15 h.h.*
POPULATION STATUS: *Rare*

According to Pakistan's Agricultural Department the original stock of the Hirzai breed is said to have been derived from a mare owned by the Rind chief named Shol, by an Arab stallion belonging to a European military officer who accompanied the contingent of Shaw Shujaul-Mulk through Shoran in the first Afghan War of 1839. Representative animals are still owned by His Highness Khan of Kalat.

The predominant color of the Hirzai is white or grey. The head is handsome with a broad forehead; the neck is medium in length, muscular, and arched; the body is compact with a short back and well-muscled loins; the croup is level; the shoulders are well sloped and powerful; the forearms are strong, but the legs lack bone. This is a horse of strength, good conformation, and stamina and it can be used for hard and fast work.

Hokkaido

ORIGIN: *Japan (Island of Hokkaido)*
APTITUDES: *Riding, light draft*
AVERAGE HEIGHT: *13 to 13.2 h.h.*
POPULATION STATUS: *Rare*

Generally called Do-san-ko in Japan as a term of endearment, the Hokkaido pony is an old breed. In the fifteenth century people from the mainland of Japan settled in Hokkaido, bringing horses with them. Primarily these were offspring of horses from the Tohoku breeding district. Development of the modern-day Hokkaido would not have been possible without the Do-san-ko, due to their adaptation to year-round pasture living with severe winter conditions. Most of these horses are found on the Pacific coast of Hokkaido. They

Hokkaido horse of Japan. Photo: Tadashi Yokata

Hokkaido horse of Japan pulling a "horse car." Photo: Tadashi Yokata

Hokkaido, at the pace. Photo: Tadashi Yokata

Hokkaido horse. Photo: Tadashi Yokata

Holstein
(Holsteiner Warmblut)

ORIGIN: *Germany*
APTITUDES: *Riding, sport horse*
AVERAGE HEIGHT: *16 to 17 h.h.*
POPULATION STATUS: *Common*

Holstein stallion. Photo: Werner Ernst

The Holstein gelding, Corlandus. Photo: Werner Ernst

are the most plentiful of the remaining ancient Japanese ponies, numbering about 2,000 head. Since 1979 pedigree registration has been kept for this ancient pony breed.

An interesting fact concerning the ancestry of horses in Japan focuses on the fact that the Chinese emperor Wu Ti, who reigned during the Han dynasty, occupied the Korean peninsula in 108 B.C. He established Lo-lang and three other colonies, providing a base for a strong influx into Korea of Chinese culture, which in turn spread to Japan. Wu Ti was interested in good horses and made successful raids to Ferghana to obtain the precious "blood sweaters" (see Akhal-Teke). It is possible that the modern small horses of Japan are descended in part from the hardy horses of the Chinese.

Hokkaido is the northernmost of the four main islands that make up the country of Japan. It is bordered by the Sea of Japan to the west, the Sea of Okhotsk on the north, and the Pacific Ocean on the east and south. Together with a few small adjacent islands it constitutes a territory with an area of 32,245 square miles, comprising 21 percent of Japan's land area.

Hokkaido horses are used for heavy transportation in the mountains where trucks and other equipment cannot go. They are very strong for their size and have a willing temperament. Hokkaido horses are also used as pleasure mounts.

As in other Japanese breeds, the Hokkaido is found in most solid colors, and many are roan. White markings of any kind are rare and not allowed for registration.

The Holstein, an old breed dating to the thirteenth century, originated in the German state of Schleswig-Holstein. This is the smallest but most successful horse breeding area in Germany. The breed originated when Neapolitan and Spanish foundation stock were bred by farmers for cavalry horses.

All people involved in Holstein breeding

belong to the Association of Breeders of Holstein Horses (Verband der Züchter des Holstein Pferds), with offices in Elmshorn. Breeding of Holstein horses goes back many centuries. The earliest information regarding this breed dates to the second century B.C. The horses came from western Asia and were most likely of Turkmenian stock, much smaller than the modern-day Holstein horse. Influenced by their new environment in the Holstein area, with different climate and soil conditions, they became larger and stronger.

The first written information regarding Holstein horses is from early medieval times. In the year 1285, permission was given to the monastery at Uetersen to graze horses on privately owned land around the cloister. The monastery had fine horses and in 1328 the convent gave two young horses to the landowners of Neumeunster, three and four years of age.

The lords were also interested in horses. The estate Bramstett was originally a stud farm. The village of Hingsthiede is said to have been named after the stud pastures around it (Hengst-Stud). The properties of cloisters and monasteries were transferred to private landowners immediately after the Reformation. The landlords continued the work of the monks and introduced fine stallions and mares, aware of the importance of horses for use on the farms and also for wartime. The first Holstein stud book shows that farmers and noblemen alike owned fine horses in the sixteenth century.

While the Thirty Years War to the south destroyed the land and many of its horses, horse breeding in Holstein flourished. Christian V (1646–99) introduced a law to prevent farmers from using stallions that were too small or too young. Christian, the founder of the horse farm Esserom, made great progress in horse breeding and later became famous for his white horses. The well-known white stallion Mignon, foaled in 1710, was raised on his farm.

A law passed in 1686 required each landlord who owned a large acreage to keep good broodmares and use good stallions. One hundred years later, further legislation entitled farmers who raised fine colts to a premium from the king.

Colts had to meet certain specifications in health and conformation to be considered in the selection of stallions. All colts that qualified were branded and could not be sold without permission of the premium commission of the king. Stallions under four years old and those that had obvious faults were restricted from breeding. Regulations became stricter in 1782, and the result was higher breeding fees; poor stock was difficult to sell and there was a noticeable reduction in the number of stallions. These regulations were abolished in 1786, but to encourage good breeding, gold and silver medals were given at the shows—gold for the two best stallions, silver for the best stallions in each district.

Beginning in the sixteenth century Holstein horses were sold to foreign lands, and until the nineteenth century the trade flourished. Horses were sent to Spain, Italy, France, England, Austria, and to all parts of Germany. In the seventeenth and eighteenth centuries Holstein horses were among the most desirable in Europe. In 1797 over 10,000 horses were exported from Schleswig-Holstein.

The horses established a fine record and were large and strong. They were outstanding riding and coach horses. The type is alleged to have been influenced by Oriental-Spanish blood. It is said that King Philip II of Spain used Holstein stallions at his famous horse ranch at Cordoba.

In the eighteenth century many Holstein horses were shipped to Italy. The stablemaster of the king of France, Guerinière, stated in 1730 that the Holstein horses had beautiful gaits, were well built, and made fine high school horses and coach horses.

Lord von Sind reported in 1770 that France bought 2,000 horses every three years. Professor Begetrapp stated in his remarks about English agriculture that the fmous animal trainer Bakewell recommended crossing English mares with Holstein stallions. Clarke described the Holstein horses in 1799 after a trip through the states: "The beauty and disposition of the Holstein find no equal anywhere in the world. They are of a dark color with small heads, broad nostrils, and large, dark eyes. Their brightness and sparkle seem to come from within. Besides their beauty, speed and performance, they are famous for their endurance" (translation from Breed Association booklet).

Many German states brought luxury horses, remounts, and breeding horses from Holstein. Holstein horses were seen at the courts of Europe and in German high society. Breeding with Holstein stallions was especially important at the famous horse ranch at Celle.

Considering the importance of the Holstein horse, it was only natural that George II (1683–1760) selected Holsteins for his new horse farm. Only under George III do we find imported stallions being used. Many Holstein mares were bred by fine English horses. One of the oldest Holstein farms is Dillenburg, in the state of Nassau, which had registered Holsteins as early as the sixteenth century.

In 1767 the Muenster government favored a cross between the native Westphalian horses called Kleisterpferde and Holstein stallions. In Oldenburg, Count Fredrich von Holstein bought Holstein stallions for his mares. In Mecklenburg, native horses were crossed with Holstein mares and stallions in order to get the then famous dappled horses, which had rather large heads and long necks. In addition, many Holsteins were used to replenish the number of horses on farms. Holsteins were even noticeable in Trakehnen in 1764, but they could not hold their own next to the English and Turkish horses.

The nineteenth century can be considered a period of crisis for horse breeding in Holstein. Increases in large land holdings, commerce, and travel caused an increase in horses, but at the expense of quality. Horse breeding was now in the hands of small farmers who could not afford good breeding material. The situation was different in the marsh country, where for years horse breeding had been in the hands of free farmers. The Holstein area became important for raising warmblooded horses, while in Schleswig breeding of heavy cold-bloods for plowing took precedence.

The 1820s brought several heavy crop failures followed by a destructive flood. Many families lost their belongings and were forced to sell their good breeding stock. The wars of 1840, 1860, and 1870 to 1871 complicated the situation further. It was therefore a blessing when, through the import of Yorkshire coach horses from England, the Holstein stock was once again improved and horse ranchers were again able to export horses. However, the blessing also brought a danger. Fantastic prices for geldings resulted in the reduction of good available stallions.

To resolve this situation the state of Prussia worked to find a new remount station in order to increase the number of good horses. Use of the breeds of East Prussia, Hanover, and Oldenburg was encouraged, but these crosses only weakened the Holstein breed. Many breeding organizations were unable to improve the breed, although they succeeded in creating interest among horse lovers.

Luckily, the Holstein Breeding Association in the marsh country came to the rescue. The foundation was laid for a new association on March 15, 1883, at the Krempen Marsh. In the first year of existence the association selected one hundred mares for breeding, and in 1891 all associations united and the name was changed to The Organization of Horse Breeders in the Holstein Marshes. The goal in breeding was to preserve the Holstein breed and to raise a good, strong coach horse with strong bone and high action, one that could at the same time serve as a heavy type of pleasure horse.

The Holstein is a somewhat heavier riding horse than the Hanoverian. Since World War II more Thoroughbred blood has been added, and the Holstein today is a large, all-around saddle horse, especially noted as a superb show jumper.

This is a powerfully built horse with good depth of girth, short legs, and plenty of bone. The quarters are of exceptional strength. Usual colors are brown, black, and bay. The Holstein is good-tempered and shows a keen willingness to learn and work.

During the nineteenth century the importation into Germany of Cleveland Bay and Thoroughbred blood from England added elegance to the Holstein breed, producing the dual-purpose characteristics of a riding and carriage horse. Since World War II further refinements have been made. Thoroughbred blood was again introduced, resulting in the modern Holstein's being a successful competitor in three-day events, show jumping, and dressage.

Rigid control by the German government assures strict minimum standards for the breeding of Holstein horses. No stallion may be used for breeding unless it has been licensed and has a current service permit. Only stallions whose pedigrees are validated by the breed society can receive such permits.

The most critical age for a Holstein stallion is two to two and a half years. Rigorous tests and assessments are then made to decide whether the horse can continue toward a career as a breed sire. The young horses are shown in hand before a panel of judges in walk, trot, and stand. Each is then classified according to type, movement, and conformation. Of the 600 Holstein colts born each year, only ten will be approved for breeding. This rigid control of breeding stallions has resulted in superb stock.

On every annual approval there are seventy stallions from which the ten will be approved. These stallions are all examined by a veterinarian. Approved stallions are used to cover thirty mares as three-year-olds. From August to November they are required to go through a hundred-day test. Only the successful stallions are then registered in the stud book. Results of the test are recorded so that breeders of riding horses can consider the field of ability when selecting stallions.

Holstein mares used for breeding are registered in various mare books. Mares are usually shown as two-year-olds to a committee of four people who select mares for the premium mare show, where the best mares of Schleswig-Holstein are shown. As a three-year-old, the mare becomes registered. A point system is used in determining her qualifications, with criteria in seven areas: type, top line, depth and width, front legs, hind legs, correctness of movement, and elasticity and suspension. Each area is scored on a ten-point range. At 35 to 39 points a mare is entered into the mare book and at 40 to 47 points she becomes a Premium Mare.

At the Olympic Games in 1984 in Los Angeles, six of the eleven horses of the German team were Holsteins. In 1988 at the Seoul Olympics, the Holstein gelding Corlandus, ridden by Margit Otto-Crepin of France, won the individual silver medal in dressage. Corlandus is the successor of Granat, who won the European Championship, the world championship, and the individual Olympic gold medal with Christine Stückleberger of Switzerland.

Hungarian Coldblood
(Magyar Hidegverü)

ORIGIN: *Hungary*
APTITUDES: *Heavy draft and farm work*
AVERAGE HEIGHT: *15 to 17 h.h.*
POPULATION STATUS: *Common*

Hungarian Coldblood. Photo: Dr. Csaba Németh

The coldblood horse is not native to Hungary, but in the eighteenth century the inhabitants of the eastern borderland (counties of Sopron, Vas, and Zala) brought draft horses with them when returning from Austria. Because of the relative proximity to Vienna, saleable goods, mostly cereals, were carted there on a regular basis. Farmers returning brought Noric (Noriker) and Pinzgauer horses as these breeds were more massive and better suited to hauling heavy loads than the native Hungarian horses.

In the hands of the people, two good types of heavy draft horse emerged—the Pinkafö or Pinkafeld, a large, massive type; and the Murakoz, a smaller, more agile type. Until the beginning of the twentieth century in the southwestern Transdanubia (Dunántúl) region—a relatively advanced area economically—intensive animal breeding activity was conducted, developing the coldblood horse ideally suited to the area.

The official state stud staff, as the governing body of horse breeding in Hungary, fought by fire and sword against expansion of the coldblood, but the inhabitants of the region battled back for continued propagation of the breed. A veritable politico-economic struggle raged between the coldblood promoters and detractors.

In spite of official opposition, in the years following World War I the horse population of the counties of Zala, Vas, Somogy, and Baranya was already predominantly coldblood. The farmers united in associations which started in 1922 to register the stock, organized horse shows and sales, awarded breeding prizes, and, to improve the quality of their horses, imported stallions in great numbers from Belgium.

Following World War II there was a severe lack of draft animals, and for a time the shortage worsened. To replace stock that had been lost and to regenerate existing stock, between the years 1948 and 1950, fifty-nine Belgian and sixteen French Ardennais stallions were imported. With these animals and their descendants, an expediently planned breed improvement program was carried out for numerous generations. From the previous mixed stock of Austrian horses and native Hungarian mares, the distinctive Hungarian Coldblood developed.

The Hungarian Coldblood has a heavy head, usually with a slightly convex profile, proportional to the body mass; the neck is short and muscular; the back short and wide. The chest well muscled and deep; the barrel well sprung; the croup wide, furrowed, and sloped. The legs are short and brawny with good joints; the hooves tend to flatten. A striking feature of this breed is the rich mane and tail of thick hair, the fetlocks also being covered with long hair.

Colors seen in the Hungarian Coldblood are bay, black, and chestnut, often with a lighter, flaxen mane and tail. Grey is also often seen. Previously the hoary, dun, and roan were also

quite frequent, but these colors are rare today.

This breed is an excellent horse for hauling heavy loads, serving farms, or slow draft work. Coldbloods continue to expand even among small landholders because of their quiet temperament, eagerness to work, and easy manageability. The Hungarian Coldblood horse develops quickly, is a good food converter, and thus is also used increasingly for meat production.

Original breeding grounds for the Hungarian Coldblood were in southwestern Transdanubia, but today the breed is dispersed all over Hungary and constitutes about one-third of the country's horse population. The best registered mares are to be found in the counties of Somogy, Baranya, and Zala. Their best male foals are raised from weaning at the Békáspuszta colt-rearing farm of the Bóly agricultural combine, and the best ones are then transferred for breeding purposes to stallion stations which serve the whole country.

Hungarian Dun

ORIGIN: *Hungary*
APTITUDES: *Riding horse*
AVERAGE HEIGHT: *15 to 16 h.h.*
POPULATION STATUS: *Rare*

Hungarian Dun. Photo: Betty Kovacova

In May 1971 the Hungarian Ministry of Agriculture and Nutrition made the decision to preserve all of the ancient native breeds of farm animals. One of the goals of this tremendous task was to reestablish the aboriginal dun-colored horse. Among other animals to be preserved were Grey Steppe Cattle, black and white Racka sheep, the dun and swallow-belly Mangalica hog, the Cikta sheep, and some native poultry strains.

The first step was to travel the country and collect all animals eligible for breeding purposes. At this period, the requirements given for the breeding stock were not stringent, as the ancient dun horse of Hungary was virtually extinct. The only limiting factor was the true color, which had to be in the dun range. Mares collected were of various types, ranging from rather primitive, thickset animals to elegant ones of halfbred origin.

The reality eventually had to be faced that it would be impossible to restore the original dun horse of the early Magyar tribes, brought into Hungary from Asia in the ninth century. But it did seem possible to reestablish the native horse of the nineteenth century. Breeders decided to develop a versatile breed, suitable for both harness and saddle, preserving the ancient dun color. The old Hungarian horse was very healthy and had great hardiness, stamina, and speed. Typical colors were the various shades of dun and grullo, with dark hooves and a dark dorsal stripe.

The original plan in 1971 was to keep a basic stock of seventy broodmares and three sire lines. To reestablish the Asian traits, imported stallions of the Akhal-Teke breed were used, especially on the more primitive type mares. According to Dr. Csaba Németh of the Mezögazdasági Minísítö Intézet, although the experiment has been carried on for years, restoration of the ancient type has been largely unsuccessful. The dun-colored horses are still being bred at two stud farms, Somogysárd and Szalkszentmárton; however, they are not considered an independent race, but rather specially colored sport horses.

The Hungarian dun has good conformation of warmblood type.

Hungarian Sport Horse

ORIGIN: *Hungary*
APTITUDES: *Riding, sporting events*
AVERAGE HEIGHT: *16 to 17 h.h.*
POPULATION STATUS: *Common*

Hungarian Sport Horse. Photo: Dr. Csaba Németh

Breeding programs for the Hungarian Sport Horse can be divided into four main groups: racing, trotting, jumping, and dressage. The performance of Thoroughbred and Trotter horses is tested according to international practice on Hungarian and international race tracks. As regards accuracy of the evaluation of breeding value, it should be emphasized that there is only one race track for Thoroughbred racing and one for Trotters in Hungary. Individuals are compared according to performance and progeny groups. In testing Hungarian Sport Horses, the goal is to improve performance in show jumping, three-day events, and dressage.

The four-year-old mare must undergo a field test. For the light riding horses, successful completion of this test is considered a precondition for becoming a broodmare. The test requires a small program of dressage, a cross-country 5,000-meter trot, 1,000-meter walk, a second 1,000-meter trot, and a 3,000-meter gallop with five to six obstacles. Length of the stride is measured in walk and trot, and there is a jumping course of ten obstacles not higher than one meter.

This program is not difficult and most healthy horses can do it. It is designed to ascertain the willingness of young horses and the possibility of placing them in line for the first selection to the stud, a sports career, or for selling. Scoring is done by an official commission.

Mares are officially classified into three categories according to their results in field tests and international or national sporting competitions; the quality of their dams is also considered. Classification is preliminary because it is always changed later according to the quality of progeny.

For stallions the breeding programs are more sophisticated. Each year only the best young stallions are selected for a performance test based on their pedigree and conformation. The others go into sports clubs and can prove their performance capacity without the official test.

The official performance assessment for young stallions is carried out in three training centers and begins at three and a half years of age, lasting one year with two tests. The preliminary test consists of judging of conformation and type; dressage, in which the elasticity of movement and ridability of the horse are scored; measuring length of stride in walk, trot, and gallop; a course of jumping with easy obstacles; and free jumping.

This test yields a preliminary ranking, and development of the training can be officially controlled. During this period of the test willingness of the horse can be judged. After a year of training, the performance test for four-and-a-half-year-old stallions consists of dressage; a 6,000-meter cross-country course; free jumping up to the jumping capacity of the horse; high jumping under a rider; and a course of jumping with obstacles at 110 centimeters (3.58 feet). The jumping is recorded on film, which gives some complementary information and can be viewed repeatedly for more accurate assessment and to allow independent evaluation.

The final ranking of stallions is made by a system of scoring by points. Addition of the points gives the ranking, but evaluation of the young stallions and their use in reproduction is always decided by considering the details along with the sum of scores.

Forty to fifty young stallions are tested yearly in this way, and approximately the best one-third begin service in the sport horse studs. The others, according to their qualities, can be used in sports clubs or sent to other studs where sports performance is not the main breeding goal. During the sports careers of the best-performing stallions, their semen is collected and used for insemination.

Sport horse stallions are also tested by the performance of their progeny, and in this way a continually improving breed is developing.

Icelandic Horse

ORIGIN: *Iceland*
APTITUDES: *Riding horse*
AVERAGE HEIGHT: *12.3 to 13.1 h.h.*
POPULATION STATUS: *Common*

The Icelandic Horse at the *tølt*. Photo: Fridpjofur Porkelsson

Icelandic Horse. Photo: Fridpjofur Porkelsson

Icelandic horses in a storm, showing the "ice tail." Photo: Fridpjofur Porkelsson

The Icelandic horse is one of the oldest and purest horse breeds on earth—a very special riding horse—and regardless of its size, those wishing to keep their Icelandic friends should refrain from calling the Icelandic a pony. This is primarily an adult's horse, usually far too spirited for a child unless the child is an experienced rider. Not that the Icelandic horse is unmanageable or difficult; far from it—but this is a lot of horse inside a small package. The label "pony" simply does not apply to the Icelander.

The Icelandic is an ancient five-gaited horse. To be able to ride the five gaits one must be an experienced rider. Many people visiting Iceland for the first time have been quite surprised when mounting for their first ride on an Icelandic horse, especially people accustomed to thinking of the Icelander as a pony. The Icelander has a gentle nature and is very intelligent, but its riders often find themselves sitting upon a time bomb of energy.

Several people in Iceland expressed their delight in the fact that, rather than borrowing information from other books, I was collecting details directly from breeding authorities. Perhaps, they said, the truth about the Icelandic would finally be told instead of the untruths often published. They reacted by sending a bounteous amount of information.

It is not true that the Icelandic horse must spend the severe Icelandic winters outside, pawing through the ice and snow for food and surviving on dead fish washed up on the beach. Both riding horses and young stock are in stables during the winter, from December until the beginning of June. They are fed good quality hay, usually twice a day. Breeding mares are usually out of doors all year round—but are fed hay, and shelter is provided for severe weather. Protective sheds are available to shield the horses from winds that may come from any direction, but often the Icelandic horse prefers to remain outside.

The horse was introduced to Iceland around the ninth century by the first settlers from Scandinavian countries. They brought horses from western Norway and the British Isles. Those strong and enduring equines were of great value to their owners both in peaceful times and in war.

The thousand-year isolation of the Icelandic horse has preserved in the breed some of the peculiarities that have been lost in other European races over the passing centuries, particularly the five gaits for which the Icelandic horse is famous and which the European horses once had in abundance through the Spanish Jennet and Asturian. It is probable that the Icelandic horse and the Asturian of northern Spain had common ancestors in the extinct Celtic pony.

The gaited horses persisted longer in Scotland and Ireland than in many other European countries due to delayed economic growth. Good roads and the carriage came late to Scotland, so riders continued to use horses that both paced and did a running-walk, called a *tølt* in Iceland. Horses of this nature were usually called "hobbeys" or "palfreys" (see Asturian and Andalusian) and were loved for their easy gaits. At one time horses possessing the comfortable gait with many names (tripple, stepping-pace, rack, *tølt, paso, rahvan,* and so forth) were popular throughout Europe. The advent of roads and speedy trotters spelled the decline of the gaited horse in Europe just as it bore the doom of the little Narragansett Pacer in the United States.

During the period from 874 to 1300, the farmers of Iceland bred their horses according to special rules. Mares and stallions were selected according to color and conformity. During the following centuries (1300–1900), less attention was paid to selective horse breeding. During this period the climate of Iceland was often very severe, and both people and animals frequently died from starvation. From the beginning of this century and especially since 1920, the breeding of Icelandic horses has been organized.

One of the most noteworthy facts about the Icelandic breed is that during the last 800 years no horses have been imported to Iceland, and this accounts for the purity and uniformity of the breed. Natural selection has played a major role, as often great numbers of stock were killed by cold and lack of food in severe times. In the most severe years from 1782 to 1784, approximately 70 percent of the horses stock was killed by starvation and poisoning from volcanic ash.

Since about 1959 upgrading has influenced the quality and appearance of the breed. Breeding societies have worked diligently in handling all matters concerning horse breeding and markets. The Icelandic horse is very popular in mainland Europe, especially in Germany, Holland, Scandinavia, France, and Switzerland, as well as England. The United States and Canada have also been enthusiastic purchasers. There are approximately 60,000 Icelandic horses outside Iceland and about that same number on the island.

The Icelandic horse has always been used as a saddle horse, providing the only means of travel and commerce in a land where roads and bridges were unknown until the early twentieth century. This surefooted little horse will carry a man all day across the mountains of sand and through cold, swift rivers. Icelandic horses weigh from 800 to 1,000 pounds and have been bred to carry adults. The most popular gait of the Icelander is the natural tølt, a four-beat gait that is easily ridden by quietly sitting in the saddle. There is no need to use artificial aids to train the horse to do this gait.

Depending on their breeding, some Icelanders also do the "flying pace," a very fast two-beat lateral gait used for racing short distances at speeds up to forty-five kilometers (almost twenty-eight miles) per hour. The horses are ridden at the pace rather than driven.

The Icelandic horse is not usually ridden until the age of four years. Generally, these horses are not completely grown until the age of seven. Their best years are from eight to eighteen, but even in their twenties they retain their strength and stamina. The oldest horse on record of any breed was an Icelandic mare in Denmark that reached the age of fifty-six years.

The horses of Iceland have never had to worry about predators, as there were none larger than the arctic fox when the Vikings arrived. Consequently, the horses do not tend to be spooky.

All horse colors are found within the Icelandic breed. Disease in this breed is unknown, other than two or three kinds of intestinal worms. Protection of the horses is assured by a law that prohibits any Icelandic horse taken out of the country for any reason from ever returning. Only new, unused horse equipment may be taken to Iceland. Due to its long period of isolation from all other breeds, the Icelandic horse has no antibodies to fight diseases found in other horses. An outbreak of disease in Iceland could easily destroy the breed.

The fertility of the Icelander is high. Both mares and stallions reach sexual maturity at two years of age. As a rule, stallions are fit for use up to the age of twenty-five and the same applies to mares. Examples are known of mares having foals at the age of twenty-seven.

Traveling on horseback is very popular in Iceland. Parties travel across the country through a magnificent landscape. The horses graze on the wild pastures during the night and do not need concentrates. During the day they can travel thirty to fifty miles carrying a weight of from 155 to 220 pounds.

The Icelandic people are justifiably proud of their plucky little horse. Most villages have large privately owned stables housing hundreds of animals. Activity is very lively when the riding season begins.

Icelandic horses are bred for use and not for show. The highest priority is given to the benefits of the horse as a riding animal. It must be courageous and resourceful, cooperative and willing, with good forward motion. The typical robustness must always be preserved. This is an extremely versatile riding horse for adults and is, under certain circumstances, also suitable as a child's mount. The Icelandic can be used for long-distance riding as well as five-gaited competition, as a family horse, and in dressage.

Ideally, the horse should be somewhat rectangular in shape and well proportioned. The head is clean-cut and expressive with a straight profile; the neck is long, supple, and well set; the shoulder is long and sloping with good muscling; the back is flexible; the croup sloping, wide, and well muscled. The legs are strong with well-defined joints and hard, strong hooves.

International Striped Horse

ORIGIN: *United States (Colorado)*
APTITUDES: *Riding horse*
AVERAGE HEIGHT: *Varies*
POPULATION STATUS: *Rare*

The International Striped Horse Association was founded in 1988 due to interest in the odd and interesting striping patterns seen on many horses in various breeds. One of the registry's goals is to catalog as many striping patterns as possible. Each breed will be given a code and will be kept separate from other breeds.

Dun-factor striping is the most widely recognized pattern. This is considered a primitive coloration and is seen most often in breeds of that type or those closely related to them. Often, foals display the striping patterns more obviously than do adult horses. The brindle dun is probably the most exotic and scarcest kind of striping phase. In addition, research is being done into the striping patterns seen on many roans. Stripes may encircle the neck and appear on different areas of the body. There are sorrels and chestnuts with dorsal and leg striping.

This new association welcomes research reports and all input that will contribute to understanding of this characteristic of genetic interest.

Iomud
(Iomudskaya)

ORIGIN: *Turkmenistan (former Soviet Union)*
APTITUDES: *Riding horse*
AVERAGE HEIGHT: *14.2 to 15.2 h.h.*
POPULATION STATUS: *Rare*

Iomud. Photo: Nikiphorov Veniamin Maksimovich

This is an old breed descending from the ancient Turkmenian breed. It was developed by the Iomud tribe in the Tashauz oasis in southern Turkmenia. As this breed occupied the margin of the Turkmenian breeding area, it was strongly influenced by the steppe breeds. In the fourteenth century the breed was influenced by Arab stallions. At a later period Mongolian, Kazakh, and local Turkmene stock was introduced. Unlike the Akhal-Teke, this breed is kept in herds in the desert and semidesert.

In 1926 large, noble mares of strong eastern expression were found in *taboons* of the area, as well as lesser numbers of small horses resembling the Mongolian and Kazakh breeds. Recently crosses have been made between the Iomud and Kazakh, and Akhal-Teke stallions were used up until recent years.

In general the Iomud has been bred pure, used for many years as a saddle horse. With the development of agriculture the Iomud has become a horse of many uses, well adapted to the conditions of Central Asia with its sharply continental climate and peculiar conditions of feeding and raising of young stock.

The Iomud is extremely resistant to the arid heat of its desert habitat and can survive without water for longer periods than most other breeds.

Less rangy in build than the Akhal-Teke, the Iomud is solidly built with a compact shape. This breed has a large, clean-cut head which is often Roman-nosed; a medium to long neck; medium to high withers; a nicely turned and usually sloping croup; a shallow chest; clean, fine legs (sometimes bowed but very strong); sparse mane and tail; and delicate skin. The colors are grey or chestnut, rarely golden chestnut or black. The coat is short and thin in the summer, growing long and dense during the cold season.

The Iomud is a healthy, long-lived horse. The action is soft and floating and easy on the rider. This breed is well suited to the sport of cross-country racing, where its abilities as a jumper, combined with its great powers of endurance, make it a strong competitor. The breed is known to have a very fast walk.

The purebred population of the Iomud has declined substantially, and stud farms were set up in Turkmenia in 1983. They were charged with protecting the breed and restoring the breeding nucleus to a size of 240 to 250 mares from the then present number of 140. Population of this rare breed is shown at 616 purebred in the Soviet Union, with 964 total, according to the 1989 FAO publication *Genetic Resources of the USSR*. A conservation farm is being established in the Kyzyl-Atrek District. A close relative of this breed is called Yamud in Iran and is included with the Akhal-Teke in horses known as Turkoman. For additional information regarding the Turkmenian breed, see Turkoman.

Regular training of Iomud horses for racing began in 1925, after the opening of a racecourse in Ashkhabad, but this was stopped due to the difficulties of transporting young horses through the Kara Kum desert. The races were reestablished after the organization of Tashauz stud. Race results revealed considerable racing ability in the Iomud, although it did not achieve the results shown by the Akhal-Teke horses.

During races and endurance tests the Iomud has demonstrated the high adaptability of the Turkmene horses for work under saddle in difficult conditions of long-lasting work with few breaks for rest. It showed less loss of weight than many others and had better scores for recovery of respiration and temperature. Animals were given limited amounts of food on race days, and little water.

These characteristics of the Turkmenian horses were developed through the process of

raising and selection for many centuries under conditions of long rides through waterless deserts where forage is sparse. The Iomud breed was used for a long time as an improver in Turkmenia, the southeastern part of Kazakhstan, and Iran. One improved type of Kazakh, the Adaev, was noticeably changed and improved through the use of Iomud blood.

Within the Iomud breed there are three types. The first are the largest horses, very near to the Akhal-Teke in size and type. The second type is the most widespread, large and similar to eastern horses influenced by the Thoroughbred. The third type includes smaller horses with shorter legs, a wide chest, and a massive build with visible influence of Mongolian and Kazakh breeds. This is the rarest type in modern times.

Breeding authorities are working to gather the best remnants of the Iomud breed to save the gene pool and prevent extinction of this ancient descendant of the Turkmenian horse.

Irish Cob

ORIGIN: *Ireland*
APTITUDES: *Riding horse, light draft*
AVERAGE HEIGHT: *15 to 15.2 h.h.*
POPULATION STATUS: *Common*

Like the English Cob, the Irish Cob cannot be considered a true breed as it varies in type although it has been bred since the eighteenth century. Irish Cobs are produced by crossing the English Thoroughbred, Connemara, and Irish Draft breeds. The Cob should have a compact body and sturdy frame with short legs. The musculature should be well developed; the action should be agile and lively. This horse was developed to produce energetic animals suitable for riding or harness, with good stamina. Previously used for light draft and farm work, the Irish Cob is now used for pleasure riding, trekking, and shows.

The typical Irish Cob has a well-shaped head with convex profile, small ears, and lively eyes; the neck is usually short and well muscled; the withers are well pronounced; the back is short and strong; the croup is rounded and gently sloped; the chest is broad and deep;

the shoulder is muscular and sloping; the legs are short and sturdy with a good hoof.

Ireland has long been known for excellent horse stock, and care is taken in the selection of good breeding animals.

Irish Draft

ORIGIN: *Ireland*
APTITUDES: *Draft, riding horse*
AVERAGE HEIGHT: *15.2 to 17 h.h.*
POPULATION STATUS: *Common*

Irish Draft Horse. Photo courtesy Irish Draught Horse Society

The name "Irish Draft" is misleading, as when we think of the draft horse we envision a Shire, Belgian, Percheron, or one of the heavy horses. The Irish Draft, however, is a lighter horse which moves more freely than does the traditional heavy draft horse. This breed is the root foundation of the Irish Hunter, which needs no formal introduction as it is known around the world for its excellence over fences.

The Irish horse earned worldwide respect centuries ago. Buyers traveled to Ireland from many foreign lands for hundreds of years, and despite the competition from other horse breeding nations, the Irish Draft has held its popularity.

In the technical sense, the Irish horse is not a true breed in that there are no identifiable foundation sires, no early breeding records, no stud book, and no preserving society or pioneering breeders to whom credit may be given for developing the breed. The Irish Draft was the horse of the countryside, not a distinctive breed within the Irish horse population but the product of its environment, evolving from the adaptation of the native horses on small Irish farms in the Connemara region.

Needs of Irish farmers differed greatly from those of their contemporaries in Britain. Demands of British agriculture stimulated the breeding of heavy horses such as the Suffolk, Shire, and Clydesdale. The Irish farmer, however, needed and bred a lighter, more adaptable animal that could be yoked, driven, or ridden. The title "Irish Draft" was brought in by army purchasing commissions and English dealers. The description "draughter," or draught horse, was used in England and applied to Irish farm horses even though in England it implied horses of very different characteristics.

The horse has enjoyed an honored place in Ireland since before recorded time. The Celts brought their dun ponies to the area 2,500 years ago, and racing was the sport of the people at horse courses or curraghs throughout the country. Today's great center of Irish racing, the Curragh of Kildare, dates to pre-Christian times. Horse and chariot races were featured at the summer harvest festivals of Lughnasa and the Tailteann Games.

Among the most perfect early Irish horse remains are those found in Craigiewarren Crannog in County Antrim during excavations by George Coffey of the National Museum. They are believed to date from the tenth century or before. Skulls found were similar to that of the modern Arab horse. Other similar remains were found at Loughrea and Shandon Cave in County Waterford.

Scientists have established that the early Irish horse stood from 12 to 14 hands, and historians believe that it began to increase in size with the arrival in 1172 of the Normans, who brought with them the larger, heavier horses needed to carry men at arms. The medieval Irish horse was, of course, the famous Irish hobby, a riding and cavalry horse which found its way to Scotland in the thirteenth century and was in demand all over

Europe until the end of the sixteenth century. Many of these early "hobbies" were Asturians from the northern tip of Spain. Exportations of the well-liked Irish Hobby were banned on several occasions in the fifteenth, sixteenth, and seventeenth centuries, so great was the drain on Ireland.

The modern Irish horse is much different from the hobby. The arrival of heavy horses with the Norman knights, and later the Andalusian or Spanish Barb which came with the Spanish trade, and finally, the Thoroughbred, eradicated the hobby and led to development of the Irish Hunter or halfbred horse as we know it today.

The first authenticated mention of the Irish Draft dates from the close of the eighteenth century, when there was increased demand for a larger, stronger horse for use not only as a pack horse but also for tillage work and riding. Heavy horses from Britain were imported, but because of the feathering on their legs, which protected against dampness but collected burs and stickers, they did not prove successful for Irish use. The farmers selected from the heavier of their own horses to produce a tillage horse with the ability to work as a roadster or jumper, and with an excellent temperament. It was found that the resulting horse crossed exceptionally well with the Thoroughbred.

Development of the Irish Draft continued until 1901 when a stud book was suggested, and in 1904 the Department of Agriculture introduced a premium scheme for approved stallions; thirteen were granted premiums. In 1911 the department received a grant to introduce a register of females, and 264 mares were identified as suitable for selected matings to the thirteen stallions. Unfortunately, due to the distances involved between stallions, the scheme failed.

In a way, World War I favored the breed. Horses required by the military for their artillery had to be steady, clean-legged, and economical to support yet have a turn of speed. The requirement was for a "cart horse that could trot, eat less, endure more and be placid and fast enough for artillery."

Numbers in the Irish Draft breed continued to grow steadily until the 1940s and 1950s,

when the impact of tractor power brought decline of the draft horse. In 1967 the Royal Dublin Show reintroduced classes after a lapse of forty-two years, and in 1976 the Irish Draft Horse Society was established with a view to halting the decline. It was affirmed that crossing the Irish Draft to English Thoroughbreds produced excellent hunters and competition horses, and Ireland became an exporter of partbred Irish Drafts.

The Irish Draft is an active, short-shinned, powerful horse with substance and quality. Standing over a lot of ground, this horse has a proud bearing and is deep in the girth and strong in the back, loins, and quarters. It has an exceptionally strong, sound constitution and is known for intelligence and a gentle nature.

The head is well shaped with a wide forehead and long, well-set ears; the forehead is broad with great width between the eyes, and the profile is straight; the neck is of average length, set in high and carried proudly. The withers are well defined, not coarse; the chest should not be too broad or beefy; the back is strong; the quarters are powerful; the croup is long and gently sloping. The forearms are long and powerfully muscled; the cannon bones are short and bone in the legs is never round or coarse; the legs should be clean and hard with a bit of silky hair at the back of the fetlock as a necessary protection. The pasterns are strong and sloped; the hooves are round and hard.

Any color is allowed in the Irish Draft, but obvious "Clyde" markings—that is, bay with white legs above the knees or hocks—are not permitted.

Despite its size, the Irish Draft horse has a natural talent for jumping.

Irish Hunter

ORIGIN: *Ireland*
APTITUDES: *Riding, sport horse*
AVERAGE HEIGHT: *16 to 17.1 h.h.*
POPULATION STATUS: *Common*

Ireland is traditionally a country known for outstanding horses and even today most Irish farmers keep a few mares for breeding. The

Akhal-Teke stallion in Germany. Photo: Evelyn Simak

The American Bashkir Curly stallion, Colonel's Fiesty Fella. Photo courtesy Carolyn Joy

Capricoso S II, a thirteen-year-old imported Andalusian stallion owned by El Canto Andalusians. Photo courtesy Ann B. Ohrel

Princess Alia Al Hussein at the Royal Stables in Jordan with two lovely Arab mares. Photo courtesy Princess Alia Al Hussein of Jordan

The Arab stallion, Mashour, of the UMM ARGUB strain. Photo courtesy Princess Alia Al Hussein of Jordan

Argentine Criollos, dun in color. Photo: Luciano Miguens

Belgian Draft. Photo: L. Rottiers

The British Spotted Pony stallion, Wantsley Ariel, at Parnham House. Photo: Mr. and Mrs. Makepeace

The Canadian is an excellent breed for use in driving competitions. Photo courtesy Paul E. Belauger

The Caspian stallion, Zeeland, owned by Louise Firouz. Photo: Louise Firouz

English Hunter. Photo: Carol Gilson

English Thoroughbred. Photo courtesy UNIC, Paris, France

Exmoor pony. Photo: Sally Anne Thompson

A Friesian stallion in Germany, showing the typical regal bearing seen in this breed. Photo: Evelyn Simak

Guanzhong stallion of China. Photo: Weigi Feng and Youchun Chen

Haflinger. Photo: Elisabeth Brotherton

The Holstein gelding, Corlandus. Photo: Werner Ernst

Icelandic Horse at the *tølt*. Photo: Fridpjofur Porkelsson

The Mangalarga Marchador stallion, Mar Diorito. Photo: Lauro Antonio Teixeira de Menezes

The Mangalarga Marchador stallion, Recado De Passa Tempo. Photo: Lauro Antonio Teixeira de Menezes

The Marwari stallion, Raj Tilak, recently deceased. Photo: Raja Bhupat Singh

Chinese Mongolian stallion. Photo: Weigi Feng and Youchun Chen

The Namib stallion, Impuls, at five years of age. Photo: Frans van der Merwe

A young Namib mare leading the others to water at Garub. Photo: Frans van der Merwe

National Show Horse, Key West. Photo: Grand Kid Acres

Northeastern. Photo: Antonio Eurico Vieira Travassos

Persian Arab. Photo: Caren Firouz

A champion Rottal mare in Germany. Photo: Evelyn Simak

The rare Schwarzwälder Füchse. Photo: Evelyn Simak

Sini. Photo: Weigi Feng and Youchun Chen

A team of South German Coldbloods. The ear coverings are for protection from flies. Photo: Evelyn Simak

The Yamud Turkoman mare, Shahnazar, in Iran. Photo: Caren Firouz

Irish Hunter is known worldwide for its brilliance over fences—the produce of the Irish Draft horse and English Thoroughbred. Although a halfbred, the Irish Hunter is usually included in breed books as this horse exhibits a remarkable degree of uniformity in type. Bred for the hunt and well suited to all kinds of terrain, the Irish Hunter is surefooted and a favorite for show jumping and eventing. While formerly the main market for Irish horses was in England, the Irish horse is now in demand internationally and often at astronomical prices.

The Irish Hunter is found in all colors except piebald or skewbald. The head is well proportioned with a straight or slightly convex profile. The eye is large and expressive; the ears are long and nicely shaped, pointing forward. The neck is long and muscular with a slight arch; the withers are prominent and long; the back is short; the croup broad, muscular, and slightly sloping; the chest full and deep. The legs are solidly built and well muscled with broad, clean joints and well-defined tendons. The Irish Hunter has substantial bone, inherited from its Irish Draft parent, combined with the refinement and fire of the Thoroughbred. This is a solidly built hunter that is agile as a cat over rough terrain and known for its endurance at the gallop.

Israeli
(Israeli Local Horse)

ORIGIN: *Israel*
APTITUDES: *Riding horse*
AVERAGE HEIGHT: *14.1 to 15.1 h.h.*
POPULATION STATUS: *Common*

Horses in Israel are an interesting combination of many breeds, seasoned with varying amounts of Arab blood.

The horse was first introduced to Israel in very early times; these animals were probably ancestors of the present-day Arab. The Holy Land has suffered the rigors of many wars through the centuries and the horse, being the war machine, was brought from various areas. All have left their stamp on the local Israeli horse.

Israeli Local Horse. Photo courtesy Ministry of Agriculture, Israel

During World War I the Australian cavalry stayed in Palestine and the famous Waler breed had a major influence, still apparent today, on the local Arab strains. Under the subsequent period of English mandate in Palestine, many hunters and Thoroughbreds kept by army officers made their contribution. During the first years of Israeli independence, the Jewish Agency bought around two thousand horses from Europe and the United States from various breeds for agricultural purposes. These included Hungarian Shagya, Norwegian Fjord, Yugoslavian local horses, Morgans, Quarter Horses, and almost any kind of horse available in Europe in those days. These breeds were all mixed with the Australian Waler/Arab types already in the area. As no clear records were kept at the time, one can only guess at the influence exerted by various breeds. The underlying fact is that Israel had long had Arab strains of varying quality and type, and Israeli horses have always shown this influence to at least some degree.

Purebred Arab horses were first brought to Israel from England in 1968, and since that time Israel has obtained additional Arab stock from England, France, Germany, and the United States. The first volume of the Arab Stud Book in Israel was published in 1974.

The first registered Thoroughbreds brought to Israel came from Kenya in 1974, and since that time Thoroughbreds have come from England, France, Holland, and the United States. The first Thoroughbred Stud Book was published in 1984.

In 1960 two registered Quarter Horse stallions were imported from the United States, and these left their influence on the local horse as well. There are also some Dutch Warmbloods, one Trakehner stallion, two Hanoverian stallions, and two Tennessee Walking Horses. In addition to the light horse breeds found in Israel, there are many ponies, mostly Welsh Mountain and Shetland.

Some years ago an attempt was made to establish a local warmblood horse, but there was not enough support at the time so this dream is still one for the future. Certainly, the ingredients are there to develop a superb competition horse. Since 1968 the main source for improving the local Israeli horse has been the Thoroughbred.

Common colors seen in the Israeli horses are bay, grey, and chestnut, with the majority being grey. The horse emerging from this mixture of good bloods has varying type but many common characteristics such as invariably good strong feet, a short back, a relatively long head, and narrow chest. The temperament is good, and the Israeli Local Horse is easily handled and full of energy.

Israel is a country with many problems and a hard life; therefore, development of horse breeds has not been a priority on a large scale. Still, individual breeders are making continual efforts to improve local horses and their own stock.

Israel only has one breeders' association at this time, the Arab Breeders Association, which deals with all aspects of purebred Arab horses. All other stud books are held by the Ministry of Agriculture in a section within the Department of Animal Husbandry of the Extension Service. The Israel Stud Book consists of the Thoroughbred Stud Book, which is a recognized stud book and a member of the European and Mediterranean Stud Book Liaison Committee; the N.T. Stud Book, which consists of all offspring of registered Thoroughbreds crossed with local mares (halfbreds); the Halfbred Arabian Stud Book, consisting of offspring of purebred Arabs crossed to local mares; and the Local Horse Stud Book, started in 1972, consisting of all local horses whose pedigrees cannot be identified.

Among activities for horses in Israel, regular show jumping and dressage competitions are held by the Israel Horse Society and amateur races (without betting) are held in some places, especially in summer in the Valley of Israel.

Due to the fact that the Israeli Local Horse is being recorded and efforts are being made to improve the type, it is conceivable that one day this horse will be the basis for a very good warmblood breed that will make itself known in equine sporting events.

Italian Heavy Draft
(Rapid Heavy Draft)

ORIGIN: *Italy*
APTITUDES: *Heavy draft*
AVERAGE HEIGHT: *15 to 16 h.h.*
POPULATION STATUS: *Common*

A young Italian Heavy Draft stallion. Photo courtesy Dott. Pigozzi Giuseppe

The Cavallo Agricolo Italiano da Tiro Pesante Rapido (CAITPR), or Italian Heavy Draft, is the only draft breed of Italy. Origin of the breed dates to 1860 when the Deposito Stalloni of Ferrara (which was the Department of War territorial branch for horse production in the northeast of Italy) began crossbreeding, using Breton stallions from France on the population of local mares reared on the farms of the northeastern Italian plain. Mares under consideration were of varied origin; Hackney, Belgian Draft, and Breton. Through continued crossbreeding of these lines, the Italian Heavy Draft emerged, a breed with fixed and heritable characteristics.

The official stud book was started in 1926, and the breed was selected and used for agricultural and draft purposes as well as for transport of artillery and engineers for the army. Some Italian Heavy draft mares were also used to produce the heavy mules required by the army.

Until the fifties and sixties this breed was a popular draft horse, not only powerful but very fast. Mechanization of the army and agriculture reduced the importance of the Italian Heavy Draft, as it did all other draft breeds. Numbers declined steadily.

In the seventies an extensive selection effect was begun with the goal of improving the meat production characteristics of the breed. Today the Italian Heavy Draft is raised primarily as a meat animal. In hilly districts where mechanical equipment cannot be easily used, this horse is still considered a valuable asset, and some mares are still used today in the production of large, strong mules.

The Italian Heavy Draft most resembles the Breton, not only in morphology but in lively temperament and gaits. The most common color is liver chestnut with a flaxen mane and tail, but some are roan or bay. White markings are commonly seen on the face and legs. The head is long but fine and elegant with a straight or slightly convex profile, a broad forehead, small ears, and large, lively eyes. The neck is short, muscular, and broad at the base with a heavy mane; the withers are well pronounced; the back is short and straight; the loins are powerful; the croup is rounded and sloping; the tail is well set on. The chest is broad, muscular, and deep; the ribs are well rounded; the shoulder is long and well sloped. The legs are short with some feather on the lower parts, and the joints broad and strong; the forearm is long and the cannons are short. The hind quarters of this breed are powerfully muscled. The feet tend to be rather straight and boxy.

Original breeding ground for the Italian "Rapid" Draft was the northeastern Italian plain, but today it is raised in many areas of central and southern Italy, thanks to the breed's fitness and adaptability to various environments. The climate ranges from subcontinental to Mediterranean. Italian Heavy Draft horses are raised in the intensive conditions of plains farms or in the sometimes harsh conditions of hilly and mountain areas. Throughout Italy the stallions are also used on local mares to improve meat production animals.

Jabe
(Dzhabe)

ORIGIN: *Kazakhstan (former Soviet Union)*
APTITUDES: *Draft, milk*
AVERAGE HEIGHT: *13.3 to 14 h.h.*
POPULATION STATUS: *Common*

A distinct type within the Kazakh breed, Jabe horses differ from the Kazakh in their massiveness and large measurements. Well adapted to year-round living in *taboons* (see glossary) under severe climatic conditions, the Jabe is greatly used as an improver of taboon horses in many parts of the Soviet Union. This breed thrives in the most severe conditions—from the Aral desert to the northern areas of Yakutia. Pure breeding of Jabe horses and their crossing to local horses gives great economic benefits, producing high quality meat and milk yield.

Since 1960 pure breeding work has been done to produce massive animals with meaty forms. The aim has been to improve breeding and productive qualities through use of the best mares and stallions according to pedigree, appearance, productivity, and adaptability to year-round taboon living. Strict culling of inferior animals has been conducted. The horses have been evaluated on a ten-point system, and the careful selection has resulted in development and concentration of desirable characteristics. Thanks to the long application of selection within the breed, a valuable nucleus of Jabe horses has emerged, highly productive and showing an increase in size during the period from 1936 to 1984.

Jabe horses have rugged heads. The neck is thick, the body wide, and the chest deep. The back is straight and the croup is well muscled; the legs are correctly set and strong.

Tested for adaptability to more severe regions, Jabe horses were highly successful. Five four-year-old Jabe stallions were taken to Yakutia and pastured with three four-year-old Yakut stallions near the Lena River. Over a three-year period, the Jabe horses increased in weight and their coats, which had originally been only half as long as those of the Yakut horses, increased in length dramatically, growing longer than the Yakut's. Pulse and respiration were almost identical for both breeds.

The Jabe breed is widespread, and presently there are three subtypes found within the breed. The best type will be given a separate status and breeding type which will be called Emben. A second type, created by crossing to local mares of the Dzezkazgan region, is called the Betpakdalin strain, and the third type, called Kulandin, was produced at the Kulandin stud farm from mares with Adaev blood crossed to Jabe stallions. The development of these three new types resulted in horses having their own typical appearance, genotype, productivity, and best adaptation to their respective regions. The new types were developed as multipurpose animals for both meat and milk production. However, it will be necessary to specialize them for pure milking horses. This work is being done now to produce a new, special breed of milking animals.

Java

ORIGIN: *Indonesia (Isle of Java)*
APTITUDES: *Riding and pack pony, light draft*
AVERAGE HEIGHT: *12.2 h.h.*
POPULATION STATUS: *Common*

Java pony. Photo courtesy Dr. Soehadji, Department of Agriculture, Indonesia

Java pony. Photo courtesy Dr. Soehadji, Department of Agriculture, Indonesia

Resembling the Timor, the Java pony is larger and stronger than other Indonesian breeds. This pony is used for pulling the *sados,* a two-wheeled cart used as a taxi, or to haul goods. The Java has a willing and gentle nature and is very tough, appearing tireless in spite of the stifling tropical heat. The breed is also known as the Kumingan pony.

Perhaps the least improved of the Indonesian ponies, the Java may be found in any common color. The head tends to be heavy with a straight profile; the neck is short, very muscular, and has a thick mane; the withers are quite pronounced. The back is straight and longer than in most other Indonesian ponies; the croup is slightly sloped; the tail set quite high; the chest deep and well developed; the shoulder often quite sloped. The legs are long and well muscled.

Arab influence is apparent in the Java pony, being of lighter build than some other breeds on the islands.

For additional general information regarding Indonesian ponies, see Bali.

Jianchang

ORIGIN: *China (Sichuan Province)*
APTITUDES: *Riding, light draft*
AVERAGE HEIGHT: *11 h.h.*
POPULATION STATUS: *Common*

Jianchang horse of China. Photo: Weigi Feng and Youchun Chen

The Jianchang horse on the way to the field to work. Note the hand-made plow. Photo: Tiequan Wang and Youchun Chen

The breeding area for the Jianchang breed is the Liangshan mountain region in the province of Sichuan.

The Jianchang is a very old, native breed, and official horse ranches were established in the Xi-chang area of Sichuan Province between the years 947 and 950 A.D. The best stock was selected through races and events combining riding at high speed and arrow shooting.

The Jianchang horse is used for mountain transport and can easily travel eighteen to twenty-five miles carrying a pack of 150 to 220 pounds. Ridden, the Jianchang travels fifty miles in a day. In tests this breed has proven its ability to pull 120 percent of its own body weight! This is a powerful horse in a small package.

Jianchang horses are very good at mountain work in subtropical areas. It has a low growth ability after one year of age, so care during the first year and especially in winter is important.

This is a light breed, although compactly built. The head is small and dry with small ears and great width between the eyes. It tapers to a small muzzle; the neck is of medium length and straight; the withers are well defined; the croup is sloped; the legs are slender with well-defined tendons and strong joints. The hoof of this horse is small, well shaped, and very hard.

Jianchang horses are mostly bay. Some are black and other colors are seen less frequently.

Jielin

ORIGIN: *China (Baicheng, Changchun, Sipling Districts)*
APTITUDES: *Riding, draft and farm work*
AVERAGE HEIGHT: *15 h.h.*
POPULATION STATUS: *Common*

The region of development for the Jielin horse was Baicheng, Changchun, and Sipling Districts. The central zone of the area includes the counties of Fore-guorlos, Nongan, Shuanliao, and Huaide. This is a semiagricultural area, very suitable for horse breeding.

Horses in this region were originally of Mongolian type, small in size, and their performance did not meet the demands of animal power for agriculture. Since 1950, based on local-bred Ardennes and Don blood, crosses

Jielin horse. Photo: Weigi Feng and Youchun Chen

were made to native horses. Both heavy and light types resulted, and these were interbred by stud farms under the common program directed by specialists from the staff at Jielin Agricultural University and Jielin Academy of Agriculture. Since 1962 a Su-Yi stallion from this system was used and the new breed of draft and riding type was established. In 1966 a linebreeding program was practiced to keep the percentage of local blood at 25 percent, light horse blood at 25 percent, and heavy blood at 50 percent. This developed breed was officially recognized in the late 1970s.

The Jielin horse has massive body construction with medium-sized head and good overall conformation. The main coat color is bay, fewer are chestnut, and black is rare.

Jielin horses are still found in two types, heavy and light, but draft is the main use in the countryside. This horse is widely used for improvement of local breeds.

Jinhong

ORIGIN: *China (Fujian Province)*
APTITUDES: *Riding, pack horse*
AVERAGE HEIGHT: *12 h.h.*
POPULATION STATUS: *Common*

The Jinhong originated along the coast in the province of Fujian. In southern Fujian an agricultural area was developed and there was great need for powerful horses. The Jinhong

Jinhong mare and foal. Photo: Tiequan Wang and Youchun Chen

Jinhong horse of China. Photo: Weigi Feng and Youchun Chen

breed was isolated from all other breeds for many years and is very pure. It is a native, indigenous breed. Recorded breeding of the Jinhong horse dates back one thousand years.

This is an attractive horse with a glossy coat, clean, strong legs, and a nicely shaped head. The main color of Jinhong horses is chestnut, and the average weight for males is about 650 pounds.

The Jinhong is a good riding horse, and it is resistant to disease. It is well able to tolerate rainy, damp coastal conditions.

Jinzhou

ORIGIN: *China (Jin)*
APTITUDES: *Riding, draft*
AVERAGE HEIGHT: *14 h.h.*
POPULATION STATUS: *Common*

Jinzhou of China. Photo: Weigi Feng and Youchun Chen

Jin County, where the Jinzhou horses were developed, is located in the southern region of Liaodong peninsula (one part of Dalian City). In 1982 there were approximately 10,000 horses in this area.

For centuries, local horses were of Mongolian blood. During the period from 1926 to 1941, the Japanese brought Hackney, Anglo-Norman, Orlov Trotter, and other breeds to cross with local strains. Later, for military purposes, the Percheron was introduced to increase power, and this was an important step in developing the Jinzhou breed. In 1948, to enhance the farming system, the local authorities selected twenty-seven stallions from three farms to enlarge horse breeding operations. In 1956, 85.5 percent of horses in Jin County were of the improved variety and 2,346 were officially recognized as Jinzhou horses. Kabarda blood was introduced at one time, but no influence from that breed was found in 1956 and this cross was discontinued completely in 1964.

The maximum power for pulling in the Jinzhou breed is 82.8 percent of the body weight. These are excellent draft horses with impressive endurance qualities, able to work hard on a daily basis. Ridden, they are fast and energetic.

A horse of harmonious conformation, the Jinzhou is solidly built and attractive. The head is of medium size, sometimes ram-headed or with a "rabbit-like" (see glossary) profile; the neck is high and arched; the withers are long and high; the back is short and strong; the chest is deep; ribs are well sprung; the croup is muscular and sloped; the legs are dry with well-defined tendons and are clean of surplus hair.

Jinzhou horses are used to improve the animals in other provinces. They are mainly bay in color.

Jutland

(Jydsk)

ORIGIN: *Denmark*
APTITUDES: *Heavy draft work*
AVERAGE HEIGHT: *15 to 16.1 h.h.*
POPULATION STATUS: *Rare*

Jutland of Denmark. Photo courtesy Niels Busk

The Jutland horse is the national draft horse breed in Denmark. Selective breeding for the Jutland was commenced about 1850. From that time on, authorities and breeders have aimed to produce a breed especially suitable for draft work in farming. A great contribution to success of the breed was the stallion Oppenheim, imported from England in 1862. Oppenheim was Shire and partly Suffolk. Six generations from this stallion, Aldrup Menkedal, the stallion considered the true foundation for the breed, was born. Almost all of the breeding Jutland horses of today can be traced back to his two sons, Høvding and Prins af Jylland.

The first stud book appeared in 1881, and there have now been 22,000 horses registered. Results of breeding work in the Jutland are a good example of the importance of having a strictly defined aim. The first association of breeders was founded in 1887 in Høver, close to Århus in Jutland, and a year later the Cooperative Jutlandic Breeding Association was established. The first judging of stallions took place in 1888, and this has been a yearly event ever since.

The most common color is chestnut, usually with a light mane and tail. Brown or black horses do appear in the Jutland breed as well, but are rare. The aim of breeders is to produce a medium-sized horse with great depth and width. The head is medium in size with sharply marked features; the neck is carried high and well placed; the back is short and strong; the limbs are strong and correctly placed with delicately hairy cannons. The Jutland breed is energetic yet calm and ready to work.

Since 1928 there has been a close association between the breed management for Jutland horses and the Carlsberg brewery, which uses Jutlands to haul brewery wagons. At one time there were 210 Jutland horses with Carlsberg, and today about twenty are still used for beer transportation in Copenhagen. In this way the Jutland horse serves as live publicity in the streets of the city and brings vitality and beauty to the cityscape. Every year the Carlsberg Brewery horses take part in many shows, festivals, and films. Wherever the Carlsberg horses show up they evoke pleasure and enhance the cultural value of the Jutland breed.

Kabarda

(Kabardinskaya)

ORIGIN: *Caucasus (former Soviet Union)*
APTITUDES: *Riding and packing*
AVERAGE HEIGHT: *14.2 to 15.1 h.h.*
POPULATION STATUS: *Common*

Kabarda or Kabardin. Photo: Nikiphorov Veniamin Maksimovich

Found mainly in the Kabardino-Balkar Autonomous Republic and in the foothills of the Stavropol area, this is an old native breed with strong links to the famous horses of Turkmenistan. The name Kabarda is also spelled Kabardin.

During the process of this breed's development by nomadic tribesmen around the sixteenth century, it was influenced by the Karabakh, Persian, Arab, south Russian steppe breeds, and Turkmenian. The Nogai horse (now extinct) also had some influence. Kabarda horses are kept in *taboons* (pasture management with a herdsman), being transferred from mountain pastures in summer into the foothill areas for winter.

This is primarily a saddle horse. In bulk it is not large but has a clean, solid build. The head is clean, sometimes a bit coarse, with ram profile and relatively long ears. The neck is well muscled and of medium length; the withers are medium in height; the back short and solid; the croup slightly sloping with heavy muscling; the shoulder long and sloped; the chest deep and long-ribbed. The legs are correctly set with the hind legs often bowed but with well-developed, clean joints and hard hooves.

Although the Kabarda has a moderate coat of hair, the mane and tail may be quite thick and the legs may have some feather at the fetlocks. The predominant color of this breed is cherry bay, often bay-brown. Black is rare.

Kabarda horses are very surefooted and have good gaits. The walk is even and cadenced, the trot and canter light and smooth. Because of their high action they do not show great speed as a rule. Some are natural pacers. They are greatly appreciated for their ability to travel on sharp, narrow mountain paths, finding their way through heavy mists in the dark, and their willingness to ford rivers and go through deep snow. The breed is known for remarkable endurance and an unerring sense of direction. This is considered one of the best horses to use in the mountains.

During the sixteenth century the Kabarda was widespread in the northern Caucasus region. Outside its natural habitat the Kabarda shows ability as a sporting competitor. It works well both under saddle and in harness, displaying a lively, energetic temperament that is at the same time obedient and tractable.

In earlier times this was a small breed with a very strong constitution and light, free movement at the walk and trot. However, many Kabarda horses died during and after the Revolution and in the 1920s restoration of the breed was commenced at the Kabardino-Balkar and Karachaevo-Cherkess studs. A stronger, even more efficient type was produced, suitable both for agricultural work and as an army mount. Today the Kabarda is still the principal breed in the Kabardino-Balkar Autonomous Republic and Malokarachaevski stud. Kabarda stallions are used to improve stock in Dagestan, Osetia, Georgia, Azerbaijan, and Armenia.

Population of this breed has dropped considerably in recent years. This is attributed to insufficient speed in hippodrome (racing speed) tests. There are four main bloodlines in the breed, giving it a limited breeding nucleus, and steps are being taken to protect the Kabarda from extinction.

Kalmyk
(Kalmyskaya)

ORIGIN: *Astrakhan and Volgograd (former Soviet Union)*
APTITUDES: *Riding and harness horse*
AVERAGE HEIGHT: *14.2 to 15 h.h.*
POPULATION STATUS: *Rare*

In 1923 Professor M. I. Pridorogin mentioned the Kalmyk horse as a member of the Mongolian group, bred in the territory along the Volga and Ural Rivers. In general appearance the breed is similar to the Kirgiz horse but is taller with a coarser head, less slope to the shoulder, and longer legs. Kalmyk horses are slow to mature, reaching full development at six years. Many pacers are found in this breed. At one time it was estimated that there were about one million Kalmyk horses. Today the breed is nearly extinct.

The Kalmyk people, who came into Russia from Dzungaria in the seventeenth century, were of Mongolian origin. They brought with them sheep, cattle, and horses. The Kalmyk horse was described as a not very beautiful, medium-sized horse with light, speedy gaits, able to withstand extreme cold and heat, starvation rations, and long-distance rides.

In 1986 an expedition was organized by workers of the Kalmyk University of Cattle Breeding to locate and determine the numbers of local horses of the ancient Kalmyk type. Since 1943 no selective breeding had been done and the entire population was considered crossbred. A goal was set to select the remaining individuals typical of the Kalmyk breed. The best remaining representatives were located in the eastern regions, where the Kalmyk was more isolated and had been best preserved. A total of 2,078 horses were examined and 522 were selected as typical of the old breed. Four breeding farms were established to preserve and increase this ancient and nearly extinct horse.

Kalmyk horses have always been bred under semidesert conditions and dry steppes. They fend for themselves year round and do not receive additional feeding in winter. Under these conditions several valuable biological characteristics have developed, such as a firm constitution, wonderful soundness, endurance, and adaptability to severe climate. Kalmyk horses are able in good conditions to put on weight quickly and during periods of insufficient feed they slowly utilize their fatty resources. Their skin is thick and firm. In winter they develop a dense coat with a heavy undercoat which protects them from the cold. They thrive even in deep snow.

The Kalmyk is a saddle horse suitable for work in harness. It has a great potential value for meat and milk production also.

Kalmyk horses have a coarse head with Roman nose; the neck is short; the back is short and often carp-like; the croup is strong; the legs are lean and strong and the hind legs are often cow-hocked. Predominant colors are bay and sorrel.

Karabair
(Karabairskaya)

ORIGIN: *Uzbekistan (former Soviet Union)*
APTITUDES: *Riding horse*
AVERAGE HEIGHT: *14.2 to 15 h.h.*
POPULATION STATUS: *Common*

Karabair. Photo: Nikiphorov Veniamin Maksimovich

This is one of the most ancient breeds of Central Asia. Developed in what are now known as Uzbekistan and northern Tajikistan—areas known for high quality horses for 2,500 years—the breed was established through influence of southern and steppe breeds. Well adapted for use under saddle and in harness *(arba cart)*, it has the typical build of a saddle and harness horse.

In appearance the Karabair bears strong resemblance to the Persian, Arab, and Turkmenian as well as the steppe breeds. It has a medium-sized, clean-cut head with a straight or ram profile, wide jaw, medium to long poll, and high-set, medium to long neck. Some have a shorter, heavier neck. It has medium high and long withers; the back is short and wide; the loin is wide and well muscled;

the croup is sloping and sometimes drooping; the shoulder is often in need of more slope; the chest is well developed. Often the fore quarters of this breed are more developed than the hind quarters. The legs are clean and strong with well-defined tendons.

Historical records show that for centuries prior to the birth of Christ the area of Uzbekistan was famous for outstanding horses. Chinese documents testify that both diplomatic and war expeditions were made to this area and to Fergana to obtain exceptional horses. During the eighth century the Arabs seized many Central Asian horses for breeding on the Arabian Peninsula (until the sixth century A.D., horses were not bred on the Arabian Peninsula). After obtaining horses from Central Asia, stud farms were established there. The development of ancient Central Asian horse breeding was due to the high degree of agriculture for that time, producing valuable foodstuffs in abundance. Outstanding horses such as the Parthian, Dawan, and ancient Turkmene were products of this culture.

As a result of insufficient feeding, the knee joints in the Karabair may be occasionally underdeveloped and hind legs may be cow-hocked. Usual colors are bay, chestnut, grey, or black. Like other breeds related to the Turkmenian, the Karabair has sparse hair in the mane and tail.

This breed shows good endurance and is a versatile working horse, characterized by sound health, average longevity, and normal fertility. The Karabair is also used in sporting games, especially *kok-par,* in which riders fight over possession of a dead goat.

There are three types within the breed—basic, heavy, and saddle—and it has eight sire lines and five mare families. The breed is zoned for breeding in all regions of the Uzbek Republic. Population of the Karabair was shown in 1990 at 28,223, with 25,499 purebred. Most widespread are the horses of the basic type, well adapted to work under saddle and in harness. Their trunk is well developed, wide and deep, and their constitution lean; the temperament is lively and alert. Some horses have retained southern traits with elegant heads wide in the forehead and having a high-set neck. Horses of the massive type are larger, have a long trunk, and are near to the harness type, but are suitable also under saddle. This type of horse is desirable in cotton-growing regions where strong horses with endurance are needed. In recent years little attention has been paid to the breeding of the massive type and numbers are low. This is the most valuable type for work in agricultural regions of Uzbekistan.

Horses of the saddle type are of a lean constitution with well-developed musculature and have great speed for short distances. Some of them have a trunk that is considered too narrow and a weak skeleton, but due to their success at the race track this type is increasing in number.

The main breeding studs are Jizak, Gallyaaral state farm, and Navoi stud in the Jizak region. This breed is improved through selective pure breeding with no outcrossing to other breeds.

The rare Karabakh of Azerbaijan. Photo: Nikiphorov Veniamin Maksimovich

Karabakh

(Karabakhskaya)

ORIGIN: *Azerbaijan (former Soviet Union)*
APTITUDES: *Riding and pack horse*
AVERAGE HEIGHT: *14.1 to 14.3 h.h.*
POPULATION STATUS: *Rare*

The Karabakh is an ancient saddle breed developed in Nagorny Karabakh in Azerbaijan between the Kura and Araks Rivers.

Prior to the nineteenth century the Karabakh khanate was the breeding center for the best horses to be found in Transcaucasia, and the Karabakh had a strong influence on improvement of horse breeds in neighboring countries. This breed was developed by crossing the native Azerbaijan horse with Persians, Turkmenians, and Arabs. The Arab influence was most pronounced, and there are many important similarities in the appearance of the Karabakh and the Arab.

Having been bred in the mountains for a long period of time, the breed has developed characteristic features. This is not a large horse but the build is clean and thickset. The musculature is well developed and the ten-

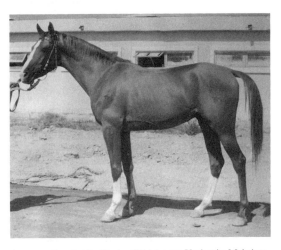

A young Karabakh. Photo: Nikiphorov Veniamin Maksimovich

dons are well defined. The head is small and clean with a straight profile and alert, expressive eyes. The neck is set high and is average in length; the loin is flat, short, and wide; the croup is of average length, well muscled and wide; the chest is deep. The legs are properly set with hooves that are small but hard and solid. The skin of this breed is thin with soft, gleaming hair that is thin in the forelock, mane, and tail. The coat color is usually chestnut or bay with a characteristic golden, metallic shimmer. This kind of skin, coat, and color are found in nearly all breeds with strong ties to the ancient Turkmenian breed.

The Karabakh is very surefooted and has no fear of harsh terrain, proving itself to be a good choice for use in rugged, mountainous country, able to gallop and maneuver, changing directions easily. The breed is not easily excited.

The Karabakh was used in the breeding of Don horses and was taken into the Ukraine area for use in *taboons* of Caucasian breeders. Karabakhs were used to create many breeds and types of horses and were exported in large numbers to other countries; they had great influence on Polish horse breeding. For many years all horses having the typical golden sheen born in the Karabakh region were considered to be of the Karabakh breed. Indeed, this characteristic shows the presence of Karabakh blood. The quality of Karabakhs was so impressive that they were considered as important to horse breeding in Asia as the Thoroughbred was in Europe.

The typical Karabakh at the end of the nineteenth century had a tough, lean constitution. The head was small and light with a well-developed forehead, large eyes, and small muzzle; the neck was of medium length and set high. The withers were well pronounced; the back was straight; the legs were straight with good joints; and the hooves hard. The skin was thin and tender with short, silky hair. Often the areas around the eyes, ears, muzzle, and inside of the legs were without hair. The general appearance was harmonious and graceful, the temperament alert but good. Colors within the breed were sorrel, grey, or lemon with a golden or silvery sheen. The best ones were the lemon-colored horses. From living in severe conditions year around, the Karabakh developed resistance and soundness.

Decline of this outstanding breed came about through changed economic conditions and as a result of the raid by Iran in 1826, when many *taboons* were driven into Iran (Persia). Another contributing factor was the small size of the Karabakh, unable to meet the needs of buyers, especially the army. The Karabakh was a specialized saddle breed, but its working abilities were not tested in races. There are many reports of its good qualities and endurance on long-distance travels. The action of this horse was calm and comfortable for the rider. The best horses had spacy, free gaits with a high-set head and were able to gallop in rough, mountainous terrain and change direction easily and quickly. Pure Karabakh horses were never hotheaded and were not known to stop before obstacles.

After the raid by Iran in 1826 when the stud farms were destroyed, Karabakh breeding began to decline. Attempts to reestablish the breed were unsuccessful. In 1949 thirty-eight mares and two stallions were collected at a specialized stud farm in Azerbaijan.

Today breed numbers are very small. The Karabakh is bred at Agdam stud, yet the total herd is composed of Arab-Karabakh crosses of various grades. Efforts are currently being made to regenerate the breed but, unfortunately, few purebreds remain.

Karacabey

ORIGIN: *Turkey*
APTITUDES: *Riding horse*
AVERAGE HEIGHT: *15.1 to 16.1 h.h.*
POPULATION STATUS: *Extinct*

A short treatise seems in order regarding the Karacabey horse. This breed is presented in most recent books published on horse breeds as a current breed, yet the Karacabey has been virtually extinct since 1979. Also, it is said to be the only "true" breed to be found in Turkey—yet Turkey has thirteen other breeds with a history of at least 800 years.

After the establishment of Turkey, the Turks bred horses they called Karacabey at the Karacabey Stud. They were developed by crossing the Turkish Arab race horse with the native Anadolu. The Turkish Arab Race Horse is not of different blood from other Arabs, but this horse has been bred specifically for racing. As a consequence, the Turkish Arab Race Horse is larger and faster than other Arabs.

The Karacabey was a very sound breed with good conformation, showing Arab influence but with more body mass. At one time, Turkey sent a fine Karacabey horse to England as a gift to the Queen, and this horse became a champion polo horse. The Karacabey was also a notable jumper.

The Karacabey Stud also bred a horse called the Karacabey-Nonius, a larger breed than the Karacabey, and now also extinct. The Karacabey-Nonius was bred as a harness horse and show jumper and was produced by crossing the Karacabey to the Anadolu and Nonius breeds. This breed easily jumped over five feet and averaged between 15 and 16 hands. Breeding of Karacabey-Nonius horses ceased entirely in 1970.

High quality show jumpers began to be imported from France and Germany, and the Karacabey breed was forgotten in favor of the exotic breeds. Another reason for its decline was competition from motor vehicles; the government decided there was no longer a need for this horse.

The Karacabey Stud ceased all breeding of the Karacabey horse in 1980, and approximately 3,000 horses were sold at auction to the public. In a short time, crossed with other Turkish breeds, the Karacabey virtually disappeared. Today it not possible to find one Karacabey horse in Turkey.

Ironically, almost as soon as the breed had vanished, the sport horse came to the fore as never before throughout the world, and Turkish officials deeply regretted the loss of this fine breed.

Karacabey Stud breeds the Arab, Haflinger, and halfbred Haflinger.

Karachai

(Karachaevskaya)

ORIGIN: *Northern Caucasus (former Soviet Union)*
APTITUDES: *Riding horse*
AVERAGE HEIGHT: *14.3 to 15 h.h.*
POPULATION STATUS: *Common*

The Karachai breed belongs to the horse breeds of the northern Caucasus. Developed at the beginning of our millennium with the blending of mountain horses and various others such as southern, steppe, and eastern horses, the Karachai is very similar to the Kabarda (Kabardin). A survey done by specialists during the first ten years after World War II revealed that in the territory of northern Caucasus only the Kabarda and Karachai breeds had survived. The Balkar horse was defined as the high mountain type of Kabarda.

Work was started after the war to reestablish both the Kabarda and Karachai breeds as saddle and saddle-harness horses for agricultural use and for the army. Soon, the result of careful selective breeding was evident. Some use of Thoroughbred blood was allowed on a limited basis, and stallion swapping took place between the Kabarda and Karachai breeds.

In 1935 the first volume of the stud book was established for the mountain breeds, registering 1,026 Kabarda and 204 Karachai mares. The Karachai at that time was smaller, averaging 14.1 hands, a tough horse of lean constitution. It differed from the Kabarda in having a more elongated and massive body with less expression of saddle type. The Karachai had an especially elegant head with curved ears, and had a wavy mane and tail; prevailing colors were black and dark brown.

In 1942 official literature dropped the name Karachai, and all horses belonging to the breed were registered as Kabarda. During this period a considerable amount of stallion swapping took place. Recently, however, the Karachai, refusing to succumb to extinction, is once again being considered as a separate breed. These horses are widespread—there are some 10,000 head—in the Karachay-Cherkess Autonomous Oblast (Province), where their best qualities and genes have been preserved. Mainly they are bred at the Malokarachaevski stud and on eight stud farms in the region.

From 1980 to 1984, workers from the department of horse breeding with specialists of the region did a survey of the Karachai horses. This study allowed authorities to determine accurately the correct standard and description of the modern breed.

The Karachai horse has a medium-sized

head with a slightly Roman-nosed (ram) or straight profile and alert, mobile ears; the neck is of medium length and well muscled, not high but sufficiently long. The shoulders are slightly slanted; the back is straight and firm, often long; the croup is of medium length and well muscled; the chest is deep and wide; and the ribs are rather long. The front legs are often found to have faults such as straight pasterns and the hind legs are often crooked. The hooves are well shaped and hard; the mane and tail are abundant, often wavy.

The prevailing colors are dark brown and black. When crossed to the Thoroughbred, the breed shows this influence but retains its tough constitution and saddle-harness type. The study revealed three types: basic, saddle, and massive.

The Karachai horses of basic type are the most widespread. They are of saddle-harness type and exhibit the main characteristics of the breed. Most often they are used under saddle and are indispensable for mountain tourism, national mounted games, and trail rides. The saddle type includes horses having a small amount of Thoroughbred blood—large horses with a lean constitution, used for sport and in breed purposes to improve horses beyond the borders of their main region. The horses of massive (blocky) type are smaller, with an elongated trunk. Often they are coarse and gain weight quickly. These are widely used in transport and rarely under saddle. Horses of this type are very desirable for development of productive horse breeding.

The breeding goals for the Karachai horse are to improve its working ability; increase average measurements; and improve the neck, shoulders, and legs while retaining the firm constitution, high fertility, and resistance to disease.

The Karakaçan of Turkey. Photo: Ertuğrul Güleç

The Karakaçan breed emerged from crosses of the Trakya horse with horses brought to Turkey by immigrant Turks coming from Tuna, Bosnia, and Kirim in Romania, Hungary, and Bulgaria, respectively. At the time those countries were part of the Ottoman Empire. Their horses had some blood of French and Hungarian draft (coldblood) horses.

The Karakaçan is a light draft horse. The head is heavy with a slightly convex profile and large, expressive eyes which are spaced widely apart. The neck is of medium length and quite muscular; the withers are well pronounced; the back is short and strong; the croup gently sloped; the shoulder muscular and sloping. The legs have good bone with strong joints and hard hooves. Karakaçan horses are very strong and active.

Today remnants of the once numerous pure breed are crossed with the Trakya horse, and it is difficult to find purebred individuals. There are about 1,000 pure specimens currently living in Turkey. Nearly all Karakaçan horses are bay in color.

Kathiawari

ORIGIN: *India (Kathiawar)*
APTITUDES: *Riding, light draft*
AVERAGE HEIGHT: *13.3 to 14.3 h.h.*
POPULATION STATUS: *Rare*

Karakaçan

ORIGIN: *Turkey*
APTITUDES: *Harness, light draft*
AVERAGE HEIGHT: *14.1 to 15.1 h.h.*
POPULATION STATUS: *Rare*

Precise information regarding ancestors of the Kathiawari breed is unavailable. The

Possessing the most extremely curved ears of *any* horse breed, the Kathiawari has ears that touch and often overlap at the tips, and purity of the breed is identified only when the ear tips at least touch one another. The Kathiawari breed (also called Kathi, Cutchi, or Kutchi) has a broad forehead, short muzzle, large eyes and nostrils, and concave profile. The neck is short and the head is held high; the body is proportional and the tail is carried high; black is the only color never seen in the Kathiawari, and piebalds are extremely rare. In temperament the Kathiawari is high-spirited but free from vice, considered very intelligent and affectionate.

In early times the local Kathi and Rajput chiefs and princes of the peninsula of Kathiawar (now called Saurashtra) bred and maintained this breed especially for their requirements. They took great pride in producing horses that could withstand extreme temperatures and were nimble and swift as the hot wind yet at the same time strong and hardy enough to carry a man, his armor, shield, sword, and lance all day through the most difficult terrain. They delighted in horses like themselves — wiry and enduring, able to keep sleek and fit on minimal rations. In the Kathiawari they produced the ideal war horse. It must have taken generations of selective breeding to produce the particular ear characteristics seen in this breed. In Lady Wentworth's book *Thoroughbred Racing Stock,* she mentioned that one can occasionally see in Arab horses the turned-in ear. In the Kathiawari this is a standard. Early breeders continued to maintain the breed until India became independent and feudalism came to an end.

Today the government of Gujarat maintains a stud at Junagadh especially for preserving the Kathiawari breed, maintaining about twenty broodmares and five stallions. The government also has stationed purebred stallions in about ten different centers for improving local horses. Other than these, there are only about fifty additional purebred Kathis in the hands of private breeders.

Within the Kathiawari breed there are approximately twenty different families. Each noble house was noted in earlier times for

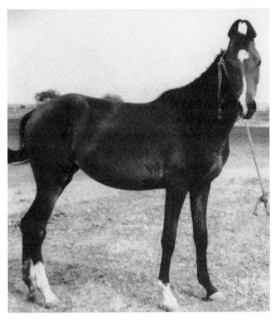

A Kathiawari youngster, showing the extremely turned-in ears typical of this breed. Photo: Krishi Bhavan, Paldi

Kathiawari mare. Photo: Krishi Bhavan, Paldi

breed takes its name from the province of Kathiawar, where it originated many years ago. It is thought that the breed descended in part from local horses which crossed with Arabs rescued from a shipwreck off the west Indian coast by the imperial guards of the Mogul emperor at Veraval Port.

specializing in breeding a particular "Kathi family." Regionally also, certain families were renowned, and the Panchaal region was most famous.

The Panchaal region became the home of the most beautiful horses Kathiawar had to offer. Here, the Kathiawari blood dominated. One should not make the error of confusing the good quality Kathiawari with the malformed, undernourished, undersized horse frequently seen in local villages. While many of these have a small percentage of Kathi blood which gives them endurance, most are of mixed breeds and do not show characteristics of the true Kathi.

The Kathi chiefs were proud of their mares and choice animals were never sold. Often a man owed his life to the strength and speed of his horse. It is said the Kathiawari horse refused to fall—even when grievously wounded—until it had carried its rider out of danger, and often stood over a wounded rider, biting and kicking at any who attempted to approach. Of the indigenous breeds of horses to be found in India, the Kathiawari is said to be the finest.

The Kathiawari is an aristocratic, well-proportioned breed that exhibits grace and style. Kathi horses were used for polo when the height limit for polo ponies was set at 14.2 hands. The cavalry maintained many of this breed until the end of World War I.

Often portrayed by authors as a pony, the true Kathiawari is a small hotblood, an excellent horse in the desert and more reminiscent of the Oriental breeds of the Near East than the Mongolian, from which some say it descended in part. There is little difference between the Kathiawari and Marwari breeds, descending from the same stock but selectively bred for slightly different type.

Kazakh
(Kazakhskaya)

ORIGIN: *Kazakhstan (former Soviet Union)*
APTITUDES: *Riding, pack horse*
AVERAGE HEIGHT: *13.2 to 14 h.h.*
POPULATION STATUS: *Common*

Kazakh. Photo: Nikiphorov Veniamin Maksimovich

The Kazakh horse in Russia is certainly related to and had the same origins as the Kazakh of China. However, having been bred along different lines, the two are different today. The Kazakh has been known since the fifth century B.C. Since that time the breed has been influenced by several breeds—Mongolian, Karabair, Arab, and Akhal-Teke. In the late twentieth century the Kazakh horses have been improved by the Thoroughbred, Orlov Trotter, and Don breeds.

Kazakh horses are kept on pastures throughout the year. Concentrated mostly in western Kazakhstan, they developed into two types throughout this vast territory, the Jabe and Adaev.

Generally, Kazakh horses fall short in performance in the Soviet Union. Their gaits are poor with short strides and the trot gives a jolting ride. At the same time, this breed is extremely hardy. Usual colors are bay, dark bay, or red, with an occasional grey. This breed numbers over 300,000. The best breeding farm is the Mugojar stud. Kazakh horses are used for work as well as for meat and milk.

Kazakh horses have strongly developed jaws as a result of feeding on harsh artemisia, grasses, and salty plants. The winter coat of this horse is dense and has a complicated structure, containing a thick undercoat and longer outer coat which sheds water. Horses living on sandy pastures and deserts have a thick brush of hair on the upper and lower lips, helping to clean the plants. Well adapted to its harsh environment, the Kazakh horse stops growing during hard times and growth resumes in rapid bursts during times when feed is abundant.

Ke-Er-Qin

ORIGIN: *China (Keerqin steppe)*
APTITUDES: *Riding, farm and draft*
AVERAGE HEIGHT: *14.3 to 15.1 h.h.*
POPULATION STATUS: *Common*

On the Keerqin steppe, Zhelimu Meng, in Inner Mongolia, there are both agricultural and semiagricultural areas. One part of this Mongolian plateau is considered a horse base for the country, providing good grasslands.

Starting in 1950, breeding experiments were carried out using Sanhe, Don, Soviet "highblood" (improved Thoroughbred), and Ardennes horses crossed to local Mongolian stock. Using complicated combinations, the Sanhe upgrade, Soviet-Sanhe upgrade, Soviet highblood, and Don grade were tested. The Ardennes-Sanhe upgrade proved the best and most effective. The Ke-Er-Qin still has a variety in type ranging from light to heavy, but the average in height is about 14.3 to 15 hands.

Most horses of this breed are bay or chestnut. The Ke-Er-Qin is fast as a riding horse and strong before a plow or in pulling heavy loads. The Chinese have proven well able to produce a horse that is versatile in all fields where horses are needed.

Mestaño, senior stallion of the Kiger mustang herd. Photo: Burns District, Bureau of Land Management

A two-year-old Kiger filly, showing the remarkable resemblance to the Sorraia of Portugal. Photo: Burns District, Bureau of Land Management

Kiger Mustang

ORIGIN: *United States (Oregon)*
APTITUDES: *Riding horse*
AVERAGE HEIGHT: *14 to 15 h.h.*
POPULATION STATUS: *Rare*

At one time vast herds of pure Spanish Mustangs roamed the open spaces of the United States just as they covered the South American Pampas. As civilization encroached and range stock took precedence over the feral horses, their range diminished in size and many of the horses were captured or shot. There are still many feral mustangs roaming open country in the United States but no one can claim them to be pure Spanish any longer, or even that many of them carry any Spanish blood today. Many have wondered, however, if a few small pockets of the pure descendants of the horses of the conquistadores might still exist in remote, isolated places, and indeed a few such bunches have been located. The Cerbat (see Cerbat) mustangs from the Cerbat Mountains in Arizona are such a group. Another group of Spanish horses was located on an old ranch in Arizona near the Mexican border (see Spanish Colonial Horse).

Imagine the surprise and delight Ron Harding of the Bureau of Land Management in Hines, Oregon, must have felt when in September of 1977 a herd of mustangs that looked to be of pure Spanish descent was brought in from the remote, rugged Beaty Butte region in Lake County. All animals in the bunch were some shade of dun, ranging from brown-dun to the nearly white claybank. All had dorsal stripes and zebra stripes on their legs. Two subsequent groups of horses were gathered from the Beaty Butte range, producing a total of twenty-seven animals very uniform in type, size, and color. The classic Barb head was there. The size was right, averaging between 700 and 800 pounds. There is every indication that this herd of feral mustangs have inhabited the Beaty Butte range for many years without infusion of other blood.

Ron Harding and Chris Vosler, the BLM district manager from Burns, Oregon, and others immediately began to take steps to protect the herd so that the distinctive animals would not be dispersed and lost. The herd was divided, twenty of the animals being turned out on the East Kiger Herd Management Area and the remaining seven released in the Riddle Mountain Herd Management Area, which lies approximately five miles to the northeast. Care has been taken to assure that other feral horses do not range near the preserves so that contamination will not occur.

Isolated in eastern Oregon and not mixing with other feral herds due to the remoteness of their range, the Kiger Mustangs apparently reverted over the years to the primitive dun color. The horses are so much alike that it would be difficult to distinguish one from another if it were not for the occasional white marking.

Noted wildlife photographer Wolfgang Bayer of Jackson Hole, Wyoming, produced a film called *America's Wild Horses,* which was aired on television nationally in the spring of 1987 and again in 1988. As a result of this film, the public saw the Kiger Mustangs for the first time. Each time the film has aired, the Burns District has received a barrage of telephone calls from people desiring to obtain a Kiger horse.

On July 15 and 16, 1988, a group of individuals from the states of Oregon, Washington, California, and Michigan met in Hines, Oregon, and formed the Kiger Mesteño Association. Its main thrust is to protect the Kiger Mustang through preservation of its gene pool in the wild and in captivity. The association aims to have an influence on the disposition and well-being of the adopted animals; adopt the Kiger Herd; act as an advocacy group concerning adoption compliance; and educate the public in reference to the BLM Wild Horse Program. They also desire to be involved in the planning process and future management decisions pertaining to the Kiger Herd Management Area so the public can rest assured that the Kiger Mustang will remain intact for posterity. The association hopes to establish a registry so that offspring of the Kiger Mustang can be tracked and pedigrees established.

The Kiger Mustang exhibits typical Spanish Mustang conformation. The head is small and wedge-shaped with a subconvex profile. The eye is placed rather high and is usually almond-shaped or Oriental in appearance. The ears are small and curved inward. The neck is arched and well muscled, and the throat-latch is usually clean. The withers are well pronounced and long; the back is short; the croup gently sloped with a low-set tail; the chest narrow but deep. The shoulder is very sloped and muscular; the barrel is rounded; the hind quarters are rounded and smoothly muscled. The legs are long and fine but strong with plenty of bone and broad, strong joints; the feet are invariably black, well shaped, and extremely tough.

Several of the Kiger horses have been adopted through the BLM's Adopt-A-Horse Program and have shown willingness to learn and good working qualities. The Kigers are natural at working cattle, a strong characteristic of the horses of Spain. For additional information see Spanish-American Horse.

Kirgiz
(Kirgizskaya)

ORIGIN: *Kirgizia (former Soviet Union)*
APTITUDES: *Riding, light draft*
AVERAGE HEIGHT: *12.3 to 14 h.h.*
POPULATION STATUS: *Rare*

Milking a Kirgiz mare. Photo: Nikiphorov Veniamin Maksimovich

Very rare today in its original, pure form, the Kirgiz horse—foundation of the New Kirgiz—was at one time widespread and plentiful, the purest and most numerous of the Mongolian group in the Soviet Union. In 1923 there were an estimated 2.5 million head. The Kirgiz was healthy, long-lived, modest in care and feed requirements, speedy, and had a firm skeleton, lean constituion, and great endurance. The hooves of this breed have always been hard and healthy. Kirgiz horses are slow to mature, not reaching full development until the age of five years. Kept outdoors year round in severe conditions, only the strongest horses survived, and the Kirgiz was an outstanding treasure of hardy genetic material. Presently the pure, original Kirgiz horse survives only in very small numbers, for excellent though the breed was, authorities felt there was need for improvement, especially in size, and Kirgiz horses were crossed to other breeds.

The Kirgiz is bred in Kirgizia and some neighboring regions of Kazakh, Tadzik, and China. This is a horse adapted to work under saddle and for packing in the mountains, and it is also used for light agricultural work. The small size limits its pulling power, however.

As are other local breeds, the Kirgiz is of ancient origin. History of the breed is bound with history of the Kirgiz nation, people who have always been involved in nomadic cattle breeding and using horses for transport and war purposes. Remnants of ancient horses have been found in tombs in the territory of southern Siberia and Altai. This early culture was breeding horses 1,500 to 2,000 years before the birth of Christ and long before the Mongols came to the area.

A great part of Kirgizia consists of high mountains with many ridges reaching over 16,000 feet in altitude. The mountain valleys with their comparatively mild slopes form a huge area of natural pastures used year round for cattle, horses, and sheep. During the summer the *taboons* go up to the alpine zone, where various nutritious grasses grow in abundance.

The Kirgiz is a typical mountain horse, well adapted to its habitat with rarified air. It moves easily in steep terrain, calmly traveling on narrow paths above deep abysses and swimming through savage mountain streams.

The most difficult time of the year for the horses is winter, with severe frosts, cold winds, and pastures covered with snow. Horses that do not go into the winter with plenty of body fat are usually lost. This breed tends to gain weight easily in summer, preparing for the long winter.

The Kirgiz is small and low-set, massive of trunk and short in the leg. The head is rather large with a straight profile; the neck is short and muscular; the withers are not well developed; the back is straight, often carp-like and medium in length; the croup is sloped; the trunk is barrel-shaped, deep and wide. The legs are short, lean, firm, and have well-developed joints; the hind legs are often cow-hocked. One outstanding feature of Kirgiz horses is the extraordinary firmness of the hooves, enabling them to travel on stony terrain without shoes.

Prevailing colors in the breed are bay and grey, but sorrel and other colors also occur.

Research has shown that six-year-old Kirgiz horses given good feed started to grow again for a time, an adaptive trait that is nonexistent in most other breeds of horses.

Kisber Halfbred
(Kisberi Felvér)

ORIGIN: *Hungary*
APTITUDES: *Riding, harness horse*
AVERAGE HEIGHT: *16 to 17 h.h.*
POPULATION STATUS: *Common*

The Kisber of Hungary. Photo: Dr. Csaba Németh

The Kisber Halfbred was developed at the Kisber Stud Farm, which was founded in 1853. The aim in breeding was to replace the often too nervous, too fine English Thoroughbred horses and to develop a more economical, serviceable halfbred to upgrade local stock. Mixed Kisber stock was crossed for several generations with Thoroughbred horses, then the most excellent strains of the resulting "highbred halfbloods" were crossed among themselves, resulting in the new breed. To increase body mass, Furioso and East Prussian stallions were also used occasionally. In 1961 the breeding stock was transferred from Kisber to the Dalmand farm where it is still bred today.

The goal in breeding the Kisber Halfbred has been production of a handsome, attractive saddle horse as well as lighter harness horses of solid constitution, useful for equestrian sports also, and near to the character of the Thoroughbred but with a larger frame and more massive structure.

The head of the Kisber Halfbred is noble and dry with small, active ears, protruding eyes, and long, straight or slightly arched neck; the withers are medium-high and long; the back is often sloped forward and hollowed; the croup has mediocre muscling; the barrel is deep and well sprung; the shoulder is well sloped. The foreleg is muscular; the legs joints are gaunt; cannons and pasterns are of medium length; the hind leg is moderately muscled, long but less wide, often with slightly open hocks. Kisber Halfbreds are usually chestnut or bay, rarely grey or black.

Because of its lively temperament, solid constitution, and qualities of endurance, the Kisber Halfbred can be used for long periods. It is good as a saddle horse, for touring or three-day eventing, as a military horse, or in equestrian sports of any kind. Some individuals as well as certain strains are very successful in sport jumping.

Recently, the principal aim in breeding has been to increase the frame and body mass while maintaining the elegant appearance of the breed. Selection is directed toward shaping a uniform type with serviceability for sports purposes. Besides maintaining traditional strains, blood of the English Thoroughbred is still used.

Candidates for breeding stallions and mares must pass a working capacity test of military character. Only those individuals passing the test successfully may be used for breeding stock.

Kisber Halfbred stallions are also tested by the performance of their progeny, and in this way a continually improving breed is developing.

Kiso

ORIGIN: *Japan*
APTITUDES: *Riding, light draft*
AVERAGE HEIGHT: *13.2 h.h.*
POPULATION STATUS: *Rare*

The Kiso horse has inhabited Japan for about one thousand years and has in the past been an indispensble aid for farm use, transportation, and power. According to research from 1946 through 1949 by a Tokyo University agriculture professor named Okabe and a Ministry of Agriculture official named Ishikazi, the ani-

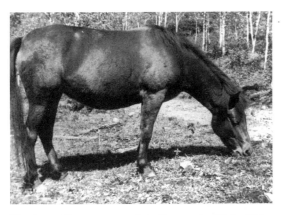

Kiso horse. Photo courtesy of the Preservation Club of Kiso Horse

mal was mentioned in a sixth-century document about a region then called Kirihara-nomaki—now the Nagano Prefecture, where the Kiso horse inhabits the area around the Kiso River. This ancient breed of Japan has since officially been named the Kiso breed.

Exact origin of the Kiso horse and other ancient horse breeds of Japan is uncertain. They are believed descended from either the plateau horses of Central Asia or the Mongolian horses of the grasslands (see Hokkaido).

Japan used horses for military purposes as well as in agriculture and transportation. In the twelfth century, the warrior Yoshinaka Kiso reportedly had 10,000 horse soldiers.

In the Edo era (1600–1867) there was again emphasis on military use. Kiso canyon belonged to the Owari feudal clan. Records from this time regarding the ancient types have been a valuable aid to modern horse breeders. The government of the Kiso area considered the Kiso horse a strategic material, and produced many; numbers again reached more than over 10,000.

During the Meiji period (1868–1903), Japan fought against several foreign countries. Finding Japanese horses inferior in size and strength to western horses, the military of Japan crossbred the Kiso horses to exotic breeds. Since World War II, machinery has replaced horses on the farms and roads.

Fortunately, however, some purebred Kiso horses escaped improvement. They are preserved now in Kaida village at the foot of Mt.

Ontake. Since 1976 pedigree registration has been kept. There are approximately seventy head living today.

The Kiso horse has a heavy head, thick through the chin, with a wide forehead; the neck is short and thick; the top line is level; the trunk is long. The legs are short and sturdy, with good bone; the hooves are well formed and very hard. The mane and tail are thick and heavy. Many individuals of this breed are knock-kneed and it is said this enables them to climb steep slopes more easily. They accustom themselves readily to different climates and have a mild, easy-going disposition.

Kladruby
(Kladrubský)

ORIGIN: *Czechoslovakia*
APTITUDES: *Sport driving*
AVERAGE HEIGHT: *16.2 to 17 h.h.*
POPULATION STATUS: *Rare*

Kladruby. Photo: Marie Goldbergerova

The majestic Kladruby in harness. Photo: Marie Goldbergerova

The Kladruby horse sprang from old Spanish and Italian blood in the sixteenth and seventeenth centuries, having a common origin with the Lipizzan. This breed was improved in the Kladruby imperial court stud on Labe peninsula, founded in 1579 by Rudolf II. Rudolf was a great lover of Spanish horses, and this is one of the oldest studs in Europe. The Kladruby was used as a majestic ceremonial carriage horse by the imperial court in Vienna. The blossoming of this stud farm was from 1712 to 1748, when the stud had 1,000 head of horses.

The Kladruby of the eighteenth century had a high, Roman-nosed head with small ears, high-set neck, and a wide, muscular croup. Colors found in the breed at the time were not only black and grey but also some less common ones, such as palomino and appaloosa. Unfortunately, breeding records for the breed's first 200 years were destroyed by fire in 1757. Later the Kladruby was bred in two color varieties, white and black, as it is again today.

The white Kladruby is still bred at the Kladruby stud. Founder of the modern breed was the black Spanish-Italian stallion Peppoli, foaled in 1764, whose descendants Generale and Generalissimus founded lines still bred today. The blood is refreshed by use of genetically allied breeds such as the Lipizzan, mainly the line of Favory.

Two stallions of the same name founded the herd of black Kladruby horses; the line of Sacramoso, foaled in 1800, lives to the present day. Black Kladruby horses were used mainly as carriage horses for clerical dignitaries. However, in 1930 the black herd was liquidated and most of the horses were sold for meat, including the stallion Sacramoso XIX. Some of the black mares were saved. Later, a bold plan was conceived to regenerate the black Kladbury from the remaining individuals. This aim was fulfilled with success in the Research Institute for Horse Breeding in Slatinany, where this herd is being bred today. This undertaking was highly appreciated by the world organization UNESCO and is acknowledged as a model for rescue of endangered breeds.

Considering that the original old Spanish horses have died out in the world, the Kladruby horse is hippologically unique. The breed has already been bred for four hundred years. In Czechoslovakia it is the only indigenous breed. This is one of the rarest of horse breeds today with only about ninety mares remaining.

The Kladruby is a vigorous, majestic carriage horse of Baroque appearance with a large frame, pronounced convex-shaped head, and vaulted neck. A cadenced trot with high knee action is typical of this breed. Kladruby horses have always been known as tough, late-maturing horses with great endurance.

Kladruby horses were also bred at the Mezöhegyes stud farm in Hungary, and were one of the bases of the Nonius breed.

Presently, the Kladruby is used for sport driving. The horses of Kladruby stud have taken part many times in world championship four-in-hand driving classes, as have the black horses.

Knabstrup

ORIGIN: *Denmark*
APTITUDES: *Riding horse*
AVERAGE HEIGHT: *15.1 to 16 h.h.*
POPULATION STATUS: *Rare*

Knabstrup stallion. Photo: Ellen Bendtsen

The association for Knabstrup horses is only nineteen years old, but the Knabstrup is claimed as a national breed, tracing back to the age of the Vikings. A fresco in Skibby church on Zealand in Denmark, dated at about 1000 A.D., shows a beautifully spotted horse.

In Oslo, Norway, a wall tapestry from about 1200 A.D. depicts a brown Knabstrup horse with fair and rust-colored spots, and this motif has been used on a Norwegian postage stamp. In Scotland the monks had spotted horses as early as 800 A.D. The spotted horse can be traced back 20,000 years from cave paintings in France showing a group of spotted horses with foals.

The original size of the Knabstrup horse was about 14.3 hands. It had clean, dry limbs; large, strong hind quarters; and a small, refined head. Basic qualities included an easy and tractable temperament, and these horses were known for their speed and endurance. Since 1100 A.D. the principal lines of distribution of the spotted horse can be closely traced. The spotted horse was very common in Asia and its distribution extended rapidly when China opened its borders for trade. The Chinese used spotted horses to transport silk and other articles. The main routes crossed Russia, through France to Spain, through Austria to Poland, and even into Scandinavia and North Africa. The main route through France to Spain is interesting as spotted horses were among the earliest to be taken to the Americas.

In 1671 there was a Knabstrup stallion at the royal stud farm in Fredensborg on Zealand referred to as a "tigery horse." The stallion became the progenitor of an entire stud. The Danish king, Christian V, gave his sister several teams of fine horses, among them a team of Knabstrup horses. Today a 300-year-old stuffed Knabstrup, which won a race between Fredensborg Castle and Copenhagen, is displayed at Christiansborg Castle.

In 1804 one of Napoleon's soldiers left a spotted mare in Denmark. Known as Flaebehoppen ("the snivel mare"), she was accepted at the estate Knabstrubgård, close to Holbaek on Zealand, and became the ancestor to many Knabstrup horses.

In 1933, approximately sixty years after the stud farm at Knabstrupgård had been dissolved, a veterinarian named Ahlstrand, who lived in Bornholm, established an association in order to preserve the spotted horses. Many fine Knabstrup horses were bred there, including the stallion Max, who in 1938 knelt to the Danish king, Christian X, at a fair on Bellahøj in North Zealand. A son of Max, Bodilsker, was later sold to a newly started association in the county of Holbaek. In 1947 an association with twenty members was established in Holbaek. In 1970 they recruited members from all over Denmark, making a nationwide effort to preserve the spotted horse.

The aim in breeding Knabstrup horses today is to preserve the horse as a Danish riding horse, no matter what size. This is unique, for few breeds can present horses in all sizes, but the Knabstrup is found today in both ponies and full-sized horses.

The Knabstrup has a beautiful, spotted coat, occurring in many colors and variations. The breed is known for its calmness and friendly temper, intelligence, and aptness to learn. Frequently this horse is used in circuses. Knabstrup horses are strong with solid, dry limbs and strong hooves.

Knabstrup horses have been used in the cavalry, as post and milkwagon horses, and as ordinary riding and work horses. Presently the breed is promoted as a riding and carriage horse. Recently there has been a good deal of crossbreeding and some breeders have concentrated more on color than on conformation. As a result, many different types are represented within the breed and the original type is nearly extinct.

Knabstrup horses are beginning to make themselves known in the show ring and in competitive sports.

Konik

ORIGIN: *Poland*
APTITUDES: *Riding, light draft*
AVERAGE HEIGHT: *12.3 to 13.3 h.h.*
POPULATION STATUS: *Rare*

A direct descendant of the wild Tarpan, the Konik, whose name means "little horse,"

Konik. Photo: Evelyn Simak

Kushum
(Kushumskaya)

ORIGIN: *Kazakhstan (former Soviet Union)*
APTITUDES: *Riding, harness*
AVERAGE HEIGHT: *15 to 15.3 h.h.*
POPULATION STATUS: *Common*

A *taboon* of Kushum mares and foals. Photo: Nikiphorov Veniamin Maksimovich

Kushum mare. Photo: Nikiphorov Veniamin Maksimovich

appeared as a native, primitive horse in Poland in the first half of the eighteenth century. At the turn of the nineteenth century, with intensification of agriculture, the Konik could not compete with the large draft breeds of western Europe and its breeding began to decline. In the years between the two World Wars, efforts were made to maintain this primitive horse and these efforts were resumed after World War II. A reserve, which still exists, was founded in 1954 by the experimental department of the Polish Academy of Science in Popielno (Suwalki County).

This area includes more than 5,000 acres where the horses roam wild all year. They are fed hay only during the winter. A group of Konik mares is also kept in the stables at Popielno.

The Konik has a rectangular, low-set body and a rather large chest and cannon circumference. Like all primitive types it is extremely hardy, is capable of living on scant feed, and is very healthy. The main deficiency in this small breed is lack of good action. The Konik is mostly mouse-colored (grullo) with a stripe along the back, and often stripes on the legs.

Koniks are good working horses, valued especially in gardening farms. Because of their good nature and docility they are also used as riding horses for children. They are in great demand from dealers abroad.

The most substantial breeding of Koniks is done at the National Stud Farm in Racot, Sierakow, and Dobrzyniewo. The population in Poland comprises about 250 mares.

The Kushum breed developed in the Urals region of Kazakhstan at the Pyatimarsk and Furman studs from 1931 to 1976. The goal in producing this breed was to provide a good army mount suitable for keeping in herds the year round. Using native Kazakh mares, crosses were made to Thoroughbreds and halfbreds as well as with trotters to obtain good size and gaits.

Care was taken to retain the high adaptability of the Kazakh while maintaining and improving size and action by mating the first crossbreds to Don stallions. The three-way crosses were then interbred. The result was the new breed characterized by high adaptability, large size, and good, versatile working qualities.

The Kushum is suitable for meat and milk production because, due to its high adaptability, it shows increased weight gain in spring and autumn. The large size guarantees a high yield of horse meat.

The Kushum is a horse with a solid build of saddle-harness type. The head is large but not coarse; the neck is average in length and fleshy; the withers are pronounced; the back is long and flat; the croup is well muscled but not very long; the chest is broad and deep; the legs are properly set. The main colors of the breed are bay and chestnut.

This is a versatile horse of high endurance. It has tested very well in distance tests, trotting in harness, and as regards walking time. Mares are high in milk yield. The breed has a good fertility rate and sound health.

Kushum horses comprise three types within the breed; the heavily muscled, basic, and saddler. Six lines are being formed. The main breeding is done at Pyatimarsk and Kransnodon studs.

Further development of the breed will be done with selective pure breeding.

Kustanai
(Kustanaiskaya)

ORIGIN: *Kazakhstan (former Soviet Union)*
APTITUDES: *Riding horse*
AVERAGE HEIGHT: *15 to 15.2 h.h.*
POPULATION STATUS: *Rare*

Kustanai. Photo: Nikiphorov Veniamin Maksimovich

The Kustanai was developed on the steppes of western Kazakhstan at the collective farm and state farm studs. The breed dates to 1887 when the state-owned stud, the Turgai, was established. Kustanai stud was established in 1888 and Orenburg in 1890. In 1951 the breed was officially recognized.

This new breed was developed by crossing the native steppe horses with Don, Astrakhan (improved Kalmyk), Strelets (now extinct), and halfbred Thoroughbred stallions. At the onset the crossbreeding was not successful. It was not until a nucleus of local broodmares, improved by pure breeding and regular creep feeding, was formed at Kustanai stud that crossbreeding with Thoroughbreds produced a positive result.

In the 1920s work was begun with two systems of management directed at developing two distinct types of Kustanai horse, saddle and steppe. The first involved keeping in pastures and stables, winter grazing in good weather, abundant hay and concentrate feeding, hand mating, and weaning of the foals at six to eight months. The other system involved year-round grazing and keeping in sheds in bad weather, free mating, hay, and concentrate creep feeding.

The saddle type included horses with a high percentage of Thoroughbred blood, while the steppe type was a mixture of Thoroughbred, Don, Kazakh, and other crossbreds. All saddle horses were put to speed tests at the hippodrome.

The Kustanai is found in Kustanai Region, in the south of Chelyabinsk Region, and in southern Kazakhstan. In 1930, the breeding herd at the Kustanai stud numbered 1,000 mares. In 1981, the Kustanai and Krasnodon studs had 726 purebred mares. Pedigreed stallions numbering 746 were used in pure breeding and general improvement.

The modern Kustanai is a massive horse, combining the best characteristics of a saddle horse and the pronounced basic steppe lineage. Conformation features of this horse include a straight, medium-sized head with wide jaws and long, straight, low-set neck; the withers are wide and of good height; the back is wide and short; and the loin is solid and well muscled with a short, nicely rounded croup. The shoulders of the Kustanai horse are long and set high; the chest is wide and deep; the legs are set correctly with well-developed joints, hard hooves, and strong tendons and ligaments. The breed is considered to be very hardy. The Kustanai shows remarkable fitness in a continental climate, also showing excellent speed. Horses of the basic type show good action in the Russian harness.

The breed consists of three types, five sire lines, and six mare families. Three volumes of the stud book have been published. The main breeding centers are the Kustanai regional experiment station (formerly a horse stud) and Krasnodon and Saryturgai studs. This breed has good prospects for pure breeding with limited corrective crossbreeding to the Thoroughbred.

Kuznet
(Kuznetskaya porodnaya gruppa)

ORIGIN: *Western Siberia (former Soviet Union)*
APTITUDES: *Draft work*
AVERAGE HEIGHT: *15.3 h.h.*
POPULATION STATUS: *Uncommon*

Organized horse breeding in Siberia developed in the second half of the nineteenth century. Prior to that time, however, local breeds had emerged which were rather large horses. The most valuable and widely known was the Kuznet (also spelled Kuznetsk).

This horse established the fame of Siberian horse breeding, had its own history, and greatly influenced other breeds. Data on the history of Kuznet horses is incomplete and often contradictory. A book written early in the nineteenth century said these horses originated near the Ob and Tom Rivers and are similar to the Bashkir, with many good trotters. Professor M. Priodorogin, on the other hand, said in 1928 that the Kuznet horse was developed from crossing the Kalmyk and Bityug.

The Kuznet horse of the early nineteenth century had many conformational faults. The trunk was long, the back soft, the legs crooked, the pasterns short, the hooves flat. Even so, this was a valuable horse with remarkable endurance, undemanding, with good movement and pulling power in agricultural work and as a pack horse.

Professor Priodorogin described the Kuznet as a type of local breed with a long body; massive skeleton; good joints; long, wide croup; deep although rather flat ribs; and strong muscles.

In the second half of the nineteenth century, private stud farms appeared in Siberia and the Kuznet horse was strongly influenced by Orlov Trotter, American Standardbred, and Thoroughbred horses. The Tomsk stud farm was established in 1884, and stallions at this farm covered about 3,000 mares annually. Up until 1917, in addition to trotters and Thoroughbreds, there were Belgian and Anglo-Norman stallions in use at this farm. After twenty years of crossbreeding the old Kuznet horse changed its type, preserving, however, its endurance, free movement, and strength. Stud farm breeds, especially the Orlov, upgraded the appearance of the Kuznet and improved its action at the trot.

Development of the Kuznet breed resulted from addition of improving breeds and natural conditions of the Kuznet basin, guarded on all sides by mountains. Formation of a draft type of horse resulted from the needs of industry in gold mining and transport.

It is impossible to evaluate the great role of the Kuznet horse for its tremendous contribution in the past. Many specimens of this breed were taken into nearly all parts of Siberia. The breed served in Yakutia in development of the Megezh type of Yakut horse. In Tuva, Kuznets were involved in development of the Verkhoyansk (Middle Kolyma) horse, and in the Altai region, the Chumysh breed.

Due to lack of railways or roads for transport, the horse was indispensable. The *Journal of Horse Breeding and Hunting,* in 1845, recorded that merchandise was transported by Kuznet horses beyond the arctic circle numbering 20,000 horses per year. All transport of Chinese merchandise was accomplished with Siberian horses, and tea export alone used about 5,000 horses. The breed, widely known, was also popularized at the West Siberian show in 1911. Each army located in Siberia used Kuznet horses. A good Kuznet mare was valued ten times higher than those of other breeds.

During the First World War the breeding of these horses suffered greatly when 50,000 horses were taken into the army. In 1916 a breeding farm for pure breeding was established, but it was later closed. The numbers of horses declined again during World War II, when 50,000 more went to the front. After the war, work began to reestablish and improve the breed.

The stock varied in type. In the north, where in addition to trotters the Belgian and Anglo-Norman were used, the horses were quite massive. In the southern regions the influence of Thoroughbreds was stronger. A breeding farm was established in the Novosibirsk region, with the task of consolidating the breed by adding the blood of lighter Belgian and massive trotter stallions. The offspring were then bred in purity. The desired type for the Kuznet horse was the light draft type, and two bloodlines were established: Ciklon, foaled in 1934, and Volkon, foaled in 1927. The first examinations of working ability showed good results.

Unfortunately, in the 1950s the stud farm was closed and after that time no breeding work was done with the Kuznet until a breeding farm was established in 1983, once again in the region of Novosibirsk. More than 500 horses were concentrated there, collected from various areas. In 1986 another breeding farm was established for this valuable horse in the Ubinsk region with forty mares.

It would seem that the establishment of two breeding farms saved the breed from extinction, yet there exists a lack of good, typical sires. For this reason some of the mares are put to Latvian stallions.

The Kuznet horse has a great importance to horse breeding in Siberia. This horse has the same live weight and working ability as the Russian Heavy Draft but is more adapted to severe climatic conditions.

Landais
(Barthais)

ORIGIN: *France (Landes)*
APTITUDES: *Riding, light draft*
AVERAGE HEIGHT: *11.1 to 13 h.h.*
POPULATION STATUS: *Rare*

Landais pony. Photo courtesy UNIC, Paris, France

The Landais pony originates in the region of Landes near Barthais de l'Adour in the southwest of France. This is a fine and elegant pony, resembling a miniature Arab. The breed has existed on its native breeding grounds since prehistoric times. Arab blood was infused at the time of the French victory over Muslim horsemen at Poitiers in 732 A.D. Additional Arab blood came through use of an Arab stallion in 1913.

In 1946 the Landais pony was crossbred with several other breeds in an effort to produce a larger, heavier animal, so today it is difficult to find many pure specimens. The French government is taking steps now to preserve the breed and using Arab or Welsh Section B stallions occasionally to help restore the breed to its original type.

Landais ponies have run wild on the Barthes plains and Landais marshes along the Adour River for centuries, numbering around 2,000 at the beginning of the twentieth century; however, lack of interest nearly resulted in extinction of the breed. This is a tough, hardy breed, living outdoors year round. It has a robust constitution.

Principal colors found in the Landais are

black, bay, or chestnut. The ponies are elegant and intelligent with an Arab-like head, large eyes, and small, pointed ears. The neck is long and arched; the back short; the croup gently sloped; the shoulder long and sloping. The coat is fine with a full, long mane and tail. The Landais always has a hard, well-shaped hoof.

Landais ponies are used as chlidren's mounts or for harness work, and these ponies are excellent trotters. Their temperament is gentle, making them a good choice for children.

Latvian
(Latviiskaya)

ORIGIN: *Latvia (former Soviet Union)*
APTITUDES: *Riding, draft*
AVERAGE HEIGHT: *15.1 to 16 h.h.*
POPULATION STATUS: *Common*

Latvian. Photo: Nikiphorov Veniamin Maksimovich

Developed in Latvia from the beginning of the twentieth century up to 1952 by crossing native horses with west European harness and saddle breeds, the Latvian is a good all-around utility horse. The native Latvian was probably closely related to the same breed or breeds that spawned the North Swedish Horse, Døle Gudbrandsdal, and other heavy draft horses.

Oldenburg, Hanoverian, and to a lesser extent Holstein stallions had the greatest influence in developing the modern Latvian

breed. Between 1921 and 1940, sixty-five Oldenburg stallions and forty-two Oldenburg mares were imported from the Netherlands. These became the core of the developing breed. The Oldenburg horses were imported from Holland, obtained at the Groningen stud. At this farm some East Friesian blood was also used, along with breeding horses in purity. Besides the purebred Oldenburgs, Oldenburg crosses and Hanoverian, Norfolk Roadster, Ardennes, and East Friesian horses were widely used. A special role in breed formation was played by the Okte stud in the Talsa region.

Two types are seen in the Latvian; the harness and the sport horse. Prior to 1960 the harness type was given special emphasis. Later, as equestrian sports developed in popularity in Latvia, the number of sporting horses was increased with infusion of Hanoverian and Thoroughbred blood. Offspring obtained from massive Latvian mares and Thoroughbred stallions are outstanding sport horses, successful in dressage and show jumping. Latvian horses have shown good results in performance tests in both saddle and harness, particularly in competitions.

The modern Latvian is a successful combination of the features of both saddle and utility breeds. Tall, with heavy muscling and good bone, it is an intermediate between the saddle and harness types. Strong and having good endurance qualities, the breed is popular with both farmers and riders.

Latvians have a well-proportioned and solid build; the joints are sometimes coarse; the musculature is well developed; the chest is broad. The withers are moderately pronounced or high and long; the back and loin are long, flat, and well muscled; the croup is long and has a gentle slope; the legs are properly set with good knee and hock joints. Defects seen in the breed include short and ringboned pasterns and cow hocks. The usual colors are bay, brown, and black. Chestnut is seen less frequently.

Pure breeding and limited crossing with Hanoverian and Oldenburg stallions are the main methods used for improvement. The best stud farms are the Burnieke state farm, Uzvere and Tervete collective farms, and the

Sigulda experimental farm of the Institute of Animal Breeding.

Lewitzer

ORIGIN: *Germany (eastern)*
APTITUDES: *Riding pony*
AVERAGE HEIGHT: *11 to 14.1 h.h.*
POPULATION STATUS: *Common*

Lewitzer pony. Photo: Prof. H. C. Schwark

A new breed developed in the German Democratic Republic, the Lewitzer, originated in 1976. Pony mares of various breeds including Shetland and Fjord were used with Arab, Trakehner, or Thoroughbred stallions. The aim has been to produce a quality riding pony for children of all ages and sizes and therefore there is a wide range in size.

There are several categories. Category 1 is the smallest type, measuring 10.3 to 12 hands, resembling a Shetland. Category 2 is 12 to 13.2 hands, a strong pony for field work. Category 3 measures 13.1 to 14.1 hands and is used for larger children.

All of the Lewitzer ponies are said to be good-natured, and indeed this was a criterion for breeding stock. The larger ponies have gaits more like the horse than the pony and give a smooth ride.

Primary colors in the Lewitzer pony are piebalds and skewbalds. Solid colors and leopard-spotted animals are also allowed. The breed is not well consolidated at present due to its short history and the variety of breeds used. The present population is around 2,500 mares and 100 stallions. A small population of the Lewitzer pony, numbering about twenty mares and one stallion, is also found in the former West Germany.

Lichuan

ORIGIN: *China (southwest Hubei Province)*
APTITUDES: *Riding, pack horse*
AVERAGE HEIGHT: *12 h.h.*
POPULATION STATUS: *Common*

Lichuan broodmare. Photo: Weigi Feng and Youchun Chen

The Lichuan horse originated in the mountainous part of southwestern Hubei Province in Lichuan County, an area having an ancient relationship with the horse. This region has long been known for horse breeding, and horses have always been important for the transport of goods in the mountainous terrain. The Lichuan is the largest of the breeds in the southwest of China. Colors are grey, bay, chestnut, and black.

In 1958 insemination stations were established and played an important role in development of the present Lichuan breed. This horse carries a load of from 80 to 165 pounds for eighteen to thirty miles, without prolonged rest, at an average speed of six miles per hour. This is an impressive performance. The horse has a good, amiable temperament.

Plans have been made to increase the numbers of this important and useful breed for agricultural purposes. Presently there are about 10,000 of the breed. Lichuans are compactly built, small horses with a remarkably strong constitution. The head is somewhat heavy with a straight profile, large eyes and nostrils, and upright ears; the neck is moderate in length; the withers are set high. The chest is well developed; the ribs round; the loin and back short and level; the croup is sloped; and the tail set low. The joints are strong with obviously developed tendons; the hoof is very solid. The front legs are straight; hind legs are often sickled. Hair of the mane and tail is long and thick.

Many Lichuan horses are sold to lowland areas, where they are highly appreciated for their endurance and working abilities.

Lipizzan stallion. Photo: Dr. Jaromir Oulehla, Bundesgestüt Piber

Lijiang

ORIGIN: *China (Lijiang District)*
APTITUDES: *Farm, draft*
AVERAGE HEIGHT: *11 to 11.2 h.h.*
POPULATION STATUS: *Common*

Lijiang horses are a newly developed breed, restricted to the Lijiang District. This is an area of high altitude with a greatly varied climate. The economy required a more powerful horse than was found in the area, so in 1944 the Arab, Yili, Hequ, Kabarda, and small type Ardennes were introduced. Crossbreeding and interbreeding of the progeny has developed a horse ideal for the needs of the area. Today there are approximately 4,000 Lijiang horses.

Lijiang horses were crossed with Arab and Arab-Kabarda and the offspring were increased in size to about 12.2 hands, but additional work to increase the size of this breed has not yet been successful.

A Lipizzan performing in Germany. Photo: Evelyn Simak

Lipizzan

(Lipizzaner, Lipitsa)

ORIGIN: *Austria*
APTITUDES: *Riding horse*
AVERAGE HEIGHT: *15 to 16.1 h.h.*
POPULATION STATUS: *Rare*

One can hardly think of the famous Lipizzan horses without being reminded of the Spanish Riding School of Vienna. This famous school is the oldest riding academy in the world. That it has remained in existence despite wars and revolutions is the best proof that the Spanish Riding School is a seat of art of the highest degree.

The Lipizzan breed takes its name from the village of Lipizza, near the northeast border of Italy. Although now in Yugoslavia, Lipizza belonged to Italy before World War II; even earlier, when the breed was being developed, it lay within Austrian territory. For this reason Austria is credited with originating the Lipizzan breed in the sixteenth century.

The Lipizzan faced great peril during the Second World War. Literally at the very last moment—in the middle of March 1945 and in the face of great difficulties, the school and its full complement of horses and irreplaceable traditional equipment were taken to St. Martin in upper Austria. Here, the riding school awaited the arrival of the victorious American forces and presented itself to General George Patton. Patton granted the request of the school director, placing the Spanish Riding School under the immediate care of the United States Army in order to return it safely to the new Austria, and had the Lipizzan Stud, which had been transferred to Czechoslovakia, brought back to Austria under military escort. In this manner the Spanish Riding School was rescued, for the school and the stud are inseparable. One cannot exist without the other.

Austria has guarded this home of classical riding as a precious jewel of the highest equestrian culture. The great Riding Hall in which the Lipizzan has been trained for over 250 years lies in the center of the city of Vienna. The structure, commenced in 1729 and completed in 1735, is a work of art of the first magnitude—a masterpiece of architecture.

Only stallions of the Lipizzan breed are used for the work done at the Spanish Riding School. They are the successors of the elegant Spanish breed whose representatives were once never allowed to be absent from any of the splendid performances and colorful carrousels (tournaments).

The Iberian horse was famous in the days of the Romans and much sought after, and "Caesar's snow-white steeds which Hispania did him send" were well known all over the world. The Carthaginian colonies on the European continent favored the introduction of Oriental blood, thus helping in the improvement of horse breeding. Although the breed suffered under the Vandals and Goths, it attained a considerable status during the seven-hundred-year occupation of the Moors because the caliphs were able to bring over the best breeding material. After the Moors were driven out of Spain, attempts were made at a better class of horse breeding. Various European countries attempted to preserve the Spanish blood in their own studs. The Spanish horse was transplanted with great success to Italy, and there the Polesinian (horses from Polesina stud) and Neapolitan breeds reached great fame. In the Danish stud at Frederiksborg and the studs of Prince Lippe-Bückeburg of Denmark, there were always excellent examples of Spanish horses to be found.

Maximilian II, son of Emperor Ferdinand I, introduced the Spanish horse to Austria about 1562, founding the court stud at Kladrub. His brother, Archduke Charles, the governor of Styria, Carinthia, Carniola (Krain), and Istria, also established a stud of similar breeding, and caused a mews to be erected in 1580 at the village of Lipizza, near Triest. Each of these studs and, later, the royal stud at Halbturn (1717–43) furnished valuable horses to the Spanish Riding Stable at Vienna. There was close collaboration and continual exchange of breeding stock among the three studs. At a later period the stud at Halbturn was closed down, not having proved favorable for the breeding of Spanish horses. Kladrub continued to breed the heavy coach type, and Lipizza the riding and light carriage horse.

Unfortunately, the stud books have been kept only since 1701 and then incompletely, so we can only speculate regarding the original type of horse from various descriptions which have been found. It is known that Archduke Charles ordered the purchase of both stallions and mares from Spain during the year of the foundation of Lipizza. The horses were endowed by nature not only with beauty and elegance but also with hardiness, endurance, and courage.

The Lipizzan breed is based on six dynasties of stallions which date back to the following sires: Pluto (grey), origin Dane, foaled 1765; Conversano (black), origin Neopolitan, foaled 1767; Neapolitano (brown), origin Neopolitan, foaled 1790; Favory (dun-colored), Lipizzan breeding from Kladrub, foaled 1779; Maestoso (grey), sire origin Lipizzan, dam origin Spaniard, foaled 1819 at Mezöhegyes; and Siglavy (grey) origin Arabian, foaled 1810.

The Lipizzans distinguish themselves by their proud appearance, smooth, elastic gaits with high knee action, and lively temperament, combined with a wonderful character and docility. They have a compact but well-formed body and are classified today as a heavy warmblood. The head is long with a straight or slightly convex profile, well-pronounced jaws, small ears, and large, expressive eyes. The nostrils are flared and flexible. The neck is muscular and arched; the withers are somewhat flat; the back is long and sometimes hollow; the loins are strong; the croup is slightly sloped, short, and broad; the tail is set high. The chest is deep and wide; the shoulder sloping and powerful; the legs are strong, muscular, and hard with broad, clean joints and well-defined tendons.

Today the predominating color of the Lipizzan breed is white (grey). Very few Lipizzans are brown. The grey is born dark—bay or black-brown—becoming snow white only much later, some at four and others near ten years of age.

Training of the Lipizzans is done slowly and with infinite patience and kindness. Great emphasis is placed on ending the day's work when the horse has performed his task well so that he is praised and rewarded, returning to his stall with a good feeling. Traditionally, it is the old stallions that train the young riders, and only older, experienced trainers may work with the young stallions. The Lipizzan stallions are brought to the height of perfection in the difficult high school movements and this can only be achieved over time with great kindness and patience. The stallion can never be brought to a state of perfection by force or abuse, but only by a carefully arranged system which combines cajoling and commanding with much praise and little punishment. Consequently, the best stallions are the oldest.

In Austria the state has a monopoly on the breeding of all Lipizzan horses: the only place where they may be bred is at the Bundesgestuet Piber in Styria.

Lipizzan horses are bred in many other countries and yet the estimated world population is around 3,000, making this a rare breed. There are about 420 purebred animals in the United States.

Lithuanian Heavy Draft
(Litovskaya tyazhelovoznaya)

ORIGIN: *Lithuania (former Soviet Union)*
APTITUDES: *Heavy draft and farm work*
AVERAGE HEIGHT: *15 to 16 h.h.*
POPULATION STATUS: *Common*

Lithuanian Heavy Draft horses. Photo: Nikiphorov Veniamin Maksimovich

Lithuanian Heavy Draft. Photo: Nikiphorov Veniamin Maksimovich

Developed by crossing Zhmud (Zhemaichu) horses with the Swedish Ardennes, the Lithuanian Heavy Draft is a solid horse of large size and harmonious body proportions. The breed was developed in Lithuania beginning in the early twentieth century and was officially recognized in 1963. In 1964 there were 62,000 head in Lithuania, making this the most widespread breed in that area.

This breed has a coarse, large head; the neck is short and well muscled; the back is long and sometimes a little dipped; the loin is flat and broad; the croup is broad, long, and well muscled; and the legs are solid and well set. Defects within the breed include its coarse head, dipped back, pin toes, and sickle-hocked hind legs. The most common colors are chestnut and bay.

The Lithuanian Heavy Draft is a long-lived horse with high fertility. The solid build of this breed enables it to withstand extreme and harsh weather conditions. It is used for heavy draft and farm work and to improve meat and milk production in native Altai horses. Currently, crossbreeding of the native Altai and improved horses with the Lithuanian Heavy Draft is being used to develop a new breed.

There are nine lines within the Lithuanian Heavy Draft. It is bred at Nyamun, Sudav, and Zhagar studs and at horse breeding units of collective and state farms.

Llañero
(Venezeulan Criollo, Prairie Horse)

ORIGIN: *Venezuela*
APTITUDES: *Riding, stock horse*
AVERAGE HEIGHT: *14 h.h.*
POPULATION STATUS: *Common*

The horses of Venezuela originated from those brought by the Spanish conquistadores, specifically from those abandoned by Don Pedro de Mendoza in 1535 near Buenos Aires.

Horses of South America developed into different types depending on the climate and other environmental conditions of their habitat. Venezuela, with its extended droughts and hard winters, shaped the Spanish horse into a small, tough horse with lively action.

The Llanos, wide grasslands of northwestern Venezuela and western Colombia, are the home of the native Venezuelan Criollo, called the Llañero. The Llanos (plains) have an area of approximately 220,000 square miles and are delimited by the Andes mountain range to the north and west, the Guaviare River and Amazon rain forest to the south, and the lower Orinoco River and Guiana highlands to the east. Elevations rarely exceed 1,000 feet and rise from the Llanos Bajos (low plains) west to the Orinoco River and Llanos Altos (high plains) below the Andes.

Most of the Llanos area is treeless savanna covered with swamp grasses and sedges in the low-lying areas and with long-stemmed and carpet grasses in the drier areas. Trees are concentrated along the rivers.

The native horses of Venezuela, bred in a hotter climate than horses of the Pampas and nurtured on coarser grasses, have become slighter and less stocky than the horses of Argentina. Venezuela has a high number of horses, and even today the Criollo is found in a half-wild state on the plains. Generally the coat is dun with a dark mane and tail, but other colors are present in the breed. Retaining resemblance to the Andalusian and Barb ancestors, the Llañero has fine legs and is overall a slimmer animal than its other Criollo cousins in South America.

Llañero horses are used primarily for working stock and have great stamina and easy gaits.

For additional information see the Spanish-American Horse.

Lokai
(Lokaiskaya)

ORIGIN: *Tadzhikistan (former Soviet Union)*
APTITUDES: *Riding and pack horse*
AVERAGE HEIGHT: *14 to 14.3 h.h.*
POPULATION STATUS: *Common*

The Lokai is classified as a saddle breed of Oriental lineage. Bred in central and southern Tadzhikistan, this breed was developed by the Uzbek Lokai tribe in the sixteenth century through improvement of local medium-sized steppe horses. Various Central Asian breeds were used, such as the Iomud, and to a lesser extent, the Akhal-Teke and Karabair. Later, the breed was influenced by Arab stallions brought from Bokhara.

Lokai. Photo: Nikiphorov Veniamin Maksimovich

Shortest in stature of all the Central Asian breeds, the Lokai is not uniform in type. The head is sometimes coarse and bulky and occasionally lacks breed character. In most cases, the head is short with a straight profile. The neck is medium in length, lean, and often set quite low with a prominent throat-latch. The withers are medium in height and broad; the back is straight, wide, and short; the loin is prominent with good development in the muscles; the croup is long, often sloping and well muscled. The chest is deep and broad, the ribs rounded, the groin short. The legs are solid with very hard hooves. The legs are not always properly set; the forelegs are often splayed and the hind legs are often cow-hocked or bowed. Coarse joints and general flabbiness are found in horses of large size. Bay, grey, and chestnut are the most widespread colors. The surface hairs of the coat are characteristically curly.

In spite of conformational irregularities, the Lokai horses are very hardy and show good action. They have great endurance under saddle and pack and in national games, especially *kok-par* (see Karabair). These horses mature late but respond readily to improved feeding and management. When purebred Lokai were raised in good stable conditions they surpassed their contemporaries raised on pastures by 2.34 inches at the age of two and one-half years. Lokai horses are fast and agile and make excellent riding horses.

A new breed of saddle horses is now being developed in Tadzhikistan by mating Lokai mares to Arab and Thoroughbred stallions.

Horses with curly coats are most often found among middle Asian breeds, especially in the Lokai. Although the curly coats of foals do not have the high quality of Karakul lamb coats, they are valued rather highly. For example, during the years from 1920 to 1930, pelts from normal foals brought five U.S. dollars, while curly furs brought ten. The breeding of curly-coated horses is of interest in the former Soviet Union, as such pelts could be used together with Karakul to make various fur products.

In the Lokai breed, the foundation sire of the curly-coated strain was a golden sorrel named Farfor, found in 1959 in the Kolchoz Lamanov in the Parchar region. He was not a large horse, measuring about 14.3 hands. Farfor was of typical Lokai breed type and curly from the ears to the hooves, with tight curls like a Karakul lamb. He was nine years old when he was discovered.

Farfor was used for breeding in the years 1955 to 1970, each year producing fifteen to eighteen foals, the majority of which had curly coats. His offspring were characterized by great endurance, resistance to disease, and good adaptation to *taboon* keeping in the mountains with changeable and unpredictable weather. Farfor was used for a long time in the Tadzik national game, *buz-kasi*.

The best curly sons of Farfor were Firuz, a grey born in 1965; Front, a sorrel born in 1958; Fakir, grey, born in 1960; and Murgab, sorrel, born in 1964. In 1970 the Farfor strain was divided into two breeding groups. In the Kolchoz Lamanov the breeding work uses the offspring of Firuz, Front, and Fakir, and there is a herd of fifty broodmares. In the Sovchov Vali Zade the offspring of Murgab are used, having forty-one broodmares.

Farfor was heterozygote, and to strengthen the heredity of the curly coat inbreeding was used in certain generations. As a result of many years of careful selection, a group of curly horses was obtained. Horses of this group are curly from ears to cannons, especially during the spring and summer periods. Currently there are ninety-nine horses of this strain—eight stallions and ninety-one mares—

all with recorded pedigrees. Recently a small group of curlies was moved into Koktas horse farm, so presently the breeding work is organized in three places.

The selection of breeding stock is directed toward improving the appearance and increasing the size of the horses. It is planned to collect all of the best curly horses at one horse farm, most likely Koktas, to continue their development and organize a serious experimental study into the heredity of this unique characteristic.

Aside from the Lokai horses with curly coats, breeding of the Lokai has been done in purity for the last eighteen years with good conditions of feeding and raising. There are various types found within the breed: horses from the northern regions tend to have coarse heads, while horses with the lightest, most refined heads are found in the south. Those representing the massive type have short heads, wide in the forehead with either a straight or convex profile; the neck is medium in length and, often set low and in the stallions, is quite meaty. Horses expressing more eastern characteristics have higher withers, with the back straight and wide and the croup long and sloping with good muscling. The legs are hard with well-developed joints and tendons. The temperament is alert, "hot" but good. It is said that the well-trained Lokai knows *only* its master.

Bay and sorrel horses of this breed often have a golden tint or sheen due to their eastern ancestors.

Lombok
(Macassar)

ORIGIN: *Indonesia (Isle of Lombok)*
APTITUDES: *Draft and farm work*
AVERAGE HEIGHT: *12 h.h.*
POPULATION STATUS: *Common*

The Lombok is the pony of the island of Lombok, also known as the Macassar. Lombok is one of the Lesser Sunda islands lying due east of Bali, across the Selat (strait) Lombok and due west of Sumbawa across the Selat Alas.

As early as 1640 Lombok was under rule of the sultan of Makasar (Macassar). Eventually, the Balinese seized control, forming four kingdoms on the island. One of them, Mataram, entered into a contract with the Dutch which lasted from 1843 until 1872. With Mataram's oppression of the Sasaks and interference in politics on Bali, the Dutch stepped in, and in 1894 eliminated Balinese rule in Lombok and imposed direct rule themselves.

Due to Dutch involvement on the island, it is likely that the Lombok pony has the influence of some Dutch equine blood, although this is uncertain. The Lombok pony is also referred to by some as the Sulawesi pony, although separate statistics were received from Indonesian authorities for the Lombok and Sulawesi.

The Lombok is said to be predominantly "red" in color, so one may assume the breed to be usually bay or chestnut. It has a calm temperament and apparently is used solely for draft work. The majority of the island of Lombok is under cultivation. No physical description was given for this pony, but it likely resembles other ponies of Indonesia, which have Mongolian ancestors and are similar in type. Other possible breed influence is uncertain.

For additional general information regarding Indonesian ponies, see Bali.

Lundy Pony

ORIGIN: *England (Lundy island)*
APTITUDES: *Riding pony*
AVERAGE HEIGHT: *13.2 h.h.*
POPULATION STATUS: *Uncommon*

Lundy pony mare and foal. Photo: A. J. Martindale

Lundy is one of several islands in the Bristol Channel with Scandinavian names, as they were used as bases for pirate fleets during the age of the Vikings. Since that time none of the islands have been continuously inhabited. The Lundy pony is a recent innovation by a proprietor of the island.

Lundy is from the Norse *lunda-ey,* meaning "Isle of Puffins," and a large colony of them still inhabits the island. The resultant guano makes the grass grow abundantly but with too much nitrogen content to be an ideal food-stuff. This is the highest of the islands and the one nearest the English shore, only a mile off the coast of Devonshire. Foundation stock for the new breed consisted of several New Forest mares and an Arab stallion. The last private owner of the island, Martyn Coles Harman, introduced thirty-four New Forest mares of the "old fashioned sort." Experiments with a Thoroughbred and later with Welsh riding pony sires that had recognizable traces of Arab blood predictably proved unsuccessful as their progeny could not stand the harsh, insular climate. Influential "outside" sires have been Welsh Mountain, New Forest, and one Connemara, and essentially there has been no other blood infused into the 117 ponies existing in 1991.

When the island was sold in 1969 on the death of Martyn Harman, it passed into pos-session of the National Trust but was adminis-tered by the Landmark Trust. Three years later the National Pony Society assumed re-sponsibility for the ponies. The Society intro-duced another New Forest stallion, Green-wood Minstrel, with successful results. Administration of the National Pony Society lasted until 1980. The entire herd was moved to Cornwall, and it is there and in North Devon that most Lundy ponies have been bred since then.

In the 1980s and 1990s Lundy ponies have been increasingly prominent at shows in the south of England. The present breed society was set up in 1984, and since that time some mares have been moved back to the island, where foals have again been born.

Conformation of the Lundy pony should show true pony character, combined with the substance and hardiness that enables them to live out all year on relatively rough pasture. The head is neat and pony-like with a broad forehead, sculpted face, fine cheekbones, and a wide, strong jaw. The foreface should be slightly convex or Roman from below the eyes to the muzzle. The ears should be of medium size and neat in appearance. The eye should be large, gentle, and alert. The neck is of medium length, strong and muscular. It should have a shapely crest and show no signs of coarseness. The shoulder should be well laid back. The body should be in proportion, strong, deep, and well ribbed up. The back should be short, strong, and not slack over the loins. Ideally, when viewed from the side, the body and legs should describe a square.

The preferable color of the Lundy pony is dun with dark points. Roan, bay, palomino, and dark liver chestnut are all permissible. Blue-eyed creams are permitted, but will only be registered as mares or geldings. Piebald, skewbald, grey, and bright chestnut are not acceptable colors.

Lusitano

(Lusitanian, Bético-lusitano, National, Peninsular)

ORIGIN: *Portugal*
APTITUDES: *Riding horse*
AVERAGE HEIGHT: *15 to 16 h.h.*
POPULATION STATUS: *Common*

Lusitano. Photo: Sally Anne Thompson

Lusitania is the Latin name for Portugal, the home of the Lusitano horse. The Lusitano has been called Portugal's Andalusian, and indeed the breed is based on Andalusian blood. Differences noted between the two breeds today can be attributed to selective breeding, which has brought about some differences in type, rather than to infusion of other blood. This is possible within any breed; doubters should consider the Egyptian Arab versus the Polish Arab—both equally pure Arab yet different in type.

The Lusitano breed displays more of the convex profile of the old Andalusian or Iberian type, whereas recent selection in Spain has favored a more Oriental head shape. Some will insist that the Lusitano has been kept more pure than the Andalusian, as Portugal did not use Arab blood in breeding programs.

The history of the Lusitano breed parallels that of the Purebred Spanish Horse or Andalusian and will not be repeated here. This has been the primary riding horse of Portugal for many centuries, and breeding of Lusitano horses is bound to the breeding of fighting bulls. Lusitano horses have been popular with the cavalry, for high school work, with farmers, and for pleasure, but the endeavor for which the breed is most famed is work in the bullring. Mounted bullfighters, called *rejoneadores*, use this breed to fight the formidable bulls and in between charges of the maddened bull will turn to doing intricate high school movements. In Portugal the bull is not killed in the bullring. The horses must be extremely agile to escape the hooking horns with their risk of maiming or death, yet even under these circumstances the Lusitano remains calm. It is considered a great disgrace to have a horse injured in the bullring.

The predominant colors in the Lusitano breed are grey, bay, and chestnut. The head is well proportioned, of medium length, narrow, and dry with a pronounced jaw and slightly convex profile; the eyes are large and lively, almond-shaped or Oriental in appearance, and the ears are small and curved inward at the tip. The neck is quite thick, of medium length, and arched, and the junction of head and neck is narrow and clean. The withers are well defined and long with a smooth transition from the back to the neck; the back is short and strong; the croup rounded and sloped; the tail set low; the chest broad and deep; the shoulder sloped and muscular. The forearm and thigh are long and muscled; the legs are sturdy but fine with broad, dry joints and solid, tough hooves. This is an extremely powerful horse.

The temperament of Lusitano horses is universally kind and willing, noble and generous. This breed remains calm and does not panic easily and is noted for its intelligence.

Malakan
(Ardahan)

ORIGIN: *Turkey*
APTITUDES: *Draft and farm work*
AVERAGE HEIGHT: *13.3 to 14 h.h.*
POPULATION STATUS: *Common*

The Malakan horse is the only heavy breed of Turkey. All other Turkish horses are light and hotblooded, of Oriental descent. This breed is about 120 years old. Ancestors of the Malakan were brought to Kars and Ardahan in the eastern part of Turkey by emigrant Turks from the Caucasus. They were raised in the Caucasus by crossing Ardennes, Percheron, Shire, Clydesdale, Biçuk, Orlov, and Danish heavy breeds with Russina native horses. (The Biçuk horse was raised in the Ukraine region, developed by crossing native horses with Orlov, Danish, and Dutch breeds.) When brought into Turkey these breeds were crossed with the Anadolu breed. Later, the government brought in Ardennes stallions to improve the heavy horse.

The Malakan breed is also occasionally called the Ardahan horse, as it is bred primarily in the area of Ardahan. Colors found in the breed include grey, black, and dun. Malakan horses have an expressive, elegant head; the neck is short, thick, and muscular; the croup is wide; the bone is heavy and strong. This is a very strong, enduring horse with a friendly, willing temperament.

Malopolski
(Little Poland Horse)

ORIGIN: *Poland*
APTITUDES: *Riding, light draft*
AVERAGE HEIGHT: *15.3 to 16.2 h.h.*
POPULATION STATUS: *Common*

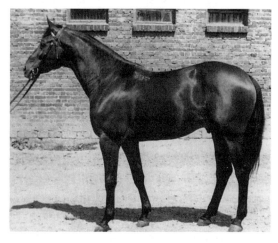

Malopolski. Photo: Maria Pazdan-Rychlak

The Malopolski is a halfbred Anglo-Arab or halfbred horse, bred in the southeastern and central regions of Poland. History of the breed dates to the nineteenth century and includes blood of the Thoroughbred, Arab, French Anglo-Arab, Austro-Hungarian halfbred, Shagya, Gidran, Furioso, and Przedswit (a halfbred Thoroughbred strain).

The Malopolski horse is an animal of general use but with distinct riding qualities. This is a beautiful horse with good conformation and good action at all three gaits. Common colors are bay and chestnut.

State breeding of these horses is performed at five stud farms having about 300 broodmares. These are: Udorz (Katowice County), Stubno (Przemysl County), Prudnik (Opole County), Walewice (Skierniewice County), and Janow Podlaski (Biala Podlaska County). In addition, Malopolski horses are used as riding horses for recreation and sport. They are good over fences, with a bold yet calm nature. They are especially good at steeplechase. Regional breeding is in the hands of private farmers and is situated in the south-

eastern and central parts of Poland. The Malopolski is still being used as an agricultural horse.

Mangalarga Marchador

ORIGIN: *Brazil*
APTITUDES: *Riding horse*
AVERAGE HEIGHT: *14.2 to 15 h.h.*
POPULATION STATUS: *Common*

The Mangalarga Marchador stallion, Mar Diorito. Photo: Lauro Antonio Teixeira de Menezes

The Mangalarga Marchador stallion, Recado De Passa Tempo. Photo: Lauro Antonio Teixeira de Menezes

Mar Diorito wrestling a stubborn cow to the ground. Photo: Lauro Antonio Teixeira de Menezes

The most popular and widespread horse in Brazil is the Mangalarga Marchador, which developed in 1740. João Francisco of Portugal settled in Brazil in Baependi, south of Minas, acquiring the lands which gave origin to the hacienda Campo Alegre, where the Marchador horses began to emerge. Francisco adopted the surname Junqueira to designate his native city. One of his twelve children, Gabriel Francisco Junqueira, the Baron of Alfenas, is credited with expansion and development of the breed. He was the representative of Minas to the Portuguese crown. His friend, Dom Pedro I (1798–1834), Emperor of Brazil and son of João IV of Portugal, gave him an Alter Real stallion named Sublime. (D. Pedro I was crowned emperor after Brazil's independence in 1822 and occupied the throne of Portugal for a short time in 1826 as Pedro IV after the death of his father.)

The stallion Sublime was the descendant of two horses brought from the breeding farm Coudelaria Alter do Chão (located in Alem Tejo in Portugal) by Dom João VI during the invasion of the Iberian Peninsula by Napoleonic troops. In Brazil Sublime was crossed with mares of the hacienda Campo Alegre, which were Spanish Jennets, Criollos, and Andalusians. It is apparent in the Mangalarga Marchador breed that most of the Spanish horses used at the hacienda in launching the breed were the famous Spanish Jennet, known to be a fast, smooth ambler. The cross of Sublime to the hacienda mares produced the first animals with the basic characteristics of the present-day Mangalarga Marchador horse: docility and smooth gaits, with a cadenced, rhythmic gait called the *marcha batida* or *marcha picada*. The horses had endurance and rusticity. The first herd of such offspring were called Sublime horses.

Origin of the name Mangalarga comes from the hacienda Mangalarga, located in Paty do Alferes, which acquired stock from the hacienda Campo Alegre and awakened interest among local ranchers, who began to go to Sul de Minas to buy Mangalarga horses.

Since that time the breed has gained in popularity, spreading throughout Brazil. In the era of travel by horseback, there was nothing better than an animal that had endurance but did not give the least discomfort to the rider. The comfortable ride resulted from the characteristic gait, the *marcha,* a four-beat lateral gait. Unlike the Peruvian Paso, the Mangalarga Marchador exhibits no *termino* in its gait.

In 1949 the first breeders association was officially founded in the Department of Animal Production offices in the Exposition Park of Gameleira (now Bolivar de Andrade) in Belo Horizonte—the Associacao Dos Criadores Do Cavalo Marchador da Raca Mangalarga (Association of Breeders of the Marchador Horse of the Mangalarga Breed). The formation of an association was the long-held aspiration of breeders of that time. Most felt the need to organize, to set standards for conformation and marketing and to distinguish the Mangalarga Marchador definitively from other breeds, principally with regard to its gait. Name changes occurred in 1967 and 1984, and the group is now called the Associacao Brasileira Dos Criadores Do Cavalo Mangalarga Marchador (ABCCMM), or Brazilian Association of Breeders of the Mangalarga Marchador Horse.

In 1989, after breeders had been organized for forty years, the ABCCMM had about 6,000 members responsible for breeding stock numbering over 150,000 animals. At its headquarters located in the center of Belo Horizonte, the diverse sectors are being computerized and linked to other systems. They occupy three floors, where the various needs of members are attended.

The Mangalarga Marchador is a beautiful horse, exhibiting classic Spanish conformation and charm. The gait is remarkably fast and smooth, a gait in which the horse moves its feet alternately laterally and diagonally with moments in which triple support can be verified. If the horse is marching on level ground at a normal rhythm, the tracks of the two hind feet will cover or pass slightly beyond the tracks of the front feet. When the horse places the feet diagonally and with moments of triple support, the gait is called the *marcha batida*. If the horse moves the feet laterally and separately and also has moments of triple support, it is called the *marcha picada*. The reason for so much preoccupa-

tion with the *marcha,* indicated by the name of the breed, is that this gait is unique in the world. The famous Spanish Jennets have died out, and the Mangalarga Marchador is probably the purest surviving remnant of that breed. No other breeds have been crossed into the Mangalarga Marchador. The best known natural movements of the horse in general are the walk, trot, and gallop, always with support on the diagonal without triple support. Due to the triple support exhibited in the Mangalarga Marchador, the *marcha* gives a very comfortable ride with little friction.

The Mangalarga Marchador is a splendid horse for working cattle, as are most horse breeds of Spanish descent. Pictured is the outstanding stallion Mar Diorito, owned by Lauro Antonio Teixeira de Menezes of Sergipe, Brazil. A video tape was sent to me to demonstrate the cattle working ability of this magnificent stallion, as well as the amazing gait of the breed. When the cattle are stubborn about going in the direction desired, this stallion has his own way of dealing with them. Grabbing hold of the back of the neck, he often forces the large cows to their knees. It was reported that Mar Diorito has never hurt a cow, and no one taught him to discipline the cattle in this manner. Needless to say, once Mar Diorito has used his special methods on stubborn cattle, they become quite agreeable about going peacefully in the direction indicated.

Another remarkable feature on the video was tape of an outstanding young bay stallion, Recado De Passa Tempo, marching champion in over seventeen expositions in Brazil. The stallion marched at a very fast pace while his rider held a glass of orange juice balanced on a tray. Amazingly, not a drop of the juice was spilled.

The general appearance of the Mangalarga Marchador is characterized by medium structure, strong and well proportioned, with agility, vigor, and soundness. The breed is light, with fine, smooth skin. The coat is smooth and silky. This horse has an active but very docile temperament. The head is triangular (not concave) with a large, flat forehead tapering to a small, fine muzzle; straight profile; and large, dark eyes which are set wide apart and extremely vivacious. The ears are proportional to the head, mobile, parallel, and well set—erect, with the tips turned inward. The mouth is of medium width and sensitive to the bit (many horses on the above mentioned video were ridden with nothing on the head, and only a rope of sorts around the lower neck, yet were perfectly obedient, even at a gallop). The nostrils are large, dilated, and flexible.

The neck, arched and well muscled, is in proportion to the head, with ideal insertion into the body; the mane is thin, fine, and silky. Withers are well defined, high and prominent, with good insertion to the neck and back. The chest is deep, long, and muscular and not prominent; the back is of medium length, straight and muscular; the loins are short, straight, and well proportioned; if the distance from the back to the loins is of lesser or equal distance to the length of the croup, it is a sign that the horse possesses excellent conformation. The hind quarters are symmetrical, well proportioned, and muscular; the croup is slightly sloping and long; the tall is set at a medium height, with a short, fine dock pointed down but lightly raised when the animal moves. The shoulders are long, broad, sloping, muscular, and well set, with a wide range of movement. The forearms are long, powerful, muscular, and straight; the thighs are muscular and well inserted. The legs are muscular on both the outside and inside, strong, long, and straight, with short cannons. The pasterns are strong and sloping.

Mangalarga Marchador horses are docile and rustic with good endurance. Training of this breed is facilitated by its intelligence, and uses for the breed are unlimited. In work with cattle, in sports, in functional trials now performed as part of the official calendar of the ABCCMM, or in cross-country horsemanship, the Marchador is outstanding and is obtaining excellent results in comparison with other breeds. This breed is extremely docile and is commonly mounted by children.

The Mangalarga Marchador is making a contribution to the current economic picture in Brazil. Confirming its slogan, "The Horse Without Frontiers," the breed is opening markets in the United States, Italy, Spain, Germany, and various Latin American countries, which have shown great interest in its importation. This is with out a doubt one of the most outstanding horse breeds in the world.

The Mangalarga Marchador neither trots nor paces, naturally going from the smooth marching gait into a canter.

Mangalarga Paulista

ORIGIN: *Brazil*
APTITUDES: *Riding horse*
AVERAGE HEIGHT: *15 h.h.*
POPULATION STATUS: *Common*

Mangalarga Paulista. Photo: Agenor Simões Neto

The Mangalarga, or Mangalarga Paulista, is a breed developed from the Mangalarga Marchador of Brazil, and the history of the breed is exactly the same. The word *paulista* means a "native of São Paulo."

When the Junqueira family moved to the state of São Paulo the Mangalarga Marchador was introduced to that state, where later it received infusions of English Thoroughbred, Anglo-Arab, and others, forming the Mangalarga Paulista, an attractive horse but of less smooth gait than the Mangalarga Marchador.

A beautiful and useful horse, the Mangalarga had 2,000 breeders in 1989, and 50,000 registered horses. This is an excellent sport horse. The influence of Arab and Thoroughbred are readily apparent in the breed.

Manipuri

ORIGIN: *India (Assam, Manipur)*
APTITUDES: *Riding horse*
AVERAGE HEIGHT: *11 to 13 h.h.*
POPULATION STATUS: *Common*

The Manipuri is an ancient breed, believed descended from Mongolian stock brought by invading Tartar tribes and later crossed to Arabs. While small in size, the Manipuri has a proportionate body, elegant appearance, and great speed.

This was one of the earliest breeds to be used in the game of polo. Manuscripts dating to the seventh century mention the introduction of polo to the state of Manipur. Polo was widespread in Asia, although virtually unknown to Europeans until the mid-nineteenth century when British colonists watched it for the first time. Manipuri horses have also been used in wartime. The cavalry of Manipur was always considered to be well mounted.

This breed is famous for its elegance and at the same time is known to be sturdy and surefooted. Resembling the Burmese (Shan) pony and likely of the same origins, the Manipuri is still in great demand in India for polo, racing, and military transport.

Usually the Manipuri is bay in color, but chestnut, brown, grey, or piebald also occur. The head is light and well proportioned with a straight profile, very mobile ears, and almond-shaped eyes; the neck is well formed with a full mane; the chest is deep and broad; the withers are prominent; the back is straight and strong; the croup usually slightly sloping; the tail set rather high; the shoulder has good slope; the legs are well built and sturdy with strong joints and very hard feet.

Maremmana

ORIGIN: *Italy (Latium and Tuscany)*
APTITUDES: *Riding, sport horse*
AVERAGE HEIGHT: *15 to 16 h.h.*
POPULATION STATUS: *Rare*

Indigenous to Italy, the Maremmana, or Maremman breed is a saddle horse which in

The Maremmana of Italy. Photo courtesy Associazione Nazionale Allevatori Cavallo de Razza Maremmana

ancient times was bred wild in the marshy areas between Pisa and Caserta.

Although precise details regarding origins of the breed are uncertain, it is generally thought that the early ancestors came from Numidia in north Africa. (In the Italian campaign, Hannibal had cavalry from Numidia.) The breed has many similarities to the present Barbary horse in structure, rusticity, and resistance to prolonged fatigue.

Until the middle of the nineteenth century the Maremmana horse distinguished itself from other breeds by several morphological features: a rather heavy head with ram-like profile; a drooping back; sturdy limbs with short pasterns and extremely hard hooves; and mane and tail of long, very thick hair.

Selective breeding has mainly involved crossbreeding, initially with the Arab and more recently with the Thoroughbred. Continued matings between the resulting halfbreds, although limited, combined with natural environmental selection, have brought about a gradual modification of the morphological features in accordance with market demands. At the same time the characteristic rusticity of this old breed has been upheld. A stud book was established in 1980.

The Maremmana is a saddle horse well

suited for equestrian tourism and for use as an endurance horse, but it excels also in jumping competitions. The Maremmana horse Ursus del Lasco, ridden by Graziano Mancienelli, won the 1977 Italian jumping championship. Some of the best jumping horses are those derived from the cross of a Maremmana mare to a Thoroughbred or Anglo-Arab stallion. This is an example of the classic halfbred, which combines the maternal virtues of rusticity and frugality with the spirit of the sire.

The Maremmana horse is still bred wild in the traditional Maremman area of Tuscany and Latium. However, the number of stud farms is increasing, largely due to the possibility of exploiting marginal terrain as pasture. Small, half-wild breeding operations, based on one or two mares, are catching on well, while in the wild breeding sector, reproduction is carried out almost exclusively through natural selection with a free stallion in a group of fifteen or twenty five mares. In small, family-managed stud farms, crossbreeding is often practiced with the goal of producing jumpers.

In conformation the Maremmana horse generally has a somewhat long and sometimes heavy head with a ram profile; the neck is muscular and of good length, quite broad at the base; the withers are high and well muscled; the back is short and straight; the loins short; the croup sloping; the chest full; and the shoulder well sloped. The legs of this breed are solid and sturdy with good joints and a correct natural stance. The hooves are well formed and extremely hard.

Defects preventing inclusion in the registry are: ears tending to a horizontal position; bluish shading in one eye; excessive extension of special peculiarities and skin depigmentations; disharmonious trunk; accentuated back curvature and length of loin; a faulty plumbline; presence of any serious hereditary defects.

Generally the Maremmana, even though not as fancy as some like a horse to be, is an animal of considerable substance and strength, accentuated by superb health, longevity, and ability.

Marwari

ORIGIN: *Indian (Marwar)*
APTITUDES: *Riding horse*
AVERAGE HEIGHT: *14 to 15.2 h.h.*
POPULATION STATUS: *Rare*

The Marwari stallion, Raj Tilak, recently deceased. Photo: Raja Bhupat Singh

Marwari mares belonging to B. K. Gadhvi in Jodhpur, India

Although it has often been reported in books on horse breeds to be a generally wretched little pony, thin, weedy, and narrow with the front legs "coming out of the same hole," even a casual glance at the accompanying photographs will reveal the truth regarding the famous Marwari horse. I can only assume that some visitors to the Marwar area have seen partbred horses, perhaps Marwari and native pony crosses, and thought this to be the extent of the Marwari breed, returning home with erroneous tales.

When the Rathore rulers lost their kingdom of Kannaju, they withdrew to lands of uninviting weather and physical features to continue their war of independence without drawing attention from their enemies. Their survival demanded that they breed a horse which would be an indispensable part of their existence; a horse with speed and the ability to withstand intense heat and cold; a horse that could gallop for long distances without faltering. Special emphasis was put on speed and hardiness.

Throughout the Middle Ages horse breeding remained the chief occupation in Marwar. During the period of the Mogul emperor Akbar, the entire Raiput population formed an imperial service cavalry of over 50,000 horses.

Bardic literature speaks volumes of the heroic exploits of the Marwari breed, leaping up to the *howdahs* of elephants during battles and across high barrier walls of cities and forts.

It is difficult to trace the origin of the true Marwari horse with precision. It is thought, however, that the breed is closely related to the Kathiawari and has similar origins with some Arab blood.

With careful selection the breeders of the past produced a horse with speed and stamina. It also possessed the remarkable characteristic of refusing to go down even when seriously injured, until it had carried its rider out of danger—and for standing near its wounded rider, biting and kicking at those who attempted to approach. These are also characteristics observed in the Kathiawari breed when it is used as a war horse.

When finally the great Raiput warrior hung up his sword and armor, the Marwari horse lost its place of pride and necessity. The scale of breeding dropped to an alarmingly low level. This lack of interest lasted until the 1930s, when the late Maharaja Umaid Singhji took steps to prevent this famous breed, which had taken centuries to evolve, from dying out. He called his sardars, who were interested in the breed, and sent them to scout around the state for top-class stallions and mares.

In his effort to save the Marwari breed, Maharajah Umaid Singhji purchased several outstanding stallions and had them sent to places where good Marwari mares were located. The stallions Mor, a dark bay of 15 hands; Kanaiya, a liver chestnut of 15.2 hands; and Pratap, Peelo, and Rajhans, all standing 15 hands, were involved in the rescue mission. Obviously the Marwari is not a pony, as many have erroneously reported.

The pure Marwari breed remains rare today and the Indian government is working to preserve the breed. The Marwar Horse and Cattle Show was organized to promote awareness of the breed, and handsome prize money is offered to encourage participation.

The stallion Rajtilak (now deceased) whose photograph is presented, was foaled in 1982 and was considered the best Marwari stallion living. His owner, Raja Bhupat Singh of Jodhpur, and several other breeders have collected some of the very best Marwari horses to help preserve the breed. He also furnished information for this volume.

No longer necessary as the superb war horse of yesterday, the few Marwari remaining are mainly used in wedding ceremonies or to pull *tongas* (taxi carriages). Some are used simply for travel, and these are taught the *rehwal* gait, very akin to the pace. Some have thought this gait to cause unsoundness, but many Marwari horses are born with the gait. Unsoundness seen in the legs of many Indian horses is caused by the fact that Indians today are not the horsemen of the past. Afraid to wait until a horse has matured before riding it, they often put a very young horse to work, causing problems in the legs (personal communication, Raja Bhupat Singh).

Whorls play an important role in selection of Indian horses. If a horse has a whorl below the level of the eyes, it is known as an *an-*

usudhal, and few will buy such a horse. The long whorl down the neck is known as *dev-man*, and this is considered lucky.

Marwari horses have good feet and, except in very stony areas, are seldom shod. The Marwari is an elegant breed with much presence and quality. The head is refined, wide between the eyes and usually with a straight profile, the ears curved inward at the tips, often touching; the neck is clean at the throat, slightly arched, and of medium length. The withers are well pronounced; the back short and strong; the croup gently sloped; the tail set high; the shoulders are well sloped and muscular. The hind quarters are muscular and strong; the limbs are long with long, smooth muscling and strong joints; the pasterns are sloped; the feet are extremely hard. The Marwari is free of hair on the legs and has a fine, silky coat. Colors are bay, brown, chestnut, palomino, piebald, and skewbald.

Many Marwari horses bear a striking resemblance to the Turkmenian type of horse.

Megezh

ORIGIN: *Yakutia (former Soviet Union)*
APTITUDES: *Draft horse*
AVERAGE HEIGHT: *14 to 15.1 h.h.*
POPULATION STATUS: *Rare*

The Megezh is one of several distinct groups found within the Yakut breed and is the largest in size. Bred in the southwest part of Yakutia in the Leninsk region, this breed was established in the latter part of the nineteenth century when gold mining began and a need for heavier horses arose. The Megezh horse, possessing great endurance and power with the typical conformation of the draft horse, was developed from crosses of the Yakut with the Kuznet breed. Planned work in breeding the Megezh horse began in 1950, and the resulting horses were well adapted to local conditions and highly suitable.

Predominant colors of the Megezh are grullo, roan, grey, dun, sorrel, and bay. The breed is considered to have great genetic potential for production of both meat and milking animals. The mares have high fertility and usually foal in May or June.

There are approximately 2,742 Megezh horses.

Megrel
(Megrelskaya)

ORIGIN: *Georgia (former Soviet Union)*
APTITUDES: *Riding, agricultural, pack*
AVERAGE HEIGHT: *12.2 h.h.*
POPULATION STATUS: *Rare*

Smallest of the Caucasian horses, the Megrel is bred in the western region of Georgia called Kolchida. Like all of the horses of the Caucasus, this is a saddle-pack breed, but it is also used in agricultural work. The Megrel is also known as Mingrelian.

The Megrel may be considered one of the most ancient Caucasian breeds, a direct descendant of the ancient Kolchida horse which was famous for centuries. Horses of Megrelia were used in large numbers in Georgia in the twelfth through seventeenth centuries, as horse breeding in the area was well developed.

Breeding of Megrel horses is now mainly on the Kolchida lowlands, which are not uniform in topography. This is primarily a humid, subtropical area known for almost constant seasonal winds; in winter the wind is from the northeast, dry and warm; in spring it is hot and dry; summer winds are humid, out of the southwest.

The Megrel has an almost square body with a massive trunk. This horse has a lean, firm constitution and is muscular. The head is large with a straight profile, ears of medium length, and somewhat small eyes; the neck is short and often thick; the withers are straight and set low; the back is straight; the croup sloped and round; the tail set low. The shoulders are slanting and of medium length. The pasterns are usually short; the hooves are hard and well shaped. The front legs are normal; the hind legs are usually cow-hocked.

The Megrel is quite long-legged. Prevailing colors are bay and nearly black, with reddish color around the muzzle and legs.

Some greys and blacks also occur. There are no white markings.

Although small, the Megrel is strong and able to carry 35 to 40 percent of its own weight—a feat usually only possible for mules and some mountain horses. The breed is enduring and able to work at high altitude; it is surefooted and calm on high mountain passes. The horses are raised in *taboons*.

Attempts were made to improve the size of the Megrel horse by using the blood of saddle breeds, but good results were not obtained, or perhaps it could be said *no* results were obtained. The crossbred animals, raised under primitive conditions by the Georgian Scientific and Research Institute, did not differ from the local horse. Crosses with the Kabarda and raising in better conditions produced a larger horse with better draft, harness, and saddle use. However, these crossings should not be carried out to the degree that the Megrel is in danger of losing its valuable adaptability to the subtropical climate.

Mérens
(Ariègeois)

ORIGIN: *France (Ariègeois Mountains)*
APTITUDES: *Riding, light draft*
AVERAGE HEIGHT: *13 to 14.1 h.h.*
POPULATION STATUS: *Uncommon*

The Mérens pony of France, also known as Ariègeois. Photo courtesy UNIC, Paris, France

A group of Mérens ponies. Photo courtesy UNIC, Paris, France

This pony has apparently existed in the Ariègeois mountains of the Pyrenees since prehistoric times. A breed with marked Oriental features, the Mérens pony is distinguished from other equine breeds in the area due to strong influence from the Arab invasion.

The remarkable homogeneity of this breed may be explained by its isolation during several millennia. Standing between 13 and 14.1 hands, the Mérens is black, and white markings are the exception. The head is fine and distinguished, with a flat forehead, straight jaw, small ears, and wide nostrils. The neck is strong (although sometimes rather short) with a thick mane; the withers are well shaped; the back is long but well coupled; loins are well set; the croup round; the breast open. The limbs are solid with wide knees and hocks, which are sometimes rather light. As a whole, the Mérens is harmonious in conformation and built low to the ground.

Although most of these ponies are still to be found in their native breeding grounds, the Mérens has overflowed the Riviera-Pyrenees region to reach the natural park of the Cevennes, the Alps, and the central region.

Hardy and disease-resistant, the Mérens is still used in its native territory for some agricultural work. It is a very surefooted pony over rough terrain and, because of this, is ideal for trekking in the mountains. Easy to school because of its gentle character, it gives confidence to young or inexperienced riders.

Very solid, this pony can also easily carry adults.

Selective breeding of the Mérens began about 1908. In the past it was used in the southwest of France for hauling minerals and timber. With the advent of mechanization, its use changed to that of a riding and recreational pony.

In general appearance the Mérens strongly resembles the Fell ponies, also of ancient origin, and bears a striking similarity to the famous black Friesian.

Messara

ORIGIN: *Greece (Messara region)*
APTITUDES: *Riding, light draft*
AVERAGE HEIGHT: *12.2 to 14 h.h.*
POPULATION STATUS: *Rare*

The Messara pony of Greece. Photo courtesy A. M. Zafracas, Aristotle University, Thessaloniki

The Messara was developed in the Messara region on the island of Crete by crossing native mares (mountain type) to Arab stallions imported during the Turkish occupation. The pacing gait is usual in this breed, making it easy and comfortable to ride.

Messara horses are used for light farm work and transport. The stallions are often used for breeding hinnies.

The main coat colors found in the Messara are bay, brown, black, and grey. They have some characteristics of their Arab ancestors.

Mezen
(Mezenskaya)

ORIGIN: *Former Soviet Union*
APTITUDES: *Draft work*
AVERAGE HEIGHT: *15 h.h.*
POPULATION STATUS: *Rare*

The Mezen breed developed in the northeastern part of Archangelsk region in the valleys of the Mezen and Pineg Rivers. This is a draft breed of the northern forest type.

The climate is harsh in this area, with winter cold ranging to −60°F with deep, dense snow. Annually, there are 180 to 200 days a year with frost. In summer, myriads of insects swarm, but the horses have developed resistance to them.

Various breeds were brought into the area over the years, including Estonian, Danish, Holstein, Mecklenburg, Finnish, and trotters. The specific conditions and selection for a suitable type of working horse melted these breeds together and the Mezen breed emerged.

Mezen horses have been bred pure for many years and only in the last fifty years has some Ardennes blood been used.

The Mezen horse has a long body, wide chest, and comparatively narrow and sloping croup. The hooves are hard; the hind legs are often cow-hocked. This breed is similar to the neighboring Pechora but is more massive and bony, with a deeper and wider trunk. Both breeds had a similar origin but developed into their own breed types. Colors of Mezen horses are bay, black, sorrel, brown, or grey.

Two distinct types were found in the Mezen breed at one time; a lighter horse with a shorter body and lean, often dished face; and the heavier type, less elegant in appearance.

Mezen horses were used extensively in agriculture, forest work, and transport. The main fault in the breed was that its small size meant less stregnth than in heavier, larger breeds. Today the Mezen horse is nearly extinct.

Mezöhegyes Sport Horse

ORIGIN: *Hungary*
APTITUDES: *Riding, sporting events*
AVERAGE HEIGHT: *16 to 17 h.h.*
POPULATION STATUS: *Common*

From a genetic standpoint, the Mezöhegyes Sport Horse is a synthetic breed composed of Thoroughbred, Holstein, and Hanover horses combined with the most talented Hungarian halfbred individuals.

Breeding studs may make their own decisions and some individual programs have been developed for selection of sport horses. The most advanced of these has been conducted for twenty-five years at the Mezöhegyes state farm, producing a good jumping type. Performance testing is nearly the same as the official test for sport horses, though requirements are higher and the detailed information from individuals is used in a more sophisticated manner in the selection process.

Minusin

ORIGIN: *Krasnodar region (former Soviet Union)*
APTITUDES: *Riding, light harness, pack*
AVERAGE HEIGHT: *14 to 14.3 h.h.*
POPULATION STATUS: *Rare*

The Minusin breed is raised in the Minusin valley in the southern part of Krasnodar region. This horse was developed by the native people, who have had an intense interest in horse breeding for centuries. The Minusin is similar to the steppe breeds but is larger and more imposing in appearance.

The development of such a horse was facilitated by climatic and topographical conditions in the region. The Minusin valley is a level area 1,300 to 1,600 feet in elevation, surrounded by mountains. The climate is dry, the summers are hot, the winters almost without snow. The soil is black, typical of the steppes, and pasture is good for every kind of farm animal.

In appearance the Minusin horse has a large head, often Roman-nosed with a thick neck, massive trunk, and wide chest. The legs are firm and bony. Predominant colors are bay, grey, dun, grulla, palomino, and buckskin.

The best Minusin horses are concentrated at the Chakass stud farm. These horses are healthy and fertile and respond well to improved conditions of feeding and keeping. Unable to meet the demands of agriculture, they have been improved by strong selection and crossing to Don, Thoroughbred, and trotter stallions.

Misaki

ORIGIN: *Japan*
APTITUDES: *Riding, light draft*
AVERAGE HEIGHT: *13.2 h.h.*
POPULATION STATUS: *Rare*

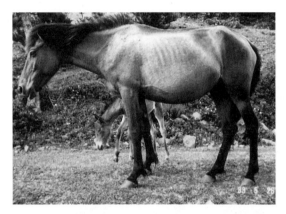

The Misaki horse of Japan. Photo: Yujiro Kaseda

The Misaki pony inhabits the meadow of Cape Toi, or Toimisaki (*misaki* means "cape"), on the south side of Miyazaki Prefecture. Cape Toi is one of the best known and most important spots for tourism in Kyushu, and the Misaki ponies run wild in the area.

All of the native horses of Japan are believed descended from horses brought from China about 2,000 years ago and are separated now into eight distinct breeds.

The ranch at Cape Toi was established by the Akizuki family of the Takanabe feudal clan in the Edo period in 1697 for the purpose of raising horses for use in war and agriculture. Farmers used the ponies to cultivate narrow gardens and rice fields on the steep slopes. After the Meiji Revolution the ranch was nationalized. In 1874 the government sold the ranch to the local Misaki Union as a joint (common) ranch.

The Misaki pony has rarely been controlled by man except for a time when almost all male horses were removed and only a few were left as stallions. In the early Taisho era a stallion was introduced which had western trotter blood, but its breed was not identified. In any event, the blood of this horse did not appear to have much influence on the Misaki ponies, which have retained their primitive, Mongolian-like characteristics.

After World War II the number of this breed was reduced drastically, and there are only approximately 100 head remaining. Today Misaki horses and their habitat are designated a natural monument of Japan because the ponies have retained the characteristics of the old Japanese breeds.

Misaki ponies are mostly bay or black, with a few chestnuts. White markings on the face or legs are very rare in Japanese horses.

Missouri Fox Trotter

ORIGIN: *United States*
APTITUDES: *Riding horse*
AVERAGE HEIGHT: *14 to 16 h.h.*
POPULATION STATUS: *Common*

Missouri Fox Trotter. Courtesy Missouri Fox Trotting Horse Breed Association

Wishing to preserve the type of horse that had long been bred selectively in the Ozarks, a group of horse breeders incorporated in 1948 and secured an organization charter for the purpose of maintaining an accurate stud book. Until 1982 the association kept its register open and accepted horses that could qualify as characteristic foundation stock. Since January 1, 1982, all additional horses allowed into the registry have been required to have both a sire and dam registered.

The Missouri Fox Trotter was developed in the rugged Ozark Mountains during the nineteenth century by settlers who needed smooth-riding, durable mounts that could travel at a comfortable, surefooted gait for long distances.

Missouri achieved statehood in 1821 and the pioneers who streamed across the Mississippi River to settle in the Ozarks came mostly from Tennessee, Kentucky, and Virginia. They brought saddle horses popular in those areas. It soon became apparent that horses able to perform the easy, broken gait called the "fox trot" were the most useful in the rocky, forest-covered hills of the Ozarks, and selective breeding for the fox trot gait began.

Easy-gaited stock imported to America's shores during the colonial era left its genetic imprint on the fox-trotting horses in the Ozarks, the American Saddle Horses of Kentucky, and the walking horses of Tennessee. Some nineteenth-century greats, such as the Canadian stallion Tom Hal, made sizeable contributions to the easy-gaited horses of all three regions.

The distinguishing characteristic of the Missouri Fox Trotter is the fox trot gait. The horse walks with the front feet while trotting with the hind. This extremely surefooted gait gives the rider little jar as the hind feet slide into place. The fox trot is a rhythm gait, and the horse can maintain it for extended periods with very little fatigue. The Missouri Fox Trotter also performs a rapid flat-foot walk and a delightful canter.

Missouri ranks second in the nation in cow/calf operations, and Missouri Fox Trotters are historically linked to the grazing cattle industry of the Ozarks. When the automobile nearly caused the horse to become obsolete in the everyday lives of most Ozarkians, Missouri Fox Trotters survived largely because the cattlemen of the region continued to use and breed them. Old Fox, one of the breed's most influential sires, was a chestnut stallion who spent his adult life trailing cattle in southern Missouri and norther Arkansas early in this century.

Stamina, soundness, and gentle disposition were serious considerations in the breeding of fox-trotting horses by pioneer families. Missouri Fox Trotters make excellent mounts for children and beginning riders because of their quiet dispositions and willingness to please. Their smooth gaits eliminate the bouncing that inexperienced riders suffer when riding hard-trotting breeds.

Today there are more than 30,000 registered Missouri Fox Trotters in the United States and Canada. Trail riders across the nation who participate in treks through mountain regions are rapidly discovering what U.S. forest rangers have known for years: Missouri Fox Trotters have no equal when it comes to delivering an easy, surefooted, willing ride on hazardous terrain.

The Missouri Fox Trotter is not a high-stepping horse, but an extremely surefooted one. The head and tail are slightly elevated, giving the animal a graceful carriage. The rhythmic beat of the hooves, accompanied by the nodding action of the head, give the animal an appearance of relaxation and poise.

The Missouri Fox Trotter may be chestnut, black, bay, grey, piebald, or skewbald. The head is well proportioned with a straight profile; the ears are pointed and the eyes large and expressive; the neck is of medium length and well formed. The withers are pronounced; the back is short and straight; the croup muscular and rounded; the tail set rather high; the chest broad and deep; the shoulder sloped and muscular. The legs are sturdy, well muscled, and have good joints and clearly defined tendons; the hooves are well formed and in good proportion to the size of the horse.

This breed is shown at the fox trot, flat-foot walk, and canter.

Miyako

ORIGIN: *Japan (Miyako Island)*
APTITUDES: *Riding, light draft*
AVERAGE HEIGHT: *14 h.h.*
POPULATION STATUS: *Rare*

Miyako Island in Okinawa Prefecture has been known as a horse breeding area for

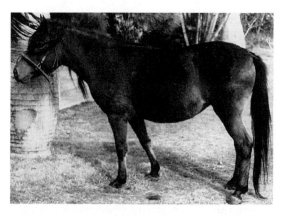

The Miyako pony of Japan. Photo courtesy of the Preservation Club of Miyako Horse

centuries, and small horses have always been found in this area. During and after World War II they were crossed with larger stallions to increase their size to around 14 hands for farming purposes.

About 1955, population of the breed peaked at around 10,000 head, but with the increase of motorization they began to decline. Since 1975 great efforts have been made to preserve the remaining few Miyako ponies, as the breed is of great antiquity, but only eleven head were living as of December 1988.

Miyako poinies are mostly bay or dun in color and resemble the Mongolian horse.

Mongolian

ORIGIN: *Mongolia*
APTITUDES: *Riding, pack, draft*
AVERAGE HEIGHT: *12 to 14 h.h.*
POPULATION STATUS: *Common*

The Mongolian is an ancient breed (possibly *the* oldest true breed) and influenced many other breeds of Asia and Europe in early times. Spread by the Mongol invasions to many areas, the Mongolian horse includes several types, some found in Mongolia and some in China (see Chinese Mongolian).

It was reported in 1985 that there were approximately two million Mongolian horses in the Mongolia People's Republic. There are four basic types found in Mongolia: forest, steppe, mountain, and Gobi.

Mongolian horses are raised in the open the

entire year without supplemental feed and are well adapted to cold climate and harsh living conditions. In general appearance the Mongolian has a heavy head, short neck, and wide body; short legs, well-developed tendons and joints, and sound, hard hooves; sloping, muscular croup; and heavy mane, tail, and hair on the legs. The winter coat is especially dense. The most widespread colors are bay, brown, sorrel, grey, dun, buckskin, and black.

The forest type of Mongolian is the largest and heaviest, standing about 13.2 hands. The steppe horse is smaller but built more for riding. This horse has a dry constitution and plenty of speed. The mountain horses average about 13 hands and are similar to the Soviet Altai breed. They are often piebald and skewbald. The desert horse, the Gobi type, is the smallest. Gobi horses are usually light in color—palomino or buckskin. They are never used for milk, while all others are milked regularly.

All of the Mongolian horses are used for draft, meat, national festivals, and entertainment. During World War II many were sent to the Soviet Union, greatly reducing the number of horses in Mongolia.

A riding school in Ulan Bator has horses which are crossed with Don, Budyonny, and Thoroughbred. They average 14 to 14.3 hands. These crossbreds resemble the foreign breeds.

The main method of breeding is pure selection of the Mongolian horse and is directed toward increased weight, milk yield, fertility, and improved appearance while retaining adaptability to *taboon* breeding and endurance.

It is planned to establish a breeding center for each distinct type of Mongolian horse.

Morab

ORIGIN: *United States*
APTITUDES: *Riding horse*
AVERAGE HEIGHT: *14.1 to 15.2 h.h.*
POPULATION STATUS: *Common*

The Morab is an American breed developed in the early 1800s. The breed is entwined in the early pedigrees of the Arab, Morgan, and

Morab. Photo: Nathan Traux

Quarter Horse. Provisions for registration of the Morgan-Arab were made in both the Arab and Morgan registries until their stud books closed to outside blood in the 1930s. In 1854 L. L. Dorsey, interested in breeding a national carriage horse for the United States, bred a daughter of the imported Arab stallion Zilcaddie to Vermont Morgan 69 in the Morgan registry. This cross produced one of the most beautiful stallions of that time, whose golden color resulted in his name, Golddust. Golddust matured to a height of 16 hands and for trotting speed was the peer of anything ever bred in Kentucky.

In 1861 Golddust defeated Iron Duke in a match race for $10,000. He was an animal of great beauty and refinement and was noted for endowing his offspring with extreme speed. Although his career as a stud was curtailed by the Civil War and his untimely death, he sired 302 foals and left 44 trotters of record.

The 1920s marked the birth of the name Morab, when William Randolph Hearst became involved in a breeding program for his herd. Crossing his Arab stallions Ksar and Ghazi to Morgan mares, he found the produce to be an excellent type for work on his San Simeon ranch in California.

Another interesting breeding program was run by the Swenson Brothers near Stamford, Texas. Purchase of the Morgan stud colts Red Bird 6775 and Gotch 5979, along with a band of broodmares, marked the beginning of their herd. Arab blood was added to the stock, and fine cutting horses evolved from the cross.

Recent years have brought new interest in the Morab. No longer used to build or improve other breeds, the Morab is appreciated for itself. Selective breeding and good training have produced top animals for nearly any activity. Whether it be English or Western riding, parade, driving, or trail, Morabs have proven to be an excellent mount.

Morab horses are tractable and easy to train. The breed is late to mature, not reaching full potential until the age of seven. Like other slow-maturing breeds, they are long-lived.

This is a powerful, muscular horse which displays grace and distinct refinement. It may be any color without spots on the body, with black skin and dark eyes. White markings on the face or lower legs are permissible. The disposition is calm, affectionate, intelligent, and dependable; the head is refined with a straight or slightly concave profile and large, expressive eyes; the cheek is broad, the muzzle narrow, and the nostrils large and flared. The neck is heavy but with refinement and good length; the shoulders are well muscled, long, and sloping; the withers are moderately high; the back short and strong; the croup level and muscular; the tail set high. The chest is broad and deep; the legs are sound, with flat bones, large joints, a broad forearm, and short cannon bone; the hooves are round and broad with tough horn.

In action the Morab has a free-flowing gait, working off its strong hind quarters, and carries itself well collected. Morabs may possess a naturally high action or the lower, quieter pleasure horse action, depending on the breeding of the animal.

Morgan

ORIGIN: *United States*
APTITUDES: *Riding, driving, light draft*
AVERAGE HEIGHT: *14.1 to 15.2 h.h.*
POPULATION STATUS: *Common*

One thing we know for certain about the little bay stallion named Figure, foaled in 1789 in Springfield, Massachusetts, and later known as Justin Morgan, is that he was an enigma.

An enigma, as no one can say for certain

Morgan. Photo: American Morgan Horse Association

The Morgan Stallion Castleberry's MCM Saratoga, owned by Geovani and Trish Capetta. Photo by the author

the name of his sire or dam. An enigma, because he turned out to be a superhorse when due to his small size it was generally thought he would be more or less useless, and because in his case, in one of very few instances in recorded history, an entire breed, stamped in his image and still enduring after two hundred years, was founded on the blood of only one stallion.

Figure has been called a "sport," a "freak of nature," a "mutant." But consider those labels: A sport is an individual that varies markedly from the normal type. Was Figure abnormal? A freak of nature is described in Webster's dictionary as "any abnormal animal, person or plant; monstrosity." Was this little bay horse a monstrosity? Mutate means "to change," a mutant "an animal or plant with inheritable characteristics that differ from those of the parents." Figure did not have characteristics he passed on that were not inherited from his sire or dam. Did he have five legs? Two heads?

Justin Morgan, the horse, was not a sport. He did not vary markedly from the normal type. He certainly was not a freak or a monstrosity. He was a beautiful, small bay stallion, perfectly normal in every way. It is doubtful that he was a mutation in any sense of the word. He wasn't *abnormal* (unless one considers that he bore no resemblance to the Thoroughbred or Arab, which were claimed *after the death of his owner* to be the breeds of his sire and dam). Had he been a mutant, he would have possessed genes he could pass on to his get that did not exist anywhere in his pedigree and pass them on prepotently, a theory I consider virtually a genetic impossibility. While there is no dispute over the fact that mutant genes may suddenly appear in an individual of any species, it is also true that sudden prepotency for those genes is remote to impossible.

The truth about Figure, or Justin Morgan, may never be known. It was lost by the time the man Justin Morgan obtained the horse. The little bay was considered too small, and it is a miracle he was left entire; perhaps it was thought he would grow larger if left a stallion for a longer period. I personally doubt whether anyone even considered his pedigree until after he was taken to Vermont and revealed the miracle he actually was. By that time, the horse was between three and five years old and people had become accustomed to shrugging him off as unimportant. Once he became famous, it was natural for people to want to give him a pedigree worthy of his achievements, and the English Thoroughbred was the rage of the day.

Figure did not carry the name of his owner, Justin Morgan, until after this man's death. Justin Morgan, blacksmith, music teacher, one-time tavern owner, and jack-of-all-trades, moved to Vermont from Massachusetts but returned to his old home in West Springfield in 1792 to collect a debt. He took as payment a big two-year-old gelding and an undersized three-year-old bay stallion. The gelding, large, well built, and strong in appearance, was sold quickly, but due to the small size of the bay stallion, Morgan could not find a buyer, even at a cheap price. Finally, he rented the horse to a neighboring farmer for a few silver dollars a year.

To the complete amazement of everyone, the little bay outperformed every horse in the state with such ease and gaiety of manner that he seemed to make light of every effort. He was soon the pride of Vermont and may well still be today. After spending a hard day working in the fields or dragging logs out of the woods, he would return to the barn fresh and bright-eyed, as though he had spent the day lolling about in the pasture beneath a shade tree.

A famous story about this horse tells that one evening after a hard day of work, he pulled a huge log with three men sitting on it that a team of heavy horses had not been able to budge. At the time, he was in charge of a man by the name of Robert Evans. For months Evans had been using the little horse in the woods to pull heavy logs, and from experience he knew the surprising power of the little bay.

One evening at dusk after work, Evans stopped at the local tavern to wet his dry throat before supper. According to eyewitnesses, he was told about a pine log lying some ten rods from the mill. Other teamsters had not been able to move it, even though some of the horses that had been hitched to it in teams had weighed as much as 1,200 pounds each.

After looking the situation over, Evans, by now perhaps feeling his drink, challenged the mill company authorities to bet him a gallon of rum that his horse could not pull the log onto the landing in three starts. After a quick acceptance of the bet and another round of drinks, they all went merrily out to the log.

Perhaps Evans had become downright lightheaded, for by the time all had gathered around the site of the immovable log, he said he was ashamed to hitch his horse to such a *small* log, and perhaps if three of the largest men would sit on it, it would make the contest more interesting and more of a challenge to his horse. He said further that if Figure could not pull the entire load, he would forfeit the rum. Three big men straddled the log, laughing and merry-making, and Figure, who had already worked hard throughout the day, was hitched to it.

When given the command, the bay released tension on the tugs just slightly; then, squatting behind and bowing his beautiful neck, he pulled slightly to one side. For a moment the chains squeaked in protest and the log shuddered. Then, while all held their breath, the big log began to move—slowly at first—then faster as Figure's legs began to work like pistons. Halfway to the mill Evans stopped the bay to let him take a short breather, then, bending to the job again, the bay moved the log to the spot agreed upon. He did it in two pulls, not three, and astonished everyone.

Needless to say, from that day forward the pint-sized bay no one had thought worth his salt was very much in demand as a breeding stallion. The men of Vermont needed strong horses with good muscle and ability to do the heavy work required at that time. The little bay horse from Massachusetts went on to outdo every horse *ever* put against him, be it walking, trotting, running, or pulling.

Justin Morgan weighed about 1,000 pounds and stood not much over 14 hands. He was a solid dark bay with a thick, wavy mane and tail. His back was short, his body rounded and very compact with powerful quarters. He was completely free of the coarseness that would have indicated draft horse breeding.

His action was not exceedingly high, but his gaits were square and even. His legs were short and he was undoubtedly without any outward appearance of the Thoroughbred. The musculature exceeded that usually seen on a horse of his size.

There has been much speculation and controversy regarding the lineage of Justin Morgan. Generally he is thought to have been sired by the Thoroughbred True Briton, out of a mare that was predominantly Arab. Others have thought him a Dutch horse. Mares in the New England states at the time of the conception of Justin Morgan were of varied breeding; the Norfolk Trotter from England was there,

the Dutch Friesian, Norwegian Dun (Fjord), Welsh Cob, and some others often not mentioned.

My conclusion in light of the many stories I have read about Justin Morgan and the horses in the area at the time of his birth is that he could not have been half Thoroughbred and nearly half Arab. Neither of those breeds has the short legs and heavy muscling of Justin Morgan. I have never seen a Thoroughbred or an Arab with the thick, wavy mane and tail for which the Morgan horse is known. He did not produce foals that looked like either Thoroughbred or Arab. However, in more ways than one, he did very much resemble the solidly built Canadian horses, the Friesians, and in some respects showed possible influence of the Norwegian Fjord.

Further, it seems an injustice to both the Thoroughbred and Arab breeds to claim that they spawned a *prepotent* offspring who could transmit *none* of their characteristics, even though both those breeds are known to have great influence when used in a pedigree. As food for thought, consider that if you could breed one million Thoroughbreds to one million Arabs—with 100 percent production of foals—how many Justin Morgans would result? How many Anglo-Arabs? That Justin Morgan did not produce Thoroughbred-type and Arab-type offspring is, to me, the greatest proof that he did not carry such blood—not that he was a freak.

It should be noted that at the time the English Thoroughbred was relatively new to the United States and extremely popular. This new breed had created a rage of excitement. A horse that could run a mile! Human nature has a tendency to give a great horse a great pedigree when there is doubt, because "unknown" carries little prestige. It should also be known that the man who owned him, Justin Morgan, always referred to the bay as "the Dutch horse," as was attested by his son, Justin Morgan, Jr.

Few of my personal opinions will be found in this volume, as I have depended on breed authorities and organizations around the world for information and have shown no preference. However, in the case of Justin Morgan, whose ancestry is open for debate, I will take the liberty of proposing what to me is an obvious and logical solution to the "mystery horse" at the top of every Morgan pedigree.

First, consider that the supposed sire of Justin Morgan, True Briton, was a stolen horse. This automatically casts a doubt on the true identity of the horse thought later to be True Briton. The owner of this stallion never recovered the horse, so its whereabouts were obviously unknown.

Secondly, Justin Morgan bore resemblance to some of the colonial horses of the period, such as the Friesian, Norfolk Trotter, and Norwegian Fjord. I have also always had an impression of Spanish influence whenever viewing a Morgan horse. In studying these breeds, however, I have not been able to reconcile to my satisfaction the possible combination of bloods that could have produced Justin Morgan.

To avoid repetition of text, I ask you to simply consider the Canadian breed in this volume, overlooked by most breed historians. History substantiates the fact that citizens in the New England states bought *thousands* of French-Canadian horses. The birthplace of the first Morgan horse was saturated with the blood of the small, blocky, powerful Canadian horse—an excellent trotter; able to pull far more than its size and weight would indicate; sure of step; curious; bold; given to lively use of eyes and ears; predominantly bay or black; having feet of iron, outstanding strength of legs, and powerful quarters and shoulders; and noted for its crested neck and wavy mane and tail—*all qualities possessed by the Morgan horse.*

It is also known that the Canadian was often referred to as "Dutch" when in the United States. The French-Canadian, imported to Canada from the royal stables of France, was known to have the blood of the Andalusian and horses from the Spanish Netherlands, obviously the famous Dutch Friesian, a breed long noted for a profuse mane and tail and absolute perfection at the trot. Perhaps Friesian blood is the reason the Canadian was often referred to as "the Dutch horse."

Of special interest is a remark found in the *Canadian Historical Review* (vol. 27, 1947, no. 2), that, according to one Vermonter of long ago, "there was [in the eighteen-forties and fifties] a terrible fear the Morgan horse would be *found* to have French blood in him" (italics mine).

Champions of the early Morgan, the foremost being D. C. Linsley, were reluctant to admit to any French-Canadian blood in their favorites, preferring to credit their merits to the popular and exciting Thoroughbreds. Canadian blood was common. Many New Englanders owned a Canadian horse or one with Canadian blood. Given the excitement generated by the Thoroughbred, it is easy to imagine powerful people wanting to make something extra-special of this obviously special new strain of Morgans. Could claims about Justin Morgan's ancestry have arisen simply from personal aggrandizement, or some vendetta we know nothing about? Could a powerful man, or several, have been so convincing in their assertion of Thoroughbred and Arab blood as to blind horse breed researchers for over 200 years?

One other thing to consider about Justin Morgan, since I have taken the liberty of expressing my opinion, is the fact he was prepotent. This is a characteristic obtained through inbreeding and linebreeding, whereby an animal is mated to close relatives. Animal breeders have used this method for centuries to fix or strengthen good characteristics of an outstanding animal. Inbreeding and linebreeding enhance the genetic qualities of a pedigree. This works in a beneficial way, causing an exceptional sire to stamp his get as though they were poured from a mold—and the method can work just as easily toward deterioration, stamping the characteristics of a poor sire. That Justin Morgan was prepotent does not indicate he was a freak, but it strongly suggests that he could have been a linebred horse, whether intentionally or (more likely) by accident. In any event it was an exceptional "nick," and whatever his lineage, it could not have been more perfect.

Bred to Vermont mares of mostly unproven ancestry and different in type, Justin Morgan sired many fine foals that could indeed have been poured out of the same mold, so strongly did they resemble him. And some were of such excellence that they went on to found families of their own—the Woodburys, Shermans, and Bulrushes. But—are there skeletons in the closet? Bulrush, one of the most outstanding sons of Justin Morgan, was out of a Canadian mare. Bred to a black Canadian mare, Bulrush got Randolph Morgan. Randolph Morgan, mated to a Canadian mare, produced Jennison Horse (sire of Old Morrill). Woodbury, another famous son of Justin Morgan, was likely from a Canadian mare, and the dam of Black Hawk was a Canadian. Even if Justin Morgan did not himself carry a preponderance of Canadian blood, it remains that the breed he spawned most certainly did.

The majority of the first Morgans were born in the states adjacent to Quebec Province. This was also where the greatest majority of Canadian horses were to be found. Vermont did not enter the Union until 1791, two years after the stallion Justin Morgan's birth, and due to its geographical affinity with the eastern townships of southern Quebec, very strong ties existed among the inhabitants; the parents of Justin Morgan (the man) lived in Quebec. That early breeders of Morgan horses claimed to have no idea as to the lineage of their horses appears suspiciously like the ultimate horse trader's hoax.

Today there is finally an opportunity to find out more about the lineage of the Morgan and other breeds. The Universities of California and Kentucky, among others, are conducting blood testing that establishes much about the relationship of one breed to another. It will be surprising indeed if blood drawn from the Canadian does not prove this horse to be the closest relative to the Morgan breed.

Ensuing generations of Morgans gained materially in size, mostly because the demand was for greater size at the time and larger horses were selected for breeding. Although size increased, the Morgan did not change its appearance and still has not to this day. Early Morgans were used in general on farms in northern New England; they were utility, all-purpose animals.

After 1820 the urge for greater speed tightened its grip on the United States and there was great demand for small roadsters that could hold their own in a chance brush on the road with the fleet Grey Messenger.

Justin Morgan's great-grandson, Black Hawk (as indicated, out of a Canadian mare), and a great-great grandson, Ethan Allen, became bywords. In 1867 Ethan Allen was acknowledged as the fastest trotting stallion in the world when he raced and beat Dexter.

By 1870 there were Morgans in every state and section of the country. Blood's Black Hawk was taken to Kentucky, where his descendants did much to further the breeding of the American Saddlebred and, later, the Tennessee Walking Horse. Keokuk and Black Eagle went to California even before the forty-niners, and Oregon stock breeders had use of Vermont and Pathfinder.

It is amazing, to say the least, that a breed such as the Morgan originated from one bay stallion, Justin Morgan. This great horse died in 1821 from a kick he received from another horse while in a corral. Lack of proper care resulted in infection, causing his death at the age of thirty-two years. I suggest, however, that rather than having descended from the single animal, Justin Morgan, the Morgan breed should be viewed as having developed as an American-bred Canadian horse with a bit of mixture to other New England breeds, and with Justin Morgan as the foundation sire.

The Morgan is a bold, alert horse with deep curiosity, and is very intelligent. Morgans have prominent withers; a sloping shoulder; long, well-sprung ribs which angle toward the rear; and the distance from the withers to floor of chest is greater than from the floor of the chest to the ground. This great depth gives the Morgan horse exceptional staying power. Morgans have beautiful heads and finely shaped, arched necks.

Morgans excel as family horses, mounts for children, roadsters or hunters, show horses, and working horses. This is one of the very few breeds in the world that is shown both as a riding horse and in pulling contests.

The Morgan is not a work horse—yet it has worked larger horses literally into the ground. It is not a race horse—yet it can outrun many of the fastest horses. It is not just a pleasure horse—yet it is one of the most pleasurable horses to ride. Whatever a good horse anywhere can do, a Morgan will gaily try, and with little effort will often get better results.

One of the oldest American breeds, it is a truly great all-around horse.

Moroccan Barb

(North African Barb)

ORIGIN: *North Africa*
APTITUDES: *Riding horse*
AVERAGE HEIGHT: *14–15 h.h.*
POPULATION STATUS: *Common*

As in several other ancient equine breeds, controversy surrounds the origin of the Barb. Some believe the breed to be an offshoot of the Arab—yet the Barb does not resemble the Arab in important ways in which the Arab usually impresses itself on the morphology of

Barb-Arab cross. Photo: Ministry of Agriculture, Morocco

Moroccan Barb. Photo: Sally Anne Thompson

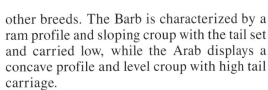

An aged Barb stallion. Photo: Ministry of Agriculture, Morocco

other breeds. The Barb is characterized by a ram profile and sloping croup with the tail set and carried low, while the Arab displays a concave profile and level croup with high tail carriage.

Some claim the Barb to be the ancestor of the Arab—but then one must wonder even more about the origin of the Barb itself, for Africa had no indigenous horse breeds. Breeds known to have Barb influence are always found to have the Barb "stamp": the distinctive shape of head, sloping croup, low-carried tail, long shoulder, and powerful front end. Probably the Barb and Arab breeds, while related, stem from two very different types of primitive horse, and herein lies the mystery and intrigue. Suppose the *same* type of horse was taken to North Africa and the Arabian Peninsula, and one, going west, became intermixed with the ancient migrants from the Iberian Peninsula, producing the Barb, while the other was selectively bred, producing the Arab—without the Iberian influence that gives ram profile. Both results had inherent desert characteristics, but they are different in appearance.

Referring to the inclusion on the Iberian horse (Andalusian) in this volume, the reader will note that many authorities today think that the ancient Iberian (with ram profile) migrated across the land-bridge of ancient times from the Iberian Peninsula into North Africa. Centuries later, domesticated horses were taken to North Africa from the east by various nomadic invaders and traders. The Barb horse emerged throughout North Africa (Morocco, Numidia, [Algeria], Tunisia, and Libya), and there are differences within the breed in various areas. There is little doubt that horses were brought into the region from Central Asia and possibly across the Mediterranean by both traders and armies.

It is also known that what is now Saudi Arabia was horseless for many years after the horse was first domesticated, so the old notion that the Arab originated as indigenous to the Arabian Peninsula should be pushed aside in light of more modern discoveries (see Arab).

The first domestic horses hailed from Central Asia—Turkmenistan and the Ukraine. These areas contained certain types or strains that are now extinct, such as the Tarpan, which had a straight and often concave profile. Also present was the ancient Caspian (see Caspian) horse, very "Arab" in appearance and known to be a much older breed than the Arab. The ancient Turkmenian horse was there, a tall, slender animal believed to be the direct ancestor of the present Akhal-Teke and Turkoman breeds. The Mongolian was also very much in evidence among early domesticated horses. Przewalski's Horse (*Equus przewalski poliakov*) was also present, a distinctly different species and not an ancestor to the modern horse.

Which of the above mentioned horses or their mixtures were taken into North Africa from the east is speculative and may never be known, but undoubtedly many of the horses were mixtures of those types. The Hyksos conquered part of Egypt in the seventeenth century B.C. using Mongolian and Central Asian horses. The Hittites conquered Syro-Palestine in 1370 B.C. The Nisaean breed characteristically had a convex profile and remains another possible origin of the Barb-type head. Many of the Hittite horses were slender, spirited, and similar to the Caspian, averaging 9.2 to 11.2 hands. They were used to pull chariots.

In the early days of domestication of the horse it is doubtful that selective breeding was practiced or even considered. It seems to me reasonable, however, that horses of the hot-blooded type such as the ancient Turkmenian and Caspian were likely the ancestors of both the Barb and Arab breeds. The Turkmenian horse was taller than most horses of that day, a swift and enduring desert horse with a delicate, tapered head. Many Turkmenian horses today possess the bulging forehead credited to the Arab breed, but with an otherwise straight profile. The Caspian also displays the bulging forehead and a concave profile and, in addition, the high tail carriage.

It is probable that the Barb horse emerged from crosses between Turkmenian and/or Caspian horses and descendants of Iberian horses along the coast of North Africa; it has desert horse attributes and retains the Iberian (or Nisaean?) profile. In the meantime, mixtures of Tarpan types and the Caspian and Turkmenian horse easily could have developed through selective breeding into the Arab as we know it today. The first Arab horses were not found in what we now call Arabia, but in Syria, Iran, Iraq, and Turkey centuries prior to formation of Saudi Arabia as a country. Those horses were obtained from Central Asia and not from the Arabian Peninsula. Later, when Arabs invaded Iran, droves of horses were taken to the Arabian Peninsula for breeding.

Formation of the Barb breed is so ancient that our view of it is dimmed to the point of oblivion. When people first began to make use of the horse other than for food, there were horses in the north of Africa, yet science has proven them to be migrants. They could only have come from two directions: northwest, from the Iberian Peninsula; or from the east.

It is safe to say that all of our present-day horses have descended from horses that once roamed Siberia (see Arab). The horse is a natural wanderer, ranging several miles each day in search of good feed. Increasing populations caused the spread of Equidae throughout the northern regions, then west to the Atlantic, and in the process, environment and natural selection led to emergence of different types; the exact number of such types is a matter of controversy. By the time of widespread domestication of the horse, most of the wild horses had already passed into oblivion, leaving only their bones for us to speculate upon. It was simply not a case of an Asia and Europe filled with wild horses waiting to be captured by the suddenly enlightened human population who had heard from their neighbors that the horse could be ridden. Small pockets of wild horses remained in various areas, and in fact Herodotus spoke in his writings of herds of wild horses. But had people not seen use for the horse as a means of transportation and work, the equine species might have died out everywhere, hunted to near extinction like the Asiatic Wild Horse or Przewalski's Horse. Refusing to be tamed, the Przewalski had only one use to humans, as food. The horse, born with a passive nature and willingness to learn, because our greatest tool and weapon and vehicle, and thus was assured of survival.

Historians refer to the Andalusian as a horse with "Oriental" blood, stating that this blood came by way of the North African Barb with the invasion of the Iberian Peninsula by the Moors in the eighth century. But to trace Oriental blood one must look to the Orient, for herein lies the true origin of our light horse breeds, including the beautiful Arab horse.

The Barb horse, swift and possessing incredible endurance, is a true desert horse, but a mountain type rather than a flat-land horse (one should not always think of the desert as being an endless horizon of sand). Most of the habitat for this horse is at high altitude, mountainous desert. The Barb is surefooted and able to gallop headlong up and down steep ravines that would give most other breeds pause. This horse is spirited and fiery and prized by its nomadic owners for its speed, agility, loyalty, and endurance. As to the profile of the Barb horse—beauty is in the eye of the beholder; I see a regal air to the profile of the Barb rather than a tendency toward ugliness.

The Barb is said to be the hardiest of all the Oriental breeds. This of course is debatable when one considers that Oriental breeds include the horses of China and Mongolia, which possess amazing qualities of hardiness. It may be true, however, that the Barb is the hardiest of the hotblood Oriental types; but here one meets opposition from pro-Arab breeders. The Arab has proven worldwide to be a champion of endurance tests. Then one encounters the Akhal-Teke, a true Oriental of ancient history known to have carried riders across nearly 300 miles of desert without water. It is safer to acknowledge the Barb to be a horse of tremendous endurance and substance, able to withstand changes in climate and the hardship of fatigue, resistant to disease—and not claim superiority over others equally enduring.

One cannot assuredly say when the Barb emerged as a true type (and many contend the Barb *still* has no true, established type), so great is the expanse of time for development of this breed. Barb blood contributed mightily to many other breeds in Europe and subsequently, the Americas. As much a "tap root" of horse breeds as any other inhabiting the earth today, the North African Barb needs no excuses for its facial profile or sloping, powerful croup and hind quarters.

The Barb is found in slightly varying models across North Africa, some strains having been crossed quite extensively with the Arab breed in modern times. The remote nomadic tribes living in the mountains and desert of Morocco consider their Barb as the purest. This is an athletic breed. The coat color may be bay, brown, black, chestnut, or grey. The head is rather long with pronounced jaws and a straight or convex profile, lively eyes, flared, open nostrils, and curved, mobile ears. The head is tapered, wide from eye to jaw, small at the muzzle. The neck is of medium length and muscular with a natural arch; the withers are prominent and long; the back is short and straight; the croup sloped with a low-set tail. The quarters are powerful with smooth muscling; the shoulder is very long, sloping, and well muscled; the chest is long and very deep, giving great depth to the heart and lungs. The legs are slender and solidly built with broad joints and well-defined tendons; the hooves are small and well shaped with tough horn.

The Barb horse has the distinctively Oriental eye seen on nearly all horses of the Akhal-Teke breed and on many horses of China.

For additional information see Andalusian, Spanish-American Horse.

Morochuco

ORIGIN: *Peru*
APTITUDES: *Riding horse*
AVERAGE HEIGHT: *12.2 to 14.1 h.h.*
POPULATION STATUS: *Common*

The Morochuco is a type of the Peruvian Andean inhabiting the region of the Plain of Cansallo in the state of Ayacucho. It is said this breed was developed by the Jesuits with stallions from the coast. This is a small horse, very fast. The majority travel with a lateral gait which is smooth for the rider, and they are gifted with great endurance.

Because of its good qualities, the Morochuco has spread all over Peru, bringing fame to the Andean horse. The conformation of the Morochuco is much like that of the well-known Peruvian Paso, except that the Morochuco is more angular and smaller in size.

Moyle Horse

ORIGIN: *United States*
APTITUDES: *Riding horse*
AVERAGE HEIGHT: *15 h.h.*
POPULATION STATUS: *Rare*

The story of the Moyle horse is fascinating, strange, and more than a little frustrating, because it is difficult to trace this unique horse in a clear line to the source of its strange little "horns." There are vague references by

A Moyle mare with little "horns." Photo: Marge Moyle

A young Moyle stallion. Photo: Marge Moyle

Flintstone, a Moyle gelding showing the odd horn-like projections on the forehead. Photo: Marge Moyle

various writers to what appeared to be small horns on horses in the past, having apparently Asian ancestors, but nothing specific is given. The only breeds of horses living today known commonly to have the small frontal bosses referred to as horns are an occasional member of the Datong breed of China, the Carthusian of Spain, and the Moyle horse. The Datong horse was the breed from which the ancient "Dragon Horse" appeared many centuries ago (see Datong).

The Moyle horses, so named for their breeder, the late Rex Moyle of Eagle, Idaho, descended from horses brought to this country by Porter Rockwell and Brigham Young of Mormon fame. It is not known where Porter Rockwell obtained his horses, except that they were brought by ship from overseas. Research is incomplete at this time, but they were famous for their endurance and were used by the Pony Express during the last days of that short-term but famous mail run.

In the early 1850s, a federal judge speeding from St. Louis, Missouri, to Salt Lake City, Utah, with the best local buckboard hitch available, admired a similar rig which was passing eastward. The judge was then on the fourth day of his journey west. He made a note in his daily journal that the passing rig belonged to the Mormon leader, Brigham Young. Changing horses the following day, the judge had proceeded for another three days when he was astonished to see the Mormon rig passing him at a fast pace, returning west from St. Louis and drawn by the *same* horses. Feats of this kind were common for the Mormon's horses, which were renowned throughout the West. The finest stock in the westward rush had contributed greatly to a strain of horses of such marked ability that the young Pony Express Company paid one Mr. Kimball $250 per head, when $25 would usually buy the best horses available.

A rancher named Chris Hansen had the good fortune of living near the route used by a Mormon messenger on an urgent mission when the messenger's mare, which he had just run for twenty-eight miles, began to stagger beneath him. He stopped at Hansen's ranch to borrow a fresh horse. He was mystified that his mare had given out, as he had run her even greater distances in the past with no distress. The rancher, knowing the value of the Mormon horses, refused to loan the Mormon a horse, stating that he would only make a trade. The mare had lain down when the Mormon dismounted, and he thought she would die, so even though he had been ordered never to let any of the special horses out of his hands, he traded with Chris Hansen for a fresh gelding and continued on his way.

The mare, merely winded more easily than usual because she was in foal, recovered and delivered a filly about thirty days later. This filly was given to Chris Hansen's daughter, a girl of about sixteen years. She married about two years later, and one of her children was Rex Moyle. Whether or not the filly had the little horns on her forehead is unknown, but when grown she produced sixteen offspring which became the foundation stock for the family ranch horses. One of the characteristics of this line of horses was the strange little knobs above the eyes, larger on some than others. Members of this strain of horses were always outstanding endurance animals, and they were used to work the cattle on the family ranch. The neighbors also hired cowboys, and a working string was usually eight to ten horses per man. On the Moyle ranch, the cowboys got one horse and were usually better mounted than on the neighboring spreads. The Moyle horses could work cattle all day without needing a rest. The blood of the old mare was diluted over the years, until by 1930 the ranch had run out of this old line and felt "unhorsed."

Around 1900 the horse importers of Utah had a law passed making it a felony to keep a stallion that was not a registered purebred. The old strain of horses the Mormons had owned nearly disappeared. After Porter Rockwell's death, the horses he had bred for the stage lines, buggy horses, and special riding horses were crossed with other breeds and bred away.

When he was still a young man, Rex Moyle went into the mountains near Salt Lake, particularly near Cedar Valley where the Mormons had grazed the best of their riding stock. By carefully combing through hundreds of captured mustangs, he was able to find a few mares that resembled the old strain the family had admired and used for so many years. Being a knowledgable animal breeder and one of the largest mink ranchers in North America, Rex Moyle was able to build his herd of special horses carefully. To avoid inbreeding to an extensive degree, some outcrosses were made to Cleveland Bay horses. Down through the years, the little knobs or horns have continued to appear on the foreheads of a majority of the Moyle horses, along with other interesting skeletal differences which one hopes will eventually be studied by scientists.

The old legend of the ancient Dragon Horse haunts those who know the Moyle horse, for there are many similarities. The Dragon Horse was kept only by the royal families, and was noted for special endurance. The Moyle horses are astonishing endurance horses, and many have done well in endurance trails in the United States. The Dragon Horses were usually bay or brown, as are those of the present Datong breed; the majority of Moyle horses are also bay or brown with few white markings, although other colors do occur.

Speculation over the origin of the Moyle horses received an interesting boost when blood samples were drawn and sent to the laboratory of Professor Gus Cothran at the University of Kentucky in 1990. Professor Cothran found that the Moyle horses showed strong relationship to the Spanish horse, and it is fascinating to note that the Carthusian Andalusian is often found to have small horns. Horses in Spain could easily trace to Asian breeds, and one has to wonder if there is a link to the ancient Dragon Horse of China. A contact of mine in Turkey wrote to say that twenty years ago he met a man with three horses he had obtained in Persia that had the same small knobs on their foreheads, and he called them "Dragon Horses." He was told that they were extremely rare. Research is continuing, and anyone with any knowledge of horses like this is invited to contact the author.

In 1961 the Moyles began to enter their horses in endurance competitions, and some amazing feats were accomplished. In August that year the eleven-year-old mare Kattawar was broken to ride in preparation for a Labor Day race in Duchesne-Heber, Utah. Ridden by twelve-year-old Dan Moyle, the mare covered the eighty-three miles to win in seven hours and nineteen minutes. The only other Moyle entrant, Copper, finished in sixth place. The two Moyle horses were the only ones completing the ride that were fit enough to be ridden into town the next day to collect their prize money.

In 1962 Copper, a seven-year-old Moyle horse, was entered in the famous Tevis Cup ride, 100 miles in one day, finishing a reputable sixth place. In 1963 the Moyles entered four of their horses. Mr. and Mrs. Moyle and two of their children finished the 100-mile test without a mishap. Rex Moyle rode Sweet Pea, a nine-year-old broodmare, who finished the distance in fourteen hours and thirty-eight minutes in a tie for sixth place.

In 1964 ranching prevented any preparation for the Tevis Cup ride until July 4. At that time three horses were selected: Copper, to be ridden by Dr. Matthew Mackay-Smith; Sweet Pea; and her five-year-old gelding son, Flintstone. Sweet Pea had just lost her foal, and Flintstone had first been broken to ride on June 8. None of the three horses had been conditioned for endurance. These three endured the most arduous Tevis Cup ride ever—in which only half the entrants finished due to the heat—to finish second, third, and fourth. Copper covered the last thirty-eight miles in under four hours, one hour faster than the leaders.

The mare Sweet Pea, one week after that grueling ride, was hauled over 3,000 miles to Pennsylvania, arriving on August 8. Less than a month later she was lightweight and grand champion winner of the Vermont 100-Mile Trail Ride. One week later, she was grand champion of all divisions in a 50-mile, seven-hour contest near Bel Air, Maryland. During this time, Sweet Pea continued to gain in condition on good hay and less than three quarts of grain per day.

Mrs. Moyle and her horse Hawk completed over 5,000 competitive endurance miles in five years. Many of these were 50 or 100 miles in a day. They completed many large and small competitions in the top percentage, and Hawk remained sound. In 1983, when Hawk was fifteen years old, they won the overall Best Condition Award for a 250-mile endurance ride (50 miles for five consecutive days). Hawk has earned from the Endurance Horse Registry its Legion of Merit Award and has also been awarded a permanent place in its prestigious hall of fame.

Besides the small horns, the Moyle horse has other conformational characteristics that would seem to warrant a close look by scientific specialists. There is unusual freedom of movement in the shoulder. The forehand is characterized by a foreleg that is placed extremely far forward on the rib cage; the first rib usually lies slightly behind the shoulder joint. The theoretical disadvantage that this might create is turned to virtue by extreme forward projection of the breastbone (sternum), coupled to the arm by exceptionally large pectoral muscles. The horizontal axis around which the shoulder rotates is decidedly lower than in other horses. This leads to a looseness in the shoulder that is controlled by the exceptional size and range of the girth muscles (latissimus). The external shoulder muscles (trapezius and deltoid) are correspondingly diminished, and the resulting elasticity is such that the rib cage can shift over a range of twenty to thirty inches laterally without moving the front feet. Due to the construction of the shoulder, traveling downhill on a Moyle horse is quite comfortable. The back is well constructed for carrying a saddle, with long muscling which is deeply seated, leaving a flatter surface for good weight distribution. The hind quarters are long and elastic. The muscling is long and smooth. The skin is very thin but tough, and covered with a dense coat. The feet are large and wide at the heel with particularly tough horn. The walking stride of the Moyle horse is exceptionally long.

This horse's remarkable power for sustained effort and quick recovery are enhanced by a liver and spleen nearly twice the normal size. The reserves of energy and red blood cells stored in these organs give a distinct advantage for prolonged exertion over horses not so well endowed.

The most interesting distinctive anatomical features in the Moyle horse are the obvious presence of horns on the forehead and the absence of chestnuts in many individuals. These features were noted in the special strain of Mormon horses. The "horns" are of course not true horns, but bony projections on the skull known as frontal bosses. They could be compared to those found on the giraffe (although smaller). They are readily apparent on the skull of a deceased Moyle horse.

Another characteristic aspect of the Moyle horse is its tendency to be a "one-person" horse. While an individual animal may be tolerant toward everyone, it is obvious that the Moyle horse prefers its "own" person above all others.

Presently there are approximately 100 horses on the Moyle ranch. Not all of the horses have the small knobs on their foreheads, but a good percentage of them do, and they are all related. Those interested in competitive endurance riding should consider this interesting horse. According to the legend of the ancient Dragon Horse of China, the horned horses were extremely special, and the Moyle horse seems also to fit into a very special category.

Murakoz

(Muraközi, Mura)

ORIGIN: *Hungary, Yugoslavia*
APTITUDES: *Heavy draft work*
AVERAGE HEIGHT: *16 h.h.*
POPULATION STATUS: *Rare*

Developed around the turn of the century, the Murakoz breed was a very popular working horse at one time, but presently its numbers are dwindling rapidly. The primary area of development was southern Hungary, around the River Mura. Foundation stock for the breed consisted of native Hungarian mares (called Mur-Insulan) and Percheron, Ardennais, Noriker, and Hungarian halfbred stallions.

The breed suffered great losses during the Second World War and new blood was introduced to reestablish the Murakoz, primarily using Belgian Ardennais stallions. Well suited to hard work, hardy, and active, the Murakoz is becoming increasingly rare these days.

The majority of Murakoz horses are chestnut with a flaxen mane, tail, and feathering on the legs, but some are bay, brown, black, or grey. The head is rather long and heavy with a slightly convex profile, large, expressive eyes, and long ears. The neck is short, crested, heavily muscled, and broad at the base; the withers are not prominent; the shoulder is sloping and long with powerful muscling; the back is short and broad; the croup is sloped and heavily muscled; the quarters are rounded and muscular; the legs are short and sturdy with large, strong joints and short cannons; the feet are rounded and well shaped.

Presently, although numbers are small, the Murakoz is bred in two types; one taller and heavier, the other smaller and of lighter conformation.

Murghese

ORIGIN: *Italy*
APTITUDES: *Riding horse, light draft*
AVERAGE HEIGHT: *15 to 16 h.h.*
POPULATION STATUS: *Uncommon*

Murghese. Photo: Dr. Alfonso Basile

The Murghese or Murge horse is bred and raised in the Murge region of Italy, bordered by the provinces of Taranto, Bari, and Brindisi. Origins of the breed date to the period of Spanish domination. The Court of Conversano imported horses of Oriental and African origin. In addition, the special environment and needs of local agriculture contributed to the formation of the Murge horse. The Murghese horse is the only one that can now be considered typically Italian, being free of any crossbreeding. In 1774 the Emperor of Austria acquired two stallions from the Count of Conversano. These two helped form one of the families of the Lipizzan breed, appropriately called the Conversano family, and this family still makes up part of that breed.

The Murghese horse is strong, has a good temperament, and lives a long, useful life. It has good limbs, solid hooves, and a black morel (reddish) coat without any white markings; there is a limited variety that is found with a metallic grey and morel shade. This breed is raised in the wild, in the dry, rocky forests of Murge, and the environment of the area tempers and enhances many superior qualities. The horses remain out in the open day and night, even during frequent snow storms. During the very dry summers when there is scarcely a trace of vegetation, the Murge horses always find some way to survive, making up for lost meals when vegetation reappears.

The Murghese can be used for any purpose due to its qualities and elegant carriage as well as its vivacious and obedient character. This horse is increasingly considered the ideal Italian horse for equestrian tourism, recreation, and country riding and for horse teams, which are currently expanding in use throughout the country. It is also well liked and increasingly employed by the National Forestry Corporation as a service horse. Murgheses, which were hardly known outside their home until a few years ago, always surprise and please those who have been able to get to know and use them.

The appearance of Murge horses at national horse shows and especially at the Mediterranean Horse Showroom "HIPPOS," which is held in the spring in Bari at the Levante Convention Center, has contributed greatly to knowledge about and appreciation and spread of the breed. The Murghese boom began in earnest in 1987. People have come to the Murge region from all over Italy, excited about the horse and wanting to get to know the breed and in particular the habitat in which it is raised. They also want to acquire specimens that demonstrate the value, qualities, and possibilities of this breed in any environment and use. Recently the Murge has crossed national borders and even the ocean, when a stallion and two mares were exported to Valencia in Venezuela.

Selection of the breed began in 1926 through what was then called the Stallion Ranch of Foggia, today the Regional Institute for Equine Increase. With the assistance of the Association of Horse Breeders based in Martina Franca starting in 1948, noteworthy results were obtained. Granduca, Nerone, and Araldo delle Murge were the foundation sires of the breed, originating three principal bloodlines that take their names. The production of Murge horses, unlike that of some breeds, is burgeoning thanks to the commitment and passion of the breeders united in their own association.

The annual horse markets and shows held in December at Martina Franca involving young stallions and foals are the best places to see the interesting results emerging from year to year through the selection and commitment of the Murge breeders. The desire for a lighter, more elegant riding horse has led to the use of some Arab, Anglo-Arab, and English Thoroughbred stallions with selected mares. The resulting offspring are very desirable for horseback riding and equestrian sports.

Mustang

(American Feral Horse, Broomtail, Range Horse, BLM horse)

ORIGIN: *United States*
APTITUDES: *Riding horse*
AVERAGE HEIGHT: *14 to 15 h.h.*
POPULATION STATUS: *Common*

A BLM Appaloosa mare and foal in the holding pen at Hines, Oregon. Photo: Burns District, Bureau of Land Management

Not so many years ago millions of feral mustangs roamed the open country of America; mostly the West. The first of these were Spanish horses or their descendants, but as settlers spread across the land there was increasing mixture of breeds into the wild herds. The Indians, once introduced to the horse and embarking on their subsequent new way of life, played a large role in the spread of horses across North America. Horses were stolen or bartered between tribes; stolen from white settlers; and obtained through battles and massacres or other means. Contrary to what some believe, the Indians found it much easier (as would anyone) to contend for gentle, trained horses than to attempt to corral and train the wild ones.

The once pure Spanish herds received continual contamination and mixture with other breeds as settlements and ranches were established across America. Sometimes wild stallions tore down the fences and made off with tame mares; sometimes tame mares went through the fences on their own; draft teams were loosed all across the country during Indian raids on farms or wagon trains. Herds of feral horses from the colonial settlements in the East were forced westward as civilization expanded, crossing the Mississippi in times of drought, and herds of Canadian horses from the area that is now near Detroit, Michigan, joined the wild bands, contributing French blood.

French blood was also infused into the American horses in the South, a fact usually overlooked by researchers of the American mustang. The French explorer, Robert Cavelier, sieur de La Salle, descended the Mississippi River in 1682 and claimed the entire river basin for France. The city of New Orleans was established in 1718 by Jean-Baptiste Le Moyne, sieur de Bienville. When Louisiana became a French crown colony in 1731, the population was about 8,000, including black slaves. Significantly, the colonization increased in the 1760s with arrival of the French-speaking Acadians (see Canadian), who had been expelled from Nova Scotia by the British. Spain controlled the territory from 1762 until 1800, when it passed once more into the hands of the French. Louisiana was acquired by the United States as a part of the Louisiana Purchase from France in 1803. Until that time, explorations and scattered settlements in the seventeenth and eighteenth centuries had given France control over the river and title to most of the Mississippi Valley. It is important to recognize the presence of the highly esteemed French-Canadian horse, undoubtedly in considerable numbers, in the Mississippi River basin. The French did a great deal of trading with the native American Indians.

Another little-known ingredient that probably contributed to the feral herds was the old-type East Friesian of Germany. In the late 1800s to early 1900s the U.S. government purchased about 150 stallions of that breed each year at the German horse sales. This practice, carried on for over ten years, brought with it the potential for a great deal of East Friesian blood in government and/or cavalry horses. The East Friesian of that time was a heavy warmblood, or coach horse, an excellent choice for pulling artillery or heavy wagons. Wherever the calvary went in those days, so probably went the East Friesian horse.

Once, the large herds of wild horses posed no particular threat to human interests, but eventually this all changed. Ranchers began to resent the horses, which ate grass needed for their cattle. The arid lands of the West cannot support an overpopulation of ungulates. Many ranches had a policy of shooting any wild horses they could to rid the range of excess animals. "Mustangers" used various methods to capture herds of wild horses; in earlier times the horses were broken in and sold as riding horses; later, large numbers were sent to meat packing plants which produced animal food, especially chicken feed.

There were an estimated two million wild horses in the United States at the beginning of the twentieth century. By 1926 this number had been halved. By 1935 the total was estimated at 150,000. Presently, numbers vary but are near 30,000. Until 1971 there were no regulations or rules to govern the treatment of wild horses, and often the methods used to capture them were cruel. A great outcry of general public opinion and the work of many dedicated people brought change to this situation, and Congress acted. Legislation was unanimously passed to protect, manage, and control wild horses and burros on public lands. The Wild Free-Roaming Horse and Burro Act (16 U.S.C. 1331–1340) described these animals as "fast disappearing" and declared them to be "symbols of the historic and pioneer spirit of the West." The Department of the Interior, through the Bureau of Land Management (BLM), and the Department of Agriculture, through the Forest Service, are charged with administering the law.

Under federal protection, the wild horse and burro populations on BLM-administered lands (where most of the animals are located), quickly grew to a point where control became a major concern. Amendments in 1976 and 1978 addressed the problem and the need to dispose of the animals being removed. Bills to provide sale authority were introduced in Congress in the early 1980s to respond to difficulties in disposing of excess animals. However, at that time no new wild horse and burro legislation was enacted.

The agencies continued to plan for appropriate management levels and to make progress toward achieving those levels by removing excess animals. These actions were taken to comply with the requirement in the act that the Secretary of the Interior "shall immediately remove excess animals from the range so as to achieve appropriate management levels." The act directs the removal of all excess animals "to restore a thriving natural ecological balance to the range, and protect the range from the deterioration associated with overpopulation."

The Adopt-A-Horse Program began in 1973 in the Pryor Mountains of Montana as a humane method of disposing of excess animals removed from public lands. Through the adoption program, these animals are made available to the public for a fee of $125 per horse and $75 per burro. Horses that have poor conformation and are less desirable for adoption are offered at a reduced fee in an attempt to avoid destroying them. All horses and burros captured by the BLM receive health examinations by qualified veterinarians and inoculations to protect them against disease. The animals are kept in large corrals where feed and water are available until they are taken to adoption centers to be offered to the public. Where in the past the wild horses and burros were at the mercy of unkind people whose idea of entertainment was to run a wild horse to death with a jeep out on the desert, the animals presently enjoy protection and good care. Anyone arrested for harassing or illegally capturing feral horses or burros is subject to severe penalties.

Adopters must meet requirements for furnishing safe transportation from the adoption center and possession of safe quarters for the animal, which remains the property of the United States government for a period of one year. At that time, the adopter is asked to submit a statement from a veterinarian, extension agent, local humane official, or other qualified individual certifying that the animal has received proper care and maintenance, and when approved, a certificate of title is issued and the animal becomes the legal property of the adopter. In this manner, a great number of horses and burros have been adopted and now serve a useful life.

Several writers have reported the mustang to be rebellious and intractable, giving the impression that these horses cannot be gentled or made useful. On the contrary, the mustang is easily gentled by experienced handlers and usually becomes as tractable as any horse raised in the barnyard. In fact, it is often the case that a formerly wild horse is *more* tractable than a spoiled colt that must be taught respect for humans.

In the past, many U.S. feral horses exhibited poor conformation or a great amount of draft blood, but in recent decades the quality of range horses has been greatly improved through the BLM's management system. Inferior stallions have been removed from many areas and replaced by stallions of good quality, resulting in better offspring, hence horses more desirable for adoption. The system seems to be working beautifully. Many people are anxious to have a true wild mustang. The American mustang is an animal that has been forged in the furnace of nature, where only the strong and fit survive while the weak are left to feed the predators. Such a horse, once gentled and trustful of its handler, is usually an excellent mount to use for endurance trials. I have never known a mustang that was not surefooted and very enduring. Many horses that were formerly feral horses under the

management of the BLM are now excellent, dependable riding horses.

Most U.S. feral horses today are of the light horse or warmblood type. Feral draft-type horses are usually separated and turned loose on a range of their own. The coat color of the mustang may be of any horse color. Type varies with conformation ranging from poor to good, and the legs and feet are usually excellent. While the Spanish blood has been diluted by that of many other breeds, many mustangs still retain some Spanish characteristics. It has been a strong belief for the last several decades that no pure Spanish horses remained in the wild in America. However, in recent years a few small herds have been found in isolated areas and blood tests have shown them to be of strong Spanish descent (see Kiger and Cerbat). For those whose heart races with excitement over the thought that one more such small group of Spanish horses roams free in an isolated area—the possibility remains. For additional information see Spanish-American Horse.

Mytilene
(Ege Midillisi, West Mytilene)

ORIGIN: *Turkey*
APTITUDES: *Riding, pack horse*
AVERAGE HEIGHT: *10.3 to 11.3 h.h.*
POPULATION STATUS: *Rare*

The Mytilene (Ege Midillisi) pony of Turkey. Photo: Ertuğrul Güleç

Mytilene pony. Photo: Ertuğrul Güleç

The Mytilene pony in action. Photo: Ertuğrul Güleç

A very small horse or pony, the Mytilene is an ancient breed known in Turkey for at least 1,000 years. The word Ege in Turkish refers to West Anatolia.

The Greek word for Midillisi is Mytilene, the name of an island near the western coast of Turkey where the Ottoman Turks raised the Mytilene pony. After the fall of the Ottoman Empire the island of Mytilene went to Greece,

and the Mytilene or Midillisi horse continued to be raised in Turkey.

The Mytilene is very much like the Caspian of Iran, is likely related to that breed, and is probably a miniature horse and not a pony. However, the Mytilene has some advantages over the Caspian, according to authorities. Naturally born with the rahvan gait (a very fast running-walk), the Mytilene can carry a large load and travel long distances much faster than the Caspian.

Mytilene horses were developed by selection of the smallest type of Anadolu to produce a small horse able to get beneath trees as a packing animal for collecting hazel nuts and olives. The Mytilene converts feed very efficiently and is able to maintain its health and weight on scant food while at the same time exerting great working effort.

There is no breed association or government stud book for this ancient breed, so appreciated by the people who use it on a daily basis.

Namib Desert Horse

ORIGIN: *Namib Desert (Namibia)*
APTITUDES: *Riding horse*
AVERAGE HEIGHT: *14 to 14.3 h.h.*
POPULATION STATUS: *Rare*

A beautiful and mysterious feral herd of horses inhabits the Namib Desert of Namibia (for-

The Namib stallion, Impuls, at five years of age. Photo: Frans van der Merwe

The Namib stallion, Impuls, just prior to being darted for capture. Photo: Frans van der Merwe

A young Namib mare leading the others to water at Garub. Photo: Frans van der Merwe

merly South West Africa) in the most inhospitable surroundings, demonstrating the ability of the horse to adapt and survive.

Dr. F. J. van der Merwe, director of the Department of Agriculture and Water Supply in Pretoria, South Africa, was responsible for bringing these horses to public attention in the early 1980s. The Namib horse is probably the only equine community living wild in Africa. Little is known about the horses, other than that they have lived in the desert for many years. The sand and gravel plains of Garub in the southern Namib Desert contain one watering hole and little feed.

The range for the horses, their numbers fluctuating between 250 and 300, stretches eastward over the desert to the mountains of the Great Escarpment and northward to the Koichab River, usually only a dry bed. The life cycle of the desert horses is governed by the unpredictable climate. Due to difficult conditions the Namib horses have never increased beyond the numbers given above. In the past a few attempts were made to exterminate them, as certain people regarded them as a threat to the habitat of the oryx. South West Africa/Namibia's Directorate of Nature Conservation has since ensured their continued existence as part of the protected desert wilderness.

Origin of the Namib horse will likely remain shrouded in mystery. There are several theories concerning their presence in the desert. One is that over a century ago Hottentot raiders moved from the south into the unmapped area north of the Orange River, bringing their hardy horses with them, likely crosses of the Cape horse and Basuto ponies. Another theory hinges on a cargo steamer with a load of Thoroughbred horses bound for Australia. The ship ran aground near the mouth of the Orange River. The strongest horses, it is said, reached the shore and made their way to the Garub plains.

The most likely theory seems to involve the German nobleman, Baron von Wolf, who built Dunwisib Castle in Namibia. When von Wolf built the castle, the stones and beams for construction were brought across the Namib from Lüderitz by ox wagon. A railway was later built linking that small harbor town with the bleak hinterland, and when diamonds were discovered during the building of the track, the way of life changed drastically in the German colony.

During the First World War, South African troops on horseback rode from Walvis Bay in the north and from Lüderitz through the desert to trap the Schutztruppe in a pincer movement and bring the Germans to their knees. It was the last time that horses played a significant role in South Africa's military history. During this campaign some horses were reportedly abandoned by the troops, while others broke free and escaped into the desert.

At that time, von Wolf sailed with his wife and a friend to Europe on a stallion-buying expedition. He left on his farm a herd of about 300 horses, on the edge of the Namib desert and about 120 miles from the Garub water hole. Von Wolf jumped ship in a European harbor and joined his unit in Germany. He was killed in action in 1916. His wife never returned and the farm was abandoned. The property was not fenced and in the turmoil following the German defeat in their South West African colony, it is quite possible that some of his horses found their way across the desert and eventually to the Garub watering point, which today is served by a windmill.

The possible link to the horses of von Wolf seems most plausible because of the quality and size of the Namib Desert horses. Namib horses do not resemble the Basuto. They do, however, bear resemblance to the Trakehner breed which was always popular with the Germans. To me, they bear an even stronger resemblance to the Akhal-Teke breed, and while this may seem a stretch of the imagination, the Trakehner breed was solidly based on blood of the Akhal-Teke horse. Of great interest is the fact that Professor Gus Cothran of the University of Kentucky and Dr. van der Merwe in South Africa are working to get clearance to ship blood samples to the blood testing laboratory in Kentucky. If obstacles concerning shipments of African horse blood (chiefly African horse sickness) can be overcome, results of genetic blood group testing of the Namib horses will be more than a little interesting.

The Namib horses have survived and multiplied in the most inhospitable circumstances and nature has eliminated the weak and unfit. The remaining specimens are not only hardy and resilient, but inbreeding has eliminated undesirable genetic material and there is little need for improvement in the Namib horse. This is a high quality animal of much refinement, and this is only possible in an environment where a group of horses such as the Namib stock is isolated from any chance of outcrossing to other strains.

In January of 1987 nine immature animals and one aged mare were darted for capture by the Namibian Department of Agriculture and Nature Conservation. With the generous assistance of Fred and Laura Burchell, they were taken to the Veterinary Research Institute at Onderstepoort near Pretoria, where they have since been maintained as a separate breeding group for comparative purposes. The first doctoral thesis on the water metabolism of the Namib desert horses has recently been initiated. The results should show whether the horses have acquired adaptive mechanisms, like the oryx or ostrich with which they share the habitat. Among the horses captured was the stallion Impuls; photographs of him accompanying this article were taken just prior to his being darted and three years later in 1990 (at five years of age). According to Dr. van der Merwe, Impuls "has the look of eagles and moves like a dream, floating and with the softest touch on the ground."

The Namib horses in the desert are separated in small family groups usually consisting of a stallion and his mares and foals, or bachelor groups of young animals. Feed is scarce and the horses must travel some distance to feed and return for water at Garub. Nature, being the best selector of breeding stock, has molded whatever the original horses in the desert were into highly specialized animals perfectly adapted to their harsh environment and, as a bonus, extremely beautiful.

According to Dr. van der Merwe, in general appearance the Namib horses are of the warmblood type, athletic, well muscled, and with straight, clean legs. The quality of bone is exceptional. The head is invariably beautiful with large, widely spaced eyes and a small muzzle. The ears are small, well set, and pointed. The back is short, the shoulders sloped and well muscled. The most dominant color is bay; some are brown or chestnut. There are no grey Namib horses.

There is a good deal of uniformity in the conformation of the Namib horses, such as the remarkably sound, dry, flat bone; the deep chest; strongly muscled, sloping shoulder; and well-defined withers. The hooves of this horse are extremely hard. The skin is thin, and the sheen on the summer coat is quite remarkable even when the horses are in poor condition. Many of the horses have a dorsal stripe, but zebra stripes on the legs are absent.

While it is generally believed that inbreeding causes a reduction in body size and height, the Namib horses average 14 to 14.2 hands.

Narym
(Narymskaya)

ORIGIN: *Western Siberia (former Soviet Union)*
APTITUDES: *Draft*
AVERAGE HEIGHT: *13 to 14 h.h.*
POPULATION STATUS: *Rare*

The Narym is similar to the Priob breed and originated near the same area in the central region near the Ob River. The two breeds live under much the same ecological and economic conditions, and may be considered two groups or types of the same breed of northern forest horse.

The Narym is larger than the Priob and is crossed with draft horses and trotters in the southern part of the breeding area. In the north, it is bred pure.

National Show Horse

ORIGIN: *United States*
APTITUDES: *Riding horse*
AVERAGE HEIGHT: *15 to 16 h.h.*
POPULATION STATUS: *Common*

National Show Horse, Key West. Photo: Grand Kid Acres

The National Show Horse Registry was conceived in the spring of 1981 by Gene LaCroix and incorporated in August of that year. A noted Arabian breeder, he was also a longtime admirer of the American Saddlebred. In the National Show Horse, he sought to combine the best qualities of the two breeds. The ideal horse had a blend of the Arab's beauty, refinement, strength, and stamina with the Saddlebred's extremely long neck, high-stepping action, and electric show horse attitude.

The new breed grew quickly. The Arabian-Saddlebred cross has emerged as a favorite show horse, combining beauty, refinement, stamina, and high action. One goal of the NSHR was to create a registry that would allow breeders a systematic means of refining and improving the cross in order to create the ideal show horse. However, founders of the NSHR were not satisfied just to create a registry for the new breed. They also wanted to produce a new atmosphere in the show ring in order to generate a broader public appeal and to develop a prize money system that would make it financially rewarding for exhibitors to show their horses.

After its official beginning in 1982, the NSHR allowed a period of open registration in order to collect a pool of horses that would be the foundation for the new breed. Today, however, the registry has established specific rules regarding the types of horses that may be used to produce National Show Horses. Eligibility for registration requires that a foal be sired by an NSHR "nominated sire." A nominated sire may be a purebred Arab, Saddlebred, or National Show Horse. Owners of one of these three types may nominate their horses, subject to approval of the NSHR board of directors.

Breeders can use three types of mares to produce National Show Horses. Arabians, Saddlebreds, and National Show Horses can be used to produce NSH foals when crossed with the appropriate stallion. Only these types of mares are eligible.

The National Show Horse Registry is dedicated to producing an atmosphere in the show ring that reflects the excitement created by the horses themselves. New concepts and formats for shows have been combined with prize money to produce new enthusiasm for both spectators and exhibitors. In its first year, the NSHR paid out more than $100,000 in prize money.

Presently, the bloodlines most consistently used are *Bask + + sire lines and Crabbet/domestic-bred mares on the Arabian side and various combinations of Valley View Supreme and Wing Commander for the Saddlebred portion. As more stallions are nominated and broodmares are introduced, additional outstanding strains are being added to the gene pool. The opportunities are endless.

Both the slow gait and rack are four-beat gaits with the horse showing extreme style and animation. Though the gaits are animated, they are smooth, and the rider sits in the saddle rather than posting as a rider would normally do at the trot. The most impressive gait is the rack, where the horse moves at speed while maintaining the four-beat gait. The horse must also perform at the walk, trot, and canter.

National Show Horses are also shown in pleasure driving, fine harness, and country pleasure classes.

National Spotted Saddle Horse. Photo courtesy National Spotted Saddle Horse Association

National Spotted Saddle Horse

ORIGIN: *United States (Tennessee)*
APTITUDES: *Riding horse*
AVERAGE HEIGHT: *15 h.h.*
POPULATION STATUS: *Common*

The Middle Tennessee region, with its fertile valleys, long growing season, and abundance of water, has long been favored by prominent horse breeders. The National Spotted Saddle Horse emerged in this area, kin to the Tennessee Walking Horse, American Saddlebred, and Standardbred. After many years of careful breeding and selection, the National Spotted Saddle Horse Association was formed in October of 1979. The association has made good progress in promoting this horse of easy gaits and kind nature. The breed has spread to many areas of the country.

The National Spotted Saddle Horse breed includes many small horses, but the average height is 15 hands. This is a rugged animal

with strong bone, intelligent eyes, and small, fox-type ears. It is popular in pleasure classes and used extensively in the original territory by hunters who appreciate its mild manners, whether hunting birds or raccoons. Surefooted in rugged, mountainous terrain, this horse gives a smooth ride — a stepping pace or rack.

Horses must qualify for registration with pinto color in addition to acceptable pedigree and exhibition of gaits.

New Forest Pony

ORIGIN: *England (Hampshire)*
APTITUDES: *Riding pony, light draft*
AVERAGE HEIGHT: *12 to 14.2 h.h.*
POPULATION STATUS: *Common*

The New Forest lies along the English coast between Southampton and Bournemouth in southwest Hampshire, comprising one of the largest areas of unenclosed land in southern England (about 60,000 acres). Once the hunting ground of kings, it is now an increasingly popular place of recreation for the general public. New Forest ponies still roam its heaths, woodland, and bogs as they have since time immemorial.

The presence of horses in the forest is first

New Forest Pony. Photo: Miss D. Macnair

indicated in the litter of a Roman-British pottery at Linwood, Hampshire. The lid of a larger water jug, identifiable as British rather than Roman workmanship and dated as first century A.D., shows horsemen approaching in a highly respectful manner a terrifying female figure recognizable as the "horrific" aspect of Epona, goddess of horse breeders. All Celtic goddesses had these three aspects: the maiden, lovely to look at but rather frigid; the matron, again very handsome but friendly and protective; and this one, the crone—bloodcurdling.

How it came to be that all of the ponies passed into private ownership is uncertain, as there have been few written references to them through the years. New Forest ponies have always been valued for their strength, hardiness, docility, and sureness of foot. Arab and Thoroughbred blood has been used from time to time to improve their appearance and increase height, but systematic efforts to improve the breed were not started until the end of the nineteenth century.

In 1891 the Society for the Improvement of New Forest Ponies was founded, offering premiums for suitable stallions to run in the forest. In 1906 the Burley and District New Forest Pony and Cattle Breeding Society started

to register mares and young stock, and the first stud book was published in 1910. At that time the current theory was that to improve the breed stallions from other native breeds should be used, and the early stud books show a curious assortment of sires.

From 1914 to 1959 registrations were recorded in the National Pony Society's stud book. In 1938 the two local societies amalgamated, and no outside blood has been permitted since the mid 1930s. In 1960 the New Forest Pony Breeding and Cattle Society started to publish its own stud book and has continued to the present.

In recent years an increasing number of New Forest ponies have been bred in private studs outside the forest and many ponies have been exported. Presently there are flourishing studs of registered New Forest ponies not only in the United Kingdom but throughout Europe, Canada, and Australia.

The breed combines the characteristics of the native British ponies such as strength, intelligence, and agility with a narrower build, tractable temperament, and speed. This is an ideal pony for the entire family and is suitable for equestrian sports, driving, or riding.

New Forest ponies may be of any color other than piebald, skewbald, or blue-eyed cream. Bays and browns predominate. White markings on the face or legs are permitted. The pony should be of riding type, with a good deal of substance. The head should be of "pony type," well set on; the shoulders should be long and sloping; the quarters should be strong and well muscled; the body is deep; the legs are straight with strong joints and good, hard, rounded hooves. The larger ponies, while narrow enough for children, are capable of carrying adults. The smaller ponies, although not up to so much weight, usually show more quality than the larger animals. The action of this pony is free, active, and straight but not exaggerated.

Most New Forest ponies are good jumpers and are naturally good at gymkhana events and mounted games. They make excellent harness ponies. New Forest ponies have been successfully trained for dressage, polo, long-distance riding, cross-country events, and carrying the disabled. They have long been raced locally and are surprisingly fast, especially over rough terrain.

Newfoundland Pony

ORIGIN: *Newfoundland (Canada)*
APTITUDES: *Riding, light draft*
AVERAGE HEIGHT: *12 to 14 h.h.*
POPULATION STATUS: *Rare*

Newfoundland mare and foal. Photo: Barbara Brown

A typical Newfoundland pony. Photo: Andrew Fraser

The Newfoundland pony is a distinct breed developed in the easternmost province of Canada—the island of Newfoundland.

Newfoundland ponies have played an important part in the rural life of the island. They were used to plow the fields in spring, helped with the harvest in fall, hauled wood in winter, and provided the people with a means of transportation. During the summer the ponies were allowed to roam freely, fending for themselves and breeding without restriction. They have evolved naturally as wild ponies would do, on the basis of survival of the fittest, and over the centuries a distinct native type has developed.

The Newfoundland is believed to be descended from pony stock which existed in southwestern England prior to 1610. Newfoundland was officially settled by English settlers in 1560. Undoubtedly some of the early settlers and fishermen brought horses or ponies with them, but their origin is uncertain. The Newfoundland fishing trade was important in the early days to the English towns of Bristol, Bideford, Barnstable, Plymouth, Dartmouth, Exeter, Weymouth, Poole, Teignmouth, and Southampton. It is known that horses were shipped to Newfoundland in 1612 by John Guy, who founded a colony in Cupids Cove, Conception Bay. It is not known what sort of horses or ponies he imported. Others may have come to Newfoundland from England between 1610 and 1630. There is little doubt that the stock was of pony and not horse type, however, as the Newfoundland bears strong likeness to British pony stock and is never considered a small horse.

With mechanization and banning of free-roaming herds, the role of the pony in Newfoundland has decreased in importance. The Newfoundland became a source of cheap meat for the meat packing industry on the mainland, shipping meat to France through Quebec, and threatening the pony's survival. The Newfoundland Pony Society was formed in November of 1980 to protect this hardy pony and prevent extinction of the breed. Less than 300 ponies have been registered, but the actual number on Newfoundland is much greater.

The role of the Newfoundland is changing, thanks to its willingness to learn and gentle disposition, from an all-around work animal to a pleasure horse. Newfoundlands are used for riding and carriage driving, and during the winter months they are used to pull sleighs.

The Newfoundland has a small, well-shaped head with small, sharply pointed ears and large, expressive eyes. The ears usually contain a great deal of thick hair which helps to keep insects out in the summer and adds warmth during the winter. General conformation of the Newfoundland is good, with strong, muscular neck and shoulders.

Predominant colors are black, bay, and

brown, with many roans. Piebald and skew-bald ponies are not accepted for registration and are not considered typical of the New-foundland. White markings are the exception rather than the rule, usually only amounting to a small amount of white on the forehead. White markings on the legs are rarely seen, and breeders see the presence of white on the legs as an indication of outcrossing.

The Newfoundland grows a thick, long coat in the winter, and the mane tends to cover both sides of the neck. In summer, however, the mane lies on one side.

This pony is virtually unknown off the island, although it has been on Newfoundland for centuries. It is hoped that with vigorous promotion the Newfoundland may be saved from extinction. This old pony breed repre-sents valuable genetic material in a world that has reduced genetic resources to a marked degree.

New Kirgiz
(Novokirgizskaya)

ORIGIN: *Kirgizia (former Soviet Union)*
APTITUDES: *Riding horse, light draft*
AVERAGE HEIGHT: *14.2 to 15.2 h.h.*
POPULATION STATUS: *Common*

Developed from the old Kirgiz breed bred by nomadic tribesmen, the New Kirgiz emerged between 1930 and 1940. The original breed was the native pony of the mountainous re-gions of Kirgizia and Kazakhstan and was derived from Mongolian stock. It was used by local tribes for many centuries. During the early twentieth century demand grew for a larger animal to work the farms, and Kirgiz ponies were crossed with English Thorough-breds and Dons.

The founder of Issyk-Kul stud brought Thoroughbreds into the area in 1907. This man, V. Pianovski, considered the local Kir-giz horses unsuitable for the army but appre-ciated their naturally healthy constitution. From his initiative, the Przewalski racecourse was established in 1908, the first one in Kir-gizia. In 1918 there were forty-eight Thor-oughbred stallions at the stud to cross to local Kirgiz mares. These improved horses were the foundation for development of the new breed—New Kirgiz.

The new breed was developed in the state and collective farms of Kirgizia, using the improved stock and also horses of Don-Kirgiz breeding. Numbers of the newly developing breed increased during the years between 1937 and 1953 to 9,100 head. The Don-Kirgiz crosses were of large size, firm constitution, and well-developed skeleton, but they were not as fast as the Thoroughbred-Kirgiz crosses. For development of a fast, firm horse of saddle-harness type, crosses were made be-tween the three foundation breeds, Kirgiz, Don, and Thoroughbred, and the resulting offspring were interbred. Three types devel-oped within the breed: basic, massive, and saddle.

New Kirgiz horses of the massive type are well adapted to highland conditions and are used for meat and milk production. These are the best type within the breed. They are short-legged, massive in build, and have a strong constitution. This is also a good horse under saddle and in harness, and is often used in the mountains as a pack horse. In many ways the New Kirgiz resembles the Don breed.

Due to lack of good conditions of raising and feeding, the number of second-class horses increased, with faults such as a short neck, small size, cow hocks, tender constitution, and poor adaptability to mountain conditions. Fertility was also notably lower.

New Kirgiz horses of the saddle type are poorly adapted to *taboon* raising, lack en-durance, and are demanding as to feed re-quirements. It has been proposed in future breeding plans to have no more than 8 percent of the breed in stallions and mares of saddle type, for use as sporting horses.

Mares of this breed are used to produce milk for the produciton of *koumiss* (fermented milk).

New Kirgiz horses have a medium-sized, clean-cut head with a low neck; good withers; and straight and level back and heavily mus-cled croup. The top line is level; the legs are clean with well-defined tendons. Sickle-hocked legs are common. The primary colors found in the breed are dark—black, brown, bay, and sorrel.

This breed has performed well at hippodrome trials and is a good all-around horse.

Fertility of the New Kirgiz mares is low as a result of excessive addition of Thoroughbred blood, with only about half of the mares delivering foals.

Nigerian
(West African Dongola, Nigerian Native)

ORIGIN: *Nigeria*
APTITUDES: *Riding, pack, light draft*
AVERAGE HEIGHT: *14 to 14.2 h.h.*
POPULATION STATUS: *Common*

The Nigerian is a member of the group of horses often known collectively as the West African Dongola or West African Barb. Horses of this classification are often referred to as "degenerate Barbs" and are known to have various admixtures of Barb blood. The Barb horse was taken to Nigeria by nomadic tribes. It is possible the Nigerian horse has genetic connections to the Poney Mousseye (see breed account) of Cameroon.

Compact in conformation, the Nigerian may be of any common color. The head is small with a straight profile and small, pricked ears. The eyes are large and expressive. The neck is somewhat short; the withers are well pronounced; the back is short and straight; the croup is sloped with a low tail-set. The chest is deep and long; the shoulder is sloping and well muscled; the legs are strong and of medium length with strong joints and well-defined tendons. The Nigerian has a strong hoof. Temperament of the Nigerian is willing and docile, and this horse is strong and enduring.

Noma

ORIGIN: *Japan (Noma County)*
APTITUDES: *Riding, light draft*
AVERAGE HEIGHT: *10.1 h.h.*
POPULATION STATUS: *Rare*

The Noma pony is native to Noma County, just outside the city of Imahari in Aichi Pre-

The Noma pony of Japan. Photo courtesy of the Preservation Club of Noma Horse

fecture. This is the smallest native pony of Japan. The breed originated in the seventeenth century and belongs to the Asian group descended from Mongolian stock.

The Hisamatsu family, the feudal lords of Matsuyama, entrusted the care of these horses to local farmers, who used the animals for transport on the steep hills around Setonaikai and on other islands nearby.

Although the government had made a policy of improving the native breeds, people in Noma County kept their horses pure to preserve them, and in 1978 the Noma Pony Preservation Society was formed. The main purpose of the society is to preserve and promote this small, ancient breed.

As of December 1988, there were only twenty-seven pure specimens living. They are being preserved as part of the local cultural heritage.

Nonius
(Nóniusz)

ORIGIN: *Hungary*
APTITUDES: *Riding, light draft*
AVERAGE HEIGHT: *Small, 14.2 to 15.3;*
Large, 15.2 to 16.3 h.h.
POPULATION STATUS: *Uncommon*

The progenitor of the Nonius breed was an Anglo-Norman stallion named Nonius Senior, brought to Mezöhegyes stud in 1816 and used for breeding for twenty-two years. His sire

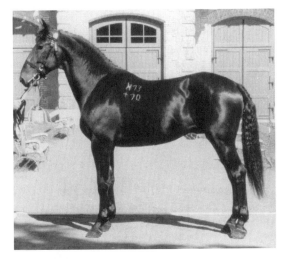

Nonius stallion. Photo: Dr. Csaba Németh

was a Norfolk Roadster. His descendants were interbred. Mares used in development of the Nonius breed were of Spanish-Neapolitan bloodlines, and characteristics of these mares are still evident in the breed today. Later, to correct faults that had manifested in the breed, blood of English Thoroughbreds was used. Initially, the mares were sorted and grouped according to their body size, from small to large Nonius. Aside from the strain bred at Mezöhegyes, a heavy local variety developed in the Municipal Stud Farm of the town of Debrecen, known as the Hortobágyi Nónius (Hortobágy Landrace), which has been bred since 1948 at the Hortobágy state farm. In 1961 the strains of Mezöhegyes and Hortobágy were united.

During the Second World War the Nonius breed was nearly destroyed, and in 1947 only fifty-two mares were found near the border region. By 1954 the number had risen to 120 at the Mezöhegyes stud farm.

Continuous breeding of this horse has produced an animal of vigorous constitution and large frame, suitable for any agricultural work. When it is crossed to the Thoroughbred as the maternal line, an excellent stock for sport horses is produced.

The Nonius is a medium to heavy harness type, powerful, robust in appearance, with a large head that is proportional to the body mass. The head is usually convex or Roman-nosed. In stallions the neck is well arched; in mares it is straight and muscular, of medium length; the withers are low and wide; the back is somewhat soft and wide; the croup is described as a wide, muscular "melon" croup; the shoulder is medium to long, very muscular, and somewhat straight; the joints are strong and large; the cannons are short and wide; fetlocks are short and tight.

Colors found in the Nonius breed are black and bay. The Nonius is a strong, steady worker with a solid constitution, a breed most adapted to heavy terrain conditions typical to the Nagyaföld (Great Hungarian Plains). This breed is easily managed but does require supplemental feeding. Nonius horses have a calm, steady nature.

Recently, the principal breeding aim is increase of the population combined with steady quality improvement. In breed selection, sorting for large frame, pleasant exterior, and good action are determining factors. Candidates for breeding stock are required to pass a working capacity test in harness.

Nooitgedacht

ORIGIN: *South Africa*
APTITUDES: *Riding pony, light draft*
AVERAGE HEIGHT: *13.2 to 15 h.h.*
POPULATION STATUS: *Rare*

Nooitgedacht pony. Photo: Dawn van Wyk

Nooitgedacht ponies of South Africa. Photo: Dawn van Wyk

One of the few indigenous breeds in South Africa, the Nooitgedacht is the only true descendant of the famous Basuto. Basuto ponies declined after the nineteenth century to a point that raised fears of extinction, and the Nooitgedacht is also one of the rarest breeds in the world, with only about 400 in existence.

Over a period of many years a uniquely South African cultural heritage was conserved with dedicated care. As a result, the Nooitgedacht pony is a pedigreed breed, a living monument to the fearless Basuto horses which were once famed far beyond South Africa's borders.

The Nooitgedacht has inherited the courage of the Basuto, its loyal heart, and the blood of its foundation Oriental horses. In its mountainous, inhospitable home in South Africa, careful breeding had brought about an entirely new breed noted for stamina and staying power; a breed that can hold its own as a farm and riding horse; and that is willing to serve and has a loving temperament guaranteed to conquer the heart of any child.

Inherent affinity for people is a legacy from the Basuto and was also characteristic of the Oriental and Boer (Cape) horses from which that breed descended. The gaits of the Nooitgedacht are natural and comfortable. Some are natural pacers and "tripplers." In spite of their small size they are quite capable of carrying a large man comfortably over long distances. These ponies are noted for their remarkable surefootedness, also inherited from the Basuto, and for their virtually perfect legs

and hooves. Natural selection in the rugged Maluti Mountains, plus careful breeding, has resulted in an animal with exceptional stamina and hardiness. It does not tire easily. It requires very little care and is capable of living off the veldt. The weight-carrying ability of Nooitgedacht horses seems completely out of proportion to its size. It is known to be able to travel with burdens on a regular basis, eating only what it can pick up on the veldt, sixty to eighty miles in a day.

Development of this breed began in 1951 when the South African Department of Agriculture bought a nucleus group of Basuto-type ponies. In 1952 a breeding project began on the Nooitgedacht Research Station. This was done primarily to save the famous Basuto from extinction but also because South Africa needed an indigenous farm and riding horse adapted to conditions of the region.

A departmental committee consisting of Professor F. N. Bonsma, Professor J. H. R. Bisschop, and a Mr. Lategan carefully selected horses to become founders of the Nooitgedacht breed. The stallion Vonk and six mares were brought from J. A. N. Cloete of Molteno, whose father had had a notable stud before the Anglo-Boer War, and five additional mares were acquired from M. G. Willemse of Memel.

Due to large-scale inbreeding in the past, the breeders at Nooitgedacht had to use extreme care. For this reason, a Boer stallion and a partbred Arab were used to a limited extent in addition to Vonk. The progeny of Vonk dominated the breeding project to such an extent that nearly every member of the Nooitgedacht breed today is a descendant of Vonk through the sire, dam, or both. The foaling percentage was exceptionally high, ranging at 90 percent. So strict was the selection that only one in four ponies was kept.

The success of the project may be attributed to the efforts of W. M. Bosman, who was personally responsible for the horses, Professors Bisschop, and Professor J. F. W. Grosskopf, who attended to technical and scientific aspects, and Drs. N. Barrie and W. P. Grobbelaar, who were at different times in charge of the research station.

Systematic breeding resulted in the gradual

crystallization of a recognizable and uniform type, and breeders were able to start talking about a "breed." It had been the intention eventually to supply breeding stock of these horses to responsible private breeders and, in 1967, eight such studs were established as part of the project.

In 1968 the Nooitgedacht Breeders' Association was formed by participants in the project. Breeding standards were established as guidelines for breeders and judges at shows. According to these standards, the Nooitgedacht should ideally be a good riding horse with outstanding stamina. Marked features are a compact build, short back, and well-sloped shoulder. The animal should be goodnatured and intelligent. This pony is extremely hardy with good bone structure, good joints, and excellent hooves which seldom require shoeing.

Satisfied that the development task had been completed, the Department of Agriculture Technical Services sold all of its horses by public auction in March 1977. The breeding project was then in the hands of the breeders' association, which had some sixty members in 1979.

Each year the breeders' association organizes a Nooitgedacht Day to allow breeders to get together and exchange views. Lectures and demonstrations are presented on aspects of breeding and care.

It was a joyous occasion for every South African horse lover and particularly for those breeders who had worked tirelessly for so many years, when the Nooitgedacht pony was affiliated with the South African Stud Book Association in 1976 as South Africa's first indigenous pony breed. This bore testimony to the fact that the breeders had succeeded in their aim to keep the good qualities of the Basuto alive and to make available a truly South African–bred farm, riding, and children's horse.

Noriker
(Noric)

ORIGIN: *Austria*
APTITUDES: *Heavy draft work*
AVERAGE HEIGHT: *15.1 to 16.3 h.h.*
POPULATION STATUS: *Common*

Noriker. Photo: Evelyn Simak

A Noriker team owned by the Austrian government. Photo: J. Griebner

Noriker horses have been bred for approximately 2,000 years in the Alpine region and foothills of Austria. In the Tyrol this horse is known as the Pinzgauer (usually the Pinzgauer is spotted), and in Germany's Bavarian mountains a lighter version, the South German Coldblood, is bred.

The original homeland of the breed was Thessaly, in northern Greece, where heavy war horses were bred; later, this was carried on by the Romans. This horse came into the region now known as Austria over the Alps during the course of the Roman conquest. Breeding has continued, and due to the influence of the landscape, climate, and soil, the heavy Alpine horse emerged. In the valleys, on the slopes, and in the mountain pastures of this folded range, the ancient Roman horse developed into an adaptable and peaceful animal of good balance with heavy-bred character.

Four hundred years ago, breeding that had been rather arbitrary was submitted to strict regulations. A stud book was founded, instructions for selection were issued, and stud farms as well as covering stations were established. Pure breeding of the Noriker was declared a principle, the environment and purpose of the breed determining the breeding goals.

Use in Alpine terrain demanded strict observance of the following objectives: Breeding of a healthy, medium-weight, heavy-bred mountain horse of harmonious measurements in breadth and depth and sufficiently strong bones and joints; with correct actions, maneuverable and surefooted; peaceful, industrious, and enduring; calm-tempered; and with good cross-country performance.

In the course of time, breeding of the Noriker has been fixed on the most valuable bloodlines: Volcano, Nero, Diamond, Schaunitz, and Elmar. Today this horse is the only heavy breed in Europe still living in an untouched and natural breeding area, centering on the Austrian federal states of Salzburg and Carinthia. At present there are approximately 8,000 Noriker horses kept in Austria, and the breeding area has over 3,000 stud book mares.

Measurements, weight, and other physical and biological properties of the Noriker today represent the horse that to a high degree meets all requirements demanded of a modern, commercial horse. The Noriker differs from other heavy or draft breeds in several ways. It has quality, nerve, and personality, and it is a qualified work horse. It has a long stride and lively trot, is surefooted, and has a very good, strong hoof.

Due to the rough environment and special requirements of use in forestry and as a draft and pack animal in the mountains, there was selection for contentedness, toughness, endurance, and tame character.

The difficult breeding conditions high in the mountains, in which horses are kept on pastures above 6,000 feet in altitude without shelter or additional feed, have given them a toughness and quality and the necessary basis for fertility and endurance. This breed is very agile, with quick reactions. Not as heavy as the Percheron or Belgian, the Noriker is best suited for work in the Alpine forest. During the past few years there has been an ever increasing demand for Noriker horses, especially for lumber work in the Alpine regions of southern Germany. At times there is no alternative, as heavy machines cannot climb the steep terrain. Working with horses is not only more economical, at least in certain areas, but also more ecologically sound. Horses neither pollute the air nor damage trees and soil as heavy machinery does.

Given their small hooves, Norikers are very surefooted in rough terrain. Plodding and snorting, they haul their loads of many tons with the bells on their collars tinkling rhythmically. They obey spoken commands in a split second. One step too many, or one movement too soon, and the horse might well lose its footing and its life.

Thanks to the wide range of possibilities for use as a work horse, parade horse, for equestrian acrobatics, and in meat production, the Noriker is still very popular in Austria. This horse also renders excellent service as a riding horse, as is demonstrated by the large number of Noriker riding groups. Interested people are invited to pay a visit to the breeding regions and see for themselves the vigor of this breed, which continues to find friends and buyers even in our technical age.

Northeastern

(Nordestino, Crioulo Brasileiro, Curraleiro, Sertanejo)

ORIGIN: *Brazil*
APTITUDES: *Riding horse*
AVERAGE HEIGHT: *13 to 13.2 h.h.*
POPULATION STATUS: *Common*

For many years scientists and breeders of northeastern Brazil have fought for preservation and selective breeding of the Northeastern or Nordestino horse, which is viewed as a special feature of the genetic heritage of the region. While small, the Northeastern is a direct descendant of the desert horse of North Africa and should not be classified as a pony in any sense of the word. The Marajoaro is a variety of this breed.

Northeastern. Photo: Antonio Eurico Vieira Travassos

Northeastern mare. Photo: Antonio Eurico Vieira Travassos

Today many feel that "bigger is better," yet long stretches of world history were written beneath the feet of small and heroic war horses—many strains of which are now on the verge of extinction. To be small is not necessarily bad—it may even be a virtue. There are many things a large horse cannot do.

The American, Frank T. Hopkins, who took part in more than 400 endurance and distance races including some of 1,800 miles in the United States, one of 3,000 miles in Arabia (winning on a pinto Spanish Mustang named Hidalgo), and several rides of over 1,000 miles, scarcely losing one of them, said

he preferred small animals that were not too fiery because the large ones were weak in comparison and the spirited ones too easily exhausted.

The Anglo-Boer War in South Africa was a veritable cemetery for English horses and their crossbreds, making it necessary to use small horses from South America. According to General Dumas, animals crossed with the Thoroughbred in the Crimean War and the Indian and Chinese Wars "melted like snow in the sunlight, ere the work began."

In England, in the March of 200 leagues, military horses succumbed, leaving the greater part of the army on foot in 1914, whereas the cheap, creole horses showed prowess, wearing out the enemy without causing the loss of human life. They had come from San Martin in the Andes during World War I.

In the seventeenth century, Brazilian armed forces, called to defend Angola against the Dutch, were to bring with them two horses with an aptitude for war. The small, rustic horses of Brazil became famous for their performance.

The phenotype of the present-day Northeastern horse as well as that of the Barb of North Africa may be "plain," but its valor is recorded in the pages of history. To lose this genetic heritage would be to trample on the past and forget about glory in the future.

It is said "the sun consumes hybrid vigor," and indeed over hundreds of years the Northeastern horse, abandoned in the semidesert of Brazil, grew away from its crossbred appearance and recovered its genetic purity, restoring the virtues of the Barb from which it descended. Today there are many of these horses in a wild, natural state.

Because of recidivist ancestors, Brazilians raise the most rustic (tough, disease-resistant) goats, mules, and horses in the Western world. The animals face death continually due to harsh living conditions, and the weakest succumb while the strongest restore their ancestral type and survive. Only the purest blood can withstand severe hardships.

In 1534 Doña Ana Pimentel, wife of Martim Afonso de Souza, received horses from Madeira and the Canary Islands. In 1535 Durante Coelho began a breeding operation in

Pernambuco. In 1549 Tome de Souza received horses from the Cape Verde Islands. Garcia D'Avila, the legendary "Senhor of the Casa da Torre" (Tatuapara) in Bahia, disseminated cattle and horses through the northeast of Brazil during campaigns to subdue the Indians. Severe drought brought about isolation of this remote region when the Portuguese population fled, abandoning many horses to fend for themselves among the caverns and mountain foothills.

The Northeastern horse was already well established many years before Dom João VI arrived in Brazil in 1807 bringing horses of the Alter breed, which later gave rise to the Mangalarga Marchador, Campolina, and Marchador. It was not an infusion of Alter blood that formed the Northeastern horse.

There may have been some Andalusian blood in the early Northeastern horse, but there was no Arab blood. The horse was a product of the pure Barb and its mixtures with the Marismenha (Camargue) breed and perhaps the Garrano and Sorraia, both of the Iberian Peninsula.

With the advent of the Brazilian sugar industry, the nouveaux riches, masters of artifice, wanting to set themselves apart from the nobles of antiquity as well as from the common people, began to favor larger, more imposing horses. From that time on, the small Northeastern horse would be used only by slaves, mountain people, the poor; it would no longer be hitched to carriages. Gone were the glory days of war and working cattle. It was condemned to anonymity.

The Northeastern is capable of thriving without human protection. It does not know the trough and stall but is born in the dew of night. The horse can extract energy from vegetation that appears to offer none. In the brush country during drought, the horses eat grasses containing less than one percent protein. Other, apparently stronger animals have died beside healthy Northeastern horses in the habitat of this noteworthy breed.

This breed is an example of the horse reduced to its simplest expression. Everything about it is economical and concentrated. It has a solid structure that is impressive, especially considering its meager diet. It con-

tentedly eats what others could not; harmful weeds, cogon grass, bitter grasses.

The Northeastern is a good swimmer, fearless, intelligent, and easily trained. This is a sociable, friendly horse, seemingly incapable of violent reactions or vices. It is bold rather than surly, and rarely rebels. Northeastern horses are capable of running forty or forty-five miles *a day* in rugged terrain and broiling heat. Additionally, this breed has developed an immunity to ticks, which abound in its habitat. The hooves are upright, small, and as hard as iron. The Northeastern horse does not use shoes.

Not very marketable due to its rusticity and small size, the Northeastern is used today to inject endurance into other breeding stock. Its image as a brush country horse is an impediment now that buyers prefer "big and pretty," overlooking the fact that the rugged, small Northeasterns will still be on their feet and traveling heartily when "pretty" has given up.

With a view to increasing the regional breeding stock, some defenders of the breed traveled through northeastern Brazil to obtain good examples. They encountered terrible frustration. In Piaui, where thousands of Northeastern horses are maintained in a free state, the broodmares are mated to the worst stallions, as the most attractive males are castrated and used as saddle horses. Furthermore, the only use for horses in this area is to furnish horse hair for businesses in São Paulo. Thus an entire breed with a splendid history is being decimated. The animals are born hungry and die hungry, their potential as endurance and working animals unrealized.

Some breeders have tried to have the Northeastern declared a national ecological heritage. The breed deserves a chance to improve and become more marketable, even to be used for modern sports. This horse is extremely agile and excellent for light pulling, work in the brush, corral work, traveling on rocky terrain, trekking, horsemanship classes, and rodeo activities. Northeasterns are excellent cow horses, having the inherent stock working ability of their famous Spanish and Portuguese ancestors.

Today breeders are looking for the best crosses to improve the breed while maintaining the many qualities that make this horse valuable. The breeders' association is seeking new people interested in selecting representatives of this worthy breed and has taken the responsibility for the quality of the initial nucleus. It is also trying to obtain backcountry animals that are good representatives of the breed, to restore the adventurous animal of yesteryear in the greatest possible numbers.

Northeastern horses are found in many colors, but white markings are minimal. There are many duns in the breed; chestnut, grey, sorrel, bay dun, black, red roan, bay, brown, white, and grullo also occur.

Presently, 40 percent of Brazilian horses are located in northeastern Brazil—approximately seven million head.

Northern Ardennais
(Ardennais du Nord)

ORIGIN: *France*
APTITUDES: *Heavy draft and farm work*
AVERAGE HEIGHT: *15.3 h.h.*
POPULATION STATUS: *Common*

Northern Ardennais. Photo courtesy UNIC, Paris, France

Belonging to the Ardennais family, the Northern Ardennais or "trait du Nord," with the Boulonnais and Belgian, played an important role in determination of the breed. The Belgian draft horse gave weight, height, and additional strength. The Boulonnais brought elegance and energy. Around 1910 the Northern Ardennais became an independent breed under the name Northern Draft Horse, which was changed in 1965 to Northern Ardennais.

Origin of this breed is French Hainaut, which in itself is a continuation of Belgian Hainaut, from which the draft horse was imported. The breed was soon to be found in the west and south of this region, in the fields of Thierache, and there gained solidity and hardiness. Soon, this horse displaced the lymphatic and heavy horse in the Flemish lowlands. In this fertile and muddy terrain it developed greatly and, thanks to good pastures and grain feeding from an early age, it soon became superior to animals raised in Belgium.

Like the Ardennais, origin of the Northern Ardennais traces to the ancient big horse of central Europe. Strongly built, rather short, powerful, with a bone structure and musculature that are important for the work it does, this breed may be bay, roan, chestnut roan, and more rarely, chestnut. The head is small, often dish-faced, with a lively eye, short ears, and wide nostrils. The neck is medium in length and powerful; the withers are deeply set in the musculature; the back is straight, short, and well muscled with wide loins. The breed is double crouped and has a wide and muscular chest. The limbs are bony and strong with short cannons and pasterns, which have abundant hair. The hooves are wide and solid.

Of a size and morphology more considerable than the Ardennais, the Northern Ardennais is strong, resistant to disease, and quite energetic. The breed is used on the Northern Plains for grain or beet culture and today has also become a meat animal, yielding a heavy carcass.

Northlands Pony
(Nordland, Lyngen Pony)

ORIGIN: *Norway*
APTITUDES: *Riding pony*
AVERAGE HEIGHT: *12 to 14 h.h.*
POPULATION STATUS: *Rare*

The Northlands or Lyngen pony is a rare breed. There is strong disagreement as to the exact origin of this ancient pony. Recent re-

Northlands pony. Photo: Inger S. Andreassen

search has revealed that it came to Norway from the east very early. For various reasons it was forced northward where it lived and developed through the centuries, but after World War II the breed was at the door of extinction.

This pony was given several local names because of varied environmental conditions and scattered, locally centered breeding districts in northern Norway. The breed developed differently in different districts through crossing with other breeds. Some districts were isolated, having little contact with the outside world, while others, such as northern Troms, had a great deal of contact across frontiers to the trading center, Skibotn.

Some are of the opinion that the Lyngen Pony is the result of a mixture of two different types of the same race—one type from Lyngen in northern Troms and one from Nordland, based on individuals from the district of Lofoten. The Lyngen type was somewhat larger and stronger, mainly chestnut, whereas the Nordland type was smaller, with a heavier mane and tail and a wider variation of color. The distinction is less evident today because of crossbreeding between the two types. However, there still are some individuals that are typically one or the other. Breeding of the race was resumed immediately after World War II, but it was a difficult task due to poor economy,

sparse population, and a shortage of animals.

When inbreeding conditions became serious in Troms, a Finnish stallion named Viri was brought in to infuse new blood. This was done with the approval of the Ministry of Agriculture and after extensive genetic blood tests of potential breeds. The testing gave proof of an opinion held earlier—that the Lyngen pony is related to the Finnhorse. The stallion Viri was used only in Troms and northern Nordland, whereas his offspring Silver 99 and Monark 115 were used in the south toward the end of the 1980s. Today the breed numbers approximately 1,500 animals, mainly in the south of Norway. In spite of difficult breeding conditions, there are a few of this breed outside of Norway.

The Northlands pony is robust and stocky with exceptionally strong legs and hooves; the head is well shaped with a straight profile, small, upright ears, and good width between the eyes; the croup is often sloped, the back line is good and strong; the rib cage is large and well sprung; the shoulder is well slanted. Conformation of this pony is generally good overall. All colors occur and all colors are accepted with the exception of dun and pied. In the area of Troms the chestnut color dominates.

Easily trained, willing to work, energetic, and good-tempered, this pony is well thought of. The breed is frugal, with its instinct for survival intact, fattening up in the summer in anticipation of the long, cold winter. The breed is robust and healthy, often reaching an age of thirty years. It retains the ability to breed until very old. The Lyngen is a splendid family horse.

Used in earlier times as a farm horse in agriculture, being small, frugal, and clever at getting around in snow and on rugged ground, the Northlands pony proved to be excellent for northern Norwegian conditions. Today this pony is mostly used for riding or driving or as a pack horse on mountain trips—and as an animal simply for enjoyment. The Northlands pony has also attracted favorable attention in jump racing and show riding.

North Swedish Horse

(Nordsvensk häst)

ORIGIN: *Sweden*
APTITUDES: *Heavy draft, farm work*
AVERAGE HEIGHT: *15 to 15.3 h.h.*
POPULATION STATUS: *Common*

North Swedish Horse. Photo: Kurt Graaf

As a breed, the North Swedish Horse is only ninety years old. This breed and the Døle horse of Norway are of the same origin, descending from the ancient Scandinavian native horse. During the nineteenth century some cross products of light as well as heavy breeds were temporarily bred into the North Swedish horse, including the Friesian, but on the whole, the breed has remained pure. Systematic pure breeding has been carried on since the turn of the century, using the population fragments that had been spared crossbreeding in the northern parts of the country and stallions of the Døle breed brought from Norway.

From its beginning in 1903, the Stallion Rearing Institute of Wången in Jämtland became very important in the history of the North Swedish Horse. This institute still raises North Swedish working stallions of high quality, though on a lesser scale than in former times. Approximately fifteen of the best colts found in Sweden are bought by Wången every autumn. More than 700 stallions, among them the leaders in building up the breed, have been reared at Wången over the years.

The North Swedish Horse is a medium-sized coldblood. The mares, according to statistics gathered over many years, usually have a height of 15 hands and the stallions about 15.2. The movements and energy are more like those of the Thoroughbred than of the heavy coldblood. A certain increase in size and some coarsening has been sought and brought about, although not at the expense of good qualities—energy, durability, stamina, and respectable longevity. All colors are represented in the breed.

The breeding area for the North Swedish Horse, which for a long time was mainly the county of Norrland, has been expanded in recent years to include the entire country. The breed's primary work is in forestry, where it has long been considered unrivaled.

Scarcely any other breed in the world has been tested in such a thorough and all-round way as the North Swedish Horse. Apart from the customary investigations into constitution, most of the stallions for the last thirty years have had their lower extremities X-rayed regularly. The draft test was introduced at Wången in 1928 and is still implemented. Stallions and mares are tested every summer at the County Horse Days, in cross-country work and sometimes in the breed association's specially built wagon, which has a measuring device. Great attention is paid to the temperament of the horses. For many years, information on the fertility results of all the stallions has been collected and studied.

The stud book for the breed was started in 1909 and has been an incentive for the undeniable and rapid success of the breeding work. *The Stud Book for the North Swedish Horse* is published annually by the North Swedish Association, which was started in 1924. The association tries to stimulate and raise the quality of the breed with various measures. Starting in 1949, it has arranged Country Horse Days every year, at which performance tests with stallions and mares completely dominate the program.

Two types are found within the North Swedish Horse; the draft horse and the trotter (see following breed account). Those interested in trotters wanted a light horse which had the capacity to trot fast for racing purposes, while others had a passion for a heavier horse able to compete with the Belgian breed.

The North Swedish Horse has clean, sound bone and is noted for endurance. In proportion to the amount of work it is capable of doing, it requires a small amount of feed. Longevity in the breed is remarkable. Diseases are seldom found in the breed. In breeding, the importance of good action is stressed. Besides correct, well-balanced action, energy, power, length of stride, and springiness in the gait are taken into careful consideration.

The North Swedish Horse has a head of average size but somewhat heavy. The profile is straight, the forehead broad; the neck is usually short, muscular, and broad at the base with a full and flowing mane; the withers are broad, low, and muscular; the back long and wide; the croup wide and slightly sloping; the tail set low; the chest wide and deep. The shoulder is muscular and sloping; the legs are short with strong bone structure, broad joints, strong tendons, and a bit of feathering behind the fetlocks; the hoof is broad, rounded, and solid.

North Swedish Trotter

(Nordsvensk travare)

ORIGIN: *Sweden*
APTITUDES: *Trotting racer*
AVERAGE HEIGHT: *15 to 15.3 hands*
POPULATION STATUS: *Common*

The North Swedish Trotter is the light version of the North Swedish Horse and has its own stud book, started in 1924. Many stallions of the breed are imported from Norway or are sons of such stallions (Norwegian Døle). As in the case of the Norwegian Døle, the North Swedish Trotter may trace its trotting ability to the Dutch Friesian breed. North Swedish Trotters are exceptionally talented for trotting races and are very fast. Breeding of coldblooded trotters is primarily practiced in the northern part of Sweden.

Norwegian Fjord

(Fjording, Nordbag, Nordfjord, Vestland)

ORIGIN: *Norway*
APTITUDES: *Riding, light draft*
AVERAGE HEIGHT: *13 to 14.1 h.h.*
POPULATION STATUS: *Common*

Norwegian Fjord. Photo: Werner Ernst

Norwegian Fjord. Photo: Judith Hegrenes

Also called Norwegian Dun, Norwegian Pony, West Norway, West Norwegian, Westland, or Northern Dun, in addition to the names shown above, this is an ancient breed and one of the few that has retained its original character. It comes as a surprise to many that initially the original color of the Fjord varied. The first purposeful breeding program in Norway for this hardy breed was begun in the mid-1880s. Prior to this time the Fjord was smaller than at present, averaging about 12.1 hands. At the end of the nineteenth century all crossbreeding was stopped, and the Norwegian Fjord has since been bred pure.

The Vikings used this pony for mounts in times of war, as is indicated by monumental stones. They also used the Fjord for farm work, and they were the first western Europeans to use the horse for this purpose. It was through their invasions of neighboring lands that plowing with the horse became widespread. It is thought that all present-day draft breeds in western Europe are descended in part from this ancient breed. The most distinctive features of the breed are its color and markings. It is one of the few cultivated breeds that is always the wild color, dun. The primitive markings of dorsal stripe (which extends through the middle of the forelock and tail) and zebra striping on the legs are always present.

The Fjord is known to have an exceptionally good temperament and excellent ability in rugged terrain; it is a perfect mount for those who need a horse they can trust. Before mechanization pushed the horse aside, the Norwegian Fjord was used in farming and forest labor, for transport, and as a pack horse. Presently the breed is a reputable leisure horse and good for all-around use, including jumping.

The mane of the Fjord is coarse and tends to stand straight up for a few inches; for many centuries it has traditionally been trimmed short to maintain this upright appearance.

The Fjord has a nicely proportioned head with a concave profile, pronounced jaw, and great width between the eyes. The eyes are large and kind, the ears small, and the nostrils flared and open. The neck is short and muscular; the withers are low and flat; the back is often slightly hollow; the croup is sloping; the quarters are rounded and muscular; the shoulder is very sloping and well muscled. The legs are sturdy and short with broad, strong joints, clearly defined tendons, and a little feather at the fetlocks; the pasterns are long and sloped; the hooves are black, hard, and well shaped.

The Norwegian Fjord has enjoyed popularity in many countries beyond its original borders, including the United States. In the mid 1950s twenty-one Fjords, all champion stock, were imported into the United States. Presently there are over 500 of the breed in North America. The Norwegian Fjord Horse Association of North America was formed in 1977 and registers only purebred Fjord horses.

Oldenburg (Modern type)
(Oldenburger)

ORIGIN: *Germany*
APTITUDES: *Riding, sport horse*
AVERAGE HEIGHT: *16.1 to 17.2 h.h.*
POPULATION STATUS: *Common*

A modern-type Oldenburg. Photo: Prof. H. C. Schwark

The modern-day Oldenburg was developed after the Second World War as breeders of this old and well-liked horse realized that need for the heavy coach horse of the past had declined. Automobiles had replaced the horse-drawn carriage, the railway had taken the place of the horse-drawn mail service, and now the tractor was rapidly replacing the working farm horse. The number of horses in the Oldenburg region fell in the forty years preceding 1984 from 55,400 to around 10,000. To prevent extinction it was necessary to re-evaluate the need for horses.

The concept of the "German riding horse" came into being. This concept, consistently in all warmblood breeding areas, calls for an elegant, generously lined, correct riding horse with spirited, ground-covering movements. People wanted an animal full of spirit and with character, suitable for any kind of riding, particularly sporting competition.

Thoroughbred and Anglo-Norman blood

were introduced to refine the breed, and today the Oldenburg is an all-purpose riding type, considerably finer than its coaching ancestors but still with a fairly high knee action.

Most Oldenburg horses are bay, brown, black, or grey, rarely chestnut. The head is of average size, with a straight or often convex profile, very pricked-up ears, and flared nostrils. The neck is average in length and muscular, well set on and carried elegantly; the withers are well pronounced; the back straight; the croup fairly flat and well muscled. The hind quarters are powerfully built; the shoulder sloping and muscular; the chest deep. For its size, the Oldenburg has short, powerful legs with large joints and plenty of bone.

This breed matures early and has a kind, yet bold nature.

Oldenburg (Old type)
(Oldenburger)

ORIGIN: *Germany*
APTITUDES: *Riding horse, light draft*
AVERAGE HEIGHT: *16.1 to 17.1 h.h.*
POPULATION STATUS: *Rare*

The old-type Oldenburg. Photo: Evelyn Simak

The Oldenburg is one of the best known German warmbloods. Foundation of the breed is credited to Count Johann the Younger, who ruled Oldenburg from 1573 to 1603 and based his horse breeding on the East Friesian horse. His successor, the legendary Count Anton Günther von Oldenburg, (reigning from 1603 to 1667), was instrumental in making the Oldenburg famous throughout Europe, having a good foundation from which to work.

He situated the main breeding area in the sandy regions of Geest, setting up stud farms and breeding stations. He built the royal stables and a riding school in Rastede, and completely controlled trade in horses in his realm. He imported stallions from Spain, Italy, and some other countries in order to upgrade the native stock. The result was a horse very much like the Kladruby in appearance. Ultimately, he presented horse experts in many countries with a "beautifully built and colored riding horse" and a somewhat heavier, yet elegant carriage horse for four-in-hands. The dappled greys, chestnuts, blacks, and ermine and dun-colored animals of Oldenburg origin were sought after and admired in all of Europe.

Anton Günther took over not only the foundation for horse breeding from his predecessor, but also the practice of giving valuable animals to important people to win their favor. For the Oldenburg count, this was a game of diplomacy that paid off well for his subjects.

During the period of Danish rule, from 1667 to 1773, breeding of horses was not a priority and became a matter of quantity rather than quality. Nearly 10,000 horses were counted in the marshlands and 6,700 head in the Geest in 1784. New breeding goals were not energetically turned to after the Napoleonic era. A certification law was put into effect in 1819, and on July 30, 1820, the first official Oldenburg stallion certification took place.

Horses in this period were described as "altogether nothing but Roman noses" out in the pastures, animals that had "one joint too many in their backs," "toe-hoppers" with necks too short, eyes and ears too small. Only strong blood from valuable sires could remedy the problem, and the Yorkshire stallions Astonishment, Duke of Cleveland, Luks All, and Prince William were imported after 1840.

The desired change did not come about all at once. Common blood and English Thoroughbred at first produced quite a few misshapen animals, and after the first stud book in 1861, many years passed before the Oldenburg horse was again attested to be representative in type. The breed continued to improve.

Around 1880 it was said that the Oldenburg was the only horse breed from which enough quality sires could be purchased. And at last the breed began to prove its true worth—in agriculture, in the Oldenburg cavalry, and in the horse-drawn mail service that ran between Oldenburg and Bremen, "one of the fastest connections in Germany" until construction of the railway in 1867.

In the *Brockhaus Encyclopedia* of 1898, the Oldenburg is described: "As a carriage and heavy cavalry horse, the Oldenburg horse is esteemed. Strong and more than middle-sized, with a well set neck, straight back, wide chest and strong thighs." A horse expert added, "The whole animal gives the impression of massiveness and power and at the same time nobility and refinement." The breed was also reputed to be easy to handle and light on feed. The word spread. Demand came from as far as North America, where the Oldenburg was known simply as the German Coach.

Up until the First World War, breeders of Oldenburg horses continued to breed the heavy-type carriage horse. The cavalry was enthused with the breed, as the Oldenburg proved itself in harness and mounted units; however, the war caused great losses in the breed.

After the Second World War, the Oldenburg once again faced a new beginning. Automobiles began to take the place of carriages; the heavy work horse was needed in lessening numbers for intensive farming; and a "new Oldenburg" was needed, a horse designed more for competition, riding, and light farm work. Today the old type of Oldenburg is nearly gone in Germany. Denmark, on the other hand, still breeds the original type of Oldenburg horse, which has had infusion only of Thoroughbred blood.

Oriental Horse

ORIGIN: *Middle East*
APTITUDES: *Riding horse*
AVERAGE HEIGHT: *Ranging from 11 to 16 h.h.*
POPULATION STATUS: *Common*

Obviously there is some confusion concerning the horses known as Oriental, and it seemed advisable to seek out the best authorities possible. Many readers will be familiar with the name of Louise Firouz of Teheran, Iran, known for her important work concerning the ancient Caspian (see breed account). Mrs. Firouz has bred Oriental horses for many years and kindly wrote an article for this book, which is presented here. Mrs. Firouz also provided valuable information regarding the Turkoman and Akhal-Teke (see breed accounts).

A VIEW ON THE DEVELOPMENT OF THE ORIENTAL HORSE BY LOUISE FIROUZ

The accompanying chart was based on the results of bone studies done on equid remains recovered from archaeological sites in Iran. They were examined and classified by archaeozoologists working mainly with faunal deposits from sites excavated by the British Institute of Persian Studies. Except for the drawings at the bottom, the graphic representations are taken from contemporary works of art from the periods shown.

For centuries, travelers from Europe to the Orient described the indigenous breeds of horses they encountered with a kind of incredulous awe. It was as if they were observing a new kind of animal. Their vision was filled with light, dancing horses whose eyes sparkled and whose tails streamed proudly behind them. The horses were quick, intelligent, and though full of spirit, tractable.

Soon travelers brought specimens of these horses to Europe and interest in the strangely beautiful creatures spread. At that time the indigenous horses of Europe were of draft type. They were known by the regions in the Middle East from which they had been bought or won in battle and they were collectively called, by some writers, the "Oriental horse."

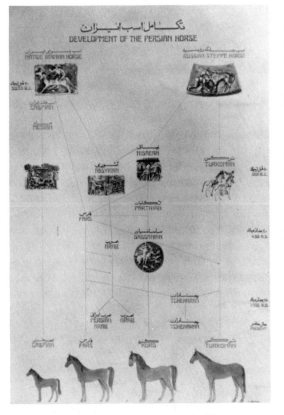

Development of the Persian Horse, by Louise Firouz

Interest did not diminish with familiarity. On the contrary, it flourished as stallions were used on the heavier indigenous European breeds and new breeds were formed based on the introduction of this blood. The Thoroughbred is the best known alloy produced by this mixing and is itself now used to "improve" existing stocks around the world. The trend came full circle when Thoroughbreds were used to improve Arabs and Akhal-Teke Turkomans [see Akhal-Teke] in Iran, once the basic ingredients for the Thoroughbred himself.

So, what is or was the Oriental horse? Lack of solid information has fostered myths and premises that were based on sheer speculation. Theses and books have been written and adherents of this cult or that have waxed eloquent in defense of their theories. Many European travelers have spent months in the desert meticulously describing the tribes and their horses. But few have actually lived there or done any scientific studies. They observed and sped back to report their findings.

I do not claim to have solved the mystery of the evolution of the Oriental horse because I too share in the limitations of having lived in only certain parts of the Middle East. But I have lived and traveled amongst the Arabs in Jordan, Syria, and Lebanon. I have traveled in Turkey and Afghanistan. And I live in Iran (Persia), which was one of the major routes for early horsemen from Central Asia and a melting pot of different civilizations. I have observed and bred horses that are direct descendants of some of the original stocks for over thirty years. What follows is based on my observations of the modern animals (which have not been mixed with European imports); the history of human migrations; ecological and geographical conditions; and working alongside archaeologists, archaeozoologists, and students of equine history.

There was a tiny wild horse that lived in the Zagros Mountains of western Iran long before the migrations of the steppe people from Central Asia began. It is possible that it also lived in other parts of what is now known as Iran, south of the great Alborz chain of mountains which separates plateau Iran from Central Asia in the west (see Caspian). The only two species of equids indigenous to the forested hills and the dry savannah were the onager (wild ass) and this tiny horse. The former preferred the open stretches of grass and desert, and the small horse flourished in the oak forests of the well-watered Zagros.

The Zagros constitute a sharp barrier separating the Iranian plateau from the Iraqi plains. Hot winds moving east strike the rocky peaks and hurl rain and snow, making lush and fertile woods and pastures. The horses roamed with deer, lions, wild boar, bear, and wild cattle. There were wolves and leopards and foxes and jackals. It was a biotype that could support a rich environment for wildlife. It could and did support a type of Neanderthal hunter-gatherer who lived in caves and gleaned a livelihood from the surroundings.

On the plains along both sides of the Euphrates River, in what is now Iraq, settled agriculture first started. The inhabitants domesticated animals as well as hunting them. In addition to keeping sheep and goats they tamed their species of onager *(Equus asinus hemipus*) and also what they called *anse-kur* ("ass of the mountains"). Evidence of use of the onagers is best known from the Standard of Ur, which shows onagers pulling chariots. The little horse (or ass of the mountains) is represented on numerous Sumerian terra cotta plaques from the third millennium B.C., showing small boys riding the tiny horse, sitting well back and controlling the animal with a rope attached to a nose ring and a stick. Evidence has also been found in southern Iran of a form of pedigree or method of counting herds dated to about the same period.

Further to the west, archaeologists have found extensive evidence for the existence of donkeys in Syria, Turkey, and Palestine but no evidence that the horse existed in those areas at that time at all. The first evidence for a horse in the western regions of the Middle East comes with migrations of Central Asiatic peoples in the second millennium B.C.

So, there was an isolated group of horses which had most likely fled the glacial Ice Age cold millennia before humans began the process of domestication. They formed their own genetic pool and fixed specific characteristics dictated by climate and geography. They were not grazers like their Central Asian steppe counterparts. They had become browsers like the deer. They developed attributes similar to those of deer, small, nimble, and quick animals with small hooves and the ability to leap fallen logs.

When the first waves of Central Asiatic people came down through the mountain passes, they brought their effects securely strapped on their sturdy steppe ponies. They overcame the indigenous populations, either mixing with them, or, more likely, conquering them or driving them out. Over the centuries they established farming communities in the Zagros and on the plains bordering the Great Salt Desert of central Iran. They bred cattle, sheep, goats, and horses.

In the meantime, the same changes were taking place in what are now known as Turkey, Iraq, and Syria. The horse had been introduced, settled agriculture permitted urban development, and civilizations began pushing against each other. Selective breeding of horses first began in Persia in the second millennium

B.C., long before areas to the west had either acquired the horse or had different enough types with which to experiment. By the first millennium B.C. in Persia, the Medians and Achaemenians had developed what the Greeks called the Nisaean, a 14-hand, massively developed horse with a vaulting of the forehead that extended down to the nostrils. But they also continued to use the tiny horse of the Zagros, as ceremonial seals show them pulling chariots in lion hunts. Evidence from archaeological sites also reveals a light horse of about 15 hands.

Raids and warfare between tribes and empires kept boundaries fluid and facilitated exchange of blood stock. The Assyrians were particularly fond of charging into the rich pastures of western Persia where the Medes kept their herds of mares, taking the mares back to the royal stables in Nineveh. They developed a light horse standing 14 to 15 hands that some believe may have formed the basis of the modern Arab after the fall of Nineveh in 612 B.C. and the dispersal of the royal herds through what is now Iraq but was then controlled by the Chaldeans and Babylonians.

In 550 B.C. Cyrus, a prince of Persia, defeated the Median king Astyages and created the Achamaenian dynasty. This was the start of the great Persian empires, which would eventually reach from the Aegean to the Indus and would last until 643 A.D. when the Arabs defeated the Sassanians. Although Alexander the Great conquered the Achamaenians in 330 B.C. and burned Persepolis, Greek influence was overcome by subsequent Persian empires created from within or by invaders from the north, such as the Parthians who spread their influence from India to Egypt.

The Parthians also introduced a new element to the makeup of the Oriental horse. They brought with them from their homeland, now called Turkmenistan, a tall, strong horse much admired by the Greeks as well as the Chinese of the Han dynasty in the first century A.D. The Turkoman, or Turkmenian horse, exerted a strong influence on the Oriental types of horses. The caliphs of Baghdad in the eleventh century had guards mounted exclusively on Turkomans, and the Ottomans (thirteenth to twentieth centuries) imported whole villages of Turkomans from Turkmenistan to breed their native horses in Anatolia for mounts for their Janissaries.

By the beginning of the first millennium B.C. the combination of chariotry and cavalry had become a menacing war machine. The success of the Persian empire stemmed largely from brilliant administration and skillful use of carefully bred horses. The Greeks were fulsome in their praise of the Persian horses and left careful descriptions of their different breeds.

The Persians moved with their horses from Central Asia and India to the northern shores of Africa and into Greece. The effect this had on local horses was far reaching and created a relatively homogenous type which eventually became known as the Oriental horse.

When the Arabs embarked on their conquests in 632 A.D., they occupied lands that were fertile and well watered and suitable for raising horses. They supplemented their camel cavalry with horses, and eventually their desert-bred mounts came to symbolize for the western eye the epitome of the Oriental horse. But a glance at a map of Arab conquests by 750 A.D. shows that they occupied an area from Libya to India and had the pick of remounts and breeding stock from that vast and varied terrain.

The Crusades, from 1096 to 1291, brought thousands of British and European people in contact with the Arabs and their horses. There is little doubt that a number of Crusaders acquired horses from the Arabs by one means or another and returned home with them. By the time of the Crusades, overland trade routes were well established from England and Europe through Russia to China, and through Turkey and Persia to China. Again, horses were moving and mixing. In the first century A.D., the Han dynasty of China had already imported quantities of the "blood sweating" horses from Turkestan.

Thus, the Oriental horse is a mixture based on Central Asiatic imports from the second millennium B.C. and bred to different degrees with the native horse of ancient Persia as well as with later importations of selectively bred horses from Central Asia. A more or less homogenous group of horses evolved from Afghanistan through North Africa. Because of environmental differences and the effects of different diet, local conformation varies according to the region: the Arab, with variations within the strains, light boned and light bodied; the Kurd, sturdy and more massive; the Turkoman, tall with dense, light bone; the Caspian, relatively unchanged from its original wild state. They all share a high degree of intelligence, stamina, and ability to withstand both severe heat and cold.

Orlov Trotter
(Orlovskaya rysistaya)

ORIGIN: *Former Soviet Union (Khrenov Stud)*
APTITUDES: *Riding, draft, trotting racer*
AVERAGE HEIGHT: *15.1 to 17 h.h.*
POPULATION STATUS: *Common*

The Orlov Trotter was developed by Count Alexius Girgorievich Orlov (1737–1808). Or-

The Orlov Trotter. Photo: Nikiphorov Veniamin Maksimovich

Racing the Orlov Trotter. Photo: Nikiphorov Veniamin Maksimovich

Russian troika pulled by Orlov Trotters. Photo: Nikiphorov Veniamin Maksimovich

lov had assisted in the conspiracy to depose Peter III and gain the Russian throne for the Tsarina Catherine the Great. For his assistance he was made commander of the Russian fleet. He soon won an important battle over the Turkish navy, and on this occasion the Turkish admiral presented him with an Arab stallion named Smetanka.

This stallion was bred to several Danish and Dutch Harddraver (Friesian) mares. The stallion Polkan was produced, and when put to another Friesian mare the first important sire of this new trotting breed was born in 1784, named Bars I.

When the breed was being established and Bars I was in use at Khrenov Stud, there were seventy-seven mares of various origins there; ten Arabs, two Persians, three Caucasians, one Don, thirty-two English Thoroughbreds, five Mecklenburgs, and one Andalusian. Bars's granddaughters, however, played the most decisive role in establishment of the breed. The number of purebred Orlov Trotters changed according to market conditions as well as social and economic factors. Significant damage was caused to the breed by uncontrolled crossing with the American Trotter during the period from 1885 to 1913, as well as by the First World War and the Civil War. After this it became necessary to reestablish the breed.

The modern Orlov Trotter is distinctive in type and conformation. The head is well proportioned and clean, the poll long, the jaws broad; the neck is long and muscular and often set high. The withers are of medium height and length; the back is long and flat, sometimes slightly dipped; the loin is of medium length and flat; the croup is straight and nicely rounded; the chest is wide, of medium depth, and the ribs are well sprung. The legs of this horse are properly set with good joints, though occasionally somewhat coarse. The forearm, cannon, and metatarsus are medium in size, the pasterns often short and quite straight. The limbs are sometimes feathered. Predominant colors are grey, bay, black, and chestnut.

The Orlov Trotter is very adaptable to either pasture or stable management, and this has contributed to the spread of the breed to many areas of the country as well as to the development of specific lines. The Dubrovski, Novotomnikov, Khrenov, and Perm stud types have emerged, all of which have distinctive features. The Khrenov is the most popular standard type, usually described in textbooks. The Dubrovski type is distinguished by smaller size, fleshiness, and a somewhat more primitive, solid build. The Novotomnikov is known by its clean build, prominent Arab-type features, and more rapid maturity. The Perm type is the largest and fleshiest type, with a coarser build characteristic of carriage horses.

The Orlov is a very fertile horse with eighty to eighty-five live births per hundred mares. The robust constitution results in a long life span, and this breed is used for improvement of other breeds.

Having gained fame as a racing trotter in the nineteenth century, the Orlov has given way to the French and American trotters, which are faster. Nevertheless, this breed still competes today on Russian tracks in races reserved for the breed.

The Orlov consists today of twelve sire lines and sixteen mare families. The best studs are at Novotomnikov, Perm, Khrenov, and Altai.

Palomino

ORIGIN: *United States*
APTITUDES: *Riding horse*
AVERAGE HEIGHT: *14.1 to 17 h.h.*
POPULATION STATUS: *Common*

Palomino. Courtesy of Palomino Horse Breeders of America

The characteristic palomino color is that of an untarnished gold coin, varying from light to dark. The mane and tail are always white. For registration, the eyes must be dark or hazel, never blue, and both the same color.

Technically, the palomino is not so much a color as the result of *lack of color* due to action of the dilution gene. The best results in obtaining palominos have been the mating of palomino to chestnut. Palomino to bay often produces a buckskin (without dun factor markings), and palomino to palomino often produces a cremello (often erroneously called albino).

Selective breeding has produced excellent quality horses in both registries for palomino horses in the United States. Many of these horses are also registered in other registries, such as the American Saddlebred, Morgan, Quarter Horse, and others.

Palomino is a horse color and not a breed, and the palomino did not originate in the United States. The golden horse with white mane and tail has existed for centuries in many parts of the world and arises in the myth and legends of many countries. Still, the United States originated the registry associations for horses of this attractive golden hue.

There are various theories regarding its name. Palomino is a common family name in Spain. Possibly, a family named Palomino owned horses of this color that were called *caballo de Palomino*—the horses of Palomino—and the name stuck. Another suggested theory is that the name came from *paloma,* the Spanish word for dove, or pigeon, and yet others feel the name comes from that of a golden Spanish grape.

Palomino horses occur in many breeds but never in certain breeds, such as the Thoroughbred, Sable Island horse, and some others. The palomino color requires that the color dilution gene be present. Although rare, palominos do occur in the Arab breed.

Called *isabella* by the Spaniards, for Queen Isabella, the palomino horse came to the Americas with the conquistadores in the sixteenth century and was particularly popular in Mexico and California. Recently there has been a resurgence of interest in horses of this color.

Panje
(Panjeskaya)

ORIGIN: *Poland, Former Soviet Union*
APTITUDES: *Riding pony, light draft*
AVERAGE HEIGHT: *13 to 14 h.h.*
POPULATION STATUS: *Uncommon*

The Panje is a descendant of the Konik of Poland and a member of the group of small horses including the Hucul, Albanian, and others of the Carpathian Mountains. When the Konik became nearly extinct, the Polish government gathered the remnants of the breed and established preserves for these closest descendants of the wild Tarpan.

The Panje may be thought of as horses with Konik blood but not in great enough percentage to earn them inclusion in the Konik breeding program. This small horse has been crossed to Arabians, trotters, and warmblood stock of Trakehner extraction as well as to Belgian heavy draft to increase size and agricultural working ability. At the same time, efforts were made to retain the valuable characteristics of frugality and endurance and the robust constitution of the Konik.

Mechanization of Soviet agriculture reduced

the need for working horses, and the Panje lost its importance. Presently the breed is primarily employed as a saddle horse or for light transport work, particularly in periods of unfavorable weather. The Panje is distributed throughout the west of the former Soviet Union, large areas of the Ukraine, and in eastern Poland.

The original Panje was about 13 hands, while the improved version is usually about 14.1 hands. Unlike the Konik, which is exclusively dun or grullo in color, the Panje is found in nearly all common horse colors.

Paso Fino

ORIGIN: *Puerto Rico*
APTITUDES: *Riding horse*
AVERAGE HEIGHT: *14 to 15 h.h.*
POPULATION STATUS: *Common*

Paso Fino. Photo: Jacqueline Garamella

Paso Finos are born, not made. For this reason, they are a breed apart. Other horses cannot learn to do the *paso fino* ("fine step")

gait, which is innately transmitted from generation to generation.

The now extinct Spanish Jennet was famous for its comfortable, ambling saddle gait and the prepotency to pass this gait to its offspring. During research for this volume, I noted an interesting possibility as regards the broken pace gait of the famous Spanish Jennet, quite different from the gait of the typical Andalusian, which is a trotting horse. Many different breeds had been taken into Spain in early times with various army invasions (including those of the Barbarians), and it was the blend of several distinct breeds or strains that resulted in the perfection of the Spanish horse. One type that stands out is the small Asturian of northern Spain (see breed account), so noted for its comfortable amble that it became a favorite of the Romans in early times—the "hobbeye horse" of Ireland and the "palfrey" of Rome and other parts of Europe. This little horse was most likely the source of the special ambling gait of the Jennet. Complaints by Columbus about "common nags" (see Spanish-American Horse) could easily point to the Asturian as well as the Sorraia breed, Garrano, or mixtures of all three.

Considering the broken pacing gait involves an attempt to investigate ancient history. An article by B. S. Dystra in the February 1968 issue of *Equestrian Journal* stated that when the Ice Ages dawned, northern European trotting horses fled south from the marshy regions and the cold, wet, inhospitable habitat to what are now France, Spain, and eventually North Africa. They met with diluvial galloping-type horses and multigaited horses which inhabited lands from the Middle East through North Africa to the Atlantic Ocean. Through interbreeding and natural selection, the prototype of Barb horse arose. Despite theories that the Barb is the result of selective breeding, the Barb is often said to be "as old as the world." There were small Barbs and large, heavy Barbs; trotting, pacing, and galloping Barbs as well as multigaited Barbs.

This article sets forth the same opinion held by some South American authorities who have researched the Spanish horse, that the Paso Fino resulted from breeding crossbreds first from the Spanish Jennet mares (pacers or amblers) to Andalusian stallions (trotters). As the mare is primarily responsible for the hereditary transmission of characteristic gait, the crossbred produce were born pacing horses. However, further inbreeding to Andalusian stallions provided the offspring with the far-famed Spanish *brio condido,* or hidden fire. The sire is primarily responsible for the general conformation of his offspring and for the style and gusto with which the offspring executes its characteristic gait. *Brio condido* means plenty of energy and spirit, held in check by intelligence, great levelheadedness and "horse sense," patience, and self-restraint. If called upon, however, the fire is very much present.

It was the "hidden fire," some contend, that was primarily responsible for the emergence of the broken pace. The impulse to move forward is given by the brain and the hind quarters promptly react—a lateral pair of legs is moved forward energetically—but movement is checked halfway by the tendency of self-restraint of the horse in question. The result is a broken pace. Whether or not the above is the case, it is offered for speculation.

Surviving difficult sea conditions, the horses of Columbus's second voyage were taken to Santo Domingo (now the Dominican Republic), where they became the foundation stock of remount stations for the conquistadores. Progeny of these hardy animals spread throughout the Caribbean as the Spaniards conquered expanding territory. Horses were first taken to Puerto Rico by Martin de Salazar in 1509. Later, Diego de Valasquez invaded Cuba with eight stallions and mares in 1511. In 1512 Spanish horses arrived at the Isthmus of Panama, and in 1517 Cortes took seventeen horses, including one foal born aboard ship, to Mexico. In 1524 the first horses arrived at Colombia, the start of that noble lineage (see Colombian Criollo). By 1550 many horse breeding and training centers had been established throughout the Caribbean.

Over time a distinctive type of horse began to emerge in the New World environment. Blood of the Spanish Jennet proved prepotent,

and in several regions horses developed that were prized for their smooth riding gaits and hence also as breeding animals. At that time long trips were common, and the demand was for a resistant and rustic horse with smooth gaits. Selection was developed by breeders, working with Spanish Jennets, Andalusians, and Barbs. The result was the breed now known by the name of its natural gait—the Paso Fino.

Different Paso Fino lines were developed in the Dominican Republic, Puerto Rico, Colombia, Cuba, and Peru. Terrain, environment, use, feed, and selection all played roles in formation of the various types found within the breed. Occasionally, these lines were crossed; however, most breeders preferred to preserve the horse of their own country in pure form.

The breeders of Puerto Rico developed a show horse with a great deal of style. This horse traditionally carried more blood of the Spanish Jennet than of the Andalusian. In Colombia a more versatile horse was produced, having in mind the work with cattle. The breeders in Peru developed a horse more related to the social culture of the time, a showy animal with an uncommon movement of the forelegs, which are excessively thrown upward and outward. In the Dominican Republic, the lineage had a less significant development regarding quantity.

Modern Paso Fino horses raised in the above mentioned countries represent breeds that differ among themselves but all have one significant common feature: the four-beat lateral gait, which is very smooth and comfortable to ride. In Peru, breeders have preferred to classify their Paso Fino horses as Pasos Peruanos, not alluding to actual relations with the Paso Fino lineage. In Colombia, many breeders preferred the more elevated and elegant trot of the Andalusian horse and increased Andalusian blood in their herds. The result was production of a type known as Trochadores, elegant horses with a great deal of pride and showing a predominance of movements in rapid triplex set diagonal (four-beat diagonal gait).

In the United States, the Paso Fino was introduced in 1950, coming from Puerto Rico.

Later some animals were imported from Colombia, the Dominican Republic, Cuba, and Peru. From the crossbreeding among these various lineages, the American Paso Fino has emerged, slightly differing from the others.

This volume has included as many as possible of the various Paso Fino horses of Central and South America, because while essentially of the same origin, they are not the same today. Selection for different attributes within any pure breed will produce horses with morphological differences. Some American and other breeds, including the Canadian, were imported to Central and South America in great numbers and had an influence on the native Spanish horses.

General appearance of the Paso Fino horse is set forth here, acknowledging that there are variances among the horses of different countries. The Paso Fino is a horse of medium port (size), natural style, and beauty, with great vivacity. This horse has retained the naturally docile nature of the old Spanish horse. The head is small with a slightly convex profile and large eyes which are spaced widely apart. The eyes do not show white sclera. The lips are firm and well shaped and the nostrils are wide and expanded; the neck is of medium length, upright and arched, and the swan neck is not rare; the shoulders are oblique and deep through the heart; the chest is moderately wide. The withers are definite but not extremely pronounced; the back varies from long to short, but an *extremely* long or short back is cause for disqualification. The legs are straight and delicate in appearance with strong tendons which are well separated from the bone; the hooves are small without excess heel; the mane and tail are encouraged to grow as long and full as possible.

The gait is a four-beat lateral gait, very smooth, with the hind foot touching the ground a fraction of a second before the front foot. The movements of the horse are absorbed in its back and loin, giving the rider great comfort. The gait may be executed in three speeds known as *paso fino, paso corto,* and *paso largo.* In the *paso fino,* the horse travels with slow forward movement but very rapid footfall. This gait is often reserved (in the United States) for the show ring. The *paso corto,* equal in speed to the trot in its relaxed form, is the ideal pleasure and trail gait. It is most elegant, snappy, and exciting to watch. The *paso largo* is the extended form of the gait, and speeds range between those of a canter and a hand gallop, which is also performed with gusto.

Paso Fino horses are gentle at hand, spirited under saddle, and endowed with tremendous individual personality. Generally good converters of feed, they enjoy the company of other horses and are especially fond of the company of their owners.

In the United States, emphasis has been placed on naturalness, and any evidence of abusive training methods results in disqualification in the show ring.

In the 1930s, the president of the Dominican Republic sent a blank check to Genaro Cautiño of Puerto Rico asking that Dulce Sueño, foundation sire of the modern breed, be delivered to the Dominican Republic. The check was returned, still blank. After seeing a Paso Fino horse ridden in a parade in the 1960s, Vice President Lyndon B. Johnson took one home. In 1986 the president of Costa Rica, on a visit to Puerto Rico, took a pair of Paso Finos to contribute to the breed he had started in his country. And the King of Spain acquired a pure Puerto Rican Paso Fino in 1987 for his private stable, bringing the breed full circle back to the land of its origin.

Patibarcina

ORIGIN: *Cuba (Escambray region)*
APTITUDES: *Riding horse*
AVERAGE HEIGHT: *14.3 to 15 h.h.*
POPULATION STATUS: *Common*

The Patibarcina of Cuba resulted from the cross of Cuban Criollo (native) horses with Andalusian and Barb stock. It is a type of the Cuban Trotter, with characteristic color. Native horses of Cuba, originally descended from Spanish stock, were crossed with horses sent from Canada prior to the American Revolution. Records show that hundreds of Canadian horses were sent to Cuba for work on plantations. The cross produced a handsome animal with many Spanish characteristics and talent at the trot.

The Patibarcina has proud bearing and a

Patibarcina of Cuba. Photo: Dr. David W. Cantera

strong constitution, characteristic of all Spanish horses. This horse is well muscled with harmonious conformation. The head has a straight or slightly convex profile, wide between the eyes; the ears are small and pricked; the neck is arched and muscular, fitting smoothly into the shoulder; the shoulder is sloped and long; the back rather long; the croup gently sloped; the tail set fairly high. The quarters are muscular; the legs are well muscled and strong with good joints and well-defined tendons. The hooves are strong and hard.

The Patibarcina is characteristically reddish brown or bay in color, with a dorsal stripe (called *veta de mulo,* meaning "mule's seam"), and zebra striping on the legs and shoulder, called *cebraduras.*

This breed has a good temperament and is easily handled while at the same time possessing great *brio* and pride. Patibarcina horses have a very fast start and show great speed for short distances, reminiscent of the American Quarter Horse. They are strong, are used in agriculture, and are outstanding riding horses, being rustic and having endurance.

Pechora

(Pechorskaya)

ORIGIN: *Former Soviet Union*
APTITUDES: *Riding, draft*
AVERAGE HEIGHT: *13.3 to 14.2 h.h.*
POPULATION STATUS: *Rare*

This horse was widespread at one time in the regions around the polar circle and in the basin of the Pechora River. Living in severe climatic conditions, the Pechora developed hardiness, endurance, and working ability, which made the breed valuable to inhabitants of the region. No breed could compete with this tough horse in its far northern habitat. The Pechora horse is able to work in deep snow and is resistant to severe cold ranging from −40° to −60°C (−76.9°F), even with sparse food. In summer this horse is resistant to the millions of mosquitoes that abound in the area.

The Pechora breed was influenced by horses from the Urals, western Siberia, and Estonia that were imported in the sixteenth century. During the nineteenth century the breed was well-known and was heavily exported to other areas. At the end of the nineteenth century breeding of the Pechora began to decline.

After the Revolution, stud farms were established and the breed was improved through pure breeding, selecting the best and most beautiful specimens. The Pechora is bred near the border of the taiga, which passes into tundra. The climate is severe and there are only about eighty days in a year without frost. Snow covers the ground 190 to 220 days.

This horse is born hardy. Foals are very sturdy and able to follow their dams across rivers at two or three days of age. The breed is rather thickset, reminiscent of trotting horses. Predominant colors are black, dark brown, bay, buckskin, and sorrel. Grey is rare in the breed, and local people always considered grey horses less adapted to the harsh weather conditions.

The head is thick; the withers are low; the shoulders are well slanted; the body is generally long; the croup is sloped; the legs are hardy. The mane, tail, and hair on the legs are thick. During the winter the horses are covered with a dense, fur-like coat.

Other types found within the Pechora breed are similar to the Tavda and Estonian Native. The goal of breeders is to increase size in the breed while retaining the valuable traits which make the horse adaptable to the northern climate. The Pechora is bred pure and is also crossed with Ardennes and trotters.

Peneia

ORIGIN: *Greece (Eleia)*
APTITUDES: *Riding and pack horse*
AVERAGE HEIGHT: *12.2 to 14 h.h.*
POPULATION STATUS: *Rare*

Peneia of Greece. Photo courtesy Aristotle University, Thessaloniki

Developed in the semimountainous regions of Eleia (Peloponnesus), the Peneia is nearly extinct. This is a heavier type than other Greek horses, used for farm work, riding, and transport. The stallions are often used for breeding hinnies.

The Peneia is good-tempered and much admired in northern Greece as a willing, dependable working horse. While the conformation of the breed gives no such indication, this horse is a remarkable jumper.

Main coat colors found in the Peneia are roan, chestnut, and black; rarely bay or brown. The body is broad with a very strong and heavy neck but often appears weak in the hindquarters. The hind legs are often cow-hocked and sickled, characteristics said to cause weakness in the animal, yet the Peneia is a remarkable climber in the mountains, defying its conformation. Often this breed is trained to do a gait called *aravani,* a sort of pace, making it easy to ride. Horses not thus trained can give an exceptionally rough and jolting ride.

When crossed to the English Thorough-bred, the Peneia produces a very useful and tough horse.

Percheron

ORIGIN: *France (Le Perche, Normandy)*
APTITUDES: *Draft and farm work*
AVERAGE HEIGHT: *Varies 14.3 to 16.1 h.h.*
POPULATION STATUS: *Common*

Percheron. Photo courtesy Percheron Horse Society

Percheron. Photo courtesy Percheron Horse Society

As with many good horse breeds that emerged from the "mists of antiquity," the exact origin of the Percheron breed is uncertain. All agree, however, on the great antiquity of the breed. Some say the Percheron is closely related to the Boulonnais horses, which were reinforcements for Caesar's legions in their advance on Brittany. Others claim the Percheron is directly descended from the large black horses that came from northern Europe with Celtic invaders. Still others maintain this old draft breed is the heir of Abd el Rahman's Arab stallions, or that when Charles Martel defeated invading Moors at the battle of Poitiers in 732 A.D., the captured cavalry was booty shared among the warriors from La Perch, Normandy, and the Orleanais, who formed the major part of the French forces. Then there is the theory that the breed comes from mares belonging to the Breton confederations subdued by Clovis after his conversion to Christianity in 496 A.D.

There *is* evidence that a Percheron-type horse was in existence in the area of Le Perche during the Ice Age. Perhaps this horse was simply crossed to incoming stock.

In any event, Le Perche is on the borders of Normandy, Maine, and Beauce, and in this area Arab stallions were mated to "native mares" around the eighth century. Oriental blood was again introduced during the Middle Ages by stallions imported by the Comte de Perche when he returned from the Crusades and expeditions into Spanish territory. Spanish stallions were also imported from Castile by the Comte de Rotrou. Several types of Percheron have been grouped together since that time, from lightweight to heavyweight, which still exist today.

The Percheron is often referred to as an Arab enlarged by the climate and agricultural nature of the work for which it has been used over the centuries.

The breed is usually grey or black. The gaits are supple and light; the head is fine with a square, wide forehead, long, thin ears, a lively eye, straight nose, and wide nostrils; the neck is long with an abundant mane, and the throat thin. The withers are well set; the shoulder is well slanted; the chest is wide and deep with a somewhat prominent breast bone; the back is short and straight and very strong; the ribs are rounded; the girth is low with full flanks; the hip is long as is the croup; and the tail is set high. The limbs are sound and clean, well set with a powerful forearm and wide and muscular thighs. The buttocks are low; the knees wide and straight in line with the shoulder. The cannon bones are wide, flat, and strong; the pasterns are clean and strong and the feet are good.

The "Good Horse Perch" is bred between Mortagne (Orne) and La Ferté–Bernard (Sarthe) in southern Normandy between the Orne, the Eure, the Eure-et-Loir, the Loiret-Cher, and the Sarthe. Its center is the Huisne valley. The area is divided into valleys of rich pasture lands, with limestone subsoil, clay soils, and a temperate climate. The environment has given the Percheron great bone structure, strength, and hardiness.

Breeding of the Percheron has spread far beyond its original areas, including many foreign countries. A wonderful draft horse, able to provide an enormous effort, the Percheron, like other draft breeds in France, is also directed toward meat production.

The Percheron stud book was created in 1893. Since 1966 it has regrouped not only the original Percheron but also the other related branches—the Berrichon draft horse, Nivernais, Marne draft horse, Augeron, Bourbonnais, Loire, and the Saône-et-Loire. This regrouping thus includes all of the breeds influenced by the Percheron. The horses are registered according to ancestry, or for the first time, since 1966. They are then branded with the letters "SP" intertwined on the neck.

The first importation of Percherons into the United States was made by Edward Harris of Morrestown, New Jersey, in 1839. Of the four horses shipped, only one mare survived the ocean trip. Harris went back to France immediately and returned to the United States with two stallions, one named Diligence, and two mares, one of which died soon after leaving the boat. This importation of Percherons did not prove very successful. The venture was not entirely in vain, however, for nearly four hundred foals were credited to Diligence before he died at the age of twenty.

A stallion by the name of Louis Napoleon imported in 1851 exerted considerable influence on draft horse stock in the United States.

During the early 1870s Percherons were imported in large numbers, and the breed has enjoyed popularity here.

Periangan

ORIGIN: *Indonesia (West Java)*
APTITUDES: *Riding, draft, racing*
AVERAGE HEIGHT: *12.1 h.h.*
POPULATION STATUS: *Common*

The Periangan pony is bred on the west of Java and is a good quality animal, obviously the result of crossing the Java pony with Arabs.

Few details were furnished for this breed, but the photos reveal an attractive looking animal. All colors are found in this breed and it is reported to have a lively temperament and kind nature.

For additional general information regarding Indonesian ponies, see Bali.

Persian Arab

ORIGIN: *Iran (Persia)*
APTITUDES: *Riding horse*
AVERAGE HEIGHT: *14.1 to 15.1 h.h.*
POPULATION STATUS: *Uncommon*

Persian Arab. Photo: Caren Firouz

Information dealing with the Persian Arab was received from long-time breeder Mary Gharagglon of Teheran, Iran.

The Persian Arab is one of the earliest known domesticated breeds of horses, dating in Persia to about 2,000 B.C. This means that the Persian Arab is some 1,500 years older than the Arab of Arabia.

Most horses throughout Iran have certain Arab characteristics. In some cases, pockets of horses are to be found that look surprisingly pure, such as in the eastern city of Bam. These horses belong to the Ameri family, originally Arabs, who were sent in the times of the Caliph Osman (the third Islamic caliph) to rule in the area of Bostan, near the present city of Shahrud. Mention is found in some old travelogues of fine horses in the Bostan area. Part of the Ameri family were later punished for some misdemeanor and banished from Bostan to Bam, a small desert city. Today around Bam there are many horses of definite Arab type. Traced, such horses are usually found to be from the Ameri stock. According to members of the Ameri family, until recently it was necessary to import stallions from Iraq or the southern part of the Iranian province of Khuzestan, located near the border of Iraq. Unfortunately, in the recent past stallions of Arab type but unknown lineage have been used in some areas, and the percentage of proven pure blood has lessened.

In the southern province of Khuzestan, the situation was different. The movement of Arabs and their horses was done in large bodies, entire tribes settling in various regions. About 500 years ago the last large group to move in was the Kaab tribe, sections of which still exist across the border. Prior to that time, other tribes had settled and were ruling in various parts of the province, among them the Mir, who had come long before the Kaab, and who were of special importance in the history of two of the better known Arab horse strains of Khuzestan. According to local authorities, the population of parts of the province was of Semitic origin even prior to the influx of the Muslims. There is no doubt concerning the long occupation of this region by Arab tribes for many centuries, along with their culture, customs, traditions, and horses (see Arab).

The strains and substrains of the Arab horses of Khuzestan parallel those of Iraq, Saudi Arabia, Najd, and other Arab nations. The main five are the Koheilan, Saglawi, Hadban, Obayan, and Hamdani. These have substrains, usually similar to those of other countries in which the Arab horse originated, occasionally lacking some substrain or having a new one, as substrains develop through some incident or are designated by the name of a tribe or person. Until African horse sickness hit Iran in the early 1950s, all strains found in other countries of Arab horse origin were also found in Iran (also a country of origin).

In Khuzestan, although there may be several different groups of the same strain, certain people are known for a particular strain and animals that trace back to these families are the most highly regarded. Examples would be the Koheilan of Mojadami, the Saglawi of Ziareh, the Hamdani of Al-e-Kassir, and the Khersan and Obayan Sharak of Mir.

Prior to the horse sickness epidemic and the appearance of motorized vehicles, horses were the sole method of swift travel. Fine horses added to the prestige of the owner, and many horses went from Khuzestan to the stables of the affluent and powerful. Some were taken by force and others were given in exchange for liaisons of friendship, but the horses were never sold. Such were those that were obtained by tribal chiefs of the Bakhtiari, Kurds, Lurs, and Quashqai, all inhabitants of the Zagros range of mountains. Their summer quarters bordered at some point the province of Khuzestan. Unfortunately most of them did not adopt the tradition of breeding Arab horses except for the leading family of the Bakhtiari, who became known for some of the strains they produced. Among them are the O Bayan Sharak, Wadnan Khersan of the Koheilan, and the Nasban, mentioned by Lady Ann Blunt in volume two of *Pilgrimage to Najd;* she called it Nusban, and this was apparently the first time she had encountered this strain.

Bakhtiari breeding began in the times of Nader Shah Afsgar, in the first half of the eighteenth century, with Ali Saleh Khan Bakhtiari; however, it became well-known in the times of Hossein Gholi Kahn Bakhtiar and was mentioned in the same book by Lady Ann Blunt. The breeding of the Bakhtiari continued, following Arab tradition for the breed,

with the exception that the Bakhtiari, who needed stallions as mounts, did not do away with the colts as was the custom with Arabs.

In the beginning of the twentieth century, the stud of Sardar Mohtashem was considered the best in Bakhtiari breeding. After his death, his stud was divided between his sons. Those that went to the son named Majid Bakhtiar, an ardent champion of the Arab breed, gained the highest renown. At his death they were donated to the government to form the nucleus of the stud for the preservation and breeding of the Asil of Khuzestan. In 1976 they were recognized by the World Arabian Horse Organization. Presently, due to upheavals within the country, they are suspended until an inspection team is able to check their continuity during the last decade.

Concerning the horses belonging to tribes of Khuzestan, though the animals were greatly diminished after the horse sickness and have never fully recovered their numbers, recent years have seen a revival in the interest of local people. During the years after the epidemic and up to the beginning of the 1980s, the incentive for the Arab tribes to keep mares was tradition. Thus a family known for a certain strain would keep a single mare or possibly two as a symbol of the past, although they were only bred infrequently and hardly used at all. Within the new generation, interest is increasing in riding as a sport, maintaining a breed originally theirs, and local racing of Arabs.

Presently, Khuzestan owners are conducting a search for the strains still existing, aiming to preserve all of the strains if possible. Contrary to views in the West, they believe in the importance of strains and the differences found among them. Western breeders have disregarded the strain structure as nonexistent or irrelevant from the very beginning, starting with Lady Ann Blunt, and have viewed the Arab horse as essentially only one type. In truth, aside from substrains, strains were to differentiate between the various types, not only in looks but in performance. It is understandable that for those importing Arabs to countries in the West, it would be nearly impossible to maintain strain breeding. The minimum number of horses necessary for such a procedure would be about 100. There are many other overlooked facts in connection with the original Arab in most western writings, which unfortunately were not challenged in the beginning and formed a standard in the West for the Arab that has drifted far from the original.

Some very strange misconceptions were also to be found, again by Lady Ann Blunt. She is used as a reference here because, of the horse-writers, she is the only one to ever have entered Persia. She claimed that the Motafigh (a tribe bordering Iran) had tampered with the purity of their horses for the sake of size as they were selling animals for export to India, and that they had mixed their horses with the Persian Turkoman horse. This comment has little merit, for very shortly afterwards, when defending someone's theory that the horses of the Anazeh (another Arab tribe) had been mixed with infusions of English blood, she valiantly defended them, stating that the finest horse ever run at Newmarket would appear as a mere *kadish* (a horse of little quality) compared to the Anazeh, and that she believed this held true for *all* Arab horse breeding tribes, from Najid, Khuzestan, or elsewhere.

The general rules for breeding in Khuzestan are that mares from the first five strains may be bred to a stallion from any one of the other five strains. They should not be bred to a stallion from the remaining strains with the exception of substrains. In the case of the Jelfan, the Jarjari, originating from the family of Seid Karim, are acceptable. There are some other exceptions. A mare of the lesser known lines can be bred to stallions of the major five, and also to stallions of the lesser known lines. Within a line, a mare must be bred back to her own line every second or third generation, maintaining the specifications of that line. It is preferable for a stallion kept for breeding to come from a sire and dam of the same strain, thus being pure in his strain.

Foals born at exactly eleven months cause suspicion as to their purity as Asil horses. Those born over eleven months, even by one day, used to be done away with immediately. It is expected that a pure Asil mare will foal several days under eleven months, fifteen days being considered correct. In forty years of observation and breeding of Asils by Mary Gharagglon, all but two births have confirmed this feature.

Several offshoots have been developed from the Asil Arab of Iran. One is the Dareshuri, bred by the Dareshuri subtribe of the Turkish-speaking Quashqai tribe in the province of Fars. This breed was made famous by the leaders of the Dareshuri, namely Ziad Khan and Hossein Khan. Toward the end of the 1970s, Ayaz Khan was also important to the Dareshuri. At one time there was interest in propagating and classifying this breed. Claiming Mongolian descent, the Dareshuri people descended from a tribe that had passed through Iran to Syria during the time of Genghis Khan, later returning to Iran. They brought Arab horses of the finest strains from Syria. However, they had not followed the mare strain and were unable to prove Asil blood on the sire's side beyond three generations. One mare was traceable for eighteen generations. The average assured Asil blood was between 75 and 85 percent.

In appearance, the Dareshuri has a good Asil croup and tail and is higher in the leg and shorter in the body than the Khuzestan Asil. The main difference is in the head, where the eye is sometimes small, the jaw bone tight, and the protrusion of the forehead (not a must with every strain) existing but continuing too low. With the death of the two men mentioned above, those once interested in propagation of the Dareshuri are losing interest and this horse may die out.

According to Mary Gharagglon, who has traveled to many of the countries breeding Arab horses in the West and has done a great deal of research on the Arab breed, the Asil of Iran is probably one of the purest small pockets of Arab horses in the world, although it is hardly known outside the country. In hundreds of years, no army accompanied by horses has conquered or occupied their territory. No foreign or, indeed, native cavalry having stallions has been stationed in Khuzestan. During the days of the Indian Army, when horse dealers traveled over three different routes collecting horses to be shipped for sale in

India, not one of the three known horse dealer routes passed through Khuzestan. The Arabs of this region have never had races for their horses, so there has been no reason to mix them to increase their speed. There is no record whatsoever of the *sale* of Arab horses, even within the country. Sale, or money taken for a horse, and especially a mare, is considered extremely dishonorable.

The dishonor of selling a good horse, an Arab taboo in *all* Arab societies, is a puzzling point in view of supposed sales of superior stock to western buyers. And there are other questions concerning the breed, as the Arab owner, who has bred his prized horse for centuries, sees it.

Peruvian Paso. Photo: Evelyn Simak

Peruvian Paso

(Costeño de paso, Pasos Peruanos)

ORIGIN: *Peru*
APTITUDES: *Riding horse*
AVERAGE HEIGHT: *14.1 to 15.1 h.h.*
POPULATION STATUS: *Common*

The diversity of horse breeds throughout the Americas began with the Andalusian, Spanish Jennet, and small pack horses, which spread after the second voyage of Columbus

The Peruvian Paso in Germany. Photo: Werner Ernst

in 1493. The horses from Spain were divided and sent to Santo Domingo, Cuba, and other islands in Central America. The horses multiplied rapidly in those lands and were taken to Peru and Mexico.

The first horses to arrive in Peru landed in 1531 with Governor Don Francisco Pizarro. The Spaniards knew that horses were a prerequisite to the continuation of the conquest. Horses in Peru came from Spain, Jamaica, Panama, and other areas of Central America. From among them were selected prototypes brought by the viceroys from Spain. Two types were established immediately; one quite luxurious with which they practiced the equestrian sports in the riding style of Spain, and the other an "ordinary" type used in the expeditions of the conquest.

After Peru's independence from Spain in 1823, horse imports were recorded of pure Arab, Hackney, pure Thoroughbred, Friesian, and other breeds. The judicious fusion of several Old World breeds provided the foundation for the Peruvian horse, which developed into three main types: the Coastal horse or Costeño de Paso, here called the Peruvian Paso; the Coastal horse developed in high altitude (see Costeño), and the Andean (which has two subtypes).

Unfortunately, the practice of crossbreeding became quite common and modified the characteristics of the national horse so that the Spanish type was not preserved. For a time there was great diversity in type, with undesirable factors appearing in successive generations. It was recognized that improvement would only be possible through good selection in which thought was given to the morphological as well as functional aspects such as toughness and *brio*.

The word *paso* simply means "step" or "gait"; hence, Peruvian Paso is a term commonly used for this breed, indicating a Peruvian gaited horse (see Costeño).

The qualities that make the modern Peruvian horse outstanding are its spirited and soft gait, elegant step, good mouth and sensitivity to the rider, sobriety, rusticity, and the power to adapt to various climatic conditions.

The Peruvian Paso has become quite popular in the United States, supported by a large and growing association. The breed transmits its smooth gait to all purebred foals, and no artificial devices or special training aids are necessary to enable the horse to perform its specialty—a natural, four-beat footfall of medium speed that provides a ride of incomparable smoothness and harmony of movement.

In addition to the easy gait, the people who developed the Peruvian horse desired the breed to retain brilliant action typified by high lift and flex of the knee and fetlock combined with *termino,* a movement of the front legs similar to the loose outward rolling of a swimmer's arm when doing the crawl.

The head of the Peruvian Paso is small and thin with a straight or slightly convex profile. The eyes are rounded and well separated, the nostrils long with a pronounced external side; the mouth is small; the jaws are pronounced and strong, widely separated at the throat. The neck is short, thick, arched, muscular, and flexible, with an abundant, silky mane. The withers are prominent and rounded; the back is muscular and slightly hollow; the loin muscular and long; the croup sloping with well-developed muscles; the abdomen rounded; the chest wide and deep with good muscle structure; the shoulder is long and sloping. The legs are muscular with a short forearm; the cannons are long in the front and short in the back; the tendons are straight and well defined; the pasterns are sloped and flexible; the hooves are short and strong.

Predominant colors in the breed are chestnut, brown, sorrel, black, grey, buckskin, palomino, roan, or dun—animals of solid color are the most in demand in Peru.

For additional information on the horses of Peru see Andean, Costeño, Marochuco, and Spanish-American Horse.

Petiso Argentino

ORIGIN: *Argentina*
APTITUDES: *Riding pony*
AVERAGE HEIGHT: *10 to 12 h.h.*
POPULATION STATUS: *Common*

Petiso Argentino. Photo: Luciano Miguens

The Petiso Argentino is a new pony breed developed from Shetland and Welsh ponies imported by Argentine farmers in the first half of the twentieth century. There is undoubtedly also a bit of Criollo blood in this pony, judging by the color and conformation. As time passed, the Shetland and Welsh breeds were no longer distinguishable and the new breed, Petiso Argentino, was created.

The photograph accompanying this article is of ponies belonging to Luciano Miguens, a Criollo breeder and president of the association for the Criollo breed.

Pindos

ORIGIN: *Greece*
APTITUDES: *Riding, light draft, pack*
AVERAGE HEIGHT: *11.2 to 12.1 h.h.*
POPULATION STATUS: *Rare*

The Pindos is the most representative pony of the mountainous type in Greece. It is bred in all of the mountainous and semimountainous

Pindos pony. Photo: Penny Turner

A Pinto of the *overo* color pattern. The *tobiano* pattern may be seen under National Show Horse. Photo courtesy Pinto Horse Association of America

regions of Thessaly and Epirus. Unfortunately only a small number of these useful animals remain.

This is an extremely hardy breed, able to exist on a minimum of feed. It is surefooted in the mountains and is used for transport and light farm work. This pony is a very good climber. The mares are often used for breeding mules, which are useful in the rugged terrain.

The Pindos is an ancient breed, and one whose origin is uncertain. The infusion of Oriental blood is obvious in its appearance. The breed is known to be extremely hardy and enduring. It is *unknown* for a Pindos to have any problem with its legs or feet. Longevity is also a characteristic of this hardy horse, with many still working at the age of thirty while appearing to be much younger. Pindos are usually of dark, hard colors, mostly bay and brown. An occasional grey is found. They are never piebald or skewbald.

The body of this horse is narrow and lacks muscle in the hind quarters and neck, giving the impression of weakness—a notion soon dispelled when using this horse in the mountains. The head tends to be slightly heavy but is well shaped. The back is short and strong; the croup is sloped but the tail is set high. The legs of the Pindos are fine boned but strong; the hooves are narrow and very hard.

Pinto

ORIGIN: *United States*
APTITUDES: *Riding Horse*
AVERAGE HEIGHT: *Varies; all sizes*
POPULATION STATUS: *Common*

The Pinto is usually associated with the American Indian, who, according to legend, considered the loudly marked horses to be magical and especially good for use in war. History of the Pinto has earlier chapters, however, than those in the New World.

In many cultures of the world the horse was an important tool of civilization, primarily as transportation. Horses are well represented throughout history in art, and even in prehistoric times ancient humans painted them on the walls of caves. A study of art reveals that all the horse colors and patterns commonly known today were also present in the past. Interestingly, these patterns do not seem to be geographically limited.

The common pinto patterns came to this continent from Europe through Barb stock used by Spanish explorers. These Barbs were hardy horses from North Africa, imported to southern Europe and crossed with native stock. The pinto patterns may also have arrived in Europe through some Arabian strains, as pinto markings appear in ancient art throughout the Middle East. The *tobiano* pattern was common among the wild horses of the Russian steppes, suggesting that this coloring existed in Europe prior to the arrival of Arabian horses.

In North America, horses turned loose by or escaping from European conquerors ran wild and developed into great herds. Due to natural selection the type changed in the feral herds and the qualities that endured included compact size, hardiness, and a high spirit. Indian use of horses faded as tribes were subdued, and white settlers continued to import most foundation stock from the well-established and popular breeds of Europe. However, on the frontier the feral horses were more available and better suited to the strenuous working conditions than the "fancy" breeds from the East. Well-bred horses were often crossed to mustang stock to increase size and attractiveness. The western horse became a fixture of America, especially when the "Wild West" was glamorized in early movies. Mexican and western saddles appeared with silver trappings in parades and fairs. Appaloosas, pintos, palominos, and other attractively marked or colored horses were popular for these flashy displays. Exhibitors crossed their colored horses with other breeds to produce specific types with beautiful coloration.

The Pinto Horse Association of America (PtHA) was founded in 1947, becoming officially incorporated in 1956. Today PtHA is strong, growing, and much more than just a color registry. It is an association that involves people and horses from all over the world. The association maintains a large and rapidly growing registry for Pinto horses, ponies, and miniature horses.

Pintos may come in any size. Horses must be at least 14 hands (56 inches) or larger. Ponies are over 34 inches up to but not including 56 inches. Miniatures are 34 inches or under. Horses, ponies, and miniatures each have their own show classes that are separate from one another, and the size of the animal determines the category; horse, pony, or miniature.

Horses and ponies are divided by type— Stock, Hunter, Pleasure, and Saddle—and each type is shown in its own respective class. Breeding is encouraged to remain with a specific type. Stock-type horses and ponies are described as being of predominantly Quarter Horse breeding and conformation. Animals double-registered Paint (see American Paint) usually fall into this category. Conformation generally is associated with a Western breed. The Stock-type pony is western in type, displaying conformation similar to that of the Quarter Horse. In the Hunter type, the horse is of predominantly Thoroughbred/Jockey Club breeding. A few Paint horses will fit this category. Conformation is generally associated with that of an English riding horse. The Hunter pony type reflects the conformation associated with the Thoroughbred horse and Connemara-bred pony. In the Pleasure type, the horse is of predominantly Arabian or Morgan breeding. The Pleasure pony type reflects carriage and conformation associated with the Arabian, Morgan, or Welsh pony breeds. The Saddle-type horse is of predominantly Saddlebred or Tennessee Walking Horse breeding, displaying high head carriage and animation. In the Saddle-type pony, conformation reflects the modern-style (gaited) Shetland or Hackney breeds.

Many people confuse Pintos with Paints. As indicated in the American Paint Horse breed account in this volume, registry with the Paint association is limited to animals exclusively from Quarter Horse, Thoroughbred, or Paint bloodlines. It has nothing to do with the color or pattern, only bloodlines. Most Paints can be double-registered as Pintos, as either Stock-type or Hunter-type horses. PtHA also registers horses with other breeding such as Arabian, Morgan, Saddlebred, and so forth, which are placed into their respective type divisions for showing. Most Paints are eligible to be registered as Pintos, but not every Pinto horse qualifies as a Paint since Pintos may come from a wide range of bloodlines.

There are two recognized Pinto color patterns, *tobiano* and *overo*. For a description of these, see American Paint.

Piquira Pony

ORIGIN: *Brazil*
APTITUDES: *Riding pony*
AVERAGE HEIGHT: *12.2 to 13.1 h.h.*
POPULATION STATUS: *Common*

The Piquira breed was founded by crossing Brazilian national (Crioulo) mares of small size with pony breeds, mostly Shetland. Since the beginning, selection has been based on two fundamental characteristics: small size and smooth gait. This is still a breed of great genetic variation due to recent pony cross-breeding and the fact that several mares of unknown origin have been registered. Consideration was given to these mares due to their small size and good gait.

There are already many animals of excellent and well-defined lineage, completely fitting the breed standard. Piquira breeders are members of the Pony Breeders Brazilian Association (Associação Brasileira dos Criadores de Cavalo Ponei), founded in 1970, with headquarters in Belo Horizonte in the state of Minas Gerais.

The main purposes of Piquira ponies are for children to ride, for leisure, and for work on farms. The Piquira is an extremely docile animal, of calm temperament and smooth gait.

Piquira ponies may be any color but albino (that is, white with blue eyes). They are well proportioned with a strong, fine skeleton, good tendons, thin skin, and silky hair. The head is expressive, proportional, and harmonious and has a flat forehead and straight profile. The eyes are large and active and have thin eyelids; the ears are small, mobile, and well directed; the nostrils are wide and flexible; the neck is light and muscular, of medium length, with harmonious connections to the head and shoulders. The withers are prominent and well implanted; the ribs are arched, long, and defined; the back is short, straight, and muscular; the croup is well linked, horizontal, and smoothly attached; the tail is set high with silky hair; the shoulder is sloped and muscular. The legs are sturdy with strong joints, short cannons, and strong tendons; the pasterns are strong and sloped; the hooves are round, solid, and hard. The gait of the Piquira is a four-beat lateral gait *(picada)* or four-beat and two-beat diagonal gait *(batida)*.

Pleven

(Plevenska)

> ORIGIN: *Bulgaria*
> APTITUDES: *Riding horse, light draft*
> AVERAGE HEIGHT: *15.3 to 16 h.h.*
> POPULATION STATUS: *Common*

The Pleven horse is a halfbred produced in Bulgaria and having a number of valuable,

Pleven horse. Photo: Dr. P. Chakarov

universal characteristics which made it a basic breed. Development of this breed was started in 1898 at the former state stud Klementina, near Pleven. It is now called the G. Dimitrov state farm.

Initial breeding animals for production of this horse were improved local horses, Arab, Anglo-Arab, and halfbred mares crossed at first to Arab, Anglo-Arab, and halfbred stallions. Later, Gidran stallions were used. The breed was officially recognized in 1951.

Horses of this breed are of uniform type with good conformation, robust constitution, light, airy action, and beautiful gaits. Most are a good chestnut color. A characteristic feature of the Pleven breed is that in addition to good economic properties it has excellent ability in dressage and jumping.

Pleven horses have a well-developed genealogical structure of seven family lines and thirteen families, which serve as a basis for the breeding work. Work with this breed is directed toward retaining good working abilities, increasing size, and improving competition qualities.

Poitou Mule Producer

(Poitevine mulassière)

> ORIGIN: *France (Poitou)*
> APTITUDES: *Heavy draft and mule production*
> AVERAGE HEIGHT: *15 to 17 h.h.*
> POPULATION STATUS: *Rare*

The Poitou is perhaps the only equine breed used predominantly to produce good mules. Although it probably existed before, the Poitou mule-producing breed is said to stem from Flemish horses imported at the end of the sixteenth century to cleanse the Poitou and Vandée marshes. The local mares bred to these horses from the north gave birth to the breed called Poitou.

In principle, any mare bred to a donkey will produce a mule, but this breed has produced stronger, more developed offspring than mares of any other breed. The mule industry, known since the tenth century, flourished during the thirteenth and eighteenth centuries and attained world fame in spite of some ups and downs.

The Poitou is portrayed by some as being of generally poor conformation, lethargic and slow in both movements and mentality. The breed varies in color and may be grey, black, bay, or palomino. The head is large, rather long with a straight profile and wide jaws; the ears are big and long; the neck is long with abundant mane; the withers are well set; the chest is wide and deep. Often the back is long and wide and well attached to wide loins. The hips are spread out and wide, and the croup is drooping. The ribs are long and slanted; the shoulder is slanted and very long; the limbs are powerful with wide, strong joints and wide and well-shaped hooves. The thighs are muscular. The coat of this breed is thick, often curly around the knees and hocks.

The original breeding area for the Poitou was the Melle region in the Deux-Sèvres. But breeding now extends to all of the Deux-Sèvres—that is, all of the Saintes and La Roche-sur-Yon stud areas.

Work and meat are two other uses for this breed, which presently exists only in small numbers. Although the output of meat is quite small, the quality is highly appreciated.

The mule industry has declined significantly since 1950 but some resurgence seems possible, given the strong demand for mules from other countries. The Baudet du Poitou is the donkey mainly used in mule breeding with the Poitou mares. The two species are in effect tightly bound, and one cannot develop without the other. Crossbreeding between the two species produces a large and sturdy mule.

Polesian
(Polesskaya)

ORIGIN: *Byelorussia and Ukraine (former Soviet Union)*
APTITUDES: *Draft*
AVERAGE HEIGHT: *13.2 to 15 h.h.*
POPULATION STATUS: *Rare*

The Polesian breed originated in the Pripet Marsh area of the Ukraine and the adjacent Byelorussian lowland (Belaruskaye Palesse), a region of marshes and forests. Grasses growing in the meadows are acidic and low in nutrients. The aboriginal horse of this area belongs to the forest type, similar to the Zhmud (see Zhemaichu).

The Polesian was bred pure for a long time under conditions of poor feeding and keeping and exhausting work. Horses in the eastern part of the region were crossed with other breeds and pure breeding of the Polesian ceased. The western section, near the Polish border, is remote, closed country with few good roads, and foreign horses seldom entered the area. Thus Polesian horses in the western zone have preserved more characteristics of the native forest horse. This ancient breed descended from the forest Tarpan and is native to the marsh region of Byelorussia. Skeletons of horses of the forest type were found in this area, dating to the second century B.C.

Breeding of the Polesian horse began in the first century A.D., but study of the breed did not occur until after the Revolution. The smallest horses came from the central regions.

The Second World War caused great loss in the Polesian breed, as they were used by partisans. During the years from 1950 to 1960 the breed was studied and more than twenty stud farms were established. Most recently, Polesians have been bred in the Brest and Gomelsk regions.

This horse is well adapted to a moderate, mild climate with high humidity, and to marshland habitat. It withstands severe cold but is intolerant of hot, dry weather. When the weather is sultry the horses hide in the forest and graze only at night.

Polesian horses have a working life span of twenty-five years and remain fertile until eighteen to twenty years and more. The mares are good milk producers.

In appearance, the Polesian has a head of medium size with a straight profile or slightly dished face; the neck is of medium length; the withers are set low; the chest is long, deep, and wide; the ribs are rounded; the shoulder is well slanted; the back is long and straight; the croup round, often sloping. The legs are clean and sturdy, and the breed characteristically has hard, strong hooves. Predominant colors are bay, sorrel, black, grullo, buckskin, dark brown, and grey.

There are two types found today in the Polesian breed; a massive type with heavy muscling, and a lighter, more refined type less suitable for agricultural work. The breed was improved by crossing with small Arabs, Gudbrandsdal, and Russian and Soviet Heavy Draft. The best results were obtained from crosses with lean Soviet Heavy Draft stallions of medium size.

Polesian horses are important in their area as working horses and are used for agricultural and transport work. They represent a valuable gene reserve and should be preserved.

Polish Draft

ORIGIN: *Poland*
APTITUDES: *Heavy draft work*
AVERAGE HEIGHT: *14.3 to 16 h.h.*
POPULATION STATUS: *Common*

Polish Draft horse. Photo courtesy Dr. Stanislaw Deskur

Beginning in the middle of the nineteenth century, expansion of agriculture launched the import of coldblooded horses of various breeds into Poland.

Based on local native mares, the breeding of several types of draft horse developed in various regions of the country. After World War II, Ardennes and Belgian draft horses were used in breeding, so that today the heavy horses bred in Poland have a great deal in common. Polish Draft horses mature early, make good use of roughage, and are docile and willing to work.

Prevailing colors in Polish Draft horses are chestnut, roan, and bay. These animals make up about 60 percent of the Polish population of horses. They are used for farming, but when useless as working horses they are destined for a slaughter house. Draft horses are common throughout Poland, but their population is declining with increasing mechanization in agriculture. Presently there are about 8,000 registered mares. State breeding is carried on at two stud farms: Nowe Jankowice (Torun County) and Bielin Nowy (Szcecin County).

Polish Draft horses are known by various names depending on the area of their breeding and certain differences, although, as noted, these dissimilarities are lessening. The Lowicz is similar to the Sztum but somewhat lighter. Main foundation stock for the Sztum were the Ardennes and Belgian breeds. The Sokólka, was also based on Ardennes stock but with some Breton and Døle blood, is mainly chestnut. The Sztum, which originated from local stock crossed with Belgian, Rheinish, Jutland, Ardennes, and Døle stock, is usually chestnut but may be bay or roan. The Garwolin is similar to the Sokólka except that some Boulonnais blood was used, so many individuals of this strain are grey in color. A smaller variety of Polish Draft is the Lidzbark, descending from the Oszmian horses, primitive animals derived from a mix of West European coldbloods and the North Swedish Horse. The smallest Polish Draft breed is the Kopczyk Podlaski, descended from a stallion foaled in Podlaska in 1921. Crossed to local mares, he produced a strain that is active, small, and usually bay or chestnut.

Pony Mousseye

ORIGIN: *Cameroon (Gobo)*
APTITUDES: *Riding pony*
AVERAGE HEIGHT: *12 h.h.*
POPULATION STATUS: *Uncommon*

Pony Mousseye of Cameroon. Photo: Ebangi Achenduh

True origin of the Pony Mousseye is uncertain. Small ponies are found throughout western and central Africa and all bear similarities. The pony is known by other names: Pony Kirdi, Pony Sara, Pony M'baye, and Lakka (or Laka) horse. This pony was first noticed in the sub-Sahel region of Cameroon in 1826, along the banks of the Logone River. It was not until 1926 that a description was given to the breed. Since then little has been written about this small horse in Cameroon, and its existence is threatened by a continuous decline in population. This has been the result of faster and easier means of transportation, little care by breeders, and random crossbreeding resulting in undesirable offspring.

The Cameroon government, conscious of the role played by each animal species in the stability of an ecosystem of a given area, created the Equine Program in 1986, with the principal objectives of safeguarding the Pony Mousseye from extinction and developing the horse industry in Cameroon.

Mousseye, or Musseye, is the name of the tribe that inhabits the small administrative unit of Gobo in the Mayo Danaye Division of southeastern Cameroon. These people raise mostly ponies, and this is the main center for Cameroon ponies. The ponies have traditionally been used for war and hunting, as exchange units for matrimonial presentations, as compensations for grievous offenses, and as a symbol of social prestige and wealth.

Some authorities believe the Pony Mousseye to be a distinctive, degenerate type of Barb. Officials in 1972 attributed its existence as a pony to deterioration in size due to scant feed. The deterioration has followed a succession of generations in a somewhat unfortunate condition. This pony may also be derived from other small, ancient types.

The animals are said to have been brought to Gobo by the Mousseye or Banana Hoho people, who had moved to Nigeria to work in the Bauchi tin mines. Between 1942 and 1943, their work was interrupted and they returned home with ponies from Nigeria.

The Pony Mousseye stands out as the only existing Cameroonian breed, raised by the Massa, Sara, M'baye, and Mousseye peoples in the Mayo Danaye Division some 155 miles south of Ndjamena in the southeast of Cameroon. The distribution of the Pony Mousseye is limited to the banks of the Logone River, a humid zone infested by the tsetse fly (*Glossina*). The existence of other horse breeds in this area has not been possible, but the Pony Mousseye is resistant to the trypanosomiasis (sleeping sickness) carried by the tsetse.

This breed has not been found in northern Africa or on the paths of migration. With its limited geographical distribution, the pony has undergone the pressures of isolated, intensive, and conservative selective breeding and the typical conformation has been preserved. A typical Pony Mousseye is described as a small, solid horse. The head is usually heavy and massive with a rectilinear forehead profile; the eyes, which are obliquely placed, form a sharp angle with the forehead. They usually express intelligence. The ears are erect and small; the muzzle is voluminous; the neck is thick and short; the legs are short and strong but show disproportion to the body weight. The hooves are thin. The body color is variable but predominant colors include white, chestnut, and grey. White markings are not common. Typical Cameroonian ponies have an astonishing rusticity. They are able to pass through difficult places and are capable of traveling eighteen to twenty miles a day consecutively for four or five days before needing a day of rest.

Despite nutritional deficiencies, the Pony Mousseye remains resistant and enduring. This breed is docile and easily managed, even by children.

The trypano-tolerant nature of this pony may warrant research and offers interesting scientific opportunities.

Pony of the Americas
(POA)

ORIGIN: *United States*
APTITUDES: *Riding pony*
AVERAGE HEIGHT: *11.2 to 14 h.h.*
POPULATION STATUS: *Common*

Pony of the Americas (POA). Photo courtesy Pony of the Americas Club

Not until 1954 was a western pony breed cultivated in the United States for a particular purpose. The development of the Pony of the Americas is the result of a study made by Leslie L. Boomhower, a Mason City, Iowa, attorney who acquired an Appaloosa horse mare and her foal. The sire of this foal was a Shetland pony. Mr. Boomhower was particularly impressed with the characteristics of the offspring, now known as Black Hand #1. This stallion became the foundation sire of the Pony of the Americas breed.

Boomhower, along with two other breeders having similar interests, set up a nonprofit organization to establish a breed association for an in-between size pony, 46 to 52 inches in height, large enough for older children yet not too small for adults to break and train. In 1955 the nucleus of the club and registry consisted of twenty-three members and twelve ponies.

The Pony of the Americas Club is one of the largest and most active youth-oriented breed associations in the equine industry today. Along with more than forty affiliated state and regional chapters, the POAC annually welcomes new individuals to its membership of over 2,200.

The POAC provides a complete program for youths under eighteen years of age, allowing them to compete against others like themselves. Many classes and events are provided.

Because of the distinctive coloration and basic characteristics of the Appaloosa horse, such as quiet disposition, stamina, strength, and speed, it was inevitable that some of the smaller ones be crossed to ponies to provide a good riding mount for children. In time, some breeders obtained foundation stock by breeding the POA to the Appaloosa horse; others imported, at great expense, herds of small ponies with Appaloosa color and characteristics from Mexico, Central America, or South America.

The newest western pony breed in the United States thus comes from one of the oldest known breeds, the Appaloosa, which is traced back to ancient days in Central Asia. POAs have also been crossed with Arabians, Quarter Horses, and Shetland and Welsh ponies. Breeders have kept in mind the height limitation of 46 to 52 inches for the mature POA and the necessary crossbreeding to get the color, characteristics, and disposition of the Appaloosa; the natural beauty, refinement, and graceful carriage of the Arab; and the added endurance, muscle, bone, and strength of the Quarter Horse. The result of such crossbreeding and careful selection has produced an outstanding pony, more like a small horse in structure. The POA is agile, fast, graceful, a good jumper, a good dressage horse, friendly and easily handled, yet spirited.

Characteristics include Appaloosa coloration with white sclera around the eye, mottled skin, and often striped hooves. The POA is a finely made pony with a good head, slightly dished profile, medium-sized, pricked ears, and large, expressive eyes; the neck is well proportioned and slightly arched; the withers are prominent; the back is short and straight, the loins are broad and well muscled; the croup is long, muscular, and rounded; the chest is deep and wide; the shoulder is well sloped. The legs are solid, strong, and well shaped, with good muscling and strong tendons. The pasterns are sloped. The hoof is well proportioned, broad, and high at the heel.

Pottok
(Basque)

ORIGIN: *Spain (Basque Province of Navarra), France*
APTITUDES: *Riding pony*
AVERAGE HEIGHT: *12 to 14.2 h.h.*
POPULATION STATUS: *Common*

The Pottok is an ancient breed found in the Pyrenees Mountains. Early cave paintings have been found that clearly depict this small horse, which may have been used by the Visigoths in early times. *Pottok* is a Basque word meaning "small horse," and the breed is known equally well as either Pottok or Basque.

Originally wild in its ancient habitat, all Pottoks today have owners. They are rounded up traditionally on the last Wednesday in January, to be branded and sold at local markets or released to roam the hills once again.

The Pottok is a rugged, tough animal that has developed in inhospitable surroundings. This small horse is considered an integral part of the Basque cultural heritage. Living in the open and finding its own food, the Pottok has fantastic endurance. When caught and tamed,

it was used in the past as a pack pony and as a pit pony for the mines. This little horse may have been among the horses taken to the Americas during the Spanish conquest for use as a pack animal. Until World War II, the breed was used by smugglers to carry packs across the Pyrenees.

Use of the Pottok as a riding horse has only occurred in the last twenty years. It is in demand as a mount for children due to its naturally gentle nature. Some breeders have sought to improve the breed by introducing Arab and Welsh blood into some lines.

The French government officially recognized the breed recently, opened a stud book, and has taken steps to assure the continued purity of the breed.

Bred today outside the mountainous regions, its results in horse shows have quickly put the hardy Pottok on the pony map. A breed association was founded in 1971 and a riding school specializing in breaking in was opened in Espelette in 1979. Soon the qualities of the Pottok were obvious, as a team from Espelette went through to the French National Finals in 1981. In 1983, Kuzko, a pure Pottok, was the French National Champion in a discipline combining dressage and show jumping. Since that time, Iago, Maitagarria, and Lhuan have been French National Pony one-day event champions. The Pottok excels in all riding pony disciplines and is also a very good harness pony. This pony is strong and tough for its size and has excellent jumping talent.

The head has a straight profile, slightly concave at eye level. The ears are short and active; the eyes are large and expressive; the nostrils are flared and open. Often the upper lip is overhanging, and when eating spiny plants during winter months in the mountains this pony often develops a thick set of whiskers for protection. These are shed during warmer weather. The neck of the Pottok is short and often ewe-like, well set on with a full and shaggy mane; the withers are well pronounced; the back long and straight; the croup rounded; the chest wide and deep; the ribs are well rounded; the shoulder is somewhat straight. The legs are clean and strong with good joints and hard feet. The tail is set very low and is usually long and full.

Priob
(Priobskaya)

ORIGIN: *Western Siberia (former Soviet Union)*
APTITUDES: *Draft*
AVERAGE HEIGHT: *13 to 14 h.h.*
POPULATION STATUS: *Rare*

Similar to the Narym, the Priob, also known as Ob, is bred in the Khanty-Mansi national district near the lower areas of the Ob and Irtysh Rivers. The climate is severe with extreme cold, deep snow and lack of grain foods, creating difficult conditions for working horses. They are used chiefly as pack animals in winter and also work in forests. During the summer months the horses do not work and are left free to graze the marshes.

Long-lived and fertile, Priob horses work until the age of eighteen or twenty. In general appearance the Priob is similar to the Yakut, although more of draft type. Small in stature with a long trunk and well-developed skeleton, Priob horses are hardy and enduring. The legs are short. The head is of medium length, coarse, and often Roman-nosed; the neck is short and thick; the shoulder is slanted and short; the withers are low; the back is long and often carp-shaped; the legs are sturdy. The hooves are wide and flat, an adaptation to the muddy terrain.

Colors are various in the Priob breed, but most typical is dun. Dun, grullo, and bay Priob horses usually have dun factor markings—dorsal stripe, shoulder stripes, and zebra striping on the legs.

The first research on this breed was done in 1936. The Priob is rare today and should be bred pure and upgraded, as it is invaluable in the northern forest regions.

Przewalski's Horse
(Asiatic Wild Horse, Mongolian Wild Horse, Taki)

ORIGIN: *Asia (western Mongolia)*
APTITUDES: *Wild, undomesticated*
AVERAGE HEIGHT: *12 to 14 h.h.*
POPULATION STATUS: *Rare*

The Przewalski's Horse *(Equus przewalski poliakov)* is the only known surviving wild

A group of Przewalski's Horses. Photo: Annette Groenveld

Przewalski's Horse. Photo courtesy A. E. Huff

species of Equidae left on earth since extinction of the last wild Tarpan *(E. p. gmelini)* in 1879. All other horses are either domesticated or feral (descended from once domesticated horses). Until very recently, the Przewalski's Horse was extinct in the wild, exterminated by hunters.

In June of 1992, through the herculean efforts of the Przewalski Foundation in Holland and the breeding reserves in Askania Nova, Ukraine, two combined breeding groups of Przewalski's Horses were released in Mongolia. Initially they are being kept in large pastures surrounded by electric fence and patrolled by armed game wardens to keep wolves and other predators out. The ultimate plan is to release the Przewalski's Horse to the open steppe, possibly by 1994. It is planned to add additional animals soon from the same sources, and the year after that, some animals from the San Diego Zoo in California.

Controversy surrounds the Przewalski's

(pronounced "sha-val-ski") Horse. Some authorities have believed strongly that this horse is the direct ancestor to modern domesticated breeds. Others contend this is not possible, as Przewalski's Horse is a different species, having sixty-six chromosomes while the domestic horse has sixty-four. It is possible to cross the Przewalski with the domestic horse, and the resulting hybrid is fertile; however this offspring has sixty-five chromosomes. When crossed again to the domestic horse, the new generation returns to sixty-four chromosomes and little influence of the Przewalski's Horse is evident.

Formerly it was believed that Przewalski's Horse was discovered in 1881 by the Russian explorer Colonel Przewalski in the northwestern part of Mongolia. He found a small herd in the Tachin Schara Nuru Mountains near the edge of the Gobi Desert, and the breed was named after him. More recent information that has come to light through the Przewalski Horse Foundation indicates that two Europeans observed these animals in the wild at much earlier dates. A Scottish doctor in the Russian service, sent on an embassy to China by Peter the Great, published his experiences as *Journey from St. Petersburg to Pekin, 1719–1723*, including a good description of this Asiatic Wild Horse. Earlier yet, a Bavarian nobleman named Hans Schiltberger was taken prisoner by the Turks and sold to the famous Tamerlane of the Golden Horde, who in turn gave Schiltberger to a Mongol prince named Egedi. Schiltberger spent several years in the Tien Shan mountains, where he observed these wild horses and wrote of them in his memoirs, "Journey into Heathen Parts." The unpublished manuscript was written in 1427 and is housed in the Munich Stadtbibliotek (municipal library).

It is difficult to imagine how Przewalski's Horse could have been the ancestor to the modern domestic horse when it seems impossible to tame this wild species. Experienced handlers have failed even to halterbreak this horse. In 1902, when the first imported pair of Przewalski's Horses arrived in New York, W. T. Hornaday, director of the New York Zoological Society, vowed to tame them. As far as he was concerned, wild horses they might be, but they were still horses and he proposed to handle them as such. Keepers made strenuous efforts to halterbreak the pair so that they could be led around the grounds and trained to pull a carriage. After a few weeks, Hornaday had to acknowledge that it was no use. The longer the men worked with them, the wilder and more obstinate the animals became.

Considering the limited knowledge and ability of early humans to control wild animals, it stretches the imagination to believe that people could have contained, controlled, and gentled this horse, which even today remains aloof. With most "rules," however, there is an exception. In Erna Mohr's book *Das Urwildpferd* (The Wild Horse), (1959) is a photograph of a Przewalski stallion ridden by a keeper. The animal seems quite tractable. However, the general belief is that the Przewalski's Horse has remained untamable and therefore unusable by people.

According to the Foundation for the preservation and protection of the Przewalski's Horse, in Rotterdam, the Netherlands, only a few Przewalski's Horses are tamable, similar to zebras. If the most human-friendly types are carefully selected and isolated as foals so that they may be handled, they may submit.

There have been several experiments of breeding the Przewalski's Horse to domestic horses, mostly in the Soviet Union.

Blood group testing of the Przewalski's Horse has revealed several genetic markers that are not found in the domestic horse, in addition to genetic markers that are common to all species of Equidae.

The Przewalski's Horse has a large, heavy head with a straight profile. The forehead is broad, the ears are large and heavy, the eyes are small. The jaw is heavy and the teeth are large. The neck is broad and short with a short, upright mane; the withers are flat; the back is straight and quite long; the croup is sloped and short; the tail is set low. The chest is deep; the shoulder is straight and short; the legs are short and sturdy with short pasterns; the hoof is narrow and elongated in shape. The tail has a distinct tuft at the end, especially noticeable during winter months. The color is yellowish dun with a mealy muzzle and dark mane and tail. There is a dorsal stripe and zebra markings on the lower legs. White markings are absent.

This horse has strong herd instincts and is very protective of its young.

Rahvan

(West Black Sea Rahvan Horse)

ORIGIN: *Turkey*
APTITUDES: *Riding, racing*
AVERAGE HEIGHT: *10.3 to 15.1 h.h.*
POPULATION STATUS: *Uncommon*

The West Black Sea Rahvan horse (usually simply called Rahvan) has been known in Turkey for at least 800 years. *Rahvan* is the word used in Turkey to describe a very fast lateral walking gait, and the proper Turkish name for this breed is Bati Karadeniz Rahvan Ati. While some horses may be taught this gait, the Rahvan horse is born with it and has been bred selectively for racing at this gait. Raised in the Black Sea region (Bartin, Kastamonu, Daday, Taşköprü), this breed has been raced at the *rahvan* gait for centuries, but the races have been more organized in the last century.

Rahvan walking horse, possessing a very fast, smooth gait. Photo: Ertuğrul Güleç

Rahvan horses racing at the running-walk. Photo: Ertuğrul Güleç

The West Black Sea Rahvan Horse is very fast at all gaits. It is descended from crosses of the Anadolu and Canik breeds. Due to their easy, fast gait, these small horses can carry a great deal of weight while using less energy than many other breeds. They are able to cover long distances without tiring and give a smooth, comfortable ride. When traveling at the rahvan gait, they hold head and tail high, reminding one of the Icelandic Horse. The gait is essentially the same as is found in several other gaited breeds throughout the world and termed, according to the region: rack, tripple, stepping pace, amble, *paso fino*, running-walk, or *tølt*. Some breeds show more action, but the beat of the gait is basically the same.

The Rahvan horse is small and eats less than larger horses, while it is able to carry the same weight. The breed is friendly and obedient. In the region of Bartin and Kastamonu, normal height of the Rahvan horse is 13.1 hands. Every common horse color is found within this breed.

The head of the Rahvan horse is very beautiful; the neck is thick and muscular; the back is short; the chest is wide and deep; the legs are strong with hooves "like iron."

Rhineland Heavy Draft

(Rhenisch-deutsches Kaltblut)

ORIGIN: *Germany*
APTITUDES: *Heavy draft*
AVERAGE HEIGHT: *16 to 17 h.h.*
POPULATION STATUS: *Rare*

Rhineland Heavy Draft. Photo: Werner Ernst

The Rhineland Heavy Draft, also called Rhenish-Westphalian, Rhenish-Belgian, Rhenish-German, or German Coldblood, was once the most numerous of all horse breeds in Germany. The breed evolved during the last quarter of the nineteenth century to meet the needs of growing agriculture and industry. Although experiments were conducted in formation of this horse using various bloods such as Clydesdale and Percheron, it was the Belgian Draft horse that contributed most to development of the Rhenish-German coldblood. The main breeding center was the Wickrath regional stud in the Rhineland. Heavy horses in Germany today make up only about 2 percent of the total equine population, whereas early in the twentieth century they were the majority. As in the case of all heavy draft breeds, when motorization began to take their place the breed declined.

This is a handsome draft horse, massive in build. The head is relatively small and well shaped with pronounced jaws, a straight profile, and small ears. The eye is large and kind. The neck is short, broad, muscular, and quite arched, set well into the shoulder; the withers are low and broad; the back is short, wide, and very strong, often slightly dipped. The loins are broad and muscular; the croup is wide, muscled, and gently sloping; the chest is wide and open; the abdomen is large and rounded. The legs are short, powerful, and well muscled with broad joints, short cannons, and broad, tough feet. The legs are feathered. This breed has a docile temperament and willing nature.

Colors in the Rhenish-German Coldblood are chestnut or red roan (with flaxen or black mane and tail), or bay.

Rocky Mountain Horse

ORIGIN: *United States (Kentucky)*
APTITUDES: *Riding horse*
AVERAGE HEIGHT: *14.2 to 16 h.h.*
POPULATION STATUS: *Uncommon*

Rocky Mountain Horse. Photo courtesy Rea Swan

The Rocky Mountain Horse originated in eastern Kentucky early in the twentieth century and is a close cousin to the American Saddlebred and Tennessee Walking Horse; all three are descendants in part of the Spanish horses. This is a medium-sized horse with a gentle temperament and an easy, ambling four-beat gait that is very comfortable to ride.

The gait made this horse the choice for the

farms and rugged foothills of the Appalachians. A horse for all seasons, it could pull plows in the fields, work cattle, be ridden bareback by four children to the fishing hole, or take its owner to town comfortably on Saturday. It even performed well when hitched to the buggy. Fancy barns and stalls were unnecessary, as the coldblooded nature of this horse enabled it to tolerate the winters of Kentucky with a minimum of shelter.

The new breed was preserved in small groups and gradually increased in number. Even though some outcrossing did occur, the basic characteristics of the Rocky Mountain Horse continued. The foundation sire of the Rocky Mountain Horse was a stallion named Old Tobe. He was owned by Sam Tuttle, who had the concession for horseback riding at the Natural Bridge State Park. Sam used these horses for many years to haul green, inexperienced people over rough and rugged trails. Old Tobe, his most treasured stallion, who sired fine horses up until the age of thirty-seven, was surefooted and gentle. This stallion was often the one used to carry the young or old or uncertain over the mountain trails of Kentucky without faltering, even though he was an active breeding stallion. He had a perfect gait and temperament. All of the present-day Rocky Mountain Horses carry his blood.

This breed is known for its gentleness. It is an easy keeper and wonderful riding horse with good endurance qualities. Today this horse is being used as a pleasure horse, for trail riding, for competitive trail and endurance riding, for local horse show classes, on cattle farms, and as 4-H projects. Due to its easy, natural gait, the breed can exert a minimum of effort to both horse and rider and cover a great distance without tiring.

The haphazard and unorganized maintenance of the breed would have led to its eventual dissipation and loss, so in the summer of 1986 the Rocky Mountain Horse Association was formed. Its purpose is to maintain the breed, increase the number of horses, and expand the area of acquaintance with the breed. The association has grown quickly, and as more people have become acquainted with this horse many have been sold to other areas of the country and even to Canada.

The Rocky Mountain Horse has a wide chest, a very well-sloped shoulder, bold eyes, and nicely shaped ears. The gait is a natural four-beat gait (single-foot) with no evidence of pacing or trotting. When the horse moves one can count four distinct hoofbeats which produce a cadence of equal rhythm. Each individual horse has its own speed and natural way of going, traveling at from seven to sixteen miles per hour. The Rocky Mountain Horse is not trained with any aids or action devices but has a naturally occurring gait. The preferred color is chocolate (dark chestnut) with flaxen mane and tail. Any solid color is accepted. White markings may not extend above the knees or hocks.

Romanian Saddle Horse

ORIGIN: *Romania*
APTITUDES: *Riding horse*
AVERAGE HEIGHT: *15.2 to 16.2 h.h.*
POPULATION STATUS: *Common*

Romanian Saddle Horse. Photo courtesy Emil Petrache

The Romanian Saddle Horse was created by crossing well-known saddle horse breeds such as the Thoroughbred, Arab, Gidran, Furioso, and sometimes the Trotters. Anglo-Arabs are currently also used, for sporting purposes. Breeding is carried out with selection based on aptitudes for sport such as dressage and jumping.

The character of this breed is docile and goodnatured, sociable, and easy to educate, while at the same time lively. This is a beautiful horse. The head is well proportioned and expressive; the neck is muscular and well connected to the body; the withers are of medium height; the back and loins are strong and flat; the croup is well proportioned and muscular; the chest is well developed and deep; the legs have very good bone with clearly defined tendons and strong, clean joints; the feet are hard and well shaped. The most predominant colors are bay, chestnut, and black. Roans are seldom seen.

Size, conformation, aptitude for jumping, and character make the Romanian Saddle Horse an excellent performer for all forms of equine recreation. The breed is in demand by sport clubs throughout the country. Some horses of this breed are also suitable for two- and four-in-hand carriage work.

Rottal
(Rottaler Warmblut)

ORIGIN: *Germany (Bavaria)*
APTITUDES: *Riding, light draft, coach*
AVERAGE HEIGHT: *16 h.h.*
POPULATION STATUS: *Rare*

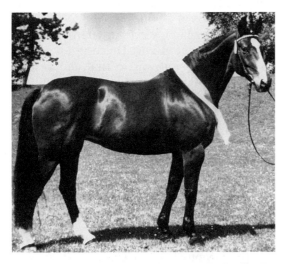

A champion Rottal mare in Germany. Photo: Evelyn Simak

A rare breed today, the Rottal had its origin in the year 909 in lower Bavaria. This is the only indigenous breed of Bavaria, which is located in the southeast of Germany. The Rott River gave its name to the area as well as the horses bred there.

The Bavarian people were raided repeatedly by the Huns before finally managing to drive them out. They left many of their horses behind, and the Bavarian farmers used these to restock their own herds. The Celtic pony ranged throughout Germany at one time, having evolved from the wild Tarpan. The Celtic resembled the Exmoor pony but was heavier in type and with more bone. The horses of the Huns were fiery and quick, of the Turkoman type. The combination of these breeds produced the Rottal horse.

The first stud book for the Rottal breed was started in the eighteenth century, and at that time the breed was a very elegant and quiet cart and military horse. During World War II and later, a heavier farm animal was needed, and Oldenburg stallions were used to produce a stronger animal. The Oldenburg was then Germany's heaviest warmblood breed. For thirty years the Rottal was used as a versatile farm animal, for work, coaching, and riding.

In the 1960s the breeding aim for warmblood horses changed again in favor of the high performance dressage horse and show jumper, and the Rottal was bred almost out of existence by the Hanoverians.

Today, Rottal breeders receive a financial award from the government as motivation to continue with the breed. The Rottal is somewhat smaller than the Bavarian Warmblood. The heavier conformation and quiet temperament make it easy to handle. This breed is a good all-around family horse and is easy to keep.

The Rottal is found in all colors except piebald or spotted. Today there are few pure specimens left.

Royal Canadian Mounted Police Horse

ORIGIN: *Canada*
APTITUDES: *Riding horse*
AVERAGE HEIGHT: *16 h.h. and over*
POPULATION STATUS: *Rare to Uncommon*

A Royal Canadian Mounted Police horse. Photo courtesy RCMP

Royal Canadian Mounted Police horse. Photo courtesy RCMP

While not specifically a breed, the horses ridden by the Royal Canadian Mounted Police (the Force) have been bred with a specific type in mind for many years. Today's technology has eliminated the horse as a viable mode of transportation, and wearing of the traditional scarlet tunic has been reduced to ceremonial occasions.

The Force's Equitation Branch serves a modern and vital role in the ongoing public relations program of the RCMP by incorporating both the horse and the traditional scarlet uniform in creating the spectacle known to millions throughout the world as the Musical Ride.

Members of the Musical Ride are first and foremost police officers who, after two or three years of active police work, volunteer for duty with the Ride. Most are nonriders prior to their equestrian training; however, once they complete the thorough courses of instruction, they become not only accomplished riders but also ambassadors of goodwill. Working through a unique medium, they promote the Force's image throughout Canada and the world.

From 1670 to 1870, the territory stretching west from the Red River for nearly 1,000 miles to the Rockies was known as Rupert's Land, granted to the Hudson's Bay Company by King Charles II. In 1879 a new era began. The territory was purchased by the Dominion of Canada. Settlement along the Red River was organized as the province of Manitoba. The limitless prairies and the Arctic north were called the North-West Territories.

On August 30, 1873, Governor-General Lord Dufferin had approved an order-in-council to form the North-West Mounted Police as a semimilitary body. Immediate objectives were to stop liquor traffic among the Indians, gain their respect and confidence, halt tribal warfare and attacks on white settlers, collect customs dues, and perform the normal duties of a police force.

The Great March West started from the Red River settlement at Fort Dufferin on July 8, 1874. The mission was to find and destroy Fort Whoop-Up, a western stronghold for whiskey traders. By 1875 the NWMP had effectively controlled the whiskey trade, but there still remained problems of Indian unrest. During the period 1874 to 1885 the NWMP's main task was to establish and maintain good relations with the Indians, but this relationship was strained by the influx of both Sioux from the United States and white settlers.

In 1876, following the Custer massacre, the rebellious Sioux fled into Canada. Their presence placed additional strain on the already dwindling buffalo herds and disturbed the peaceful relations Canada was in the process of establishing with its own Indian tribes. The

Sioux were watched closely by the NWMP, who assumed a peace-keeping role to assure there would be no revival of intertribal warfare. It was not until 1881 that the last of the Sioux, led by Sitting Bull, returned to the United States.

During this time the NWMP had succeeded in establishing a good rapport with Crowfoot, Chief of the Blackfoot Confederacy. This trust and confidence in the police resulted in the Blackfoot signing of Treaty No. 7 in 1877.

Another continuing threat to the Indian way of life was the increasing number of white settlers migrating to the vast expanse of the Canadian northwest. In 1874 the North-West Territories had a white population of only a few hundred, which grew to 23,000 by 1885. Construction of the Canadian Pacific Railway increased the flow of settlers to the west. The Force was responsible for policing railway construction as well as providing settlers with many essential services such as mail delivery, recording vital statistics, collecting customs, providing medical treatment, and acting as Indian agents. The NWMP encountered its first experience with labor unrest while policing the railway's construction. The railway indirectly precipitated the North-West Rebellion of 1885.

In 1896 the Canadian government began a vigorous campaign to attract settlers to the west. Assisting new settlers placed an additional burden on the Force. In addition to their other many duties, they now became land agents, agricultural experts, welfare officials, and immigration officers. The NWMP was also the sole police force in Canada's arctic. Members of the Force maintained law and order during the Klondike Gold Rush and later played a prominent role in opening up Canada's arctic frontier.

In 1904 King Edward VII granted the Force the prefix "Royal" in recognition of its service to the Crown, and it thus became known as the Royal North-West Mounted Police. In 1920 the Mounted Police absorbed the Dominion Police, taking over responsibility for enforcing federal laws from the Atlantic to the Pacific. In keeping with this new role, the force was renamed the Royal Canadian Mounted Police.

While the "Mounties" are more likely to be seen today in helicopters, snowmobiles, automobiles, or other vehicles, horses still play a large role in publicity events and in the renowned Musical Ride which tours from early May until late November each year.

In the early 1940s, the commissioner and senior officers of the Force decided that suitable horses were no longer available through purchase and a police horse breeding station would be established. Furthermore, it was agreed that members of the Musical Ride would be mounted on black horses, favorably complementing both the uniform and equipment. With the purchase of Fort Walsh in 1942, the first breeding ranch was established, encompassing 4,300 acres of ranch land. Located in Saskatchewan's Cypress Hills, it was the former site of the NWMP Fort Walsh, built in 1875 and sometimes referred to as the "cradle of the Force." Here the breeding program that produced the present-day horses began.

Development of the Royal Canadian Mounted Police Horse parallels that of the Canadian Hunter and is derived from a Thoroughbred stallion crossed with black mares of good conformation, size, and quality. As recently as January 1989, the equine requirements were reviewed and updated. The horses must reach 16 hands by the age of six years; they must be black in color; their temperament must be quiet and good to deal with the general public; the conformation must be eye pleasing with good substance; the bloodlines must be predominantly Thoroughbred, although the program allows for other breeds to be used in order to maintain size and the necessary temperament for Musical Ride purposes.

Although the Force has used and continues to use Thoroughbred stallions, in the past Trakehner and Hanoverian stallions have been leased in an attempt to upgrade the size, substance, and temperament of the horses. In 1989 a ten-year program was started with the purchase of breeding stock at Verden, Germany. Twelve young black Hanoverian horses (two colts and ten fillies) were purchased for introduction into the breeding program, aimed at improving overall quality.

An organization with a long and exciting history, the Royal Canadian Mounted Police is seen far beyond the Canadian borders today, having toured much of the United States, Belgium, Bermuda, Denmark, England, France, Germany, Holland, Ireland, Japan, and Switzerland. Requests are received each year from all parts of Canada, many areas of the United States, and Europe. Most involve appearances at annual shows and fairs. These requests are filed and catalogued with regard to geographical location, and every effort is made to include them during the next visit of the Musical Ride to that specific area or country.

Russian Heavy Draft
(Russkaya tyazhelovoznaya)

ORIGIN: *Ukraine (former Soviet Union)*
APTITUDES: *Draft work*
AVERAGE HEIGHT: *14.1 to 15 h.h.*
POPULATION STATUS: *Common*

Russian Heavy Draft. Photo: Nikiphorov Veniamin Maksimovich

Development of this small draft horse — strong, sufficiently fast, easy to keep, and economical — took place concurrently with the breeding of a large, heavy draft horse. The genetic material for the breed originated from the hardy native Ukrainian breeds, the mountain Ardennes, and in part the Brabançon (Belgian) and Orlov Trotter.

Systematic breeding began in the 1860s with the main nucleus of the breed formed at Peter's Academy (now the K. Timiryazev Moscow Agricultural Academy), Chesma stud in Voronezh region, Kochubei and Chaplits studs in Poltava region, and Derkulski stud in the Ukraine. In the meantime, upgrading of the local native horses with the Ardennes breed was taking place.

In 1875 there were nine Ardennes stallions in Russia. This number rose to 597 by 1915. By the beginning of the twentieth century the Ardennes type had become the most popular in Russia. Even in areas where there was a demand for large heavy draft horses and where high-grade Ardennes were mated with Belgians, the crosses retained the characteristics of the Ardennes.

The Russian Heavy Draft was introduced at the Paris Exhibition in 1900. Unfortunately World War I, followed by the Civil War, nearly wiped the new breed out entirely. By 1924 only ninety-two Ardennes stallions were left in Russia, but by 1937 the stock of purebreds had been reconstituted and isolated again as an independent breed.

The Russian Heavy Draft has become widespread due to its high profit feeding and adaptability. It is now bred in the Ukraine and North Caucasus, in Udmurtia and Byelorussia, in Kirov, Sverdlovsk, Perm, Vologda, and Archangel regions, as well as in western Siberia.

This breed is not large yet is heavily muscled and powerful. The trunk is long and broad; the joints are well developed and the limbs solid and strong. The head is of average size, clean, with Oriental "breediness," the forehead wide and the profile straight; the neck is short, broad, and fleshy with a high crest in stallions. The withers are low and broad; the back is long, broad, and sometimes rather soft; the croup is long and drooping; the loin is flat; the chest is deep and very broad; the ribs are steeply sloping. The front legs are short, set widely apart; the pasterns are sometimes short and prone to ringbone. The knee joints are often set too far back. The most common color is chestnut, with brown and bay occurring less frequently.

This breed is generally of harmonious conformation with a lively and supple flowing action at both walk and trot. There are three basic types found within the breed. Fifty percent of the horses belong to the Novo-aleksandrov type, smaller in size with a deep, massive body, noble and harmonious with high pulling power. The Ural type, developed under more humid conditions and better feeding, is larger with a shorter trunk, less lean body, thicker skin, longer feathering on the legs (which are longer), and a larger head. The massive type is the largest of the breed, coarser in conformation and less alert. In addition to the three main types there are horses of lighter type, mostly with trotter blood.

The Russian Heavy Draft is strong, shows good speed, and has a very willing and patient nature. This breed matures early and the mares are good milk producers. The breed also has a long fertility and is long-lived.

Russian Saddle Horse

(Orlov-Rostopchin)

ORIGIN: *Former Soviet Union*
APTITUDES: *Riding horse*
AVERAGE HEIGHT: *15.1 to 16.1 h.h.*
POPULATION STATUS: *Rare*

The Russian Saddle Horse is included because, although the breed is presently essentially extinct with only a few remaining specimens, vigorous attempts are being made by horse breeding experts to restore it through use of its descendants and the crosses that produced the breed originally.

The first saddle breed that developed in Russia at the end of the eighteenth century was the Orlov, developed by Count A. G. Orlov. Two separate strains were produced at the Khrenov stud at the time, through the use of two stallions; the bay Arab, Sultan, and the grey, Smetanka. Smetanka, through his son Polkan, was the foundation sire of the Orlov Trotter. Through the cross of a granddaughter of Sultan and a son of Smetanka, a prized line of saddle horses emerged.

In addition to a great number of Thoroughbred and Arab stallions and their crosses, Danish Frederiksborg, Karabakh, and Turkish horses were used. Count Orlov used only Russian-born stock that had no previous training. During development of the breed, strict selection was used for the desired type, and mild inbreeding also occurred. Natural conditions at the Khrenov stud were important in shaping a breed adapted to severe climate. Horses selected for breeding were trained in dressage and highly valued. Orlov horses were almost never trained at hippodromes, but some were run in races which were specially organized for the breed or run together with Thoroughbreds and mountain breeds.

The Orlov saddle horse was big, classy, and had good conformation. The breed gained popularity throughout Russia for its beauty and outstanding dressage qualities, and in 1866 nearly half the stallions used for breeding were Orlov saddle horses.

In the beginning of the nineteenth century, Count Rostopchin developed another saddle breed which was given his name. In 1802 four Arab stallions were imported and crossed with Karabakh, Kabarda, Persian, and Don mares. The crossbreds were carefully selected for desired type and working ability, with special attention paid to speed. The Rostopchin horses were less beautiful than the Orlovs, smaller, with a shorter trunk and neck. The Khrenov stud and the stud of Rostopchin were purchased by the government in 1845. At first the two breeds were bred in purity, but gradually Rostopchin mares were mated to Orlov stallions. The new breed produced by this cross was called Orlov-Rostopchin. The Orlov stallions had the greatest influence and were most widely used.

In 1883 Orlov-Rostopchin horses were used to breed large, elegant saddle horses for the army. The conditions of breeding were typical for raising of army horses — all day in paddocks regardless of weather conditions. At night they were kept in sheds without heating. The severe conditions developed soundness and a tough constitution. The breed was beautiful and in great demand as an improver of other breeds, including the Don, Strelets (extinct), and steppe breeds of eastern Russia.

During the First World War and occupation

of the Ukraine, nearly all of the Orlov-Rostopchin stock was destroyed. In 1931, through the initiative of S. M. Budyonny, the remaining horses were gathered at the Limarev stud farm and work was begun to reestablish the breed. Only five stallions and four mares were located. In 1933 the entire stud was moved to Derkul stud farm and joined with twenty-eight halfbred mares, nine Anglo-Arab mares, and forty-five mares of other breeds. Breeding was accomplished thorugh suppressive crossing, gradually eliminating the blood of other breeds. After the Revolution, the breed was given the name Russian Saddle Horse.

Selection for type and appearance was very strict. At the All-Union agricultural show in 1939, it was said, the only breed that could compare with the Russian Saddle Horse was the Arabian. They had light heads with a wide forehead and dished face, long nape, high, long withers, long, well-muscled back and croup, rounded ribs, and long, lean legs. Almost all of the horses were on "high legs," and the color, without exception, was black.

During the Second World War all of the horses at the stud were destroyed, and the only survivors of the breed were those at the agricultural show in Moscow. The task of reestablishing the breed was extremely difficult. Only two purebred stallions and one mare remained. Some crossbred horses remained with added Thoroughbred and Arab blood. In the first stages of rebirth for the breed, pure and halfbred Thoroughbred, Arab, and Akhal-Teke horses were used with the small number of Orlov-Rostopchin individuals remaining. The Orlov-Rostopchin stallions were widely used to concentrate the blood, suppressing the other blood. Great concentration centered on obtaining a valuable stallion from the only purebred mare left. After eight years of work the efforts proved successful, and the breed is now commonly known as the Ukrainian Riding Horse.

Presently there is a new attempt in progress to reclaim the Russian Saddle Horse. Using lines of the Ukrainian Riding Horse that have the strongest remaining blood of the old Orlov-Rostopchin breed, work has been done since 1978 using Trakehner, Anglo-Trakehner, Arabo-Trakehner, Thoroughbred, Akhal-Teke, and Orlov Trotters. In addition, the remaining Russian Saddle Horse blood is being concentrated, and animals presently used have at least one-quarter Russian Saddle Horse blood with the remaining blood usually Thoroughbred. It was revealed in studies of the breed prior to the war that this breed was dominant in all crosses, and this gave confidence that the breed could be restored, although it was a formidable challenge. Interestingly, as the work progresses, it has been found that a greater number of fillies are born with the desired characteristics.

There are seven groups of desired characteristics for the breed, and the work is being done gradually. The first traits are size, working ability, and black-only color, or "summer black" (black with brown shading on head and legs). Later in development of the breed, more concentration will be centered on type and appearance. Stallions and mares are evaluated each year according to the quality of their offspring.

Currently, objectives are not merely to rebreed the old Russian Saddle Horse but to recall the best traits of that breed; charm, elegant appearance, leanness, black color (without markings), and good temperament. In all other characteristics the breed should surpass its ancestors.

In 1988 there were ninety-six mares and eight stallions in the breeding stock. The successful reclamation of a valued breed from such a small nucleus is a testimony to the skilled genetic knowledge at work in the Russian breeders.

Russian Trotter
(Russkaya rysistaya)

ORIGIN: *Former Soviet Union*
APTITUDES: *Trotting racer*
AVERAGE HEIGHT: *15.1 to 15.3 h.h.*
POPULATION STATUS: *Common*

The Russian Trotter, also called *Métis* Trotter, resulted from crossing the Orlov Trotter with the American Standardbred and then interbreeding the offspring.

Russian Trotter. Photo: Nikiphorov Veniamin Maksimovich

Russian Trotters. Photo: Nikiphorov Veniamin Maksimovich

This crossbreeding was started in the 1890s. Prior to 1914, 156 stallions and 220 purebred mares were used. The first imported American trotters were not of high quality — but they and their crossbred offspring were more speedy. After World War I and the Civil War, the importation of American trotters stopped.

Russian Trotters are bred at studs distributed in different zones of the country—the Dubrovski Stud of the Ukraine, Gomeiski Stud of Byelorussia, Ufa Stud in the Urals, Omsk Stud in Siberia, and many others.

The new breed was officially recognized in 1950, a combination of the best features of both of the foundation breeds. In 1960 a second cross to the American Standardbred was made to improve speed, and the crossbreeding still continues.

The modern Russian Trotter is a typical harness type. It is generally clean and well proportioned with well-developed muscling and tendons. The head is light with a straight profile; the neck is long and straight; the withers are of medium height; the back and loin are straight and very well muscled; the croup is flat, broad, and long. This breed is inferior in appearance to the Orlov, and more frequently possesses such defects as bowed legs or close hock joints. This defect causes the hind feet to move outward in a semicircular motion, known as "dishing," which becomes an asset in that it allows the horse to find its pace more easily when lengthening its stride. While faster than the Orlov, the Russian Trotter does not reach the speed of the American or European trotters.

Common colors are bay, black, and chestnut, with grey occurring less often. Russian Trotters have adapted well to various conditions and distribution of the breed in the former Soviet Union is quite broad, extending from the Baltic republics into western Siberia.

Future breeding of the Russian Trotter will include pure breeding as well as corrective crossbreeding to the Standardbred.

Sable Island Horse

ORIGIN: *Sable Island, Nova Scotia, Canada*
APTITUDES: *Feral horse, riding type*
AVERAGE HEIGHT: *13 to 14 h.h.*
POPULATION STATUS: *Rare*

Sable Island lies some one hundred miles off the east coast of Nova Scotia, an enormous sand bank with no gravel, rocks, or mud banks. In addition to the bands of feral horses roaming the island, Sable is home to many wild creatures — harbor seals, grey seals, red-

Sable Island horses at a water hole that they dug. Photo: Zoe Lucas

Sable Island stallion. Photo: Zoe Lucas

breasted mergansers, black ducks, blue-winged and green-winged teals, sandpipers, semipalmated plovers, arctic and roseate terns, and black-backed and herring gulls. Perhaps most important, Sable is the sole breeding ground of the Ipswich sparrow. A few species of insects are found on the island and nowhere else in the world. Great herds of walrus once made annual calls to Sable, but these were hunted to destruction for their teeth, hides, and oil.

Feral horses have roamed the island for so long that there is continual controversy regarding their origins. Probably no one has studied the history of the horses and the island as thoroughly as has Barbara J. Christie in her quest to identify origin of the herds. Her book *The Horses of Sable Island,* along with a good deal of personal communication, have been invaluable for the purposes of this volume.

Life on Sable is not easy and the horses have adapted as only horses can, managing to survive for many years without human assistance. Sable Island is a bit over twenty miles long and covered in large areas with low shrubs, grass, and, near the water's edge, wild pea. Converging sea currents of various temperatures subject the island to a great deal of fog, while devastating winds and storms constantly change its basic crescent shape. Treacherous shallow approaches and terminal bars reach out for many miles and, like the tentacles of a gigantic octopus, have taken hundreds of ships to the bottom of the sea—

especially during the age of sail. Justifiably, Sable Island has been called the Graveyard of the Atlantic and, by some, the Dark Island.

Usually referred to as ponies, the horses of Sable by their conformation and history are not ponies in the true sense, for while small in stature, their ancestors were all horses. Island living, with restricted food and the inability to migrate to better pastures, has more than once reduced horses to smaller size; nature's provision for survival. When removed to better circumstances, such horses increase in size steadily with subsequent generations toward their original genetic potential. More than simply size is involved in the differences between horses and ponies, but true ponies remain small whatever the case. An example is the Shetland, smallest of the pony breeds and native to the Shetland islands off the coast of Scotland. Shetlands have been taken to every country on earth, popular as first mounts for young children, yet many generations later are still small in size.

The first horse taken to Sable Island of which there is record of origin was named Jolly. Many of the horses of Sable today bear a resemblance to this stallion, who was taken to Sable in the fall of 1801. He had a fine, crested neck and strong build, a typical horse of the old Acadian breed (Canadian). Many horses had been taken to Sable before him, but he stands unique today as the first identifiable horse of the island. He survived hair-raising adventures on Sable, and was last included in the 1812 Sable Island annual inventory, at eighteen years old.

History does not record who first managed to run a boat through the roaring surf surrounding Sable Island or, surviving shipwreck and the crushing waves, first lay exhausted and gasping on those lonely shores. We will never know if the first human arrival to Sable found a temporary refuge or a sandy prison. We do know, however, that the Viking round ships, called *knarr,* made their way across the Atlantic about the year 1000, and later, Portuguese, Breton, Basque, English, and Spanish ships visited the rich fishing grounds off Newfoundland and Acadia (now Nova Scotia) in the early sixteenth century.

While Sable Island could have been known

to the earliest adventurers, they left no written record. It was left to the larger expeditions of Spain, Portugal, England, France, and the Netherlands to make the first known records of sightings and landings.

The first mention of animals on Sable is included in the accounts of the ill-fated voyage of Sir Humphrey Gilbert in 1583, when a Portuguese sailor stated that thirty years prior to that, the Portuguese had put on the island cattle and swine that had multiplied exceedingly. Some historians have insisted that horses were also put ashore at that time, but there is no historical or archaeological proof of this.

Those who insist on Spanish characteristics of the Sable Island horse are correct—but likely not from survivors of wrecked Spanish ships. The first horses sent to New France by the king of France (see Canadian) in the early seventeenth century were of Breton, Norman, and Andalusian extraction; hence, prepotent Spanish blood was extant in the horses taken to Sable by early settlers. Later, animals purchased from the New England states for use in Acadia were often of the preferred Barb breed, for they were known for stamina, which was needed for the hard life. Undoubtedly, many of the horses taken to Sable for use by the settlers were of Barb extraction. A settlement was maintained on the island for many years, the main purpose of which was assisting shipwrecked survivors until their removal to the mainland.

The horses of Sable were known to be extremely handy. Occasionally the feral herds were penned so that use could be made of them and were driven into the pens by others of their kind. Of the saddle horses of Sable, it was said that they had an incredible ability to keep their feet. They were said to trot, jump, gallop, paddle, rack, prance, shuffle, and waltz—carrying a stranger anywhere without the slightest inconvenience to themselves—and in the wild roundups, needed no urging to race uphill or down, turn quickly, or take a flying leap over a gulch. Many of the Sable horses were said to have a natural pace or ambling gait, and one has to wonder if some of the Spanish blood originally sent from France may have included that of the beloved Spanish Jennet.

One theory that has been considered is that horses swimming in from Spanish shipwrecks were the root of the modern Sable herds. This belief is also held closely by enthusiasts of the horses of Chincoteague and Assateague (see Assateague) as well as the feral herds found on the Outer Banks off North Carolina (Banker horses) and on Cumberland, off the southern coast of Georgia. The "horses from the sea" theory seems to crop up in every case of horses marooned on offshore islands of both the Americas and Europe. There is little doubt that, at least in some instances, some horses may have managed to accomplish this feat, but I am certain it would have been only in rare, isolated cases.

It is difficult to imagine how horses, crosstied and/or possibly slung so that their feet barely touched the floor, would have been able to free themselves in a rolling or perhaps sinking ship, escape the hold, leap over the side, and survive the swim to shore among high, crashing waves which slammed them into huge rocks. Horses shipped from Europe at that time were generally in sorry condition by the end of the long voyage, dehydrated from lack of water, sickened by the constant pitching of the ship, and usually almost starved, from lack of edible food or appetite.

Horses of the defeated Spanish Armada of 1588 are said to have influenced the horses of England, Scotland, and Ireland, but while this is an enjoyable speculation, it is doubtful. The great Spanish galleons, driven by storms and fleeing English pursuit, ran short of water in the North Sea long before rounding the Shetland Islands in their retreat to Spain. The fleet commander, Medina Sidonia, ordered all horses and mules thrown overboard. Hundreds of dead and still swimming animals were seen, but too far out to sea to hope realistically that any reached shore alive. All expendable horses had been jettisoned long before the last vessel passed Ireland. This was the usual conduct any time a ship met with grave difficulty.

Barbara Christie furnished me with some of the old records of horses shipped in those days, which are of great interest.

Shipments of horses from Sable Island to the mainland were driven into corrals where they were roped to a Bonapart hitch (something like a war bridle), which is a fairly thin rope running from the horse's head and jaw to a back leg. This was used to force the horse to move forward to the seashore. There, they were thrown, their two front feet tied together at the pastern and the same for the back feet, with soft hemp material between the joints. They were then tied to something that looked like a hand stretcher and taken to the surf boats, three or four to a boat. As Sable Island is surrounded by some fourteen miles of hidden sandy shoals, a larger vessel had to lie off shore. The surf boats came alongside and the horses were hoisted on board with ropes attached to the joined legs. Later, slings were used for hoisting—that is, when the seas and the struggling, terrified horses permitted.

During the days of sail, the horses were deck-penned. In the days of steam, they were stowed between decks in circular pens. Island water, hay, and grass was taken for them on the voyage. For steam vessels, the voyage was about fifteen hours; for sail, anything up to a week, depending on weather and wind. Unless an animal went down and was trampled, losses were usually light, although in a sailing vessel in bad weather losses could reach eight to twenty-one in a shipment of fifty head. Their short journey did not greatly affect their appetite.

Shipments from England to Halifax, however, were very different, and it was on the voyages that took weeks that the losses were greater. In 1799 a shipment of Thoroughbreds, Hanoverians, Cleveland Bays, halfbreds, and an Arabian were first loaded on July 17 at Sheerness, England. After a short voyage to another port, where they were offloaded for exercise, they were loaded again on September 5 with the sailing transport at last sailing on September 6. They arrived in Halifax on October 29. Seven of the horses died en route.

There is a horrific account of horses dying at sea. A shipment of thirty-nine Thoroughbreds on the *Helvetia,* destined for the American, R. W. Cameron, left Liverpool, England, on December 6, 1866, and arrived in New York on December 24. Thirty-four head had been lost during fourteen terrible, stormy days.

Regarding horses on their way to Boston from England, the Halifax *Morning Herald* (September 8, 1881) noted when their vessel, the *Caladonia,* stopped over at Halifax for supplies: "Nine horses; Suffolk Punches and Shires. They are stowed between decks and are so closely packed that they cannot lie down. Some have to be swung to rest their legs and the grooms walk over their backs to feed them."

Horses often went off their feed on long voyages and nearly all did when they smelled land, usually within a day's sail out. Attempts were made to feed them oatmeal gruel when they refused solid food. Since the days of the Romans, horse had been slung for long periods with their feet just touching the deck. They had extra ties to tail, head, and legs in rough weather, causing cuts and chafing. Deck-penned horses traveling long distances often died of dehydration and exposure. Those experienced with feeding of horses will know of the danger and difficulties often experienced when a horse has refused to eat for a twenty-four hour period.

Zoe Lucas, working with the Atmospheric Environment Service in Bedford, Nova Scotia, furnished some interesting information regarding the modern herds on Sable Island.

During surveys done in the 1970s and 1980s, the horse population on Sable fluctuated between 200 and 360 individuals. Usually cold and wet winters result in increased mortality, primarily affecting the very young and very old horses. However, during mild winters the mortality rate can be extremely low (5 percent or less). The population is divided into family bands. Band structure is quite variable but usually consists of one dominant stallion, one or more mares and their offspring, and occasionally, one or two subordinate males (the subordinate males can be either immature or sexually mature individuals). Average band size is four to eight, with bands of ten or more individuals rather unusual. The sex ratio is about 50:50, and males not in family bands form loosely organized bachelor groups, or, particularly if they are older, live as solitary stallions.

The typical Sable horse stands from 13 to 14 hands, with horses under 14 as common as those of 14 hands. The horses are thick-bodied, stocky, and short. Some have dished profiles, but straight profiles or Roman noses are more common. They are rugged animals adapted to moving up and down the sand dunes and walking and galloping in sand. Coat colors include bay, a wide range of browns, and chestnut, with mealy muzzles and dorsal stripes quite common. No zebra stripes have been observed on the legs. There are no greys, roans, duns, or spotted horses. Though some reports have stated otherwise, there are no palominos on Sable. Several individuals are so dark that they are described as black; however, it is difficult to tell for certain as there is so much weathering of the coat. About 45 percent of the horses have one or more white markings on face or legs.

About 40 percent of the island surface is covered by vegetation. Though they graze many of the 175 or more plant species on the island, the horses' primary forage is marram grass *(Ammophilia breviligulata)*, beach pea *(Lathyrus maritimus)*, and sandwort *(Honckenya peploides.)* During summer and autumn, the horses graze in the lush vegetation on the high dunes and the sheltered inland fields, putting on weight and storing energy for the coming winter. During the winter they gradually lose weight and body condition, and they begin to gain again in late May and June. First foals are born during the last week of April, and the last foals appear during October.

There are no trees on the island, but shelter is readily available in the lee of dunes and heathland hills. Drinking water is available at numerous freshwater ponds on the island. In addition, the horses dig water holes in hollows and low areas between the dunes and keep the holes open year round except in the very coldest weather. Even during winter the cold is rarely prolonged, but when ponds and water holes are temporarily frozen, the horses will eat snow.

The Sable horses are the only terrestrial mammals on the island. They have no predators (and have protected status under the Sable Island Regulations of the Canada Shipping Act). All research is non-invasive. A series of long-term research programs is under way, including studies of equine parasites, fertility and reproduction, forage quality and grazing behavior, as well as a number of other monitoring programs.

The Sable Island horses are not subjected to any kind of management, though there have been sporadic calls from various individuals and groups for management programs. These parties represent one or the other of two concerns. One is that the horses must be suffering by being trapped on such an isolated and inhospitable island (and that good horse-flesh is being wasted if it is not being trained and ridden), and they should be removed or fed by the government. The other concern is that the horses are having a negative impact on the island and its native species and should be culled, subjected to reproduction controls, or removed. However, both arguments are fraught with misconceptions and lack of information, and it is unlikely that anything will come of the controversy in the near future.

The Sable Island feral horse represents one of the few gene pools of early breeds that has not been tampered with or "improved" by selective breeding for many years and could therefore be a valuable resource of genetic material in the future.

Salerno
(Salernitano, Persano)

ORIGIN: *Italy*
APTITUDES: *Riding horse*
AVERAGE HEIGHT: *16 to 17 h.h.*
POPULATION STATUS: *Rare*

The Salerno breed, as its name suggests, developed gradually within the province of Salerno and in particular in the flat area extending from Battipaglia to Eboli. In ancient times this area was considered particularly well suited for horse breeding. During the period of maximum expansion of Italian horse breeds around the seventeenth century, great importance was accorded to the Salerno breed because of its notable versatility and usefulness for light transportation, long journeys, or for horseback riding. As a result it was suitable for both civil and military use.

Salerno. Photo: Sally Anne Thompson

The influence of excellent Oriental material, already imported during the period of the Republic of Amalfi, is apparent in the Salerno breed, notably as regards manageability. Oriental horses were used throughout Campania and, crossed with pre-existing horses, formed the foundation of the ancient Neapolitan breed. The Neapolitan was one foundation for several breeds in Italy and other countries, including the Lipizzan. Oriental influence in the Salerno breed is clear even today, although modified by environmental factors.

During the Borbonic era, Charles III founded a prestigious racecourse at Persano. This began the genetic fixation of the breed. At Persano the native Oriental horse, already introduced in Campania, was crossed with local horses and other breeds considered suitable for desired characteristics. Among these, the Lipizzan race had a major influence. The stallions Pluto, Conversano, and Napoletano are considered the founders of the Salerno breed.

Origin of the Salerno as a clearly defined ethnic group with its own morphogenetic characteristics began at the end of the 1700s following the opening of the horse establishment at Persano; the breed is also known as the Persano horse. The Salerno is genetically important, possessing the blood of very antique and prestigious breeders of the Arab (nervous and resistant) and the Neapolitan and Lipizzan (docile and elegant).

The Persano stud functioned from 1764 to 1864. Stallions were selected for breeding use as "deposit stallions" of a center at Santa Maria Capua Vetere. The center is currently known as the Institute of Equestrian Increase. The Santa Maria center had jurisdiction over a wide area taking in Campania, Basilicata, and Lazio and as far as the province of Frosinone. The Salerno horse had the maximum influence for improvement of horses in this area.

By the end of the 1800s the Salerno breed began to decline. The demand changed to one for horses that were lighter and more spirited for the calvary. In 1884 the Persano stud was closed. Thoroughbreds and some Hackneys were imported. Hackney stallions were largely used in the cenral Italian countryside, crossed with pre-existing horses which were later crossed to the Salerno breed. This produced a beautiful, elegant horse. In Calabria, the Salerno horse was used toward the selection of the Calabrian (Calabrese) breed. The Thoroughbred was largely adopted in Campania. The crossbred (Thoroughbred and Salerno) was an immediate improvement in size and spirit. This horse was greatly appreciated by the army, which bought large quantities of them.

After World War I a period of decline took place for the entire Italian horse breeding population. This was due to reduction of the cavalry and mechanization in agriculture and transport. There being little interest in equestrian sports or racing, many horses were abandoned and became feral. This crisis could have produced grave repercussions had the center at Santa Maria Capua Vetere not intervened in time with action to conserve the work done with the Salerno breed. The horse establishment at Persano was reopened as a "shelter for the four-footed," organized to host abandoned colts. The breed was saved from extinction and a field was sought for use of these horses.

During the period between the two World Wars, the authorities at Santa Maria used the utmost caution in selection of horses for reproduction. The Thoroughbred stallions Rockbridge and My First were purchased, both distinguished race horses. Four paths of breeding evolved at Santa Maria: the Salerno-Thoroughbred; the halfblood Salerno–halfblood Salerno cross; halfblood Salerno–Thoroughbred; and halfblood Salerno–Salerno.

The Salerno has a light head with a straight profile, medium-sized mobile ears, and large eyes. The neck is of medium length; the withers are short and prominent; the shoulder is muscular, prominent, quite long, and sloped; the back is short and strong; the croup muscular and rounded; the chest wide and deep. The joints are dry; the tendons well defined; the cannons slender but strong. Good structure of the hoof is considered a characteristic of the Salerno breed.

The Salerno is an extremely rare breed today with little more than twenty pure individuals remaining. Many halfbreds may be found. It is a superb sport horse with great endurance.

Sandalwood
(Sandel, Soemba, Sumbanese)

ORIGIN: *Indonesia (Sumba Isle)*
APTITUDES: *Riding, light draft, racing*
AVERAGE HEIGHT: *12.2 h.h.*
POPULATION STATUS: *Common*

The Sandalwood pony, also commonly called Sandel, is named for sandalwood, one of Indonesia's main exports. The pony is native to the islands of Sumba and Sumbawa. This is the best quality pony in Indonesia, carrying a good deal of Arab blood.

Sandel mare. Photo: Dr. Soehadji, Department of Agriculture, Indonesia

Sandalwood stallion. Photo: Dr. Soehadji, Department of Agriculture, Indonesia

A pony of great endurance, the Sandalwood is quite fast, especially considering its small size. It is ridden bareback on the island in local races, over distances of two to three miles. Crossed to the Thoroughbred, the Sandalwood has produced a very good strain of racing horses used on Cambodian and Thai tracks.

The coat color varies widely and includes most common horse colors. The pony has an attractive, light head showing Arab influence with alert ears and expressive eyes; the neck is well proportioned; the withers are well defined; the back is long and straight; the croup slightly drooping; the chest well developed and deep; the shoulder is sloping and gives good action and stride. The legs of this pony are strong; the feet are well shaped and hard.

For additional general information regarding the ponies of Indonesia, see Bali.

Sandan

ORIGIN: *China*
APTITUDES: *Riding, packing, draft*
AVERAGE HEIGHT: *13.3 to 14 h.h.*
POPULATION STATUS: *Common*

The Sandan breed was developed solely at Sandan Farm, located at the foot of the Qilian Shan mountain range at an altitude of 8,200 to 10,500 feet. The average temperature in the area is 32°F. The region includes plateau steppe, shrub prairie with some conifers, grasslands,

Sandan horse. Photo: Weigi Feng and Youchun Chen

Sandan mare. Photo: Weigi Feng and Youchun Chen

and semiarid pastures, supplying a unique place for horse breeding.

The Sandan steppe was called "large horse camp" in ancient times. During the Han dynasty it was a base for supplying large numbers of horses. Gansu-wanmashi, which means "temple of horse garden," was set up during the Ming dynasty and was located on the north slope of the Qilian range. In 1934, at the remains of this setting, the Sandan Horse Farm was established. Since that time local horses have been crossed with many exotic breeds. Earlier records show Yili, Chakouyi, Datong, Hequ, and other breeds, but Sandan horse size never reached more than about 11.3 hands. In the early 1950s Don horses were used, with each generation followed by grading up of crossbreeding, but adaptability in the horses was lost rapidly. In 1963 backcrosses were made by elite local stallions, and

crosses from them were successful. After the linebreeding program was practiced, the new breed was officially recognized in 1980.

Sandan horses are of riding and draft use. They have a good temperament and the conformation is harmonious with dry, clean legs. The head is light, the neck somewhat plain but well connected to the body; the body trunk is strong; the rump is wide; the natural stance is correct, and the legs have prominent joints and good feet. The main coat color is bay, followed by chestnut.

Performance tests of the Sandan horse at 9,000 to 13,000 feet above sea level at a rapid pace, wading rivers and climbing up and down mountains, showed impressive endurance. This breed can pull 89 percent of its own body weight.

Sandan horses are raised in large pasture herds with seasonal pasturage being popular. Simple shelter is provided only in winter, and oats are fed as an addition to the diet in winter and spring. The Sandan is bred pure, and improvement is by careful selection within the breed.

Sanfratello

ORIGIN: *Italy (Sicily)*
APTITUDES: *Riding, harness, light draft*
AVERAGE HEIGHT: *15 to 16 h.h.*
POPULATION STATUS: *Rare*

Sanfratello. Photo: Alberio Agatino

Information about the origin of the Sanfratello breed is vague and fragmentary due to lack of thorough studies on the ancestral phenotypes bred through the years and the difficult task of finding authoritative sources.

Local tradition claims that the Sanfratello horse is a descendant of the work and war horses that accompanied the founders of ancient San Philadelphia, known today as San Fratello. These founders were the Lombards who arrived in the eleventh century as part of the retinue of Adelaide, Count Ruggero's third wife. Others believe the horse is descended from the Sicilian horse. Even before Roman times the Sicilian horse was so well known that the Greeks admired and were jealous of it.

The Sanfratello horse still lives in the wild, roaming a large forested area found mostly in the province of Messina. More specifically, the area covers land within the communes of Acquedolci, Caronia, Militello Rosmarino, Santa Agata Militello, and San Fratello Mistretta, the northern slopes of the Ebrodi hills, and along the Tyrrhenian strip in the province of Messina in an area that runs between the hill and the mountain. Another small breeding area is found in the commune of Nicosia, in the province of Enna.

Climatic conditions here range from considerable heat and humidity in summer to cold in the winter. The Sanfratello is bred at altitudes ranging from sea level to about 6,000 feet at Mount Soro. The heat of summer and low temperatures of winter are mitigated by the sheltering effect of trees, shrubs, and bushes, creating an environment that is almost ideal for the breeding of Sanfratello horses. Nevertheless, the breed requires a certain amount of rusticity in order to thrive in this habitat.

This breed is completely wild, living in the Ebrodi hills and roaming in large groups, within which many impressive horses may be found. The Sanfratello is known for its morphological structure, physical health, and strength. The breed leads a wandering life, encountering every climatic and nutritional difficulty without artificial shelter. Besides surviving the cold, the Sanfratello must satisfy its hunger on scanty fare—grass stems and shrubs in pastures also foraged by cattle. Natural selection over the centuries has developed in the breed characteristics of strength, endurance, and rusticity. Obviously, these specimens do not suffer the diseases of civilization like horses raised in captivity. This background has made the breed a prime candidate for many uses.

Sanfratello horses have a strong skeletal structure that, combined with their sobriety, enables them to perform well even in the most disagreeable work and climatic conditions. They also have physical beauty, strong personality, and noteworthy ability in the athletic arena. In the world of agriculture, the Sanfratello horse has long been considered incomparable, especially in the south and in the islands. It has also been used in the production of mules. Approximately half of mule production in Italy takes place in Sicily.

Today, the Sanfratello is considered a valuable new export for sporting purposes. Breeders in the region have improved the breed's characteristics over the decades through crossbreeding, and it is seen as an animal that, with the proper crossbreeding, could raise interest in many equestrian circles. Athletic and strong, the Sanfratello is best adapted for difficult cross-country riding. In ravines and on rocky paths and steep slopes, the horse can greatly impress its rider.

Control over the quality of breeding stock turned loose to roam assures the continuance of excellent quality. Given the evolution that breeding has gone through over time, the Sanfratello horse should not be considered as a member of an ethnic group having uniform, finite genotypes and phenotypes (with standardized characteristics). Rather, the horse should be considered a polyhybrid.

Some authors have stated that the modern Sanfratello is the product of Hackney and Mecklenburg stallions as well as English, Anglo-Oriental, Maremmana, and Nonius derivatives. However, there is no documentation indicating that stallions of the first two breeds were used, although the others were certainly employed; unfortunately, some records assembled by the Equine Expansion Institute of Catania covering the years from 1864 to 1925 were destroyed in World War II.

Beginning in 1925 the decision was made to eliminate some of the fundamental problems of the breed, such as the heavy head, ram profile, large jaw, short shoulder, ram-like knee, and above all, weak pasterns. Large stallions of English and Oriental descent were used to phase out the unwanted qualities and increase size in the breed. Use of these stallions over a nine- to ten-year period partially modified the defects without refining the structure of the breed.

Population of the Sanfratello breed is about 1,500. Each year, the Equine Expansion Institute of Catania assigns nine or ten stallions to the selected stations—Fazio, Mirtoti, Sambuca, Raneri, and Sorba—located within the borders of Caronia and Cesaro. Breeding of the Sanfratello remains fragmented. Generally, every breeder has a group of six to ten mares. Because of the way the pastures are situated, the groups are assembled into groups of sixty to eighty, not including foals.

The short- and medium-term breeding objectives of the Sanfratello are to make it sober, strong, and tranquil. This will be done by using the best mares and Maremmana stallions, carefully studying the results. Offspring are targeted for use in equestrian sports.

Sanfratello horses reach maximum development at five years. The head is slightly heavy with an almost straight profile and marked development of the masseter (jaw) muscles. In many individuals the profile is ram-like. The neck is slightly short, generally straight, and crested; the withers are well formed and well attached to the shoulder; the back is of medium length, sagging in some individuals, and well attached to the croup; the croup is large and muscular, tending to a quadrangular shape, and sloped; the thorax is generally large, deep, and well formed; the shoulder is muscular, with good slope. The legs are well muscled and strong with clearly visible tendons; the fetlocks have some hair; the hooves are strong and solid. Generally the Sanfratello horse is bay or dark brown with various shades. No special marks are required. Morphologically, the horse is considered a muscular type with sensitivity and a sanguine temperament.

Sanhe

ORIGIN: *China (eastern Inner Mongolia)*
APTITUDES: *Riding, racing, light draft*
AVERAGE HEIGHT: *14.3 to 15 h.h.*
POPULATION STATUS: *Common*

Sanhe stallion. Photo: Weigi Feng and Youchun Chen

The Sanhe area is in the eastern part of Inner Mongolia Autonomous Region. Sanhe means three rivers in Chinese. The basin formed by these rivers is one of the best grassland areas in China and became the base for forming the Sanhe breed, considered the best of developed Chinese horse breeds.

During the Liao dynasty about one thousand years ago, there were horses of good quality in the area, some having been sent by nomadic tribes to the emperors. In the Qing dynasty (seventeenth century) the Soulun breed, famous for its beauty and good saddle type, was bred in the area. It was used as a cavalry horse against invaders. The Soulun nation lived in the Heihe region and were related to the Manchu.

In 1904 to 1905 Russian horses were brought through the Baikal with Zabaikal horses to cross with the local breeds. In 1917 Russian emigrants brought Orlov and Bechuk (Biçuk) improved horses and settled in the area. The Biçuk horses were derived through crosses of native Russian horses to Orlov Trotters, Danish, and Dutch horses. This horse is known as Malakan today in Turkey. From 1934 to 1945 a stallion farm was established by the Japanese with records of Anglo-Arab, Arab, Thoroughbred, American Trotter, and Chitran horses. In 1955 the Ministry of Agriculture of the People's Republic of China organized a special assessment of crossbreeds, then established two stud farms to develop a new breed under the name of Sanhe. Another breed, called Hailar after a city of that name, was also developed and is famous as a race horse in Shanghai hippodrome.

Sanhe horses have compact, strong conformation. Bay and chestnut are the prevalent colors, and other colors are rare. The light type of Sanhe is best at racing and is also used extensively for saddle and harness work.

The Sanhe is well made and the body is harmonious. The shoulder is well slanted, giving good action and a smooth ride. The neck is slightly heavy. As in all Chinese horses, the legs are strong with good muscling and joints. Good, hard feet are typical of most Chinese horse breeds.

Sardinian Anglo-Arab

(Anglo-Arabo Sardo)

ORIGIN: *Italy (Sardinia)*
APTITUDES: *Riding horse*
AVERAGE HEIGHT: *15.1 to 16.1 h.h.*
POPULATION STATUS: *Common*

The Sardinian Anglo-Arab spring from Arab and Barb horses brought to Italy when the Saracens dominated Sardinia. Their horses were crossed to native horses of small stature. Early in the sixteenth century Andalusian blood was imported to improve the breed.

The Sardinian Anglo-Arab went into decline at the beginning of the eighteenth century, but new Arab blood revived the breed in the beginning of the twentieth century. Recently Thoroughbred blood has been used, producing the Sardinian Anglo-Arab of today, a horse differing greatly from its ancestors.

This breed has shown great talent in equestrian sports.

Sardinian Pony

ORIGIN: *Italy (Sardinia)*
APTITUDES: *Riding, light draft*
AVERAGE HEIGHT: *12 to 12.2 h.h.*
POPULATION STATUS: *Rare*

The island of Sardinia has contained horses for centuries, and accurate information regarding the pony of Sardinia is not available. Throughout history Sardinia has imported horses. The island breed is said to stem originally from Arab and North African horses, but records show that Ferdinand the Catholic (1452–1516) had a stud of Andalusian horses and allowed breeding privileges to the public. When Sardinia passed to the House of Savoy in 1720, horse breeding declined. In 1908 Arabian stallions were imported, and development of the Sardinian Anglo-Arab began.

While we have fairly accurate and complete information regarding formation of the Sardinian Anglo-Arab, records are apparently silent on the small horse or pony known simply as the Sardinian pony. Although it is clearly an ancient breed, the first reliable reference to the Sardinian pony is dated in 1845.

This pony lives in the wild on a plateau on the island, about 2,000 feet in elevation. Living conditions are poor and the vegetation thick but scrubby. Pastures are sparse, flourishing only in the spring. Perhaps the Sardinian pony is a descendant of early imports of Arabian and Barb stock, and island habitat in the wild under adverse conditions and scant feed have caused the breed to diminish in size. This would make the Sardinian pony a small horse and not a pony, and, indeed, Sardinian foals are proportioned like horses rather than ponies.

The Sardinian is a tireless worker and very strong for its small size. The breed is hardy and surefooted and does well on a minimum of feed.

The most common colors are bay, brown, black, or liver chestnut. White markings are minimal. The head appears rectangular from the side but very wedge-shaped when viewed from the front. The profile is straight or slightly convex; the eyes are large and expressive; the ears are small and curved inward at the tip. The neck is of medium length, arched, and well muscled; the withers are well defined; the shoulder is sloping and muscular; the back is short and sometimes slightly hollow; the croup is sloping with a low tail-set; the chest is deep and wide. The legs are slender with long cannons and pasterns and strong joints; the hooves are small but sound. Often, the Sardinian is cow-hocked.

The breed is used under saddle and for light draft work on the land. It is slow to mature but very long-lived and healthy.

Sárvár

ORIGIN: *Hungary, Austria*
APTITUDES: *Riding horse*
AVERAGE HEIGHT: *15.2 to 16 h.h.*
POPULATION STATUS: *Rare*

A rare Sárvár stallion. Photo: Evelyn Simak

The Sárvár was originally a Hungarian breed, developed at Mezöhegyes stud farm, a blending of Thoroughbred, North Star (see Furioso), and Furioso blood. This is a halfbred type, very much like the Senne. The stud was founded by emperors of the Austrian Habsburg dynasty and later was owned by the family of the Weittelsbaches. Hungary was under the rule of the Austrian emperor until 1914, when the Austrian prince was assassi-

nated in Sarajevo. World War I was initiated by his death.

The Austrians had influenced Bavaria, and the emperors were closely related, Austria and Hungary almost being one country at the time. During the war the owner of Sárvár Stud, Prince Ludwig, had to flee with his horses. He ended his flight in southern Bavaria near Lake Starnberg, where his family owned a castle and some land. Today the stud is owned by one of his sons, His Royal Highness Prince Ludwig of Bavaria, who still breeds a small population of Sárvár horses. The greater part of this breed (fifty animals) were returned to Hungary some years ago and placed at an ancient stud farm at Pusztaberény. At the Ludwig stud no foreign stallions had been used apart from Thoroughbred, and the Sárvár is considered in Hungary to be of the old strain of Furioso blood.

Sárvárs are on the small side for German thinking, and most are bays. For tradition, one grey is kept in the German herd. This breed has the least Thoroughbred blood of the Hungarian warmbloods and numbers only about 300 today.

Sárvár horses are used for long-distance rides, cross-country races, and coaching. They are quite full of temperament. In Germany there are about twenty broodmares and four stallions. One Thoroughbred stallion is also kept at the stud. The horses are kept outside to keep them hardy.

Schleswig
(Schleswiger)

ORIGIN: *Germany (Schleswig-
Holstein)*
APTITUDES: *Heavy draft work*
AVERAGE HEIGHT: *15.2 to 16 h.h.*
POPULATION STATUS: *Rare*

Rarest of Germany's coldblood draft breeds, the Schleswig is predominantly a chestnut horse with a flaxen mane and tail. Some bays and greys also occur. The breed takes its name from Schleswig-Holstein, the northernmost region of Germany, and early ancestors were Jutland horses from neighboring Denmark.

Schleswig Heavy Draft. Photo: Evelyn Simak

Schwarzwälder Füchse
(Black Forest Chestnut)

ORIGIN: *Germany*
APTITUDES: *Heavy draft work*
AVERAGE HEIGHT: *15 h.h.*
POPULATION STATUS: *Rare*

The rare Schwarzwälder Füchse. Photo: Evelyn Simak

This breed is also known by the name Niedersachsen. In the early days, some Yorkshire Coach Horse and Thoroughbred blood was used, but this influence was not a lasting one. The main goal in breeding was to produce a draft horse of medium size, and the breed had deveoped by the end of the nineteenth century. This horse was in great demand as a bus and tram horse.

Blood of Danish horses was used up until the Second World War, but then Boulonnais, Breton, and Suffolk Punch stallions were introduced. One Suffolk stallion in particular, Oppenheim LXII, had a great influence.

The modern Schleswig is a compact, cob type which is active and has a good, docile disposition. Currently there are only about 100 registered mares and ten stallions remaining in the breed.

The head of the Schleswig draft horse is well proportioned with a slightly convex profile, pricked up ears, and small eyes; the neck is short, muscular, and crested; the shoulders are very powerful and muscular, somewhat straight; the body tends to be long and slabsided but has plenty of depth; the withers are low and broad; the back is strong and short; the croup is full and round. The legs are short, muscular, and have strong joints, with feathering on the lower leg. The feet of this horse tend to be flat and soft.

Breeders are encouraged to keep the breed alive with financial help from the government.

The Schwarzwälder Füchse is an old breed, a small coldblood intermediate in size between the Haflinger and Noriker. The breed was influenced by the Noriker in the past. *Füchse* is a German word meaning "sorrel."

This breed is indigenous to the higher altitudes of the Black Forest (Baden Württemberg), a poor area where farmers needed a surefooted work horse that would not require a great deal of food. Another name for this breed is Black Forest Chestnut. The breed also carries Breton blood.

The Schwarzwälder Füchse typically has an attractive head showing refinement, wide between the eyes and tapering to a small muzzle; the ears are small, the eyes large and expressive; the neck is short and muscular with a clean throat-latch; the withers are set low into the musculature; the shoulder is sloped and powerful and heavily muscled; the back is short; the croup is sloped with a low tail-set; the body is deep and well rounded; the chest is broad and deep; the legs are trim, strong, and clean of heavy feathering. The breed is typically sorrel with a white or flaxen mane and tail and a broad blaze on the face.

Selle Français

(French Saddle Horse)

ORIGIN: *France*
APTITUDES: *Riding, sport horse*
AVERAGE HEIGHT: *15.1 to 17 h.h.*
POPULATION STATUS: *Common*

The Selle Français, Flambeau C. Photo courtesy UNIC, Paris, France

The French have always had a passion and acknowledged talent for breeding good horses. History of the horse in France has been greatly affected by changes in breeding classification and policy, and breeds have changed in appearance accordingly, with the exception of the purebred Arab and English Thoroughbred.

When William the Conqueror crossed the English Channel to march on England, he took with him a large number of Norman horses. These were said to be powerful, enduring, and very much appreciated as war horses during those times. The horses of Normandy were the foundation of the Anglo-Norman breed. A great deal of improvement took place in the native horses of England with the addition of Norman blood.

The Norman horse eventually deteriorated through haphazard breeding with the Mecklenburg cart horse and some Danish breeds, and after 1775 Arabs and Thoroughbreds were used to improve the stock. Between 1834 and 1860 a large admixture of Norfolk Trotter blood gave origin to the Anglo-Norman Trotter, bred primarily in the district of Morleraut.

The soil of this area is rich in iron and lime, and the qualities of climate and water greatly enhance the breeding of excellent horses with strong muscle and good bone. Anglo-Norman Trotters were very enduring and hardy animals with a good reputation.

Two basic types were found in the old Anglo-Norman breed, the first of which was a draft type standing from 15.2 to 17 hands. These horses had a good deal of Percheron and, later, Boulonnais blood. Usually grey but sometimes bay, chestnut, or black, they were used for pulling mail carts because of their ability to move a heavy load at a good, fast pace. The second type was the cavalry horse, used in the army and for sports. Breeding centers in Normandy and the Bearn possess racing records dating to the sixth century. The first regular races, however, took place during the reign of Louis XIV. Toward the end of the eighteenth century Thoroughbreds were imported from England, and in 1780 racing rules were drawn up.

In 1833 the Société d'Encouragement, which still rules French racing, was founded, and two years later it created the Prix du Jockey Club (the French Derby). By 1865 Gladiateur's victories in England proved beyond any doubt that the French Thoroughbred could hold its own against its English counterpart.

After World War II, Normandy began to dress its wounds, rebuilding destroyed towns, devastated farms, and a countryside which in general appeared a desolate spectacle. Horse breeding had suffered the marks of war. As the horse was not an object of luxury and leisure during the four years of occupation but rather an instrument of labor, many breeders had their broodmares covered by Cobs to increase strength and size. In addition, along came the swift progress of motorization, increasing under the impetus of the Marshall Plan, causing the horse market to suffer considerable repercussions from 1947 onward.

Normandy had a large number of mares at its disposal, mostly cart horses and Cob breeds, as the Thoroughbred had made little impression before the war on native breeds. Half the stock of the National Stud at Saint Lô was destroyed, and most of the stallions survived the liberation by remaining at the stud farms, where only a few animals perished.

Gradually, horse shows were organized. Before the war the only existing show was at Avranches. Though the first shows were without the nicely turned out horses and freshly painted obstacles in front of an elegant audience, there was a sporting atmosphere which made the shows successful. At the same time, breeding trends changed. Breeding in Normandy once again had the means necessary for success: exceptional terrain, quality stallions, and breeders of goodwill, all working together to bring this capital to bear fruit. The Selle Français, sport horse of international excellence, was at the door.

The main regions for production of saddle horses in France are: the Hague, Cotentin, Bessin, and Avranchin. Situated at the extreme northeast of the Cotentin region, the Hague is an area of austere landscape with direct influence from the sea. The soil is not very deep here and does not retain water, which falls in abundance.

Horses of the Hague are characterized by their rustic and robust form. This they undoubtedly inherited from their distant ancestors, the Hague ponies, which still populated this area in the middle of the nineteenth century and were said to resemble the British ponies. But the horse of the Hague owes its sobriety and endurance to the infusion of Arab blood that took place during the nineteenth century. The saddle horse of the Hague has retained its own characteristics; a strong, tough animal of sober character.

The Cotentin proper, situated in the center of the Cotenin peninsula, corresponds to the perimeter of an ancient gulf that existed in former geological times. Thanks to the chalky subsoil and clay soil, the area boasts fine pastures year round. On the other hand, the bordering land is less privileged: wooded zones, the moors of the Bricquebec and St.-Sauveur-le-Vicomte regions, and the marshes of the Carentan gulf, submerged during the winter but able to carry considerable numbers of horses and other livestock at other times of year.

The Cotentin is a rich area, where the area cultivated is greater than the regional average and where horse breeding goes back to ancient times. Up to the middle of the nineteenth century, the local breed of "Black cart horse of the Cotentin" was produced here. It was valued as an important horse with strong muscles, but unfortunately often suffered from roaring and exaggerated lethargy.

Besides this breed, there were also "speedy nags," small, untiring horses that covered great distances for those who needed to move about a lot. One of the dominating features in the betterment of the local horses was the contribution of the trotting horse, which was to revitalize the breed with its well-known energy.

The Bessin is a transitional area with rich soil, divided into vast pastures; here many large estates and a few Thoroughbred breeders are to be found. Breeding of the trotting horse is also widespread in the area. The French Saddle Horse broodmares are linked to three stud farms at Isigny, Bayeaux, and Trévières. This area has always bred very good horses, which is why the Bessin coach horses of yesteryear were in demand as gala horses for the emperor. Many Selle Français stallions are products of the Bessin broodmares.

The Avranchin, a verdant, gently rolling region, benefits from the gentleness of a temperate climate favorable to the growth of grass, which now covers increasing areas of land formerly plowed.

Geologically, the area consists of dense alluvium covering ancient sedimentary terrain. The people of Avranchin have always cherished horses, especially the Thoroughbred, but the Avranchin conditions lends themselves better to the breeding of a sturdy medium-sized horse rather than well-developed horses. Using the Thoroughbred, breeders obtained horses which are distinguished in appearance, finely made, with a fluent bearing.

Several breeds usually termed halfbreds evolved in the various horse breeding areas of France, including the Vendéen, Charollais, Corlais, Angevin, Limousin, and others. Horses of the popular riding type were crossed to Anglo-Norman, English Thoroughbred, and

Anglo-Arab stallions, producing the prototype of a highly bred, somewhat tall, powerful horse of fine conformation and renowned for its jumping skills. The Charollais came from the center of France, an ancient native breed upgraded by Anglo-Norman and Thoroughbred blood. This horse had tone and quality, producing fine hunters.

The Limousin was a remarkable horse, full of quality and averaging about 15.2 hands; extremely refined, supple, and light on its feet; ideally suited to dressage work. Additions of Thoroughbred and Anglo-Norman blood made it an even more excellent animal.

With the Anglo-Norman often used to cross on native mares, the various regional breeds began to resemble each other more and more, and creation of the Selle Français was a logical action. All of these regional breeds were effectively "canceled" in 1958 and gathered into one stud book termed Selle Français, which means French Saddle Horse.

Due to its diversified origins, conformation of the Selle Français varies, but selection tends toward large size and attractive conformation. As a rule, this is a horse with powerful build and musculature and strong-jointed limbs. The neck is often long and carries a slightly heavy, although elegant head. The color is usually chestnut or bay and more rarely red roan or grey. The Selle Français may be over 17 hands. Of calm disposition, this horse usually has a good temper. The breed is classified in five sections of the stud book according to size and capacity to carry a medium or heavy rider.

The Selle Français remains today an amalgamation of breeds, with 33 percent having a Thoroughbred sire, 20 percent Anglo-Arab, 45 percent Selle Français, and 2 percent trotter. Most valuable are the competition horses. This breed has shown great aptitude for jumping, dressage, and eventing. Next in importance are the race horses. France has many race meets not for Thoroughbreds, most of these being over fences. Then there are the nonspecialized horses with a variety of uses such as riding schools, trail riding, and general pleasure.

Luckily, moral fiber and energy are not exclusive to the Thoroughbred, and old breeders have all known good cart horses that were used for long treks along the roads, keeping up a fast pace and preserving healthy limbs. Mares by such cart horses are often the base of today's excellent French Saddle Horses. The old Norman horse lives on through the Selle Français, a careful blend of today's best bloodlines and substance.

Senne
(Senner)

ORIGIN: *Germany*
APTITUDES: *Riding horse*
AVERAGE HEIGHT: *15.2 to 16.2 h.h.*
POPULATION STATUS: *Rare*

Senne horse of Germany. Photo: Evelyn Simak

The Senne is considered a halfbred type, emerging from indigenous warmblood horses now extinct and crossed with Arabs, Spanish horses, and Thoroughbreds.

Originating in the Senne, an area of heath on the southern slopes of the Teutoburger Waldes between the cities of Bielefeld and Paderborn, the Senne horse is an old breed dating to at least 1160. A private stud originally situated near the city of Detmold and then moved to Lopshorn in 1680 (and then named Lopshorn) became well-known for its fine horses. In the eighteenth century Arab blood was crossed to the horses of Lopshorn.

Senne horses are used for riding and coaching. They are hardy, as the tradition from the beginning was to keep them outdoors all year round. Easy keepers, they have good tempers and are late to mature. In the place of origin, the mares and foals were taken from pasture only to select horses needed for work or riding.

After 1870 the horses were not kept in the old way and natural selection no longer governed the type of the Senne horse. In 1935 the Lopshorn stud was closed and the horses sold to private breeders. J. M. Jmmink, from the Netherlands, brought several of the horses and bred them in Lopshorn. In 1946 the stud was finally dissolved entirely.

In the hands now of a few dedicated breeders, the Senne horses are once again bred as they were originally in several places in Germany. Today there are about fifty horses of Senne type. The main coat color is grey and some are bay.

This is an attractive, well-made horse, elegant in appearance and action. Arab influence is clearly evident in its characteristics, and the Senne horse has good endurance. The head is of Arab type, with a straight profile and often the classic dished face; the neck is long and slightly arched and fits well into the shoulders and withers; the withers are well pronounced. The back is long; the croup gently sloped; the shoulder well sloped and mucular; the chest deep; the barrel well rounded. The legs are refined yet with good bone, strong joints, and hard feet.

Shagya. Photo: Dr. Csaba Németh

An oddly marked Shagya mare. Photo: Evelyn Simak

Shagya Arabian

ORIGIN: *Hungary*
APTITUDES: *Riding, sport horse*
AVERAGE HEIGHT: *15 to 16 h.h.*
POPULATION STATUS: *Rare*

During the Turkish occupation of Hungary from 1526 until 1686, numerous Arab and Oriental horses arrived in the country. The Bábolna Stud Farm, founded in 1789, adopted the breeding of Arab horses and improved its stock with original, imported Arab stallions. Since 1870 this stud has bred Arab horses exclusively. With improved husbandry and professional breeding processes, the Arab horse increased in body mass while preserving nobility and character.

Progenitor and name-giving sire of the breed was the stallion Shagya, imported from Syria in 1836. Other stallions such as Kemir, Siglavy Bagdady, Gazal, Jussuf, Mersuch, O'Bajan, Koheilan, and Kuhaylan Zaid also became founders of the breed. The breed was simply called "Arab" until 1982, but following that year the name was officially changed to Shagya Arab after the most populous strain. This breed has become internationally recognized.

The Shagya is an easy saddle and leisure horse and is used to improve other breeds, adding good conformation, elegance, and constitutional strength. Every Hungarian horse breed carries Arab blood to some extent. The Shagya, compared to the desert-type Arab, is of more mass overall, a taller, more robust horse.

The Shagya is a beautiful animal, proportional and appealing in appearance. The head is characteristically concave in profile with high-set, small ears, large eyes with a vivid expression, and wide nostrils. The neck is long, slender, slightly arched, with a fine, silky mane; the withers are medium high and long; the back is slightly forward and a bit sunken. The loin is short; the croup is narrow, of medium length, and scantily muscled; the barrel is not as deep as in some other breeds but still well sprung; the legs are strong with good, dry joints. In general appearance the Shagya looks like the desert Arab, the only actual difference being the larger size. The breed is overwhelmingly grey in color, with some bays and rarely black or chestnut.

Primarily a light saddle horse, the Shagya is also excellent in harness to light carts. Of vivacious temperament, hard, enduring, with appealing action, yet docile and easily managed, the Shagya is used to improve the appearance of common-bred heterogenous mare stock. A rare breed, the Shagya must be protected genetically, and prevention of genetic narrowing is of great importance. In order to avoid this, crossing of unrelated Arab blood is constantly used. Selection is directed also to maintain the Shagya conformation and size as well as the natural measure of tameness, vivacious temperament, and solid constitution.

The Shagya is gaining in popularity in the United States as a small but growing nucleus of the animals are winning friends wherever they are seen.

Shan
(Burmese)

ORIGIN: *Burma*
APTITUDES: *Riding, light draft, pack pony*
AVERAGE HEIGHT: *13 h.h.*
POPULATION STATUS: *Common*

Bred by the hill tribes of the Shan state in eastern Burma, the Shan shows many similarities to the Manipuri and is believed related to other ponies of northern India—that is, the Bhutia and Spiti. Descended from the Mongolian, the Shan shows evidence of mixture with foreign bloods. This is a strong pony, slower in movement than some other breeds.

Larger and taller than the Manipuri, the Shan is considered a version of that breed. It was used in the past as a polo pony by British colonials, being the only horse available. The Shan is a surefooted mount, reliable for use on steep trails in the mountains.

Prevailing colors found in the Shan breed are bay, brown, black, grey, or chestnut. The head is small and light with a straight profile; the eyes are typical of the Oriental horse; the neck is well proportioned and muscular; the withers are somewhat flat; the back is straight and long; the croup is sloped; the chest is deep and long; the shoulder tends to be straight. The legs are slender but strong with good joints, and the hoof is tough.

Shetland

ORIGIN: *Scotland (Shetland Islands)*
APTITUDES: *Riding pony, light draft*
AVERAGE HEIGHT: *Maximum of 42" (10.2 h.h.)*
POPULATION STATUS: *Common*

Shetland. Photo courtesy Shetland Pony Stud-Book Society

A group of Shetland ponies. Photo courtesy Shetland Pony Stud Book Society

Smallest of all pony breeds, the Shetland is known as "Scotland's Little Giant." This pony is believed to trace to the same origins of small pony stock found in Iceland, Scandinavia, Ireland, and Wales. It is possible that the pony was introduced to the Shetland Islands from Scandinavia before the lands were divided by water about 8,000 B.C. and that it subsequently crossed with ponies brought over to Scotland by the Celts. In any event, the Shetland is among the oldest of known breeds.

In the eighteenth century the Shetland was bred in large numbers on the islands, and only a few were shipped out on small sailing vessels to various European countries; the remainder worked on the island crofts. This changed in 1847 when an act of Parliament prohibited children from working in the coal mines. Previously, children had been employed for pulling coal tubs along the mine workings, many of which were too low to allow even a small child to stand upright. Almost immediately, there was a great demand for the Shetland pony to replace the children, and very soon a great boom in trade for the ponies took place, upwards of 500 being shipped annually to the mines. To avoid stabling problems, only colts and stallions were taken, and in the 1850s a mature pony would fetch a very substantial price.

Conditions under which the Shetland was developed, with harsh weather, little or no shelter, and scant feed, produced through nature's reliable method of survival of the fittest a very strong, hardy animal. This pony is known for its long life span, sound feet and legs, wonderful endurance, and good eyes.

Comparatively stronger than its counterpart the horse, the Shetland can carry half its weight for many miles without ill effects.

On the islands, the ponies were put to work carrying peat from the hills for fuel. The peat was packed in "cassies," each weighing no less than sixty or seventy pounds, and two were slung across the pony's back on a pack saddle of sorts, shaped like a sawhorse and fastened by a breast collar, girth or two, and breaching. The whole load weighed as much as a good-sized man, and the pony often carried this load for several miles without a halt.

In the late nineteenth century, Lord Londonderry and various other mine owners took great interest in the breed, and many formed their own studs. They recognized together with many Shetland Island breeders that the considerable trade in male animals was reducing the number of good stallions available for use on the islands. Realizing the necessity of improving the breed, they founded the stud book.

The first volume was published in 1891 and included 457 ponies, all which had been inspected by the committee for correctness of type and conformation. The height limit was fixed at 42 inches, and all colors other than spotted were allowed.

Ponies were still being sold in numbers to the pits in the years prior to the Second World War, but during the war years there was a catastrophic drop in stud book registrations, especially in Shetland. Quality of the ponies bred on the native islands deteriorated, and this was partly due to the crofters, who had shares in the common grazing land (known as scattalds). Since crofters had the right to run any pony, male or female, on these commons, many scrub stallions ran wild, resulting in haphazard matings.

This was cause for great concern to the Department of Agriculture in Scotland, which was and still is active in promoting improved breeding in all kinds of livestock. No solution could be found until the Crofter's Act of 1955 established the Crofter's Commission, which had powers to regulate common grazing. A regulation was made that only registered and approved stallions could run on the scattalds.

Each year, twelve selected stallions approved by the Stud-Book Society were sent to the islands for the breeding season, a modest fee being paid to their owners. This stallion scheme continues to this day, but happily almost all of the stallions are now owned in Shetland, reducing the need to import them from the mainland. Results were immediately evident in the improved quality of the foals, and this prompted the society to organize auction sales in Shetland. They were a great success and took the island breeders out of the hands of dealers, who had become their only buyers.

Over the years, numerous ponies have been exported to the United States, the Netherlands, Denmark, Norway, Sweden, and indeed all over Europe.

On the mainland, in addition to their use in the pits, Shetlands were much sought after as driving ponies, and a pair was used at Windsor and at Holyrood by Queen Victoria. The Shetland has a naturally kind disposition and makes an excellent first mount for children. Presently, Shetlands are extensively ridden by children and driven in harness.

Because these ponies are "cute," some owners have had a tendency to spoil them, and the subsequent misbehavior has given the Shetland a bad reputation for being rebellious. However, the Shetland is no more easily spoiled than any other equine breed, and if trained correctly makes a wonderful mount for a child.

In the United States, different breeding practices have resulted in a modified version of the breed.

Shire

ORIGIN: *England*
APTITUDES: *Heavy draft*
AVERAGE HEIGHT: *16.1 to 17.3 h.h.*
POPULATION STATUS: *Uncommon*

Information detailing the origin of the Shire breed is varied and not always in agreement. This breed is of great antiquity and is thought descended from the "great horse" used in medieval times as a charger for jousting.

A team of Shires. Photo courtesy of the Shire Horse Society, Great Britain

Shire. Photo courtesy of the Shire Horse Society, Great Britain

Sir Walter Gilbey, an early authority on the breed, contended that the Shire was the purest survivor of the "great horse." If the Shire did not originate in England, this country at least acquired a very early and widespread reputation for producing it. The English "great horse" seems to have been a native development of the British war horse, whose strength, courage, and aptitude for discipline were praised by chroniclers of the Roman legions upon their first landing on English shores.

The history of England and that of the Shire horse are inexorably entwined. During the reign of John I (1199–1216) one hundred stallions were imported into England from the lowlands of Flanders, Holland, and the banks of the Elbe. At least some strains of England's heavy horses must owe their origin to these stallions. In the period between the reign of Henry II and that of Elizabeth I (beginning in 1558), it was the constant aim of the government to increase the size and number of horses. Large, strong horses were needed to carry horse soldiers, who, combined with their armor, weighed upward of four hundred pounds.

It is probable that the Shire was strongly influenced by the Friesian breed, as during the sixteenth century Dutch engineers were hired to drain the fens and their black Friesian horses came with them. From that period large black horses called the Lincolnshire Blacks were known in the area.

While useful in war, the Shire proved to be even more valuable in peace. The breed became nothing less than a national treasure in the 1800s. Large geldings moved the commerce of England off the docks and through the streets of the cities. Over badly paved streets and on rough roads, weight was opposed by weight. For decades there was a great demand for massive horses with great muscular strength.

The marshy fen counties of Lincolnshire and Cambridgeshire lay claim to having exerted the earliest beneficial influence upon the breed, and it was from these counties that sales were first made for the improvement of draft horses all over England.

The Shire is the world's largest horse, in some cases reaching 19 hands with a weight of 2,640 pounds. Like other draft breeds, this one went into decline in the 1950s but has experienced a renewal of interest.

Shires may be bay, black, brown, chestnut, or grey, frequently with white markings. The head is small in relation to the size of the body, with pronounced jaws and a somewhat convex profile. The forehead is broad, the eyes are large and prominent, and the ears are long. The neck of the Shire is quite long, arched, and muscular; the withers are wide and continue the line of the neck; the back is short; the croup sloping; the chest broad and muscular; the shoulder long and well sloped. The legs are short, well muscled, and feathered below the knees and hocks; the joints are broad, the cannons usually somewhat long; the hoof is sound.

Silesian
(Slaski)

ORIGIN: *Poland*
APTITUDES: *Farm and draft*
AVERAGE HEIGHT: *15.1 to 16 h.h.*
POPULATION STATUS: *Rare*

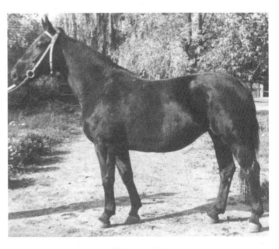

Silesian. Photo: Andrzej Turczanski

The Silesian horse emerged around the turn of the nineteenth century in Lower Silesia, the result of crossing Oldenburg and East Friesian sires with local halfbred mares. This developed what has been called the largest warmblooded, clean-legged horse in all of Europe, a powerful animal very useful in both agricultural and carriage work. The Silesian has good action at the walk or trot and great traction power. The breed is said to be free of vices, although particular in regards to its feed and living conditions.

Breeding of the Silesian was interrupted during the war years but was continued immediately following World War II, and in the process of improving the breed some Thoroughbred and halfbred stallions were introduced into the state breeding program. Most Silesian horses are bay, dark bay, or black, and there is little variation in the shades of these colors.

Today the state breeding is done on two stud farms, maintaining a population of around one hundred broodmares: Strzelce Opolskie (Opole County) and Strzegom (Walbrzych County). Since the early 1970s the Silesian horse has been in great demand among country farmers, and the area of breeding has spread into southern and central Poland.

Lately the Silesian has found a place in carriage competitions and has kept a leading position, winning a number of international carriage events.

The purebred population for registered mares is about 1,500.

Sini

ORIGIN: *China (Sini River Basin)*
APTITUDES: *Riding horse*
AVERAGE HEIGHT: *13.2 to 14.2 h.h.*
POPULATION STATUS: *Common*

Sini. Photo: Weigi Feng and Youchun Chen

Originating in the early twentieth century in the Sini River Basin in eastern Inner Mongolia in China, the Sini horse was developed from crosses of Baikal (Buryat) mares in Siberia with Sanhe (see Sanhe) stallions. Progeny were interbred for many years. Some other bloods were involved in development of this breed, but their effect was negligible and those breeds were eliminated from the breeding program. By 1982 there were 10,000 Sini horses in China. Pastoral riding is the main use for this breed, which is quite fast and extremely enduring.

The Sini is generally a goodlooking horse. The head is light and clean, with a straight or sometimes slightly dished profile; the neck is not heavy but is well muscled; the withers are of good height and well defined; the back is long, straight, and very strong; the loin is long; the croup is gently sloping and relatively short, with a high tail-set. The shoulder is generally well sloped. The legs are clean and strong with well-defined tendons and good muscling, and the hoofs are dry and strong. The mane and tail are long and silky.

The main colors found in the Sini breed are bay, chestnut, and black. Other colors are rare. Some white markings may be found on the face and legs.

Skyros

ORIGIN: *Greece*
APTITUDES: *Riding pony, light draft*
AVERAGE HEIGHT: *9.2 to 10.3 h.h.*
POPULATION STATUS: *Rare*

The Skyros is the smallest of all Greek horses. There is some debate over whether the Skyros is a pony or a miniature horse. Interestingly, between the latitudes of 30° and 45° from Greece into China, several small breeds developed that are remarkably similar in appearance, including the Mytilene of Turkey, the Sardinian pony, the Carpathian, the Caspian of Iran, and the tiny Guoxia of China. All resemble miniature horses more than ponies.

Skyros pony. Photo: Penny Turner

A team of Skyros ponies. Photo: Penny Turner

The Skyros horses developed in the mountainous and stony southeast of the Aegean island of Skyros. They have lived in this area for many centuries in a semiwild state and their exact origin is unknown. The islanders always kept a few of these ponies during the summer months to use for farm work. Turned loose for the winter months, they roamed freely in small herds without supplemental feeding, so in severe winters numbers were often reduced.

With the advent of mechanization of both transport and agricultural machinery, use of ponies has been greatly reduced. There are approximately 102 pure Skyros left on the island, including twenty-four mares and seven stallions. A small group is kept at the University Farm in Thessaloniki. In addition to those mentioned, a small number of Skyros may be found in various other parts of Greece and are kept mainly for children to ride.

Penny Turner of Thessaloniki has been breeding the Skyros for over twelve years and owns some of the best individuals of this ancient breed. She reports that the Skyros breeds true to type and has many excellent qualities. It has a delicate bone structure yet is strong, and lameness or problems in the legs

are uncommon. The temperament of the Skyros is good, spirited but friendly and tractable.

The preferred color is bay. Brown, grey, and rarely dun are also seen. The hooves are black and always very hard. The Skyros seldom needs shoeing, even when trekking in rocky areas.

The Skyros is very small with a narrow body and short neck. The head is pretty and light with a large eye and wide forehead, small ears, and large, open nostrils; the chest is flat, the back straight, and the croup gently sloping. The mane and tail are long. The legs are free of hair and are long and fine-boned but very strong.

Slovak Warmblood

ORIGIN: *Czechoslovakia*
APTITUDES: *Riding, carriage horse*
AVERAGE HEIGHT: *16 h.h.*
POPULATION STATUS: *Unknown*

Similar to the Czech Warmblood, the Slovak Warmblood is a saddle and harness horse. The primary difference between the two breeds is that the Slovak Warmblood has more blood of the Hungarian halfbred, Furioso, Przedswit (a halfbred Thoroughbred strain), Gidran, and Nonius and little of the Oldenburg and East Friesian. Recently the breed was improved through the use of Trakehner and Hanoverian blood.

The Slovak Warmblood breed is not unified yet, and both light and more massive types are represented within the breed. The nature of this horse is alert, and the gaits are good and energetic. The general appearance depends on ancestry, but it is an elegant warmblood. Feeding requirements are moderate.

The goal in breeding Slovak Warmblood horses is to fix the type and breed in purity, with emphasis on the massive type.

Sokólka
(Sokólski)

ORIGIN: *Poland*
APTITUDES: *Heavy draft work*
AVERAGE HEIGHT: *15 to 16 h.h.*
POPULATION STATUS: *Common*

A powerful draft horse originating in the northeast of Poland, the Sokólka (also spelled Sokólski) was developed over the last one hundred years. The breed has a sound constitution and economical feeding habits, making it popular as a farm worker. The Sokólka was developed from Norfolk, Breton, Ardennes, Døle, Belgian, and Anglo-Norman stock. This is not an excessively heavy animal, and the legs are nearly free of feather. The Sokólka has a free-moving gait. Its temperament is calm and docile, yet the horse is active and full of energy.

The Sokólka has a large head with straight profile, upright, nicely shaped ears, and large, expressive eyes; the neck is long and well muscled, quite broad at the base; the withers are well pronounced; the back is short and straight; the croup sloped and muscular. The chest and girth are deep; the shoulder is sloping; the legs are muscular and stout with short cannons, well-defined tendons, and a round, broad hoof.

The predominant color in this breed is chestnut, but bay and brown also occur.

Sorraia
(Sorraiana)

ORIGIN: *Portugal and Spain*
APTITUDES: *Riding and pack horse*
AVERAGE HEIGHT: *12.2 to 13.2 h.h.*
POPULATION STATUS: *Rare*

Sorraia. Photo: Evelyn Simak

Originating in the western region of the Iberian Peninsula and specifically in Portugal north of Lisbon, the frugal Sorraia is an ancient and important breed. This small horse is believed to be the primary ancestor of the famous horses of Andalusia and, both directly and indirectly, ancestor to many horses throughout Europe and the Americas. The Sorraia is found portrayed faithfully in prehistoric cave art (see Andalusian), displaying the classic Iberian profile often credited to the North African Barb. The name Sorraia is taken from the rivers Sor and Raia, small tributaries of the Sorraia, which irrigates the plains where these horses have roamed for millennia.

The Sorraia represents *Equus stenonius,* one of the few types of original wild horses known. While some believe the breed to be descended from the Przewalski's Horse, discussions with leading geneticists have convinced me this could not be true (see Przewalski's Horse). There could, however, be strong ties to the wild Tarpan, now extinct.

The Sorraia is noted for its ability to withstand extremes of climate and to survive on very little food while at the same time maintaining its health. This hardiness once made the breed highly valuable to local stockmen. Too long-legged to be seriously considered a pony type, the Sorraia is a small horse.

It is believed that Sorraias were among the small horses taken to the Americas by the Spanish conquistadores, probably for use as pack animals. Sorraia blood in the Americas would explain the emergence of several breeds in both North and South America that bear the dun coloration and other physical characteristics of this ancient breed.

Presently the Sorraia is rare, its numbers diminished and remnants to be found only in a small preserve maintained by the d'Andrade family in Portugal. A few specimens can be found in Germany. The breed has attracted the attention of zoologists and geneticists from all over the world.

The Sorraia is known to have a calm, steady nature and willingness to learn and is believed the progenitor of these characteristics so cherished in the Spanish horse.

In color the Sorraia is invariably some

shade of dun or grullo, a primitive coloration with a black dorsal stripe, black-tipped ears, and a heavy black mane and tail. Usually the mane and tail are fringed with the body color. The legs are striped with bars (zebra stripes). There is also a stripe across the shoulders. The pure Sorraia never has white markings. The breed displays no infusion at any time of either Oriental or northern European blood.

The head is somewhat large with a convex profile and wedge shape. The eyes are set high, the ears long and curved; the neck is long and slender but well shaped; the withers are quite high and well defined; the back is short and straight; the croup is sloping; the tail is set low; the chest is deep and narrow; the shoulder is rather straight; the legs are long and solidly built with strong joints, well-defined tendons, and long, sloped pasterns. The hooves are hard and dark in color.

This breed is believed to have migrated across the prehistoric land-bridge at Gibraltar to North Africa thousands of years ago (see Andalusian, Moroccan Barb) and contributed to the breeds found there today.

South African Miniature Horse. Photo courtesy P. W. du Preez

South African Miniature

ORIGIN: *South Africa*
APTITUDES: *Riding horse, light draft*
AVERAGE HEIGHT: *Maximum of 35"*
POPULATION STATUS: *Common*

In South Africa the miniature horse is represented by the South African Miniature Horse Breeding Association in Glenstantia. The South African Miniature Horse may be no taller than 90 cm. (35") and is not a pony but a scaled down model of a full-sized horse. Miniature horses in South Africa come in a variety of conformational types, from the refined Arabian to the heavy-boned draft horse type. True miniatures are not dwarves or genetic errors. By nature they are gentle, affectionate, and easily trained and make ideal companions. Because of their small size they are easily handled and have minimum requiremens for shelter, fencing, pasture, and feeding.

Origins of the first miniature horses are obscure. Some feel the first miniatures were bred for the royal courts of Europe as early as the sixteenth century and that no miniatures existed prior to that time. However, in China there is a miniature known as the Guoxia, which existed during the Han dynasty and was used as a plaything in the imperial courts. The Guoxia still exists today as a rare breed.

In Europe the remarkable ingenuity of the Renaissance period was required to produce this special breed of tiny horses. Stature of the horse was gradually reduced from generation to generation. The royal stable master was given a free hand and everything necessary was supplied, as the royal stables were considered a part of the national treasure. The horses produced there were counted part of the personal fortune of the royal family.

This life of supreme luxury was short-lived for the miniature horse. As the kingdoms of Europe began to decline, the tiny horses found themselves in traveling circuses. In the rigors of travel from city to city, selective breeding was impossible to maintain and the breed nearly became extinct. Several animals managed to survive, but were scattered throughout the world.

The earliest reports in the United States of miniature horses date to the early 1880s when such horses were included in shipments from Europe. A man in Virginia named Norman Fields purchased some of these tiny oddities and in the course of fifty years was producing miniatures as small as 75 cm. in height.

Today many breeders throughout the world are producing miniature horses. The breed is characterized by an even temperament. During the breeding season the stallions are said to be very manageable.

South German Coldblood

(Süddeutsches Kaltblut)

ORIGIN: *Germany (Bavaria)*
APTITUDES: *Heavy draft work*
AVERAGE HEIGHT: *16 to 16.2 h.h.*
POPULATION STATUS: *Uncommon*

The South German Coldblood is a close relative and considered a smaller version of the Austrian Noriker. The original breeding grounds for both the Noriker and South German Coldblood were Bavaria and the mountainous regions of Austria, once belonging to the Roman province Noricum. The German version of this breed received infusions of Norman, Holstein, Cleveland Bay, Hungarian, Clydesdale, and Oldenburg blood. This resulted in a smaller, more refined draft horse.

South German Coldblood. Photo: Evelyn Simak

A team of South German Coldbloods. The ear coverings are for protection from flies. Photo: Evelyn Simak

While in Austria all colors are to be found in this breed, most specimens in Bavaria are chestnut with a white mane and tail, or bay.

This breed has been used to improve and create other breeds in Germany. South German Coldbloods have been used extensively in the past for carrying artillery and packs in mountainous regions. Like the Noriker, this breed is active and quick, with fairly high action.

While the population has diminished with that of other heavy draft breeds, the South German Coldblood is still popular with small mountain farmers and the breed is also used in forestry work, where horses are more efficient than machinery.

The South German Coldblood is a heavy working horse. Distribution is the upper country of Bavaria and the area of the Bayerische Wald, a forest region near the border of Czechoslovakia. The breeding aim is a horse of medium size which is hardy, dry, agile, and able to work in both hilly terrain and flat country, for farm and forest work.

While South German Coldbloods have basically been bred pure since 1880, some Clydesdale and Belgian blood was used after World War II, and since 1989 a Suffolk Punch stallion has been used.

Southwest Spanish Mustang

ORIGIN: *United States (Southwest)*
APTITUDES: *Riding horse*
AVERAGE HEIGHT: *13.3 to 15 h.h.*
POPULATION STATUS: *Rare*

Southwest Spanish Mustang. Photo: Gilbert Jones

Several groups are breeding Spanish Mustangs, and there is some confusion as to why so many groups exist. Gilbert Jones of the Southwest Spanish Mustang Association kindly furnished some insight. The following is quoted with his permission from a 1984 issue of that association's newsletter.

In the beginning, there was only one Spanish Mustang registry, formed by Robert E. Brislawn, Sr., Robert E. Brislawn, Jr., and Lawrence P. Richards, in 1957. At first, their only interest was in preserving the few remaining remnants of *pure* Spanish Mustangs; however, as the registry grew and more members became involved, differences of opinion caused some individuals to branch off and form their own registries. The Spanish Barb Breeders Association, originally formed by Susan Paulton of Hot Springs, South Dakota, was formed to perpetuate breeding a particular *type* of Spanish Mustang. They pursued a breeding program whereby they came up with a Barb-appearing horse. The Southwest Spanish Mustang Association was established with a twofold purpose: To keep the purest stock in existence by keeping strict breeding records and to make the public aware of the stamina and endurance capabilities of the breed. Other registries were formed to encompass the wild mustangs being gathered by the Bureau of Land Management (BLM).

All of the officers and directors of these registries know each other well, having been closely associated for many years. Differences of opinion caused the eventual splits among them. Some

horses were inferior and never should have been allowed into any of the registries. Some breeders bred for color. Others bred for show qualities only, while others bred strictly for performance, paying no attention to bloodlines whatsoever. Some called theirs Barb-Arabian, some Indian Pony, while others called theirs Barb and Spanish-Barb.

The Southwest Spanish Mustang Association was formed by Gilbert Jones from all the foundation bloodlines, and this is the reason the Association still accepts some horses for registration that come from the other registries. Basically, the same little Spanish Mustangs founded *all* of these registries and they all emerged out of the same brew.

The SSMA maintains a very comprehensive library at Medicine Spring Ranch, which documents history of the breed, the forming of all of the registries, early photographs of foundation stock with their pedigrees, and hundreds of letters received through the years from the various officers, directors, and breeders.

The fascination Gilbert Jones has with the Spanish Mustang dates to 1924, when he was seventeen years old. His uncle gave him a Spanish Mustang mare named Susie. Some of the Mustangs now running on the Medicine Spring Ranch have the blood of this mare. Others, obtained steadily over the years, trace to authenticated Spanish Mustangs foaled in 1887. Gilbert Jones feels there is a great deal of evidence to confirm his belief that two different strains of Spanish horses developed in North America: the Spanish Mustang, derived from horses left in Florida by Spanish explorers; and the western Mustang, descended from horses left in Mexico.

"There was a time," Gilbert said, "when I was gathering them, that half the ranchers thought the Mustang had more cow sense than any horse. Then the oldtimers started dying off and the word *'mustang'* got a bad ring to it. A lot of people just naturally had it in for mustangs. They thought they was mean, hard to break, hard to handle. Now they're gaining a comeback, maybe because so many people want to be what I call old-timey."

The Spanish Mustang is not like the feral horse of today, but the Mustang breed that evolved from horses brought to the Americas by the conquistadores. Nearly all pure Spanish Mustangs were gone by 1890 due to slaughter, crossbreeding, and castration. By this same time the buffalo had been slaughtered almost to extinction and cattlemen stocked the western ranges with cattle. They brought in blooded horses from the eastern states to cross with the local mustang mares. The resulting offspring, although larger than the native horse, never had the natural herding instinct and "cow sense" the pure Mustang had, or the endurance.

Perhaps the last episode in the long and colorful career of the Mustang was the gathering of millions of wild longhorn cattle off the Texas range in the 1800s. The Mustangs were ridden by a breed of men as wild as the longhorn and as tough as the Mustang. The Mustangs swam every river from Texas to Canada, enduring stampedes, tornadoes, hailstorms, and freezing blizzards. They did it all while foraging on grass and brush without grain. They came through it with their heads up. No other breed of horse could have accomplished what the Mustang did.

The aim of the Southwest Spanish Mustang Association (SSMA) is to restore and preserve the few pure remaining specimens of this fine breed existing today and to prove their ability to perform as they did in centuries past. All crossbreeding is discouraged by SSMA due to the rarity of the breed.

The Spanish Mustang's sensible disposition, smooth way of traveling, and size make it an ideal horse for all members of the family. It is very intelligent and learns extremely fast. This breed averages from 13.3 to 15 hands and weighs 750 to 1,000 pounds. With modern feeding methods and good care, today's Mustangs are reverting back to their ancient size of 14.2 to 15 hands. This is a very attractive, smooth-bodied horse. There are two different types, one lighter in the body and slightly leggy, like its Barb ancestors, and the other blockier, like the old Spanish Jennet.

There are three types of head profile; straight, slightly dished, and convex. The ears are medium in length, thin and slim with a small base, notched at the tips; the forehead is of medium width; the eyes are large and bright, showing intelligence, often with much white sclera showing around the eye. Some of the purest have glass eyes and multicolored eyes; some have heavy bone protruding over the eyes. The muzzle is refined and small; the mouth is shallow, with thin and firm lips; nostrils are small and crescent-shaped; some of the purest in blood have a moustache. The neck is moderate in length, never heavy, with a well-defined throat-latch; the chest is medium in width to somewhat narrow; the ribs are well sprung; the heart girth is deep and the shoulder is sloping; the withers are well-defined but not too pronounced; the back is short and stout but never wide; the croup is short with a low tail-set; the hips are well rounded but never wide; the barrel is slightly tapered. The legs are strong and of medium bone with long, smooth muscles that carry down well into the knees and hocks; the forearm is long; the cannon is of medium length, never too heavy; the tendons are strong and well separated from the bone. Some of the purest have no chestnuts, but when they are present they are small and nonprotruding; ergots are small and often indistinguishable; the fetlock hair is sparse or lacking; the pasterns are medium in length and springy for easy riding; the hooves are small and very tough.

No breed of horse has the variety of colors seen in the Spanish Mustang, ranging through all the colors and patterns found in horses. For additional information see Spanish-American Horse.

Soviet Heavy Draft
(Sovetskaya tyazhelovoznaya)

ORIGIN: *Former Soviet Union*
APTITUDES: *Heavy draft work*
AVERAGE HEIGHT: *15.2 to 15.3 h.h.*
POPULATION STATUS: *Common*

Development of the Soviet Heavy Draft breed started in western Russia between 1880 and 1890; in 1952 it was officially recognized as a new breed. The breeding zone was quite extensive, including Vladimir, Yaroslavl, Gorki, Penza, Ryazan, Tula, Tambov, Voronezh,

The Soviet Heavy Draft at a performance test. Photo: Nikiphorov Veniamin Maksimovich

Soviet Heavy Draft. Photo: Nikiphorov Veniamin Maksimovich

and Orel regions and Mordovia. This was a zone with developed industry and intensive agriculture, and there was great need for a strong horse of ample size.

In the beginning, native horses of the area were improved by stallions of the Belgian Brabançon breed. In 1885 three Brabançon stallions were recorded in use in the above zone; in 1895 they numbered fifty-eight; in 1905 there were 394; and in 1945, 891. Mares of various origins and breeds (Percherons, Ardennes, Suffolk, Danish, and some varieties of saddle horse) were mated with the Belgian stallions for three to four generations and the offspring were bred to each other. At the same time, grading up was taking place on a large scale and crossbred stallions were widely used.

The modern Soviet Heavy Draft is distinguished by ample height and clearly expressed harness type. These horses are heavily muscled and boned. The head is average is size and clean-cut; the neck is average in length, often quite short, and heavily muscled; the withers are low; the back is long, often slightly dipped; the loin is of average length; the croup is broad, drooping, furrowed, and heavily muscled; the chest is broad and deep; the ribs are well sprung. The forelegs are often pigeon-toed and the hind legs are often sickle-hocked. The build is quite coarse, soft, and yet much cleaner and more solid than some other breeds. The predominant colors are chestnut, brown, and bay.

This breed matures extremely early and is a good producer of both meat and milk. It is not as disease resistant or as adaptable to extreme management conditions as some other breeds are.

The leading breeding centers for the Soviet Heavy Draft are Pochinkovsk in Gorki Region and the Mordovian Stud Farm.

Spanish-American Horse

ORIGIN: *Spain, the Americas*
APTITUDES: *Riding horse*
AVERAGE HEIGHT: *14 to 15.2 h.h.*
POPULATION STATUS: *Rare*

This inclusion is made because nearly fifty distinct breeds exist in the Americas (North, Central, and South) that were based on Spanish blood brought by the conquistadores during the conquest.

While there is some controversy (but little), it is generally agreed that the horse died out in the Americas prior to the Spanish conquest. Indians of both North and South America, upon first seeing horses, were frightened of the animals and communicated that they had never seen such beasts. Many at first thought a horse and rider to be a single godlike entity.

Further, there is no memory of horses in the ancient past among the many and various Indian tribes of the Americas—it is absent from their early oral tradition and there is neither any word for it in their original languages nor any ancient art of those cultures resembling the horse. Some people claim that the horse did not die out entirely and hold solidly to early historical references to "wild" horses, believing there already existed large herds when the earliest European explorers came to the New World. To my knowledge, no scientific proof has been found to substantiate such claims.

Generally it is understood, based upon archaeological evidence, that the prehistoric horse expired in the Americas for an unknown reason and first returned with the second voyage of Christopher Columbus in 1493.

Many have believed the Andalusian and Spanish Jennet were the only breeds that accompanied the conquistadores, and if that be true, one is forced to wonder at the great diversity found among the horses termed "Spanish" in the Americas. Nearly all are believed by their breeders to be "pure" and uncontaminated by foreign blood. Realistically, this is doubtful.

As this book approached publication after four years of preparation, proof was found that the Sorraia—a primitive, indigenous horse of Spain—did indeed come to the Americas, and possibly in great numbers. Blood type testing done at the University of Kentucky by Professor E. Gus Cothran has found in the Sorraia distinctive genetic markers that are also found in the Cerbat (see Cerbat), an isolated pocket of Spanish horses found in Arizona. This particular marker has not been found in any other horse breeds tested, and there have been many, both American and European. That the Sorraia genetic markers are intact and clearly evident in the Cerbat after so many years is an amazing revelation. Those who feel the "old blood" is gone should reconsider.

To comprehend the Spanish horse of the Americas, one must look closely at its history, and a very broad picture is quickly evident. Spain contained, as horses of its own development, the Andalusian and Spanish Jennet, descended (but not entirely) from indigenous Spanish horses; the Sorraia, Garrano (Minho), Asturian, Galician, and Basque or Pottok—all *still existing* today except the Jennet and Galician (see Andalusian). There were also various crossbreds of varying quality.

The Andalusian was produced through admixture of many different breeds taken into Spain and Portugal over the ages and crossed with the native horses—Libyan, Goth, Roman, Phoenician, and others—and, much later, with the horses of North Africa (see Andalusian). The unique combination of both hot and cold blood produced the horse that was famous throughout the known world for centuries and considered the best horse in the world.

Invasion of Spain by the Barbarians influenced the equine population, as these people brought heavy Germanic horses and small Mongolian stock of Asiatic origin. It is not known for certain how long the influence of foreign horses was felt in Spain prior to the Moorish (Arab) invasion of 711 A.D. History reveals that the great Moorish calvary would have had immense influence on equine production. Christians, recognizing the usefulness of warring with light, spirited, and fast animals, readily adopted this type of horse.

Long before the discovery of the New World and for a considerable time after, Spain was famed for having the best saddle horses in all of Europe. This was due in part to the prolonged occupation of the Iberian Peninsula by the Moors, whose fine Barb horses, though not noted for their beauty, were unrivaled for speed, refinement, patience, and endurance. The fame of the Spanish horse is largely attributed to the Barb, yet the Spanish horse was famous prior to introduction of many Barb horses. In the year 707 the Arabs conquered Barbary, and they held sway for many generations. While the invasion of the Iberian Peninsula was thus politically made by Arabs, the armies were Berber and so were their horses. History of Berber and Moorish horses reveals they were of the same type and does not speak of Arabian horses. It is pertinent to note that while Barbary is very near to Iberia, Arabia is very far from it. The total picture is lost to our view simply due to the passage of time and lack of complete records.

When Mohammed went to Medina in 622 A.D., he rode on the back of a camel, for there were no horses available. His preferred mounts were two mules. Learning that there were some feral horses in the desert, remaining from those of the legendary herds of King Solomon, he went there and obtained five mares "as swift as the wind." Legend says these gave birth to the Arab breed through selective breeding. Even if that is the case, the Arab breed could not have grown by the year 711 to the point to have horsed the thousands of warriors who invaded the Iberian Peninsula. At the time the Arab horse was still scarce, highly prized by its owner, covetously pampered, and kept for personal use; there was no surplus to horse an army. Those who insist that thousands of pure Arab horses were sold to invading armies should consider that it was (and still is) akin to a religious taboo for an Arab to sell his good horses (see Persian Arab). To sell a horse is like selling a friend or member of the family. Solomon obtained his horses from Cilicia (ancient Anatolia, now Turkey), and they were undoubtedly of the Turkmenian strain.

The Mohammedan conquests began in 634 A.D. The Muslims, by 696, had completed dominion of most of northern Africa, entering Spain in May of 711. Their cavalry was largely composed of Berbers, whose horses had come from Numidia, Syria, Egypt, Nubia, Barbary, and Africa. The great preponderance of these were Barbs. Domination of Iberia by the Mohammedans lasted nearly eight centuries. Only one region in the north of Spain resisted it, and there some Navarre Jennets still exist.

Musa arrived in Spain with some 19,000 additional Arab troops in 712, and for the next few years he and Tarik occupied more than two-thirds of the Iberian Peninsula.

The Barb (Barbo, Barbe, Berber, or Moorish horse) is said to have had its origins in Numidia, the ancient home of the Berber people. Inhospitable conditions developed its special characteristics. No breed is more temperate and rustic than the Barb, considered by many to have been the best war horse.

According to the South American researcher Rinaldo dos Santos, one breed introduced into Spain in addition to the Barb but long before — when the Romans entered the country — was the Marismenha, the horse of the salt marshes, believed to be the ancient Camargue breed. This horse was well liked by Julius Caesar. The Camargue is known as "the horse of the sea." Centuries later, while the Barb was requisitioned for the breeding of war horses in Iberia, the Marismenha was considered more and more *puduro* (an animal without value). The mixtures of Barb, Marismenha, and other small horses of Spain used as packhorses were among those sent for colonization of America, along with pure examples of those breeds and, according to some, a small number of the most valuable horses of the period, the Andalusian. It is a matter of record that later voyages carried some of the finest Andalusian stock, once accommodations had been erected for them.

Without doubt one of the most misunderstood terms when dealing with history of the horse is the word "oriental." Oriental blood does not always imply the Arabian horse, which did not exist when the old Andalusian emerged.

In the 854 years between 711 and the colonization of Spanish Guale in 1565 (now Florida), the Oriental and Barb horses were crossed and recrossed on both the native stock of Spain and its adopted imports, producing the Andalusian, the world's greatest war horse. These were the horses, among others, that were later taken to the Americas. The Andalusian was the vehicle by which Spain was able to claim vast empires and world prominence; the horse that laid claim as progenitor to many of the world's modern horse breeds.

One glaring misconception regarding the Spanish horse of the pre-conquest period is that at the time of the Moorish invasion of Spain in 711 A.D., the Spaniards, like many of their European neighbors, had only heavy, draft-type horses, and the invading Arabs conquered Spain solely by maneuverability upon their light desert horses. This is not entirely accurate. The horses of Spain at the time of the Moorish invasion were heavier than those of the invaders, but they were neither draft nor slow horses. They were muscular, agile riding horses, already renowned far and wide for their superiority as saddle animals. The slowness of the Spaniards may be attributed to their style of riding and bulky armor.

Once Spain overthrew the yoke of the invaders, expeditions were sent to the New

World, landing in the West Indies and carrying in the holds or on the decks of ships the horses destined to become the progenitors of many American breeds. The first horses to arrive disembarked in December of 1493 in Hispaniola, sponsored by Ferdinand and Isabella of Spain. It is evident that the Crown was determined to form breeding farms in the New World at the earliest opportunity. The royal edict of May 23, 1493, states: "We command that a certain fleet be prepared to send to the islands and mainlands which have newly been discovered—and to prepare the vessels for Admiral Don Christopher Columbus—and among those we command to go in the vessels there shall be sent two horses each, and the two horses which they take shall be mares."

The intention was to ship Andalusian mares of fine quality to mate with the cavalry stallions. However, Columbus wrote to the Crown in 1494 complaining bitterly that due to fraudulent dealings of his cavalrymen, substitutes had been made and he had received "common nags." This could indicate the first introduction to the Americas of mixtures of Sorraia, Garrano, or other small Spanish horses often used as pack animals.

The Sorraia, Sorraia-Andalusian cross, or other small, common horses, could be purchased for one-tenth the price of a highly bred Andalusian, and it is likely that they were the "common nags" mentioned by Columbus. No other written statements have been uncovered concerning substitutions, but if it happened once, it may have happened again. It is also probable that the Sorraia, Garrano, and other small horses of Spain were shipped intentionally for use as pack animals. Many modern Spanish Mustangs bear remarkable resemblance to the Sorraia in color, size, and conformation, as does its South American cousin, the Criollo.

Horse breeders all over Spain and Portugal bred animals that, though perhaps quite closely related to each other, were the products of individual breeders. Some were bred for war, some for work in the bull ring, some for pleasure riding, and many for other work. Though all were Spanish horses, they must have had conformational differences reflecting their uses and the tastes of their breeders. When the Crown declared its intent that stock transported to the New World should include finely bred horses, we cannot at this late date specify what those animals would have looked like, though certainly they all followed the typical conformation and description of the old Iberian horse. To me it is doubtful that Spain would have sent its best horses into uncharted waters and unknown peril, aware that their return was beyond consideration and their survival in doubt. It was these horses that were the foundation of the first breeding farms of the Americas, arriving in various colors and patterns no longer found in modern Andalusians but certainly found in its descendants in both North and South America.

The first Spanish horses brought to the Atlantic Coast mainland were landed in Florida by Ponce de León in 1521. He arrived with two ships, two hundred men, and fifty horses. The horses must have been shipped at Puerto Rico as Ponce de León put in there en route to Florida. This hurried colonizing expedition was a dismal failure, only one ship surviving to return to Cuba. The horses that did not die would likely have been taken aboard ship and not abandoned, as horses not disabled were very valuable. It is improbable that any progeny could have been left on the mainland from this lot of horses.

Royal breeding farms were established in the Indies, Cuba, and at other locations with purebred Spanish horses of the highest order provided to them. Consignment after consignment arrived, and eventually the horse population from these bases spread with the onward march of the conquistadores and the settlers that followed them. The Portuguese settlers of Brazil imported horses usually termed Lusitanos, carrying much the same blood as that of the Andalusians; there is little on record about these imports, but they certainly contributed to the horses in the New World. Eventually the importation of Spanish horses ceased, and horses from North Africa were shipped to the Americas. By then the breeding farms were well established, providing fine horses in great numbers.

During the sixteenth, seventeenth, and eighteenth centuries the explosive population of Spanish horses in North America was the result of Spanish colonization of the West Indies and Mexico. About the same period, however, other countries were making their claims on areas of the Americas, and they brought their horses with them. In the mid-seventeenth century the French, English, Dutch, and Danes struggled for control of the West Indies and North American continent. The French descended the Mississippi in 1682 and claimed the entire river basin for France, filling the area with the old French-Canadian horse.

The Indians became mounted and found that not only was the horse beneficial to them for riding (and eating, in certain cases), but it was also a means of trading with other tribes. Indian raids into what is now Mexico were frequent and fruitful, as the Mexican-bred horses were easier to capture and brought more booty than the wilder animals now running on the plains and in the mountains. Expeditions by Coronado and others provided nucleus animals for other wild herds, as every escaped or stolen animal was a potential breeder. The Spanish used no geldings.

Due to Indian raids, storms, and incidental straying of domesticated Spanish horses, by the time of the overthrow of the Santa Fe settlement in 1680 the local Indians were well versed in horse handling. When the Spanish beat a hasty retreat southward from Santa Fe, they left thousands of head of stock, including horses. Though some of these undoubtedly were taken by the Indians, many joined the feral bands on the plains, further enlarging the wild herds. Spanish missionaries contributed their share of horses to the plains and deserts, losing many along their routes. Thus the wild herds proliferated. The feral herds multiplied at an amazing speed and became known as "mustangs," derived from the Spanish word *mesteño* or *mostrenco,* meaning wild or roving.

Formation of new horse breeds began in the Americas soon after colonization and settlements from other countries, with almost immediate mixture into the Spanish horses of imports from France, England, Holland, and other countries. Some breeds retained more Spanish blood than others, and in some cases,

such as those of the Spanish Mustangs and Spanish Barbs (see breed accounts), some remained pure in isolated pockets of feral herds or in the hands of breeders (see Cerbat and Kiger Mustang). By the latter part of the nineteenth century, however, most purebred descendants of the conquistadores' horses had been crossbred or reduced to near extinction. Had it not been for a few privately owned fragments, the Spanish horse in North American would have been lost. Central and South America and the West Indies have also received the addition of horses from many other countries during the passage of time since the conquest, and of the horse breeds in the Americas based upon Spanish blood, only a few remain with no mixture in their veins.

Spanish Anglo-Arab
(Hispano)

ORIGIN: *Spain*
APTITUDES: *Riding horse*
AVERAGE HEIGHT: *14.3 to 16 h.h.*
POPULATION STATUS: *Common*

The Spanish Anglo-Arab developed from crosses between Arab-Spanish (Andalusian) mares and Thoroughbred stallions. A synonym for the breed is *Tres sangres,* meaning "three bloods." The breed is known for its courageous attitude and is often used to test the fighting spirit of the bulls chosen for the bullring in the contest called *acoso y derribo.* In this event the riders test the bull by bringing it down with a thrust from a long pole. Horses used in this event must be extremely agile, for often the bulls rise and charge at considerable speed. The Hispano is an excellent all-round riding horse and a proficient jumper, dressage, and event horse.

Conformation in the Spanish Anglo-Arab is varied, as it is a composite breed, sometimes resembling the Thoroughbred more than the Arab and often with Andalusian characteristics. Most often it has an elegant, light-framed body. The head is well shaped and proportioned, usually with a straight profile, large, expressive eyes, and small, mobile ears; the neck is of good length and usually arched; the withers are pronounced; the back is straight and short; the croup gently sloped; the chest full and deep; the shoulder long and sloping. The legs are long but strong with solid joints and prominent tendons; the hooves are small and well shaped.

Spanish Barb

ORIGIN: *United States*
APTITUDES: *Riding horse*
AVERAGE HEIGHT: *13.3 to 14.3 h.h.*
POPULATION STATUS: *Rare*

A Spanish Barb gelding. Photo: Susan Paulton

Sun, a foundation sire of the Spanish Barb breed. Photo courtesy Susan Paulton

The Spanish Barb is the horse of Iberian and Barb blood brought to and bred in the Americas by the Spanish during the sixteenth and seventeenth centuries. A superb blend of elegance and durability, the Spanish Barb was immortalized in the literature and art of post-Renaissance Europe. The name Spanish Barb is one coined in the United States, and refers to the famous horses of Spain known as Andalusian and Spanish Jennet. The roots of this title are found in the fact that Barb horses were crossed on the horses of Spain.

The Spanish Barb is a hotblooded, Old World horse, the culmination of centuries of superior breeding. It is a horse of quality, breediness, and regal character, standing 13.3 to 14.3 hands and resembling its ancestors, the North African Barb and Spanish Andalusian. The distinctive Spanish Barb head is well chiseled and refined with a convex or subconvex profile and tapered muzzle. An almond-shaped eye is set low and somewhat forward in the head; the ears are crescent-shaped. A prominent heavily crested neck, deep at the base and set on a long, sloping shoulder, allows natural flexion and suppleness necessary for proper collection and head-set. The throat-latch is deeper than in other light breeds but is well-defined behind the jowl.

Predominance of the fore quarter over the hind quarter is created through the well-proportioned neck, powerful shoulder, short-coupled back, and rounded croup, with legs positioned well under the torso.

The Spanish Barb has long, smooth muscling throughout and a long, full mane and tail with a low tail-set. The breed's correctness of leg structure and tough, thick-walled hooves impress owners and farriers. Chestnuts and ergots on the legs are small, nonprotruding, or absent. The Spanish Barb is usually a solid-colored horse of black, sorrel, chestnut, or any of a wide variety of shades of roan, grullo, dun, or buckskin. A unique *overo* strain of the Spanish Barb is the Medicine Hat Paint, a consistent coat pattern that originated with the ancient African Barb. ("Medicine Horse," or "Medicine Hat," is a coined American term for a color phase that, according to the Cheyenne and Sioux Indians, held mystical powers.)

The Spanish Barb is phenotypical of Spanish breeds worldwide; the Andalusian, Lusitano, Peruvian, Argentine Criollo, Puerto Rican Paso Fino, Mangalarga Marchador, and North African Barb, to name a few. The specific features, or type, of the head, neck, shoulders, and croup of the Spanish breeds have collectively become known as the "Iberian factor."

Honed to perfection in Spain as the world's finest war, stock, and pleasure horse, the Spanish Barb never lost its cutting edge as a performance horse once transplanted to American soil. Its agility, intelligence, courage, and endurance only sharpened as its mettle was tested in the wild interior of the continent. Never grained or sheltered, the Spanish Barb endured and thrived under the hardest of conditions. Bred to handle the agile cattle of Spain, it was a tailor-made buffalo horse and more than a match for the swift, "snaky" longhorn. To gallop a hundred miles in a day was not uncommon for the Indian, *vaquero,* or frontiersman. With smooth, effortless speed, this horse comfortably carried overseers across vast plantations and ranches. Yet by appearance and nature it was also suitable as a parade horse or ladies' mount during festivities. It is recorded that Spain accomplished the impossible: a melding of opposites of nature into one. The Spanish horse perfectly fits the description of "fire and feather."

Today the Spanish Barb continues to be a highly versatile performance horse. It is appreciated for its inherent "cow sense," sensitivity, response to rein, and incredible endurance. This breed has a natural aptitude for jumping. The gaits are noticeably smooth. The stride is long and the Spanish Barb is a surefooted traveler and truly a pleasure to ride. The impressive self-carriage and style also make it a light harness horse of distinction. This is a very sensible, dependable horse and a good working partner under saddle and in harness.

The disposition of the purebred Spanish Barb displays the unique, hotblooded Spanish temperament, combining exuberance, intelligence, and spirit, with gentleness, willingness, and ease of training. This is a highly expressive breed. The stallions are easily managed and well mannered with royal bearing.

The Spanish Barb matures slowly and has a long, useful life-span. Broodmares continue to produce well into their mid- and late twenties, and stallions remain equally fertile. It is not uncommon to find aged Spanish Barbs in their late teens and mid-twenties with the appearance, action, and ability of an eight-year-old. They are easy keepers, thriving on grass alone even under working conditions—a valued characteristic also notable in their hardy South American cousin, the Criollo.

The recently formed International Association for the Preservation of the 16th Century Spanish Barb Horse is a nonprofit institution dedicated solely to the accurate restoration of the Spanish Barb—in essence, the horse that *was*—through selection and management of four bloodlines, represented as the purest genetic material available at its inception. A purebred herd is maintained by the association, based on purity of blood and breed type, while ongoing scientific and historical research complement and direct the breeding program. The goal is to stabilize the Iberian type genetically, while retaining the inherent performance abilities and marvelous disposition and spirit of the Spanish Barb. The association is not a registry. Education of the general public is essential in guiding awareness about the unique Spanish Barb of the twentieth century. Public interest and support are crucial if the breed is to persist. Of the true Spanish Barb, it is probably safe to say that less than 250 exist in the United States.

The current effort is based on the breeding program of Susan Field-Paulton, who has been referred to as the "Lady Ann Blunt" of the Spanish Barb breed. The results of her breeding program represent a herculean effort requiring a great deal of time and purposeful selection. She had long admired the extraordinary feats and colorful past of the Spanish Barb when, in 1957, she realized that a few survivors of the virtually extinct Spanish Barb breed did, in fact, still exist under private ownership. By 1962 she had acquired some of the rare stock, and she was actively breeding the Spanish Barb by 1965. The entire concept was a genetic challenge and not without setbacks, as strains were bred, then carefully retained or rejected, as they proved or disproved their validity. Purity of blood was established through the horse's background history, pedigree, and overall Iberian appearance as well as its consistent ability to produce progeny of authentic Spanish Barb type. Conformation, disposition, and performance are meaningless without type if an unadulterated breed is to be restored.

Ms. Field-Paulton's foundation stock consisted of two stallions and two mares. Using this minute gene pool, she began the meticulous work of restoring the breed. She spent over thirty years researching the purebred Spanish Barb and selectively breeding the four strains that she believed represented the purest blood to be found in North America.

The foundation stallion, Scarface, is from a Spanish Barb strain kept intact by the D. D. Romero family since the early 1800s. The Romero land in New Mexico, an original Spanish land grant, and their horses were purchased by Weldon McKinley in 1957. Breeding programs have proven the Scarface blood to carry tremendous beneficial impact for the breed. Linebreeding and outcrossing to Scarface has established him as a sire of prepotent genetic strength, excellence, and influence.

Using the breeder's best tool, that of linebreeding to a superior individual, the Romero/McKinley bloodlines were perpetuated primarily through linebreeding and outcrossing of Scarface blood to the foundation mares, Coche Two and A-ka-wi. These mares' pedigrees represent the earliest bloodlines of the horses bred by Robert Brislawn, Sr., who began working to preserve America's Spanish horses in the 1920s. The incorporation of the second foundation stallion came in 1969 with the purchase of a rare and uniquely colored Medicine Horse named Sun, bred by Ferdinand Brislawn. Sun's pedigree is deeply rooted in Santo Domingo Pueblo in central New Mexico. Santo Domingo Pueblo, having never been displaced by the federal government, has remained in its present location since before the Spanish invasion. These people have traditionally been known for breeding fine Spanish stock, many being Medicine

Horses. Their generous custom of giving or selling only their finest horses made possible the purchase of Sun's sire by Robert Brislawn, Sr. The esteemed Medicine Horse coloration is perpetuated in the breed through the blood of this uniquely marked stallion.

While the Spanish Barb and Spanish Mustang (see breed account) share identical ancestors, separation of the breeds lies in the different goals and methods of breeding. The Spanish Barb is being restored to the horse that *was*, while the Spanish Mustang, also rare, is preserved as the Spanish horse shaped in America through hardship and natural selection. Today, due to these differences in breeding, there is a clear distinction in their appearance.

For additional information see Spanish-American Horse.

Spanish Colonial Horse
(Wilbur-Cruce Herd)

ORIGIN: *United States (Arizona)*
APTITUDES: *Riding horse*
AVERAGE HEIGHT: *14.2 h.h.*
POPULATION STATUS: *Rare*

Spanish Colonial horse. Photo: Marye Ann Thompson

Though virtually unknown outside of the immediate area, a herd of authenticated Spanish Colonial horses has existed in virtual isolation for over a century in southern Arizona. Dr. Ruben Wilbur, a physician educated in the East, came west in the mid 1800s to practice as a company doctor for the largest gold mining operation in southern Arizona. Setting up a ranch near the town of Arivaca, he married a Mexican woman from the state of Sonora and subsequently purchased twenty-six head of Spanish horses from Juan Sepulveda of Magdalena, Mexico.

Sepulveda, a horse trader, trailed a herd of approximately 1,000 Mexican horses over the border into what is now southern Arizona with the intent of selling them in Kansas City. However, due to the demand for his horses, his ultimate destination was never achieved. Legend claims the herd was sold before he exited New Mexico. Dr. Wilbur purchased a stallion and twenty-five mares and placed them on his spread with the purpose of using them as ranch horses.

Historically, horses from Magdalena trace back to Father Eusebio Kino, a Jesuit priest who in the late 1600s and early 1700s provided Mexican settlers with horses, cattle, and sheep as well as providing his Indian workers with domestic stock in his efforts to Christianize the various tribes. Spanish records show that Father Kino was an extremely astute breeder of stock, and his Sonoran headquarters of Nuestra Señora de los Dolores provided thousands of head of livestock to northern Mexico and southern Arizona. Records state that he moved a herd of 1,400 head of livestock consisting of cattle, sheep, mules, and hundreds of Spanish horses on one drive into southern Arizona.

Due to Apache depredations at the mine and the death of the owner's brother at the hands of the Apaches, the mine was shut down and Dr. Wilbur eventually became the first Indian agent at the Mission of San Xavier del Bac, south of Tucson. Upon his death around 1884, ownership of the Wilbur-Cruce ranch passed to his son, who maintained it as a cattle ranch, breeding and using his own herd of cow horses descended from the original stock purchased from Sepulveda. Upon the son's death in 1933, his daughter, Eva Wilbur-Cruce, took over operation of the

ranch. Thus it remained for many years, with the Wilbur-Cruce herd living in a semiferal condition in the mountains on the ranch, brought in and trained for cattle work as needed.

Mrs. Cruce says that no other bloodlines have been infused into the herd except for one "Morgan" stallion introduced for a period of one year or so in 1933. This so-called Morgan was a Colorado paint, and it is likely that he was in fact of Spanish descent, given his coloration and place of origin. His contribution to the gene pool of the Wilbur-Cruce herd was negligible because of the brief period of time he remained on the ranch and also because the existing range stallions of the herd would have prevented him from acquiring mares from their harems. A number of years ago the cattle ranching operation ceased and the herd ranged in a feral state with little or no human interference. The herd has never developed in large numbers due to depredation by mountain lions and possibly Mexican jaguars upon the very young and old.

In early 1990 the Wilbur-Cruce ranch was purchased by the Nature Conservancy, which in turn donated it to the United States Fish and Wildlife Service to be added to the adjacent Buenos Aires ranch, where certain species of endangered quail are bred and released. Due to prohibitions against "domestic" species on such federal lands, which can only be changed by federal legislation, the Wilbur-Cruce horses were ordered removed from their home of over a century.

Time was of the essence. Officers of the Spanish Mustang Registry became aware of the dilemma, and several trips were made into the mountains to ascertain the number of horses and determine whether they indeed appeared to be Spanish. Satisfied with what they saw but hindered by lack of time, they contacted the American Minor Breeds Conservancy (AMBC) through Dr. Philip Sponenberg, technical chairman of that organization and a professor of pathology and genetics at the Virginia-Maryland Regional College of Veterinary Medicine. Thanks to the efforts of Dr. Sponenberg and the AMBC, ownership of the herd passed from Eva Cruce to the AMBC and the horses were trapped in mid-June 1990.

Members of two mustang organizations and several private breeders volunteered to take breeding herds to perpetuate the unique, historic bloodlines, and to that end approximately sixty mares and stallions were placed in Arizona, California, Texas, and Oklahoma. The remaining sixteen horses, all stallions, were auctioned off to private persons in Arizona.

Two groups, one in Arizona and one in California, are in refuges open to the general public for viewing. The Arizona herd is at the Arizona Pioneer Living Museum north of Phoenix on Interstate 17, and individual animals will be displayed under saddle and to buggy within the next few years, perpetuating their heritage as colonial Spanish horses.

Preliminary results of blood samples taken from each horse in the herd show that they do indeed carry Spanish genetic markers. Though the analysis is not complete at the time of this writing, results show they have been a closed population (little or no outside blood introduced), but at the same time individual variation is high, indicating that due to the high number of original mares, inbreeding is low.

The primary color in the herd is chestnut, ranging from light chestnut to the rare "black" chestnut, with some individuals having flaxen manes and tails. A lesser number are bay with a few blacks and greys. A number of paints exist within the herd, including the rare Medicine Hat pattern. Though duns and grullos were present many years ago, due to natural selection and sale of horses of these colors in the past none exist today.

The average size is approximately 14.2 hands with moderately long head, rather high-set eyes, semicurved ears of moderate length, moderately long neck, good withers, fair shoulders, short to moderately long back, rounded croup, low-set tail, and extremely good feet and legs. None of the individuals observed had any type of leg or foot defect. Manes and tails are full and fetlock hair is scant. The chest is deep and well rounded and endurance and agility in their native mountainous terrain is remarkable.

The name "Spanish Colonial Horse" was chosen because these horses represent a remnant of the colonial horse of America, predominantly of Spanish blood.

For additional information see Spanish-American Horse.

Spanish Mustang

ORIGIN: *United States*
APTITUDES: *Riding horse*
AVERAGE HEIGHT: *14 to 15.1 h.h.*
POPULATION STATUS: *Rare*

A Spanish Mustang gelding, showing remarkable similarity to the Sorraia of Portugal. Photo courtesy Marye Ann Thompson

A rare Medicine Hat stallion of the Spanish Mustang breed. Photo: Marye Ann Thompson

In the mid-1950s a group of dedicated men met to form an organization for the purpose of preserving the last of the true, old-type Spanish Mustangs, a breed that once roamed the western part of the United States in great numbers but was now threatened with extinction. The efforts of Robert E. Brislawn of Oshoto, Wyoming, were the primary moving force that brought this group together.

Brislawn had long been concerned over the dwindling numbers of his beloved Spanish Mustangs — so concerned that for many years prior to the actual formation of the Spanish Mustang Registry he had personally collected individual animals he considered the best examples of the breed. Thus he formed the nucleus of his own breeding herd and foundation stock for the future registry. He was familiar with the breed, as he had used these horses during his years of service with the United States Topographical Service. He chose his stock carefully, culling out those he believed less than ideal.

The name "Spanish Mustang" was agreed upon as the most descriptive of the breed; perhaps not as prestigious as some other name might have been, but more accurate, acknowledging their Spanish origin and utilizing the name by which they had been known for many years. Descended from the Andalusian, Barb, and Sorraia — Spanish breeds first brought to the New World — they were a combination of all these breeds and no one breed name was totally descriptive.

The word *mustang* has become somewhat confusing in the past few years, connoting to many the wild horses controlled by the Bureau of Land Management (BLM) and put up for adoption. However, there is little or no relationship between these adoptive horses and the true Spanish Mustangs — not in genetic inheritance or in appearance. The BLM horses are descended from various American breeds, and few, if any, carry any trace of Spanish blood (see *Wild Horse Parentage and Population Genetics* by Bowling and Touchberry, University of California at Davis, 1988).

The Spanish Mustang Registry is the original and oldest of the Mustang registries. In recent years several Spanish horse–based organizations have been formed utilizing registered Spanish Mustangs as foundation stock, either in part or altogether.

The original intent was simple — to preserve the last remnants of the old Spanish imports and to form a breeding herd with a large and diverse gene pool to ensure that the registry would prosper and grow and would never have to resort to crossbreeding or heavy inbreeding in order to prosper. The founding directors made it very explicit that no crossbreeding of any kind would be allowed. The aim from the inception of the registry has been to preserve in purity the old Spanish horse. Currently, with 1,500 horses now in the registry, the genetic pool is varied enough that no consideration will ever be given to going outside the breed to perpetuate it.

Long noted for its stamina and toughness, both in fact and in legend, this breed truly shows its Spanish inheritance in its ability to survive and multiply where other breeds would have perished. Considered to be the finest in the known world at the time of the Spanish conquest of the Americas, the Spanish horses left in their tough, hardy descendants a legacy that endures to this day. Once the animals were left to themselves in the Americas, their heads grew larger, their ears longer, their eyesight more acute, and they developed a great sense of direction. They were greatly changed in conformation through feral life and feed, acquiring over time qualities not found in the domesticated horse.

Though the Spanish horse was not a feral animal when it arrived in the New World, once turned loose it managed not only to survive but to flourish, attesting to the versatility and strength of the breed. Fired in the crucible of war and conquest and molded by the necessity of survival, the Spanish Mustangs retain the characteristics of their illustrious ancestors and have added a few traits of their own through natural selection for survival. Genetic imperfections, if any, were culled by nature, the most critical judge of all. The result is an extremely hardy and sturdy horse possessing the ability to perform in almost any equine field, and perform well.

The staying power and endurance of these horses became notable when in 1877 Frank T. Hopkins, half white and half Sioux, purchased a small white-eyed mare (probably her eyes were blue) from the Sioux in order to prevent her slaughter by the army. Hopkins, a courier for General George Crook, was present when the army captured a band of Sioux horses and, in order to immobilize the Indians, ordered destruction of their horses. One of the Sioux convinced Hopkins to buy the white-eyed mare as he said she could carry a man all day on the run, though weighing only 700 pounds. Hopkins bought her for a few dollars. Later he acquired a paint Apache stallion, using this pair as the foundation of his famous White-y line of mustang endurance horses. Their descendants were noted for their extreme endurance.

In 1886 an endurance ride from Galveston, Texas, to Rutland, Vermont, was organized with prize money of $3,000. Hopkins entered the contest on a buckskin stallion. Fifty-six riders set out from Galveston on September 6. Allowing his horse to proceed at his own speed, other riders left him behind. Within a week he began overtaking them and on the seventeenth day passed the lead rider. Reaching Rutland on the thirty-first day and waiting another *thirteen days* for the second rider to complete the ride, and then more days for the third, he proved his mustangs were indeed possessed of exceptional stamina. No others reached the finish line, and the two that had were in worn and poor condition. All in all, Hopkins rode mustangs in some 400 endurance competitions, winning most of them.

In 1890, while in Paris with some of his White-y line of horses, he was asked to enter a 3,000-mile race in Arabia, expenses to be paid by the Congress of Riders of the World. He rode his paint stallion Hidalgo, and crossed the finish line in sixty-eight days, more than a day ahead of the second-place entry. Hopkins said (according to information from SMR): "You can't beat mustang intelligence in the entire equine race. These animals have had to shift for themselves for generations. They had to work out their own destiny or be destroyed. Those that survived were animals of superior intelligence."

Jack Best, a cowboy on a dun cow pony, raced 530 miles from Deadwood, South Dakota, to Omaha, Nebraska. The race started on Saturday morning at 6:00 A.M. In order to qualify for the $1,000 award, the riders had to reach Omaha before midnight the following Saturday. Though the smallest animal in the race and carrying at least 190 pounds, Best's coyote dun mustang won the race.

Thousands of mustangs were used as cow ponies and hundreds as cavalry mounts. The cavalry required horses of hardiness and stamina. In 1876 the Fifth Cavalry Regiment, mounted on cow ponies, was ordered from Arizona to Santa Fe, where they were required to exchange horses with the Sixth Cavalry, on its way to Arizona to replace the Fifth in a campaign against the Apaches. Though the Sixth originally balked at giving up their "blooded" mounts for rough and shaggy mustangs, they found that you must fight fire with fire—mustang against mustang—when fighting an elusive and fast-moving foe such as the Apaches. In six months they lost not a single horse, another attestation to the mustang's ability to travel hard and frequently with only forage to sustain it. Colonel Richard Dodge once took a fancy to a Spanish pony in El Paso, Texas, a small horse hardly 14 hands tall. He offered the owner what he considered to be a fairly high price, forty dollars. The owner refused, saying he could not take $600 for the animal and rightly so, for every week for the past six months the horse had carried dispatches between El Paso and Chihuahua, traveling by night and hiding by day, covering the more than 300 miles in three nights, then back again.

The ancient heritage of the Spanish Mustang is evident in today's horses, not only in their type and conformation but also in the distinctive colors and patterns. *Overo* paints are numerous, including the rare Medicine Hat pattern. Grullos, duns, and buckskins are common, again tying in their relationship to the original Spanish horse. The more exotic colors and patterns are still accepted by the Lusitano Society (Portuguese Andalusians), but the Andalusian Society of Spain does not allow these, though old paintings in many museums show Andalusians in the colors and patterns found in modern Spanish Mustangs. Indeed, a painting of Queen Isabella in the Prado Museum in Madrid shows her mounted on a Medicine Hat paint! Van Dyck painted Charles I on his buckskin Iberian charger (1632).

Many conflicting and inaccurate theories concerning the origins of the Spanish Mustang have been promulgated over the years. One of the most accurate and comprehensive texts concerning the Iberian or Spanish horse is *The Royal Horse of Europe* written by Sylvia Loch, herself a Lusitano owner and show judge. Her research was vitally assisted by the utilization of old records in the archives of Spain and Portugal while she was living in those countries (see Andalusian).

The written description given in ages past of the Iberian or Spanish horse could well be that of the modern Spanish Mustang—slightly rounded forehead, almond-shaped eyes, straight or subconvex profile, heavy mane and forelock; powerful sloping shoulder, good withers, short back, strong legs, deep rib cage, rounded hind quarters with sloping croup and low-set tail; dense bone, sloping, elastic pasterns, round, high hooves, easy keeping ability, and good temperament.

The Spanish horse has had a profound influence on many breeds, both in the Old World and the New. Today's Spanish Mustang retains those qualities that allowed the Spaniards to conquer a new world.

Spiti

ORIGIN: *India (Himalayas)*
APTITUDES: *Riding, pack pony*
AVERAGE HEIGHT: *12 h.h.*
POPULATION STATUS: *Common*

The Spiti pony is localized in the Spiti valley of Himachal Pradesh. Very hardy and surefooted, the Spiti is smaller than the Bhutia of India but strongly resembles this breed. Both breeds are believed closely related to the Tibetan.

The Spiti is bred mainly by the Kanyat Hindu tribe in the high mountains. It is used primarily as a pack animal and is said to be tireless and able to thrive under adverse conditions such as scarcity of food and long journeys. The Spiti thrives in cold, mountainous conditions only and is intolerant of the warm, humid climate of the lowlands. Evolution of this breed in the high mountains has

also resulted in a horse that performs better at high altitude.

The body of the Spiti is well developed and the legs have quite a lot of long, coarse hair. Color varies from dark grey to dun. The head is somewhat heavy with a straight profile and small, sharp ears. The neck is short with a heavy mane; the withers are low; the back is short and strong; the croup is sloped slightly; the tail is full and long; the chest is deep; the shoulder is well muscled and often rather straight. The legs of the Spiti are short and sturdy with strong joints and good bone structure. The hoof is rounded and well shaped with extremely tough horn.

Standardbred

ORIGIN: *United States*
APTITUDES: *Harness racing, riding*
AVERAGE HEIGHT: *15 to 16.1 h.h.*
POPULATION STATUS: *Common*

The Standardbred trotter, Mack Lobell. Photo: Edward Keys

When the stud book was founded in 1871 for trotting and pacing horses, the requirement for registration was that a horse must be able to trot or pace a mile in a standard time. For the trot this was 2.30 minutes and for the pace, 2.25 minutes. Hence the name Standardbred came into being. As the years passed and Standardbreds became faster through selective breeding and improved tracks, time for the mile has diminished.

Although harness racing is generally believed to be an American sport, it can be traced back some 3,000 years. And the deep roots of the sport, like those of most U.S. citizens, can be found in other lands. Horse racing while drawing two-wheeled carts long preceded horse racing with riders astride the horse.

Many reasons have been considered for the disappearance of trotters from history for a period of more than 2,000 years until they reappeared in modern history. Warfare and its needs, and the painstaking care necessary to train a trotter, probably played a large part. The Norfolk Trotter, developed in England just over 200 years ago, was bred as a road horse and not a race horse.

Modern American harness racing began about 1806. Records of sorts began to be kept, and a gelding by the name of Yankee trotted a mile in less than three minutes for the first time on record.

Some feel the Standardbred is descended in part from Spanish horses, but it was the horses that landed in Massachusetts that helped to establish the easy-going mount known, the better part of a century later, as the Narragansett Pacer. Although it no longer exists, the Narragansett Pacer was important in development of the Standardbred, American Saddlebred, and Tennessee Walking Horse.

The first horses to arrive at Massachusetts were twenty-five mares and stallions, landing in Boston in 1629 after shipping out of London. The Dutch, six years later, sailed into Salem harbor with twenty-seven mares and three stallions.

Residents of New England preferred the Dutch horses. They were larger than the English horses, more muscular, and had larger bones. They were used for general purposes and for a variety of needs and were not bred for racing, as were the horses of Virginia and the Carolinas.

The largest of the Dutch imports was 14.2 hands high, and the process of *reducing* this height began as soon as the horses started to arrive. All attempts to increase the height of the early American horses had no effect. The English horses had good saddle qualities which the Dutch did not, and as the colonies developed, people wanted saddle horses. The English and Dutch breeds were crossed and, as

more horses were brought to America, they were selectively bred to produce a stout and hardy horse that could pace long distances without tiring the rider. The pace is a natural gait, and many foals paced soon after birth. A fact not commonly known or acknowledged is that a great many horses were brought into the New England states from Canada. They were the French-Canadian breed, which was known for producing both trotters and pacers.

As the colony of Rhode Island began to take root after initial settlement in 1636, the fastest and best of the New England pacers were taken there for breeding. The settlers loved a good horse race. Race courses were established, valuable prizes offered, and improving the breed became serious business.

Careful breeding produced the Narragansett Pacer, which reigned for a century and a half as the most wanted saddle horse in the Americas. Its name was taken from the bay at Narragansett, in turn named for the Indians who once lived on its banks. This horse was not large, averaging about 14.1 hands. Its color was usually sorrel. The Narragansett was neither stylish nor handsome but was surefooted, easy riding, and dependable. Its back remained straight as it traveled, not producing a swaying or rocking motion as did other horses when they paced. It is likely that this horse traveled at a four-beat amble, or "broken pace." Many Narragansetts were shipped to Cuba for use on the newly established sugar plantations. So many were exported, in fact, that their numbers in the North American colonies were sharply reduced.

In Maryland and other colonies there was a practice of branding horses and turning them loose to roam and fend for themselves. It is difficult now to imagine the eastern coastal states as having once been the home of great herds of feral horses. The animals were allowed to overproduce and many were turned loose without brands. Such horses became a problem, and the state legislatures dealt with it directly and sometimes savagely. Maryland passed an act in 1715 that allowed a person to shoot on sight any older male horse of any size, provided he was running wild.

After more than a hundred years of great popularity the little Narragansetts began to disappear, just as their ancestors had disappeared a century before. Rhode Island was no longer a frontier settlement but had grown into a prosperous state. Broad, smooth highways had developed and wheeled vehicles had taken the place of the saddle. The style changed to a larger, more dashing type of horse.

Attempts to increase the height of the horses proved futile. Though diminutive in size, the little pacers were fine saddle horses, and until good roads and vehicles appeared on the scene they remained indispensable. Out of this vast stock of early imports to the colonies were developed race horses that both paced and trotted.

Harness racing in America began quietly in the 1800s on country roads, village main streets, and prominent city avenues. The pastime had gone well into the nineteenth century before Americans began calling it a "sport."

The Standardbred breed, representing both trotters and pacers, received its most important infusion of blood in 1788. In that year, a grey Thoroughbred named Messenger arrived in America from England to become the patriarch of the family of trotting horses. Messenger arrived on American shores five years after the Revolution.

The Revolutionary War had stopped the importation of all kinds of high-bred horses from England, and during the war's seven years a terrible destruction of stock had taken place. It was necessary to import horses once again, and England was a principal source. At first, southern planters who preferred the English Thoroughbred took the initiative. But horses began to arrive along the northeastern coast, too, and it was here that Messenger arrived, in Philadelphia.

Messenger's sire, Mambrino, was made of stout material. As one of the best race horses of his day, he was the master of four-mile heats. And Messenger's grandsire, Engineer, was also a famous stayer. Messenger's career over the race courses in three seasons found him winning eight races, losing six, and receiving forfeit in two. In twenty seasons at stud he got at least 600 foals, most of which were relatively large. He produced both runners and trotters, the latter known for their great trotting action, speed, and gameness.

While the Thoroughbred is credited with refinement and increased size in the Standardbred, the Thoroughbred is a galloper and could not have originated the trotting and pacing excellence of the Standardbred horse. Acknowledgement should be given to the Canadian horses known for those gaits.

Most American-bred horses at the time were small. Messenger's family had a history of being big and strong. His American offspring possessed these qualities, along with a superior trotting action and speed that placed them in a class by themselves.

Harness racing received another massive shot in the arm with the birth of Hambletonian, a descendant of Messenger. Foaled in 1849, Hambletonian, owned by William Rysdyk of Chester, New York, raced only sparingly. However, he quickly proved himself to be history's greatest progenitor of both speed and gait. All but a few of today's trotters and pacers trace to Hambletonian. Small wonder that the most famous harness race of them all bears his name.

The birth of Rysdyk's Hambletonian might be considered a miracle. In 1844 John Seely, an Orange County farmer, was returning home after delivering a drove of steers to the New York market. He found a crippled bay mare in the stable of a butcher named Charles Kent. The mare was no stranger to Seely: she was a daughter of One Eye, a mare that had belonged to his father. She was suffering from injuries received in a runaway and, out of family sentiment more than anything else, he brought her home.

The bay mare was a daughter of the imported Bellfounder, who had spent a season in Orange County the decade before. On the maternal side she was linebred to Messenger. Her dam, One Eye, was a daughter of a mare named Silvertail that had been bred to one of Messenger's best sons, Bishop's Hambletonian, and had produced One Eye.

Bred to Abdullah, the crippled mare produced Hambletonian in 1849. Abdullah— ungainly, rat-tailed son of Messenger's Mambrino and out of Messenger's famous granddaughter, Amazonia—had been rejected by Kentucky breeders and returned to his native Long Island. In his younger days a great future had been prophesied for Abdullah as a sire of trotters, but he had not fulfilled the prophecies. His looks and personality were both against him. Besides being an extremely ugly horse, Abdullah had a vicious disposition, was continually cantankerous, and resisted no matter what anyone tried to do with him. He was sold once for five dollars by a disgusted owner. Even at the age of twenty, Abdullah was known as a horse who would not submit to harness and spitefully kicked his master's cart into matchwood with no provocation.

But out of this unlikely mating of the crippled bay Kent mare and the intractable Abdullah, Hambletonian was born. Thinking the colt would never amount to anything, Seely sold both the mare and colt to his hired man, Bill Rysdyk, who had fallen in love with the colt. The rest, as they say, is history.

Farmers in and around Orange County, New York, soon discovered that if they bred their mares to Rysdyk's Hambletonian, the offspring were natural trotters rather than runners. It was remarkable to them, since this stallion had the blood of runners in him. All of his colts showed great speed at the trot rather than the gallop.

There have been many great racing trotters over the years, but a few will always stand out. One of the truly great names is that of Rosalind. In 1937, on September 9, a great example of the stamina and courage of the breed was seen in this mare. Sired by Scotland and out of the mare Alma Lee, she was owned by Ben White of Lexington, Kentucky.

She had previously won the Junior Kentucky Futurity and had also won the Hambletonian as a three-year-old. Now she was up against the greatest handicap ever asked of a trotter. She had to start 240 feet behind the wire, 60 feet behind the nearest horse.

People shook their heads. Some said she would break her heart to win. Other horses in the race were two bay geldings, Friscomite and Fez, 120 feet behind the wire; Calumet Dilworthy and Lee Hanover, 180 feet behind the wire; then Rosalind, at 240 feet.

That day Rosalind won the All-American handicap at Syracuse, New York, in 3.12¼ for the mile and a half, plus her handicap. She set a new record when everyone thought it was impossible for her to win.

Another great Standardbred racer was the grey gelding Greyhound. In 1938 he became the fastest trotter, setting a record that held for many years with a time of 1.55¼. Greyhound was another horse that as a youngster was not considered to have champion qualities. He was by Guy Abbey, by Guy Axworthy, and out of the mare Elizabeth by Peter the Great. He seemed awkward, and it was said that he did not have a good gait. Put up for sale at Indianapolis as a yearling, he passed from his original owner for a price of slightly more than $900.

Under the capable management of trainer Sep Palin, Greyhound came into his own with a polished and powerful gait that measured a twenty-seven-foot stride.

The Standardbred is a more heavily muscled animal than the Thoroughbred to which it owes so much, is longer in the body, and has a flatter rib cage. The quarters slope to a greater degree than do the Thoroughbred's, and the muscles of the quarters, especially in the thigh, are powerful and long. This, combined with the extra slope to the quarters, gives the Standardbred much impetus in its racing gait. The head is not as refined as in the Thoroughbred, often being straight or slightly convex. The nostrils are capable of great extension, allowing a generous intake of oxygen when racing. The ears tend to be on the long side, the neck somewhat short and straighter than is seen in the Thoroughbred. The predominant color is bay, with brown, chestnut, black, and some greys. Manes and tails grow profusely and are thick and very long. The average height is between 15 and 15.3 hands, although some horses are over 16 hands.

When trotting, the left front leg and right hind leg move forward together, then the right front and left rear. In pacing, the left front and left rear move forward simultaneously, then the right front and right rear, giving a rolling motion to the gait. It is often described as piston-like or sidewheeling.

A trotter is trained to trot, and to trot fast without breaking its gait. When one breaks its gait, going offstride and into a gallop, it must be pulled up and put back into the trotting stride. This does not disqualify the horse but usually means it loses any chances of winning. The same rules apply to the pacing horse. It must pace consistently.

Over the years harness racing has fluctuated in popularity. It was immensely popular in the 1800s, but when the automobile appeared on the scene and replaced the horse as a means of transportation, it fell from favor. It came back to prominence in the 1940s. The resurgence resulted from several factors. New York State approved pari-mutuel betting on the races; a group of intrepid investors proved that nighttime harness racing could succeed by offering races under the lights at Roosevelt Raceway on Long Island; and a man named Stephen G. Phillips developed the mobile starting gate.

Today harness racing is big business in many countries of the world, and it is here to stay. In the United States, more than 12,000 men and women are now driving harness horses in races across the land, some of them professionals, some amateurs. Over 50,000 people hold membership annually in the U.S. Trotting Association.

Standardbred yearlings have sold for as much as $625,000, while drivers like Herve Filion, John Campbell, and Carmine Abbatiello have each already driven horses to more than $35 million in earnings during their careers.

The Standardbred has been used to improve racing trotters around the world.

Suffolk
(Suffolk Punch)

ORIGIN: *Great Britain*
APTITUDES: *Heavy Draft*
AVERAGE HEIGHT: *16.1 to 17.1 h.h.*
POPULATION STATUS: *Rare*

The Suffolk is the oldest breed of heavy horse in Great Britain existing today in its original state. The breed dates to at least the sixteenth century, but precise information as to its ancestry has been lost. All animals alive today trace their male lines to one stallion, Crisp's

Suffolk. Photo courtesy Suffolk Horse Society, Great Britain

Horse of Ufford, foaled in 1768. Volume 1 of the *Suffolk Stud Book* is a classic among livestock books. The author, Herman Biddell, first secretary of the Suffolk Horse society, spent two years tracing the pedigree of all Suffolks alive at that time. These he published in the book, which he prefaced with a fascinating history of the breed and a contemporary account of the Suffolk and the people associated with it. The book was illustrated by the Ipswich artist, Duval.

Homeland of the Suffolk horse was bordered on the north, east, and south by the North Sea and on the west by the fens. Isolated from their neighbors, the farmers of Suffolk independently developed breeds of livestock to fit their special needs. To plow the heavy clay soil they needed an agricultural horse with power and stamina, naturally good health, longevity, and docility.

Suffolk farmers used their horses to till and harvest their own lands, so they seldom had horses to sell. This kept the breed localized and relatively unknown but also pure.

The Suffolk is always chestnut (meaning chestnut, but traditionally spelled without a "t" in the middle). A few white hairs well mixed with the chestnut on the body, and a star, stripe, or blaze are allowed. Seven shades of chestnut are recognized: bright, red, golden, yellow, light, dark, and dull-dark. The legs are rather short, and the impression the breed gives is that the body is too big for the legs. This shape gave the breed its nickname, the

"Suffolk Punch," and its great strength. Pulling matches were popular in Suffolk in the eighteenth century, and their recent revival has shown this horse to equal the strength of any.

Average height in mares is about 16.1 to 16.2 hands and that of stallions 17 to 17.1 hands. The legs are clean with no feather, thus making the working Suffolk an easy animal to care for. In temperament, so important in a good working animal, this horse is exceptionally good, and a long working life and economy of feeding are well-known features of the breed. Common practice on East Anglian farms was to feed the Suffolks loose in yards.

Home to the breed is East Anglia, and this is still where the largest number are to be found, although there are now also breeders throughout the country. The Suffolk can be seen at the Royal Show, the Essex, Suffolk, East of England, and Royal Norfolk shows, and many smaller shows throughout the summer. The Woodbridge Horse Show is the principal stallion show, and the Framlingham Show provides the best display of mares and foals. The Suffolk Horse Society was founded in 1877, and its office has always been in Woodbridge. The society publishes the stud book and encourages breeding by presenting premiums and grants.

The government of Pakistan uses Suffolk stallions to produce horses for the army, and the breed has recently been exported to Denmark, Italy, and Australia. At home the Suffolk is being used for breeding hunters and show jumpers.

The Suffolk in America was hard hit by the headlong mechanization of the post–World War II period. Although it had made great strides in popularity during the thirties, it still did not have the numerical base necessary to withstand the onslaught of the fifties. For a few years the American Suffolk Horse Association ceased to function. Then in the early sixties, as the draft horse market began its recovery, the few widely scattered breeders who had kept faith in their Suffolks reorganized, holding their first meeting in fifteen years in May of 1961.

The early seventies saw some outstanding importations from England to America. Inter-est in the draft horse in general was expanding, and demand for the Suffolks began to increase. The early eighties saw an encouraging increase in the number of Suffolks being registered, and once again there have been some excellent importations from Great Britain.

Sulawesi

ORIGIN: *Indonesia (Isle of Sulawesi)*
APTITUDES: *Farm work, riding, racing*
AVERAGE HEIGHT: *12 h.h.*
POPULATION STATUS: *Common*

Sulawesi is the name given to the pony of the Indonesian island by that name. It is said to be found in all common horse colors and to have a patient and calm temperament but to be quite lively and fast. This pony is occasionally used for racing on Sulawesi.

No physical description was given for this pony by Indonesian authorities.

For additional general information regarding Indonesian ponies, see Bali.

Sumba

ORIGIN: *Indonesia (Isle of Sumba)*
APTITUDES: *Riding, light draft, packing*
AVERAGE HEIGHT: *12 h.h.*
POPULATION STATUS: *Common*

Nearly identical to the Sumbawa pony, the Sumba is a very tough individual, willing to put forth hard effort. This pony is famous for its use in native dance competitions. Each ridden bareback by a small boy, and to the accompaniment of tom-toms, ponies dance with bells attached to their knees. They are docile and obedient and are ridden in bitless bridles. Judging is carefully based on lightness and elegance. The Sumba pony is also used in lance-throwing competitions, the national sport of Indonesia.

Although this pony is usually dun with a dorsal stripe and dark mane and tail, other colors are also seen in the Sumba. An ancient, primitive type, it usually has a heavy head with a straight profile; the eyes are almond-shaped; the neck is short and broad with a thick mane; the withers are not very prominent; the back is quite long and slightly dipped; the croup sloped; the girth quite deep; and the shoulder straight. The legs of this pony are slender but very strong, as are the feet.

For additional general information regarding Indonesian ponies, see Bali.

Sumbawa

ORIGIN: *Indonesia (Isle of Sumbawa)*
APTITUDES: *Riding and light draft*
AVERAGE HEIGHT: *11 h.h.*
POPULATION STATUS: *Common*

Sumbawa pony. Photo Dr. Soehadji, Department of Agriculture, Indonesia

Nearly identical to the Sumba pony, the Sumbawa also descended from Mongolian horses with little evidence of other breed influence. The Sumbawa is extremely hardy and strong, of primitive morphology, active and willing with a calm and kind temperament. This pony is so docile and obedient that it is ridden with a bitless bridle and obeys many spoken commands.

Morphology is like that of the Sumba pony. The coat color is usually bay or dun with a dorsal stripe and black points. Another Indonesian pony called the Bima, or Bimanese, is said to resemble the Sumbawa closely.

For additional general information regarding Indonesian ponies, see Bali.

Swedish Ardennes
(Svensk Ardenner)

ORIGIN: *Sweden*
APTITUDES: *Heavy draft work*
AVERAGE HEIGHT: *15 to 16 h.h.*
POPULATION STATUS: *Common*

Swedish Ardennes. Photo courtesy Jan-Sivert Antonsson

Swedish Ardennes stallion. Photo courtesy Jan-Sivert Antonsson

During the latter part of the nineteenth century agriculture increased greatly in Sweden, and this intensified the need for a heavy, strong draft horse. The native Swedish horses had proven too light for use with heavier equipment.

In 1872 Count C. G. Wrangle called the attention of Swedish horse breeders to the sturdy Ardennes horse, recommending that it be interbred with the Swedish country horse. In 1873 the count imported three stallions, Faro, Husar, and Railleur, to Värmländska Hingstbolaget. The most important import took place in 1874, when the mares La Me're and Rigolette were taken to Kinnekulle by Wrangle and the stallion Mouton was taken to Kavläs. Other imports followed, and by 1880 there were Ardennes stallions in nearly all parts of south and central Sweden. The cross of the Ardennes to the Swedish horse turned out well, and many breeding studs were formed. The horses increased in power and size while retaining their dryness, motion, hardiness, and long life. In 1901 the association for the stud book was formed.

For many years the Swedish Ardennes dominated Swedish horse breeding, and through careful selection the breed has become completely free and indepenent of imported blood. Recent years have seen a decline in numbers as mechanization has taken over, but the Swedish Ardennes is still used as a cart horse.

The object of Swedish breeding of Ardennes horses has been to produce a draft horse of medium size; a deep and broad working horse especially distinguished by its mobility, sturdiness, health, and longevity. The Ardennes is sweet-tempered and easy to handle.

Predominant colors are brown and bay, but black and roan are also seen. The head is rather heavy with a straight profile, broad forehead, and comparatively small eyes; the neck is short and heavily muscled; the withers are somewhat flat, muscular, and broad; the back is short; the croup muscular and rounded, often doubled; the chest wide and deep; the shoulder well sloped and muscular. The legs are short and strong with a bit of feather at the fetlocks; the hoof is round and broad with tough horn.

Swedish Warmblood

(Swedish Halfbred)

ORIGIN: *Sweden*
APTITUDES: *Riding horse*
AVERAGE HEIGHT: *16.2 to 17 hands*
POPULATION STATUS: *Common*

Callaghan, a Swedish Warmblood imported to the U.S., at the age of twenty-two. Photo: Ruth Sobeck

Sweden is a very long country. The climatic conditions and growing season are very different in the south and north. When spring grass is growing well in the south, snow is still falling in the north. Autumn or winter may come as early as September in the north, while animals may graze contentedly in the southern regions of the country until December. The consequence of this is that animal husbandry differs greatly between north and south. In the northern third of Sweden, there is no horse breeding at all. In the next tier moving southward, the North Swedish horse is the only breed represented. Horse breeding is primarily concentrated in the southern and central parts of the country.

In 1920 the horse population of Sweden was about 720,000 horses, of which nearly 80 percent were draft animals. As mechanization came to farming, horse populations declined steadily, reaching their lowest point in 1969 at about 82,000.

The history of the Swedish horse parallels that of the people. Archaeological findings of horses have been dated to 4,000 B.C. The ancient native Scandinavian horse was small—12 to 14 hands tall, high spirited, and had good endurance. The horses were kept outdoors in the woods and on moorland year round. Only the utility horses were kept at stud. As early as the sixteenth century steps were taken to improve horse breeding, and Friesian stallions were imported from the Netherlands to increase the size of the native horses. Importation went on, including Oriental, Spanish, Dutch, and English stallions. Prohibition of exportation of big horses was promulgated as a protective measure. The national studs at Kungsör, Strömsholm, and Flyinge got the best horse material, and their produce was usually for the royal stables.

Ignorance of the elements of horse breeding and inadequate feeding methods obstructed positive progress in Sweden for many years. Knowledge of the value of a good pedigree and the power of selective breeding was poor. Pedigrees of the stallions used were often unknown, and stallions were selected only on their appearance. The terms *phenotype* and *genotype* were unknown in Sweden. The terms *hotblood* and *coldblood* were not even used prior to 1860.

The two largest buyers of horses in Sweden, the army and farmers, could not agree on the type of horse needed, and breeding effort was divided. For breeding saddle horses, Thoroughbred and halfbred stallions of the Anglo-Norman, Hanoverian, and East Prussian (Trakehner) breeds were used. In order to direct horse breeding on the right path and to consolidate the different breeds, the government of Sweden started the stud book examination system in 1874. Examination has since been obligatory for registration in the stud book. Breeds that have been the object of examination since that time have attained a high degree of quality. These breeds are the Swedish Ardennes, the North Swedish Horse, and the Swedish Warmblood. All breeds are included by the law, and partbred stallions may not be licensed for breeding. There is no stud book for partbred horses in Sweden.

Today the Swedish Warmblood is an elegant horse well-known in Olympic and other international competitions. The breed is exceptionally good in the dressage ring, with extreme forward reach in the trot and suppleness in the shoulder.

Prior to entering the stud book, all mares must meet minimum standards as riding horses, demonstrating riding qualities in walk, trot, and canter. Stallions must meet a very high standard concerning gaits as well as jumping ability. For the final approval for commercial breeding, they also must pass a stringent cross-country test.

For more than a century, stallions have been scrupulously tested before they are licensed for breeding. The population of licensed stallions presently amounts to around 150. Each year some 100 three-year-old colts are evaluated in hand for conformation and soundness. From ten to twenty meet the minimum standards and are given a temporary breeding license for one year. They are then invited to do the performance tests. About half are selected for breeding after being tested on three different occasions for gaits, jumping ability, eventing performance, and health.

The stud book includes approximately 5,000 mares. Annually, about 10 percent of new mares enter the files. They are selected from among the 600 to 700 young fillies evaluated at regional shows every year, and to some extent also from among mares having completed a career in competitions. A mare is eligible to enter the stud book after being shown in hand with a foal at foot and ridden. The best three-year-old mares from each region are annually presented at the international agricultural show ELMIA in Jönköping in late May or early June.

The Swedish Warmblood is usually bay, brown, chestnut, or grey. The head is nicely shaped with a straight or convex profile, long ears, and lively eyes; the neck is well formed and long; the withers are prominent; the back is long and straight; the croup broad, long, and flat; the chest wide and deep; the shoulder is sloped and well muscled. The legs are long and strong with good muscling and clean, broad joints, and the hoof is solid and well formed. This horse has a good natural stance.

Swiss Warmblood

(Einsiedeln, Swiss Anglo-Norman)

ORIGIN: *Switzerland*
APTITUDES: *Riding horse*
AVERAGE HEIGHT: *15.3 to 16.3 h.h.*
POPULATION STATUS: *Common*

Switzerland has an old tradition of breeding warmblood and halfblood horses. In medieval times the monks of the monastery of Einsiedeln bred top quality horses called the Cavalli della Madonna. Until the middle of the nineteenth century there also existed, in the Emmental, the Bernese Oberland, and Les Ormonts in French Switzerland, stocks of a rather small, hardy, usually black riding and draft horse.

Around 1870 the military authorities began to breed cavalry horses but without much success. After many years of stagnation, interest in horse breeding enjoyed a revival in Switzerland in 1960 as well as elsewhere in western Europe. Along with rapid economic progress, the demands for recreational activities grew dramatically and riding sports became increasingly popular.

The horse breeding authorities did their best to take advantage of the situation and began to acquire top quality stallions each year from France, Germany, and Sweden. By systematic crossbreeding of these stallions with both imported and indigenous mares, the modern warmblood riding and sport horse gradually emerged. While at first the stallions used were mainly imports, today more and more Swiss Warmblood stallions are being used to produce this talented competition horse at the national stud at Avenches.

The breeding concept is to produce a good-tempered, rather tall, and goodlooking horse with correct paces and a great deal of class. This multipurpose riding horse can be used for leisure riding as well as for top performances.

The Swiss Warmblood may be of any color. The head is well shaped with a straight or slightly convex profile; the neck is of medium length and muscular; the withers are prominent; the back straight and strong; the croup slightly sloped; the chest broad and deep; the shoulder long and sloping. The legs are strong with well-defined tendons and strong joints; the hooves are well formed with strong horn.

The Swiss Warmblood has an elegant appearance and makes an excellent riding horse capable of competing in all levels of competition. This horse also goes nicely in harness.

Taishuh

ORIGIN: *Japan (Tsu Island)*
APTITUDES: *Riding, light draft*
AVERAGE HEIGHT: *12.2 h.h.*
POPULATION STATUS: *Rare*

Taishuh horse of Japan. Photo: Hiroyasu Ogawa

Native to Tsu Island (or Tsushima), in Nagasaki Prefecture, the Taishuh is an ancient breed. It is said that this pony dates to the eighth century. In 1920 there were more than 4,000 of them, but today only about sixty-five remain.

Since 1979 pedigree registration has been set up for this ancient breed, and attempts are being made to save it from extinction. Tsushima is noted for its extremely rugged coastline, and villages are extremely isolated. The Taishuh pony was very useful in times past for transportation.

The Taishuh is a strong and rugged animal with sturdy legs. Noted for their gentle nature and willingness to obey, these ponies were often ridden by farmers' wives and children.

Three Taishuh ponies have been sent to the national animal husbandry center in Hokkaido for special attention, and sperm and ova have been banked there.

Tarpan

(*Equus przewalski gmelini antonius*, European Wild Horse)

ORIGIN: *Poland and Russia*
APTITUDES: *Riding horse, light draft*
AVERAGE HEIGHT: *13 h.h.*
POPULATION STATUS: *Rare*

Although included here as a rare breed, the true wild Tarpan is completely extinct, and

Tarpans at the Munich Zoo. Photo: Evelyn Simak

the Tarpans of today have been "reconstructed" from remnants of the ancient breed. The last wild Tarpan was accidentally killed while resisting capture in 1879, and the last one in captivity died in the Munich Zoo in 1887. The Tarpan was one of the few wild Equidae in existence in historic times.

Tarpan in Russian means "wild horse." The Tarpan is believed to be the prototype for the light horse breeds originating on the Russian steppes. It originally inhabited all of eastern Europe and western Russia and was the basis of stock used by all the chariot-driving nations of the eastern Mediterranean from the Hittites to the Greeks and, originally, the Celtic tribes. This is the main ancestor of the small horses found in east-central Europe and the Balkans, such as the Hungarian Goral, Romanian Hucul, Polish Konik (considered the Tarpan's closest living descendant), and the extinct Prussian Schweike.

When the Tarpan became extinct, people sought, in hindsight, to recreate the breed—a feat which is not really possible. Had foresight revealed the coming extinction of the wild Tarpan, steps could have been taken to preserve this ancient horse. Such a move would have enabled modern scientists to make a comprehensive study of the roots of most domestic horses.

As it was, the Polish government collected from peasant farms animals that closely resembled the wild Tarpan and put them into forest reserves at Bialowieza and Popielno. Most of these animals were of the Konik breed, a very close descendant of the Tarpan. Careful selection was used in mating the most representative specimens, and the resulting

"Tarpan" is very much like the original, although it is acknowledged that the reconstructed Tarpan is not pure. The peasants had always captured the steppe Tarpan for use on their farms. The main reason for extinction of the wild Tarpan in the nineteenth century was widespread hunting for meat.

The Tarpan has a long, broad head with a straight or convex profile. The ears are long and point slightly to the sides; the eyes are small and round; the neck is thick and short with a full mane. The withers are somewhat flat; the back is long and straight; the croup is sloped; the chest is deep; the shoulder is sloped and long; the legs are long and slender but strongly built with broad joints.

This horse is always some shade of dun, ranging from mouse to brown dun, with a dorsal stripe, zebra markings on the legs, and usually a dark stripe across the withers. The mane and tail are black, as are the lower legs. Often there is a fringe of the body coat color lining the mane and at the base of the tail.

Like the domestic horse, the Tarpan has sixty-four chromosomes. It is not considered a separate species as is the Przewalski's Horse, nor are half Tarpans considered hybrids.

The zoo in Munich, Germany, also has a reconstructed Tarpan, slightly different from the one in Poland.

Tavda
(Tavdinskaya, Tavdinka)

ORIGIN: *Former Soviet Union*
APTITUDES: *Draft and farm work*
AVERAGE HEIGHT: *13.3 to 14.2 h.h.*
POPULATION STATUS: *Common*

The Tavda or Tavdinka is one of the northern Soviet horses of draft type. This breed is considered "natural," having developed more under the influence of natural conditions than at human hands. The Tavda is a forest type. It was widespread in the northeast part of Sverdlovsk and western part of Tjumen regions, in the valleys of the Tavda and Tura Rivers. Through the influence of severe climate and living conditions a special type evolved, able to cope with the cold and to work under poor feeding circumstances. These horses were used for work in rough terrain; in winter without any roads; and in summer in wet, muddy conditions.

In 1923 two Tavda types were exhibited at the agricultural show in Moscow. They were called Tavda and Tura and came from Krasnoufimsk, some 1,116 miles from Moscow, traveling this distance in twenty-one days. The first research on the Tavda horses was done in 1948.

The Tavda is lean with a deep trunk, wide chest, and well-developed muscles. Its legs are lean and strong, and hooves are hard. This breed is fertile, and the mares, even with poor feeding conditions, always produce adequate milk for their foals.

The primary color of Tavda horses is dun, with grullo (mouse dun), buckskin, bay, brown, and sorrel occurring infrequently. Most of the horses have a dorsal stripe and zebra stripes on the legs. Hair of the mane, tail, and legs is dense and long.

Because of its small size, the Tavda was unable to meet needs in agriculture. Some measures were taken to improve the breed—better feeding, training, establishment of stables, and also crossing to Ardennes horses and trotters. The goal has been to develop a larger horse while still preserving the Tavda's unique adaptation to severe local conditions.

This Tennessee Walking Horse, in its running-walk, demonstrates the high-stepping gait that makes the breed so popular in the show ring. The horse reaches high with the front legs while "walking" and pushing with the hind legs. Photo courtesy Tennessee Walking Horse Association

Tennessee Walking Horse mare and foal. Photo courtesy Tennessee Walking Horse Association

Tennessee Walking Horse

ORIGIN: *United States (Tennessee)*
APTITUDES: *Riding horse*
AVERAGE HEIGHT: *15 to 16 h.h.*
POPULATION STATUS: *Common*

The Tennessee Walking Horse is a composite breed that evolved from the Narragansett Pacer, Canadian, Morgan, Thoroughbred, Standardbred, American Saddlebred, and the Spanish horses descended from those brought to the Americas in the sixteenth century. All "American" breeds were based on Spanish stock. These bloods were fused into one animal in the middle Tennessee bluegrass region, resulting in one of the greatest pleasure, show, and trail riding horses.

Early breeders of the Walking Horse used gaited Spanish Mustangs brought in from Texas to Natchez, Mississippi, and trailed them up the Natchez Trace to Middle Tennessee, crossing them on pacing and trotting stock to eventually produce a new breed. To the scrutinizing eye, some of the Spanish characteristics are clearly evident.

The Tennessee Walking Horse is a light breed, originating especially in five or six counties of middle Tennessee. Here, native bluegrass abounds; the soil is rich in limestone and phosphate. The rich grass and the abundance of limestone water had much to do with building the Tennessee Walking Horse into a hardy, healthy animal with great endurance.

Some may think of the Walking Horse as a new breed, but its history reflects the fact that it had a strong influence in the daily lives of our forefathers.

Mainly used for utility and riding stock, the breed gained wide popularity for its smooth and easy gaits, speed, and the ability to stride faultlessly over the rocky hills and through the valleys of middle Tennessee terrain. Used as a utility animal for farm work as well as family transportation and recreation, the old plantation-type horse was not trained for gaits in those days—its gait was naturally inherited.

Often an entire family could be seen astride one horse traveling the roads, or one might see a gentle family horse carrying a group of school children to their daily classes. While everyone, including the horses, merited a day of rest, that day seldom came for the family Walking horse. After laboring all day in the fields in front of a plow, a farm horse might be saddled in the evening for a trip to a neighboring farm or church. Saturday morning usually found the animal in front of the farm wagon for a trip to town for supplies. Farmers from surrounding areas often gathered at the public square in town. Here, as well as at the "jock yard," the farmer's trading ground, horses were bought, sold, traded, and bragged about. Not infrequently, the outcome of bragging was a match race between the speed horses, the stakes a sack of flour or some other commodity or perhaps the horses themselves.

On Sunday the family horse was shined and hooked to the buggy to draw the family to church. One of the things a man always held dear was a good horse, and most would do without other necessities in order to keep a good mount. Many horse shows were held in Middle Tennessee before the war between the states. Horses were ridden or shipped to shows by rail. Competitors might set out to a show driving one horse to a buggy and often leading one or more behind. As the early Walking horse was still doubling as a utility animal, it might have come out of the fields just in time to make it to a show. If it did well, however, it was sometimes rewarded by being shipped home by rail.

Early breeders of the Tennessee Walking Horse were mostly farmers, and the time came when Walking horse colts were considered an additional crop in the Tennessee area.

As the automobile made its appearance in the southern way of life, many felt the days of the horse were finished. But in Tennessee communities, even when saving precious time and speed became a necessity, the Walking Horse remained essential. Roads were nearly impassable much of the time, and the easy-riding, swift saddle horse remained the better form of transportation, passing many horse-less carriages stuck deep in the muddy ruts.

In 1885 a cross between a stallion called Allandorf, from the Hambletonian family of trotters, and Maggie Marshall, a Morgan mare, resulted in a black colt with a white blaze, off hind coronet, and near hind sock: Allan F-1, foal of 1886. This horse was later chosen by the Tennessee Walking Horse Breeder's Association as the foundation sire of the Tennessee Walking Horse breed. While the bloodlines of the Gray Johns, Copperbottoms, Slashers, Hals, Brooks, and Bullett families ran thick and produced a type known as the Tennessee Pacer prior to the arrival of Allan F-1 in Middle Tennessee, it was the cross between Allan and the Tennessee Pacer that produced the Tennessee Walking Horse of today.

Early foundation animals of these breeds were brought into Tennessee from Virginia, the Carolinas, and other states by the pioneers who settled this area. North Carolina was known especially for its fine breed of Narragansett and Canadian pacers as well as some fine Thoroughbreds.

The particular area of Tennessee previously mentioned was populated mainly by pioneers from North Carolina, and early history of Tennessee refers to these horses being ridden back and forth between the two states as early as 1790. There are statements in the references of General Andrew Jackson that include mention of the horse Free and Easy (named for his gaits) being sent into Warren County, North Carolina, for breeding purposes.

Jackson also referred to the horse Copper-bottom, a Canadian Pacer and Thoroughbred brought from Kentucky as a colt, leaving

North Carolina at the age of twenty to spend the remainder of his life in Tennessee. He was the founder of the Copperbottom family and a great sire of saddle horses. Many of today's good Walking horses trace to him. One of his sons, Morrill's Copperbottom, established the Slasher family. Most notable is Mountain Slasher P-59.

Another important horse of this time was McMeens Traveler, by Stump the Dealer, by Timoleon. It has been said that Traveler never sired a "sorry" offspring, and his colts always brought fancy prices. General Nathan Bedford Forrest's cavalry possessed forty-seven horses sired by Traveler, and not one of them was lost in the Civil War. Records show that one of his sons, a grey gelding twelve years old, brought a thousand dollars in gold during those hard times.

Blood of the pacing horse from the strains named Brooks, Hals, Joe Bowers, Snow Heels, Grey John, and the impressive pacer Pat Malone left their mark on future generations. The family of Stonewalls, who came from the great Denmark and Cockspur blood found in gaited stock, contributed their outstanding characteristics to the Tennessee Walking Horse of today. These breeds were well established in Tennessee by about the 1780s.

The Tennessee Walking Horse Breeder's Association of America was formed in 1935. Its committee set out to find the stallion that had contributed most to the breed, an achievement that would make him eligible as the original foundation sire. Allan F-1 was the horse chosen, as he had done more to establish the breed than any other horse in history with the exception of Justin Morgan, whose blood was so strong he founded an entire breed. Justin Morgan was a profound contributor to the Tennessee Walking Horse breed.

Allan F-1 was by Allandorf by Onward, by George Wilkes F-54, by Hambletonian 10. His dam was Maggie Marshall, by Bradford's Telegraph, by Black Hawk 5, by Sherman Morgan. He was a double-gaited horse, both trotting and pacing, and held a record time of 2.25 at the pace, which was excessive for that time. His sire, Allandorf, was the acme of fashionable harness breeding. In his dam ran the best Narragansett pacing blood. Being bred with the hopes of being a great trotter, he wanted to pace so was never raced.

Roan Allen F-38 was by Allan F-1 and out of Gertrude F-84, a Denmark mare, out of a Canadian pacing mare. Most of the registered Tennessee Walking horses show Roan Allen F-38 in their pedigrees. It has been said that he could trot in harness and win; then do five gaits and win; and then step in and win the Walking classes. He sired many show winners and great colts that went on to make names for themselves.

Much of the fineness and finish of this breed today is credited to Giovanni, an American Saddlebred stallion. He was by Dandy Jim II, by McDonald Chief. His dam was a great Denmark mare. Giovanni's most prominent get as a show horse was Sir MacAlvanni, crowned world champion at Madison Square Garden. His blood appears in many of the pedigrees of today's world champions.

The docile temperament and keen intelligence of the Tennessee Walking Horse have made this a popular breed as a pleasure mount as well as a fine show horse. His gait gives great pleasure in the saddle, giving the rider a sensation of gliding, without shock. Standing on an average from 15 to 16 hands, the Walking Horse is refined but solidly built with plenty of substance. There is no limitation as to color or markings. In the forties the roan was popular, and later the sorrel with flaxen mane and tail. Least desirable before 1945 was the black or bay Walking Horse. With the success of the legendary breeding stallions Midnight Sun and Merry Go Boy, the taste of the general public turned to blacks and solid colors, where it has remained.

As people and times change, so do their horses. The coarseness typical of the early Walking Horse has long since disappeared and been replaced by a great deal of refinement and elegance. Excellent conformation is now also apparent throughout the breed. Selective breeding has added the polishing touches while continuing to maintain the muscular vigor, unexcelled disposition and calmness, size, and structural soundness required in today's Tennessee Walking Horse.

The flat-foot walk, running-walk, and canter are natural, inherited gaits and can easily be recognized from the time a young foal starts to amble beside its mother with rhythmic coordination of leg, head, and body movement.

The flat walk and famed running-walk are both a basic, loose, four-cornered gait, a 1-2-3-4 beat with each of the horse's feet hitting the ground separately at regular intervals. As it moves, its head nods in rhythm with the regular rise and fall of its hooves, and it overstrides with the hind foot the track left by the front foot. In general it should travel in a straight, direct motion, never winging, crossing, or swinging. The flat walk should be loose, bold, and square with plenty of shoulder motion. The running-walk should be executed with loose ease of movement, pulling with the fore feet, pushing and driving with the hind. There should be a noticeable difference in speed between the flat walk and the running-walk, but a good running-walk should never allow proper form to be sacrificed for excessive speed. Today's rocking chair canter is a high, rolling motion with distinct head motion, chin tucked, and a smooth and collected movement.

Many people have the impression that the Tennessee Walking Horse is a show horse, a pampered darling that shouldn't be used for any real work, but this could not be further from the truth. As its history reveals, this is one of the most versatile of horse breeds.

Tersk
(Terskaya)

ORIGIN: *Former Soviet Union*
APTITUDES: *Riding horse*
AVERAGE HEIGHT: *14.3 to 15.1 h.h.*
POPULATION STATUS: *Rare*

A relatively new breed, the Tersk was developed between the 1920s and 1940s at the Tersk and Stavropol studs by a team of horse breeders and veterinarians under the leadership of S. M. Budyonny. Both studs are located in the northern Caucasus, which has a continental climate.

Foundation stock consisted of the Strelets stallions Tsilindr and Tsenitel and mares of

Tersk horse. Photo: Nikiphorov Veniamin Maksimovich

the Arab-Don and Strelets-Kabarda complex. As the initial gene pool was quite limited, the Arab stallions Koheilan IV, Marosh, and Nasim were brought in for use in producing the Tersk breed. Selection was directed toward a breed as good in type as the Arab but better adapted to harsh climate and more massive. Crossing the Strelets stallions, followed by thoughtful inbreeding, created the new breed.

The Strelets breed, used as foundation for the Tersk, resembled the Arab but was larger and better suited to local conditions. This horse was produced from purebred Arab stallions crossed on high quality Orlov and Orlov-Rostopchin mares. By the early twenties the Strelets had nearly died out, with only two stallions and a few mares surviving. In 1925 the remaining horses were sent to the Tersk stud, where work was begun to increase their numbers. The Strelets was not preserved in its original form since those that survived were too inbred, and three pure Arab stallions and a number of crossbred mares (Arab × Don) were used. The result was the Arab-type Tersk, with good height and a strong constitution.

Close to the Arab in appearance, the typical Tersk has a light head with a straight profile, wide forehead and jaw, and long poll; the neck is medium in length; the back is flat and wide with a short, wide loin; the croup is rounded and well muscled; the chest is deep and wide; the shoulders are long and have good slope; the legs are correctly set. The coat of this horse is thin, including the mane and tail. The usual color is grey with a peculiar silvery sheen, with bay and golden chestnut also occurring.

Tersk horses are known for their remarkable endurance qualities. All of the horses that participated in a distance race of approximately 192 miles finished the event easily and in good condition. This breed is also known for excellent jumping ability and is becoming increasingly important for sporting events. It has excelled in show jumping, cross-country, and dressage. To increase size, Tersk horses are crossed with Trakehner or Hungarian breeds.

There are three types within the Tersk breed: basic or regional, eastern, and heavy. Tersk stallions are widely used for improvement of native mountain breeds such as the Lokai and Deliboz. This breed is also in demand for export due to its many excellent qualities.

The Tersk has inherited not only the general appearance of the Arab but also the quality of its paces. The walk and trot are very light and elegant. In addition, it is energetic with a kind disposition, easy to train and very well suited for circus work.

Breeding is concentrated at the Stavropol stud. The breeding nucleus is very small with only 250 purebred mares, and protective management is required to prevent the Tersk from becoming extinct.

Thai Pony

ORIGIN: *Thailand (former Siam)*
APTITUDES: *Riding horse*
AVERAGE HEIGHT: *12 to 12.2 h.h.*
POPULATION STATUS: *Common*

Origin of the native horse of Thailand is unclear but traces to the Ayudthya period in the eighteenth century. King Taksin the Great and his cavalry, mounted on Thai native ponies, fought off the besieged capital city of Ayudthya and captured Chanthaburi near the border of Siam. Taksin (1734–1782) then founded Dhonburi as the new capital city of Thailand. After his death, the capital was

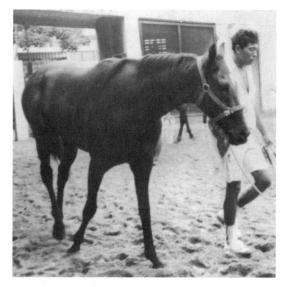

Thai "country bred" pony. Photo: Major General Panas Manjetjit

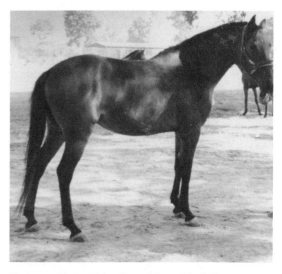

Thai pony. Photo: Major General Panas Manjetjit

moved to the opposite side of Chao Phya River and named Bangkok. A statue on the Big Circle (Wong Wian Yai) Road in Dhonburi depicts King Taksin the Great on his native Thai pony.

Native Thai ponies are believed descended from Mongolian stock. They have great stamina and are generally bay or brown. During the two World Wars, this pony was used by the military as a pack animal. Although crossbreeding has occurred in recent years using Australian, New Zealand, and some

English breeds, the pure native Thai pony can still be found in the northern part of the country, being used by people of the Hill Tribe as pack animals to transport tea leaves or fruits from the mountains.

Improvement of the native horse through crossing to the above-mentioned breeds has brought increase in size and refinement. Two classes of modern Thai pony are now found in the improved group; Class 1 reaches 14.3 to 16 hands, and Class 2 averages 13.3 to 14.2 hands. Called Thai "country bred," the Thai pony is a good quality animal.

Thessalian

ORIGIN: *Greece*
APTITUDES: *Riding, light draft*
AVERAGE HEIGHT: *14 to 15 h.h.*
POPULATION STATUS: *Rare*

Thessalian of Greece. Photo courtesy Aristotle University, Thessaloniki

Thought long extinct by many, the Thessalian still survives in small numbers. This is the ancient breed that provided excellent mounts for the Greek cavalry in days of yore. Bucephalus, the famous horse of Alexander the Great, was a Thessalian horse. Alexander's father, Philip, obtained many horses—some 20,000—from the area of Fergana to mount his troops, so the Thessalian horse is undoubtedly descended from the ancient Turkmenian breed.

Prior to the World Wars the remnants of this old breed were small, averaging about 13.2

hands. After World War II, the need was realized for larger horses, and the Greek authorities began to breed Thessalian mares with imported stallions of the Arab, Anglo-Arab, and Lipizzan breeds.

Predominant colors in the Thessalian breed are bay, brown, chestnut, and grey. This breed is of Oriental type, intelligent, and affectionate. The head is nicely shaped with a straight profile and the conformation is generally good. The legs are slim but very strong, with short cannons.

Tibetan

ORIGIN: *Tibet, China*
APTITUDES: *Riding, packing*
AVERAGE HEIGHT: *12 to 12.2 h.h.*
POPULATION STATUS: *Common*

The Tibetan horse is a breed located in the Qinghai-Tibet plateau area, mainly in Tibet and adjoining districts. This breed has always been considered very special. Remains were found in the excavation in Kanuo, considered to date 4,000 years ago. Using native stock, the Tibetan people developed this small but strong horse with very good adaptation to life on the high plateau. Lamaseries were always the owners of closely held, special breeding strains.

Although Tibet is a nation of many tribes, the Tibetan horse has been bred pure for many centuries. During the T'ang, Song, and Ming dynasties, Tibetan horses were often sent to the Chinese interior as tributes for the emperors. Blood of horses from the interior was never brought to the Tibetan plateau during those times, however. In recent decades a few attempts were made to introduce other blood to the Tibetan breed, but without success. The Tibet horse has evolved with a special adaptation to living and working in very high altitudes, and crossbreds do not survive in this habitat.

The Tibet breed was officially recognized in 1980. This is a smaller breed than the Mongolian but larger than some breeds found in the southeast of China. The head is small with a straight profile; the body is compact

Tibetan horse in China. Photo: Tiequan Wang and Youchun Chen

Tibetan horse. Photo: Tiequan Wang and Youchun Chen

and muscular; the legs are short with plenty of bone. Colors are bay and grey, and the coat is very thick.

Riding and packing are both important work for the Tibetan horse. This breed is phenomenal in hardiness and endurance qualities, and at an altitude of from 10,000 to over 16,000 feet can easily carry a load weighing approximately 130 pounds a distance of twenty miles without resting. Considering the size of this horse, its working ability is remarkable

and, in fact, unique in the world.

There are several types found within the Tibetan breed. The direct hereditary line of original stock of mountain-type Tibetan horses accounts for about 32 percent of the breed. The Plateau type is mainly in Naqu district, making up about 44 percent of the stock. The valley type is located in the south end of the region, which supplies barns for animals raised there. They account for 18.8 percent of Tibetan stock.

The Ganzhi type is located in the province of Sichuan, Abatibet Autonomous District; historically, this type was called Maiwa horse, well known for its developed chest. The Yushu type is found in Qinhai Province at an altitude of about 9,800 to 13,000 feet. The Zhongdian type is found in the province of Yunnan. This horse has large eyes, clean, open nostrils, and a compact and harmonious conformation.

All of the Tibetan types share in the extraordinary qualities of endurance.

Tieling

ORIGIN: *China (Tieling County)*
APTITUDES: *Farm and draft*
AVERAGE HEIGHT: *15 h.h.*
POPULATION STATUS: *Uncommon*

The Tieling is a draft breed developed at Tieling Stud Farm in Tieling County north of the city of Shen-Yang. The area around this city is developed for agriculture. Corn, sorghum, soy, oil plant, and even rice are cultivated there, so concentrates needed for feeding and developing strong horses are plentiful.

A survey done in 1949 revealed forty-four crossbred mares with various percentages of Anglo-Norman and Anglo-Arab blood, and mongrels in the area were of varying quality. After 1951 Anglo-Norman and Percheron crosses were made and the female progeny were crossed with Ardennes. The best offspring from the second generation were interbred. In order to overcome inbreeding depression, the blood of Soviet Draft, Jinzhou, and Orlov Trotter was introduced. The Ardennes

Tieling stallion. Photo: Weigi Feng and Youchun Chen

Tieling mare. Photo: Weigi Feng and Youchun Chen

gene base is narrow and increase in population is needed.

Timor

ORIGIN: *Indonesia (Isle of Timor)*
APTITUDES: *Riding, light draft*
AVERAGE HEIGHT: *10 to 12 h.h.*
POPULATION STATUS: *Common*

Timor pony. Photo: Dr. Soehadji, Department of Agriculture, Indonesia

played the most important part in development of the Tieling horse. Tieling horses can pull 80 percent of their body weight and show good endurance and willingness.

The head of this breed is of medium length with a straight profile, straight, forward ear, broad forehead, and strong jaw; the neck is slightly long and crested; the chest is deep and well developed; legs are muscular and have prominent, strong joints; the hooves are solid and fetlocks are free of hair. The main colors found in this breed are bay and black, with an occasional chestnut. The Tieling is a chunky individual but also shows refinement. The breed is used to improve other local breeds.

As the Tieling breed is small in number, the

Smallest of the Indonesian ponies, the Timor is extremely strong and agile and is often used as a cow pony. In spite of its small size it is even used for roping cattle.

The Timor has great powers of endurance and is lively yet makes a good mount for children, having a very docile temperament. Timors were among the first horses taken to Australia, and they contributed to the famous Waler breed as well as being very influential in development of the beautiful Australian pony. Timors were also taken to South Africa, where the famous and hardy Basuto and Cape horse developed.

The Timor may be any common horse color but is usually bay, brown, or black. The head of the Timor is usually small with a straight profile, short ears, and well-opened nostrils; the neck is short and muscular and quite broad at the base; the withers are prominent; the back is short and strong; the croup sloped; the chest deep; the shoulder quite straight. The legs of this pony are sturdy with good joints, and the feet are well shaped and very hard.

For additional general information regarding Indonesian ponies, see Bali.

Tokara

ORIGIN: *Japan*
APTITUDES: *Riding, light draft*
AVERAGE HEIGHT: *12 h.h.*
POPULATION STATUS: *Rare*

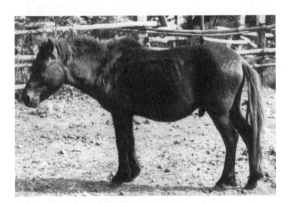

Tokara pony of Japan. Photo: Enkichi Koumoto

The original breeding ground of the Tokara pony of Japan was a village on the Tokara Islands, Kagoshima Prefecture.

Early in 1950, Dr. Shigeyuki Hayashida found a group of small horses living in the south side of the Tokara Islands and named them Tokara ponies. According to this man, around 1897 people from Kitaiga Island brought about ten native horses to Takara (not Tokara). They were used for farming, transportation of heavy objects, and in processing sugarcane grown in the area. In 1943, at the height of their popularity, the breed reached 100 head, but after World War II their numbers were reduced drastically.

Kagoshima Prefecture declared the ponies a prefectural national monument and sent them to the mainland for a time because of the difficult living conditions on the island. Today Kagoshima University, Iriki ranch, Kaimon National Park, and Nakano Island in the Tokara Islands are the centers for breeding and preservation of the Tokara pony. As of December 1988, a total of eighty-eight purebred animals remained.

Tori
(Toriiskaya)

ORIGIN: *Estonia (former Soviet Union)*
APTITUDES: *Riding, draft*
AVERAGE HEIGHT: *15 to 15.1 h.h.*
POPULATION STATUS: *Common*

Tori or Toric. Photo: Nikiphorov Veniamin Maksimovich

The Tori (also called Toric) is an all-purpose utility horse developed in Estonia at Tori stud between 1890 and 1950. Native Klepper mares were crossed to European halfbred stallions. (Today a very small number of the Klepper breed remain.) The Klepper was a predominantly dun pony, prepotent for toughness and hardiness. The Vyatka also has some Klepper blood.

The foundation stallion of the Tori breed is listed as Hetman, the son of Stewart and a hunter mare of unknown breeding. Stewart was a crossbred of a Norfolk Trotter sire and Anglo-Norman dam. Formation of the breed involved extensive use of Hetman and his sons. The result was a valuable breeding nucleus. However, by the end of the 1930s, signs of inbreeding depression were noted, manifested by deterioration of performance and robustness. Crosses with Breton stallions were used to eliminate the problem. The result was a massive-type horse which became widespread within the breed, and the quality in the gaits began to decline.

A need arose for a combination of utility

and sporting qualities in the horses, and to meet this need a limited experimental crossing of Tori mares to Hanoverian and Trakehner stallions was undertaken. The modern Tori is clearly a harness type and has a very clean and solid build. Conformation features include a large to medium-sized head; neck of medium length and fleshy; and withers of average height. The back is long and flat; the loin and croup are broad and well muscled; the legs are strong and clean. Common colors are chestnut, bay, and reddish bay.

This breed exhibits good performance with excellent gaits and jumping ability. The Tori has spread throughout Estonia. The main breeding centers are Tori Stud and the breeding farms at Pyarivere and Aravete state and collective farms.

The Trakehner stallion, Caesar (deceased). Photo: Terri Miller

Trakehner

(Ostpreußischen Warmblutpferdes Trakehner Abstammung)

ORIGIN: *West Germany, East Prussia*
APTITUDES: *Riding, sport horse*
AVERAGE HEIGHT: *16 to 16.2 h.h.*
POPULATION STATUS: *Common*

The full name of the Trakehner horse translates as "East Prussian Warmblood Horse of Trakehner Origin."

While known today as a West German breed, the Trakehner originated in East Prussia in an area now part of Lithuania. "Trial by fire" is an apt term when referring to development of this versatile breed, as the Trakehner was nearly destroyed during the Second World War. A beautiful and hardy breed prior to the war, the few that emerged from the fiery furnace in their flight into West Germany ahead of the advancing Russian army—wounded, sore, exhausted, and nearly starved to death—represented the purest refined mettle in horseflesh.

Probably it is true to say that today there are two Trakehner breeds; the one that survived the war to become the famous German warmblood known around the world, and the group taken into Russia directly from the remains of Trakehnen Stud. The two have been bred

Trakehner. Photo: Prof. H. C. Schwark

Although this is not commonly known, the Trakehner is sometimes pied. Photo: Evelyn Simak

differently, with less foreign blood being used in the Russian Trakehner.

Origin of this breed goes far beyond the Second World War to the days of the ancient Royal Scythians who inhabited the area in the fifth and sixth centuries B.C., after being driven from their former homelands by the Medes. They established themselves as rulers of an area reaching as far as what later became Hungary and East Prussia, enjoying considerable economic power until about the first century B.C.

The Scythians were superb horsemen and fierce warriors. Every Scythian owned at least one gelding to serve as a riding horse, but the wealthy possessed many mounts; most Scythians also owned oxen or rough-type ponies which served as beasts of burden. The finest riding horses were of the Fergana breed (ancient Turkmenian), but the majority were Mongolian horses. The frozen tombs of the Scythians at Pazyryk revealed many artifacts and remains of both humans and horses, the latter of the Fergana type. Thus while it is true to say that the foundation of the Trakehner breed was the indigenous, local Schweiken (extinct) breed, it should also be remembered that this native Lithuanian horse was based solidly on Turkmenian blood with various mixtures of the hardy Mongolian, a combination that could only produce an extremely hardy, fleet, and spirited animal of great endurance that was, at the same time, sensible and quiet in nature.

Early in the eighteenth century, King Frederick William I of Prussia, father of Frederick the Great, saw the need for a new type of cavalry horse for the Prussian army. He chose the best horses from seven of his royal breeding farms, and in 1732 moved them to the new royal stud at Trakehnen. Arab, Thoroughbred, and Turkoman blood was used, and careful selective breeding developed the Trakehnen horse into a fixed breed. In 1787, after the death of Frederick, the stud was transferred to state ownership under direction of a chief stud administrator named Lindenau, who set clearly defined breeding objectives and began practical work to improve the local horse population. With the basis of the hardy native Lithuanian horses, Thoroughbred, Mecklenburg, Danish, and Turkish blood was used. The final stage in development of the Trakehner was strongly influenced by a Turkmenian (Akhal-Teke) stallion and three of his sons. The Trakehner stud book was started in 1878.

Lindenau brought even stricter breeding practices than had the king, eliminating two-thirds of the stallions and one-third of the broodmares. He allowed private breeders to bring their mares for service by the royal stallions. Between 1817 and 1837, the blood of selected English Thoroughbreds and Arabs was added to the breed.

The Trakehner breed was very successful through the latter part of the 1800s and up until World War II, excelling as a military and endurance horse while also doing light draft work on the farms. In the area of performance the Trakehner also made an impression when the gold and silver medals in dressage in the 1924 Olympic Games went to Sabel and Piccolomini. In the 1928 Olympics, the Trakehner Lija won the bronze medal in the three-day event. In 1936 the famous Trakehner Kronos won the gold medal in dressage, Absinth won the silver, and the gold medal in the three-day event went to Nurmi. That same year, the German jumping team went to the United States to compete at Madison Square Garden and their Trakehner, Dedo, won the Prix des Nations. Between 1921 and 1936, the Great Pardubice Steeplechase, considered the most difficult next to the English Grand National, was won a total of nine times by Trakehner horses.

The stud at Trakehnen was forced to evacuate several times. In 1794 the stud went into exile to escape the invading Poles. In 1806 and 1812 the establishment fled when Napoleon's army defeated the Prussians. During World War I the Trakehners were again forced from their home, and they did not return until 1919. The end of this outstanding breed nearly came, however, in 1944 as World War II drew near its end. As the Russian army advanced on East Prussia, orders were issued to evacuate the horses from the Trakehnen Stud. Private breeders were also ordered to leave. Many of the best horses were transferred both by foot and rail but did not go far enough west, and most of them, along with their papers, fell into the hands of the Russian occupation forces and were shipped to Russia. The private citizens hitched their horses to wagons loaded with all their worldly goods and fled their country in the middle of winter in what came to be known as "the Trek." About 800 horses started the 600-mile journey to safety. The people were mostly women, children, and elderly people. Many of the mares were heavy with foal. Snow covered the ground. They could not stop when a mare lost her foal or if horses became lame or ill. Their feed ran out and the horses had to live on what they could find along the way. Many lives were lost on this journey of misery. Of the 800 once proud and beautiful Trakehner horses, less than 100 pitiful skeletons reached their destination, limping, wounded with shrapnel, and with frozen burlap bags tied to their sore and bleeding feet. Only the hardiest and best survived the "trek."

Scattered throughout West Germany, the horses were slowly collected about 1947 and the West German Association of Breeders and Friends of the Warmblood Horse of Trakehner Origin was formed. In 1950 the German federal government recognized the effort being made by dedicated breeders and agreed to join with the government of Lower Saxony in providing support.

Prior to World War II there were more than 10,000 breeders in East Prussia, and 18,000 registered mares. As previously noted, less than 100 survived the journey into West Germany. The majority of the breed was left behind in what became Poland. Many of the horses were sent to Russia and others formed the basis of the Mazury and Poznan breeds in Poland, later amalgamated into the present Weilkopolski breed.

Trakehner horses are now bred in most areas of Germany, and the breed has been brought back to a healthy population. Russia has approximately 2,000 purebred animals.

The Trakehner is perhaps the most elegant of the German warmbloods. A superb sport horse, this is a large, heavily muscled animal of classic saddle build. The head is elegant and refined, broad between the eyes, which are large and expressive, narrowing to a fine muzzle; the neck is long and crested; the withers are quite pronounced; the back is straight and short; the croup is gently sloped; the chest is deep; the shoulder well muscled and sloping. The legs are well muscled with broad, clean joints, clearly defined tendons, and a solid hoof.

Most commonly the coat color of Trakehner horses is chestnut, bay, brown, black, or, more rarely, grey. Piebalds also occur in this breed and, while not encouraged, may be registered if their conformation and other characteristics are good.

Trakya

ORIGIN: *Turkey*
APTITUDES: *Riding, packing, harness*
AVERAGE HEIGHT: *13.3 to 14.1 h.h.*
POPULATION STATUS: *Uncommon*

The Trakya horse has existed in Turkey for at least one thousand years in the region of Trakya (ancient Thrace), located in the far west of the country. This is a small horse, slightly larger than the Anadolu but generally of the same type. The Trakya, like the Anadolu, is very enduring and hard, able to live and work under very poor feeding and weather conditions.

Trakya horses developed from crossing the native Anadolu horse with the native horses of Trakya. They are used for riding, harness, and packing. All common coat colors are found.

A halfbred Trakya horse. Photo: Ertuğrul Güleç

A purebred Trakya horse. Photo: Ertuğrul Güleç

Trote en Gallope. Photo courtesy Marshall S. Ezralow and Associates

The Trakya has a medium-sized head; the neck is short; the withers are medium in height; the chest is narrow; the croup sloped; the tail set low; the legs are good with plenty of bone. Hair of the mane, tail, and feathering on the legs is coarse.

Presently there is no stud book for this enduring little horse, but breeders hope to form one in the near future.

Trote en Gallope

ORIGIN: *Colombia*
APTITUDES: *Riding horse*
AVERAGE HEIGHT: *14.2 to 15 h.h.*
POPULATION STATUS: *Common*

The Trote en Gallope is a relatively new breed, originating in Colombia about sixty years ago. The breed was produced by crossing of the Spanish horse (Andalusian) to the native Colombian Criollo or Colombian Paso Fino. This cross resulted in a very elegant, smooth-gaited horse called the Trote en Gallope. A special feature of the breed is the trotting gait, executed very smoothly by the fact that two feet are always in contact with the ground, so the horse is never suspended in the air as is the case with other trotting horses. The gallop is short, smooth, and very rhythmic. With these two combined characteristics, the horse presents an aura of elegance and grace while delivering a comfortable ride.

Turkoman
(Turk, Turkmene, Turkmen, Turkmenian)

ORIGIN: *Turkmenistan, Iran (Persia)*
APTITUDES: *Riding horse*
AVERAGE HEIGHT: *15 to 16 h.h.*
POPULATION STATUS: *Rare*

The Yamud Turkoman mare, Shahnazar, in Iran. Photo: Caren Firouz

The Plateau Persian in Iran. Photo: Caren Firouz

While the ancient breed known as Turkmenian is considered extinct today in its pure original form, horses bred in Persian Turkmenistan are still called by that name. The purest descendants of the ancient Turkmenian breed are the modern Akhal-Teke and Yamud; the Yamud in Iran is much the same breed as the Iomud in Russia (see also Akhal-Teke and Iomud).

Iran basically has three categories of horses: Plateau Persian, Turkoman, and Caspian. The Plateau Persian, a term coined by Louise Firouz of Teheran, includes all horses bred in the central provinces (south of the Alborz or Elburz Mountains) such as Azerbaijan, Isfahan, Kurdistan, Khuzestan, and Fars. These are collectively known as Plateau Persian and share a basic Oriental conformation and heritage (see Oriental Horse). Traditionally they were bred mainly by tribes, and the breeds take their names from tribal origins: Darashuri, Qashqai, Bakhtiari, Kurd, and some others.

Only the Western Arab no longer reflects its tribal origins. This is because its various different strains have become so mixed that the animals can only be known by the generic term "Arab." In Iran the Arab is referred to as the Asil (pure) horse. The Arab in Iran today is a mixture of horses bred by Arabs and Bakhtiari tribesmen in Khuzestan Province in the south. They are the usual strains of Ham-

dani, Kuheilan, Nesman, and so forth. Possibly some of the original stock came from Iraq, but they have been bred in Iran since before the Arab invasion in the seventh century A.D. The term "Arab" is somewhat misleading, since it lumps together a number of strains and subbreeds. This is essentially due to early British travelers in the Middle East, Lady Ann Blunt particularly, who substituted "Arab" for "Oriental" horse, thus creating one breed out of many. The purest of the Arab types in Iran is bred by a subtribe of the Kashkai (or Qashqai), the Darashuri. This has never been mixed with any other type and closely resembles the Assyrian Oriental types on the friezes now in the British Museum. The Darashuri do not call their breed an Arab and are, in fact, insulted if you make this mistake. They consider their own breed pure but do not consider the "Arab" pure, as it has been mixed with horses from Hungary, Poland, Lebanon, Jordan, Egypt, Bulgaria, and the United States (see Persian Arab), with no regard for strains.

The Turkoman horse has been largely neglected by historians. The majority of references by writers from the sixteenth to nineteenth centuries have been far from complimentary, their eyes reflecting current western fashions. The Turkoman horse is slim and narrow, angular and "long" everywhere, and often considered a weak type by those uninitiated to the breed. Sharply to the contrary, this is a horse of exceptional endurance and ability and most likely the main contributor of hot blood to the very breeds to which it is considered inferior. At present, the Turkoman horse is raised in a remote area of Central Asia and numbers for the pure breed are dwindling.

The treaty of Turkmanchai, signed by the Russians and Persians in the nineteenth century, divided Turkmenistan into two separate entities. This put an end generally to the traditional raids on agricultural communities in the plateau regions of Iran, which had provided the Turkomans with slaves and other amenities. It also decreased the need for the Persian war horse.

For centuries there had been a fluid movement of boundaries in the Aralo-Caspian region, but the Yamud and Akhal-Teke breeds kept to their respective breeding grounds and remained pure. The Akhal-Teke is bred by the Teke Turkoman tribes, which encompass an area from Merv to Ashkhabad—flat, open land divided by irrigation ditches where the Oxus, Morghab, and Tejend Rivers have been diverted to agriculture. Between these oases, however, lie miles of dry steppe where the few water holes are invariably brackish.

The Yamud is bred by the Gouklan and Yamud tribes, whose boundaries are within Iran. The topography here is varied, between the steppes beginning at the eastern edge of the Caspian Sea and the dry steppe (Gehezel Bayer) bordering the Atrek River. The Yamud is a more compact animal than the Akhal-Teke, reflecting generations of crossing with the Plateau Persian, a relic of the Indo-European invasions from the steppe country and subsequently bred on the rich alfalfa pastures of the Zagros Mountains. Conformation of the Yamud is well suited to the flat steppe and to the mountainous regions of the Alborz.

When compared to other horse breeds in Iran the Yamud has a scanty mane and tail, long back, and angular attachments of the head and neck. Nowhere, however, is the exaggerated angularity of the Akhal-Teke to be seen.

Turkoman horses are raised on the open steppe where once great herds of mares roamed from summer to winter pastures. The colts are caught when six months of age and staked out on long ropes. From that time on they are never turned loose. They are fed a protein-rich diet consisting of barley, native mountain alfalfa, and barley straw with additions such as raisins and dates. Boiled chicken is often a part of their diet, to provide protein. Traditionally, they are blanketed with felts from the tips of their ears to below the tail to sweat out all excess fat. The colts first feel the weight of a light rider at eight months and face their first race as yearlings. Incredibly, few of the colts break down under this treatment, and those that do are never used for breeding.

The Turkomans traditionally use proven methods to condition their horses. Only animals that have shown their ability for traveling great distances are used for racing. Conditioning begins slowly. The first day only the saddle is placed on the horse's back. The second day the horse is ridden to water and for one week thereafter walked slowly each day for increasing periods. After one week the horse is put to the trot and will not canter for another two weeks. Galloping is done only at night, and sweating beneath the blankets is the usual fare for the day. Only small boys are used for jockeys, while all sizes and ages are used for the slow conditioning. While being conditioned for racing and during the racing season, the horses are fed barley up to seven times a day. They are fed chopped straw and alfalfa five times a day.

During the breeding season the stallions are released with the mares. This practice makes accurate pedigrees difficult, as there are many herds of mares and many competing stallions. Once covered, the mares are allowed to roam from stark areas along the Atrek, through wheat fields, and to the Alborz forests. They must deal with cheetahs, wild boar, bears, wolves, cobras, and other snakes. The perils of this life have developed the Turkoman into an intelligent, capable, and self-sufficient animal. Any foal born with a debilitating malformation simply does not survive to pass it on.

The Turkoman breed was kept completely pure until the 1960s, when Thoroughbreds were imported for racing. Many Turkoman horses were bred to Thoroughbreds for increased speed. This practice was limited to the Yamud, while the Akhal-Teke was kept pure.

The revolution brought a dramatic change in horse breeding in Iran, and many horses were turned loose in the mountains toward the Soviet border, the breeders confused over policy. For a time anyone owning more than one horse was considered feudal and subject to execution. Today large herds of Turkoman horses roam the Ghezel Bayer near the Atrek River. They are wild (feral) now but are descended from some of Iran's finest horses. The Turkoman is an Oriental breed, and additional history can be gleaned by reading about the Oriental horse in this volume.

The coat color of the Turkoman may be bay,

grey, brown, black, or chestnut. The head is well proportioned and wedge-shaped with a straight profile, broad forehead, and large eyes; the neck is long and lean but muscular. Withers are prominent; the back is long and straight; the croup long and sloping; the chest deep; the shoulder long, sloping, and muscular. The legs are long and slender but strong, with broad joints and well-defined tendons; the hoof is small but strong. The skin is thin and the coat is fine and silky; often the mane and tail are sparse. Everything about the appearance of the Turkoman speaks of speed and endurance.

Tushin

ORIGIN: *Georgia (former Soviet Union)*
APTITUDES: *Riding horse*
AVERAGE HEIGHT: *13.1 to 14 h.h.*
POPULATION STATUS: *Rare*

The Tushin is a rare breed today, bred in the eastern regions of the Georgian Republic. It is used principally under saddle and is able to cover long distances to control herds of cattle being driven from one seasonal pasture to another. Tushins are very surefooted on the narrow paths of the Caucasus mountains.

The Tushin is described as a small horse with a long, shallow body and fine legs. The breed is well suited to high altitude and thrives on highland pastures. Many are natural pacers, a quality highly valued by the local inhabitants. The predominant color is bay, although chestnut, grey, and black are also found.

The Tushin breed descended from ancient Georgian horses that were strongly influenced by the Persian, Bactrian (Turkmene), Arab, Turkish, and other eastern breeds. Georgia was famous for its horses long ago. They were of higher quality at one time than the celebrated Persians, although smaller. Horseback riding has developed as a culture since ancient times, as life in the mountains and wars for independence made the horse a vital necessity. Arabs acquired many Georgian horses during their rule in the eighth century.

Tushin horses are bred year round on alpine pastures, and only the strongest working animals are kept in stables. Food is scarce during the winter months, and this horse developed the ability to store fat quickly during the summer.

The constitution of the Tushin horse is lean, with the general appearance being compact and round. The head is lean with a straight profile, wide forehead, and short ears. The neck is of medium length and straight, set on rather low; the withers are moderately pronounced and long; the back is straight; the croup is short and slightly sloped; the chest is somewhat narrow but very deep; the shoulders are long and sloped; the legs are lean with clean tendons, and in the majority of Tushins the hind legs are cow-hocked.

Prevailing colors are various shades of bay, with some animals being sorrel or black. Paints and appaloosas occur rarely, with the majority having no white markings of any kind.

The temperament of the Tushin horse is calm and obedient, and these horses are able to cover long distances in a short time. Enduring, they are often worked two or three days without a break and yet show no loss in their condition. They are worked hard without grain.

Tushin horses are late to mature, reaching full height by about five years. There are many amblers in the breed, which are prized for their easy gait. The trot is weakly developed.

In recent years attempts were made to improve the Tushin breed with crosses of Thoroughbred and Kabarda blood for increase in size. Interestingly, the crosses had *no effect* due to the pure, genetic strength of this old breed.

In 1976 approximately 1,000 pure Tushin horses were in existence. Work is in progress to prevent extinction of the breed.

Tuva
(Tuvinskaya)

ORIGIN: *Siberia (Tuva Autonomous Region)*
APTITUDES: *Draft and harness work*
AVERAGE HEIGHT: *12.3 to 13.2 h.h.*
POPULATION STATUS: *Rare*

The great Yenisey River of Siberia is formed near Kyzyl, the capital of Tuva, where two mighty tributaries run together: the Great Yenisey (Bolshoy Yenisey) or By-Khem, which rises on the eastern Sayan Mountains of the Tuva Autonomous Region, and the Little Yenisey (Maly Yenisey) or Ka-Khem, which rises in the Darkhat basin of the Mongolia People's Republic.

Tuva is a mountainous country with many valleys. It is far from the sea, so the climate is sharply continental. Summers are hot, winters long and very cold. While the territory of Tuva is comparatively small, there are differing climatic zones. In the south, camels are bred, and in the Todzin region, reindeer.

Currently there are 30,000 horses in Tuva. Horse breeding is widespread in each of its twelve districts, the horse having adapted to various climates. In the southern regions the *taboons* (large herds) of horses graze together with camels, sheep, and goats.

Horse breeding in Tuva has been in existence since ancient times. Archaeologists excavated a barrow/kurgan near the village of Arzan where a nomad leader and fifteen of his guards along with 150 riding horses were buried. The barrow was dated to the seventh or eighth century B.C. Horses of ancient Tuva were small, similar to the Mongolian horse. Some of these survive today in southern semi-desert areas. The native people of Tuva were nomadic cattle breeders.

Most of the Tuva horses have developed into a larger type as the result of adding light horse blood such as Don or Thoroughbred. In some areas of central Tuva a different type is bred, known as Verkhne-eniseiskaya, or Upper Yenisey horse. These average 15.2 hands and are reminiscent of massive trotters or trotter–draft horse crosses. They are raised in *taboons*, outdoors without stables. During the entire summer they are kept on the mountain taiga pastures, and in winter in the valleys. They are bred for meat, and geldings are used for work. In remote regions the Tuva breed has been saved in purity.

Curiosity must arise over development of such comparatively large horses that are well adapted to *taboon* keeping under severe conditions in the area of Tuva. Other Siberian breeds are much smaller and quite different in morphology. A look at the history of Tuva answers the question. Until the middle of the nineteenth century, all of the inhabitants of Tuva were nomadic, raising cattle and hunting. Around the 1860s, emigrants from the Altai and Minusin regions began to arrive; peasants who brought horses with them. They were Kuznet, draft horses and massive trotters.

The main activities were cattle raising and the output of minerals—gold, asbestos, and coal. For this work the people required strong working horses. By the end of the nineteenth and beginning of the twentieth centuries, large private farms were established which had their own taboons of horses. The book *History of Tuva,* published in 1964, describes some of these horse farms. The largest was the farm Vavilin, with 3,000 mares and 50 sires. The Safjanov farm had more than 2,500 horses. From the beginning of the twentieth century until 1917, some great landlords attempted to develop their own breeds. The young horses were kept in stables until grown, receiving good food and care. They were trained, and stallions were selected from the best four- and five-year-olds. These horses were named after their owners and breeders.

After 1944 when agriculture was being reestablished in Tuva, the various family taboons were mixed together. The emerging breed displayed high working ability and endurance. In 1951 a standard was established and they were officially named Tuva Harness Horse. In recent decades the Tuva Harness Horse has declined, suppressed by tractors and harvesters in the fields and as transport with horses decreased. For shepherds working with the herds of meat cattle and sheep in Tuva, the need was for riding horses. The stud farm was closed in 1957.

The Tuva, however, thanks to its high adaptability, is still bred in some areas of the central regions and is used to improve local taboon horses of some farms. There is interest today in studying the horse stock of these areas to locate the entire living population of Tuva horses. The state plan for 1980 to 1990 was to study the preservation of genetic material of each kind of farm animal. The Tuva horse has great agricultural value and is interesting also for genetic research and study of the process of developing horse breeds.

Ukrainian Saddle Horse
(Ukrainskaya verkhovaya porodnaya gruppa)

ORIGIN: *Ukraine (former Soviet Union)*
APTITUDES: *Riding, sport horse*
AVERAGE HEIGHT: *15.1 to 16.1 h.h.*
POPULATION STATUS: *Common*

Ukrainian Saddle Horse. Photo: Nikiphorov Veniamin Maksimovich

The Ukrainian Saddle Horse emerged in the studs of the Ukraine after the end of World War II by the blending of Hungarian mares (Nonius, Furioso, and Gidran) with Trakehner, Hanoverian, and Thoroughbred stallions. Special value was attached to individuals that carried the blood of the now technically extinct Russian Saddle Horse.

Crossbreds possessing the desired characteristics of good height and bone, a broad, powerful body, and a strong but light build were further interbred. Mares inclined to coarseness were crossed to Thoroughbred and Thoroughbred-Hanoverian stallions, while those of the finer, lighter type were put to Hanoverian or large Thoroughbred-Hanoverian stallions.

The horses were kept on the combined system and fed a high nutrition diet. The best of the young stock were taken for breaking and training at the age of eighteen months. Two- and three-year-olds were tested on racecourses as well as in other competitive events such as dressage, cross-country, and show jumping. The most promising of these were subsequently used at stud.

Breeding was started at the Ukraine stud in Dnepropetrovsk region and later continued mainly at Aleksandriisk, Dnepropetrovsk, Derkulsk, and Yagolnitsk studs. Ukrainian saddlers are bred mostly for sporting use, where they have proved themselves to be excellent mounts.

This is a heavy, large saddle horse. It has an attractrive head and well-made body. The eyes are expressive; the neck is straight and long, as is the poll; the withers are well developed and prominent; the back is long and flat; the loin is well muscled and broad; the croup is long and has a normal slope; the chest is deep and broad; the body is heavy; and the limbs are correct and well set. This is a solidly built horse. The most usual colors are bay, chestnut, and brown. This horse is smaller than the Thoroughbred but with more substance and a calmer attitude.

Ukrainian Saddle Horses perform extremely well in classic events and especially in dressage. Soviet equestrians competing on them have won repeatedly or have been placed high in many top-ranking competitions such as the Olympics and the European and world championships.

Breeding is mainly of pure stock with corrective crossing to the Thoroughbred. The Bespechny line consists of horses derived from the last of the Russian Saddle Horse (Orlov-Rostopchin) breed.

Unmol

ORIGIN: *India (Northwest Punjab)*
APTITUDES: *Riding horse*
AVERAGE HEIGHT: *15.1 h.h.*
POPULATION STATUS: *Rare, possibly Extinct*

The Unmol is extremely rare and is verging on extinction. In spite of all the efforts made some years ago by the Army Remount Department to save this breed, it is doubtful that any purebreds are left. Some breeders in the Punjab states have horses referred to as Unmol, but it is believed that today all of them have some blood of imported Arabs.

The various families of the Unmol breed are known as Harna, Hazziz, Morna, and Sheehan, all of which are referred to as Unmol. *Unmol* means "priceless" and indicates the great value and preference put upon this breed.

The Unmol is described as very strong, elegant, and "shapely," possessing a long mane and tail and compact body. Predominant colors are grey and bay. Tradition has it that the ancestors of the breed were brought by Alexander the Great when he invaded India. If that is the case, the Unmol was originally of Turkmenian blood.

Uzunyayla of Turkey. Photo: Ertuğrul Güleç

Uzunyayla

ORIGIN: *Turkey*
APTITUDES: *Riding, packing, harness*
AVERAGE HEIGHT: *14.1 to 15.1 h.h.*
POPULATION STATUS: *Rare*

Origin of the Uzunyayla breed dates to 1854 in Turkey. Ancestors of the breed came from the Caucasus, and it is believed that they were of the Kabarda breed. They were bred pure in Turkey until 1930, when Anadolu and Nonius blood was introduced.

The Uzunyayla has a large head with concave profile; the eyes small; the neck is of medium length; the withers are well pronounced; the legs are strong with good joints; the pasterns are sloped and very strong; the feet are well shaped and of tough horn. The Uzunyayla is usually bay in color. Feathering on the legs is coarse, and the tail grows very long.

This breed is very good for riding long distances. Uzunyayla horses have a good gallop but cannot do the *rahvan* walk like some other Turkish breeds. The normal walking speed is not fast as is seen in the Anadolu breed.

There is no association or stud book for the Uzunyayla breed, but breeders hope that one will be formed in the near future. Presently there are only about 2,000 specimens of this breed.

A good example of the Uzunyayla horse. Photo: Ertuğrul Güleç

Vladimir Heavy Draft
(Vladimirskaya tyazhelovoznaya)

ORIGIN: *Vladimir (former Soviet Union)*
APTITUDES: *Heavy draft work*
AVERAGE HEIGHT: *15 to 16 h.h.*
POPULATION STATUS: *Common*

Developed in the area for which it was named, the Vladimir Heavy Draft is a horse stemming from large native breeds crossed to various draft breeds such as the Percheron and Suffolk. The breed was developed on horse breeding collective farms of Vladimir and

Vladimir horse. Photo: Nikiphorov Veniamin Maksimovich

Ivanovo Regions. Later, Clydesdale and to a lesser extent, Shire blood was used. The aim was to breed a horse of medium draft power with rather high speed. A particular role was played for more than a century in development of the breed by the Gavrilovo-Posad breeding station, previously a stud farm and a state breeding stable. This breed was officially recognized in 1946.

The region where the Vladimir was developed has been famous since early times for breeding good horses. In the eighteenth century there was a stud farm producing riding and harness horses of large size that had great influence on the local horses.

The Vladimir Heavy Draft combined ample size, stout build, energetic temperament, and speed. Compared to the Clydesdale, the Vladimir has a more developed chest and cleaner, more solid build. Frequent inadequate size of the Vladimir is due to its having been raised in unsuitable conditions at collective stud farms.

The Vladimir has a long, clean-cut head with an often arched (convex) profile; the neck is long and well muscled; the withers are pronounced and long; the back is somewhat long and a little dipped; the loin is short and broad; the croup is long and sloping; the legs are long and properly set. The chest is broad but not deep and the ribs are lightly sprung. The hair of the mane, tail, and limbs is heavy. The predominant color is bay, with brown and black being seen less frequently. White markings on the face and legs are common. The Vladimir has excellent gaits. This is a breed with excellent working ability, good disposition, and a high growth rate.

The main breeding work is being carried out at the Yuryev-Polski stud. There are four bloodlines.

Voronezh Coach Horse

(Voronezhskaya upryazhnaya)

ORIGIN: *Voronezh District (former Soviet Union)*
APTITUDES: *Harness, draft work*
AVERAGE HEIGHT: *15 to 15.3 h.h.*
POPULATION STATUS: *Rare*

The Voronezh horse exists in small numbers today and is grouped with other rare breeds in the former Soviet Union which collectively number some 116,200 head. Among these are included the Hucul, Kuznet, Chumysh, and others. Presently there is interest in reviving this old breed which is descended from the now extinct Bityug. Present-day Voronezh horses retain many characteristics once common to the Bityug breed.

The Bityug horses were of medium size, firm constitution, and good, calm temperament. They were of transport-agricultural type and had great endurance, able to travel thirty to fifty-five miles a day pulling freight weighing 1,700 pounds and more. The Bityug horse was developed from local steppe horses crossed to various breeds, especially Danish, Dutch, and Orlov Trotter. The breed received its name from the Bityug River. Colors in the breed included grey, roan, and paint, rarely black. This horse was reminiscent of a coarse, partbred Orlov Trotter. By the 1920s the breed was described as rare and rapidly disappearing.

By the end of the nineteenth century demand arose for a more massive, larger horse, and the Bityug horses were crossed to draft breeds and trotters. As a result, they lost their uniformity of type.

The Bityug breed was highly valued by peasant farmers, who continued to breed horses of the Bityug type, and a breed nearly identical to it was eventually produced.

In 1936 a study was done on local horses of the area and the decision was made to improve the draft horses by way of pure breeding and some addition of trotter blood. A stud farm was established in 1938 for the breed. In addition, the commitment was made for improved feeding of young stock. The breed was given the name Voronezh harness or coach horse, not Bityug, and it was shown with this name at the All Union Agricultural show in Moscow in 1939.

The Voronezh horse has a long trunk and well-developed, long croup; the back is straight and strong; the front legs are straight with well-made joints and large, hard hooves. There is well-developed feather on the lower legs. The hind legs are often cow-hocked. This breed is strong on endurance while modest in requirements of feed and raising. Presently there are three types within the breed. The type smallest in number is the most massive, with a longer trunk, tracing in the sire line to draft stallions. A lighter, more harmonious type represents trotter-Bityug crosses. The third type represents offspring of local stallions of unknown origin with medium massiveness; it is slightly coarse and usually not very harmonious in conformation.

The basic goal in breeding work with the Voronezh breed is fixation of the desired type. Some highly valuable stallions have been produced recently. For improvement is is planned to use trotter and Vladimir stallions in addition to Voronezh stock.

Vyatka
(Vyatskaya)

ORIGIN: *Kirov and Udmurtia (former Soviet Union)*
APTITUDES: *Riding, draft work*
AVERAGE HEIGHT: *13 to 15 h.h.*
POPULATION STATUS: *Rare*

A native northern breed, the Vyatka (also spelled Viatka) was strongly influenced by natural conditions in the areas now known as Kirov and western Perm regions and in Ud-

Vyatka. Photo: Nikiphorov Veniamin Maksimovich

murtia. Valleys of the Vyatka and Obva Rivers provide rich meadows and quality hay, ideal for horse breeding.

Conformation of the breed was affected at various times by native Estonian horses and Kleppers brought by Novgorod colonists in the fourteenth century and by edicts from the state horse breeders during the reign of Peter the Great. Subsequently, horses of the Estonian Native breed were imported during development of the mining industry in the Ural Mountains. The valuable features of the Vyatka included good draft abilities such as speed, extraordinary endurance, and good use of feed. The breed was commonly exported beyond the limits of Vyatka Province, even being sent to Poland for improvement of local stock.

In conformation, the Vyatka has a clean head with a wide forehead and broad jaw. The neck is short and fleshy, often well arched; the withers are avearge in height; the back is long, broad, and sometimes slightly dipped at the withers; the croup is wide and somewhat short; the trunk is wide and deep; the legs are short and strong with very good hooves. The hind legs are often sickle-hocked. The forelock, mane, and tail are thick and long. One strong characteristics of the breed is its chestnut roan or bay roan color, with a black stripe along the spine and wing-shaped patterns over the shoulders as well as zebra striping on the legs.

Development of agriculture, transportation, and industry sharply reduced the demand for the breed, and most of the purebred mares were mated with heavy draft horses and trotters. There were no provisions for breed protection, and the Vyatka was considered essentially extinct in 1917. Steps were taken after the Revolution to reestablish the Vyatka, however.

The Vyatka is a strong, very usable horse, energetic, tractable, and frugal, with great stamina. It is believed that this breed is a direct descendant of the Tarpan. Aboriginal colors were red dun or brownish dun with a dorsal stripe, zebra stripes on the legs, and a cross on the shoulder. After fifty years of breeding to reestablish the Vyatka breed, new colors have appeared (above).

The Vyatka has a fast trot and was much used in the past for pulling the troika sled (three horses to one sleigh); by the first half of the nineteenth century it was the best troika horse in Russia.

This breed could become quite competitive due to development of tourism in the former Soviet Union. As draft horses, the crosses of the Vyatka and heavy draft are very strong, showing good use of feed and almost no reaction to midges and blood-sucking insects.

A 1980 report indicated that some 1,000 Vyatka horses were still in existence; a report from the Moscow journal *Konevodstvo* (no. 6, 1986) says that the Vyatka breed is in danger of extinction, very few pure individuals remaining. It is possible to find a small number in the regions of Irgiz, Koz, Debec, and Balezin. Throughout the Soviet Union about 1,774 horses can be found having Vyatka blood. The majority of these are crossbreeds, preserving the typical dorsal stripe but not having zebra stripes on the legs.

There is currently no breeding farm for the Vyatka, but a stud farm has been proposed in Kolos in Debes region to save the breed. Breeding animals of Vyatka type will be collected. It is possible that another stud farm will be established in the Irgin region. There are three mares and one stallion of the ancient type in the Udmurt region. Their offspring will be used in these farms, not only to save the breed but also to improve it.

Waler

ORIGIN: *Australia, especially New South Wales*
APTITUDES: *Riding horse*
AVERAGE HEIGHT: *15.2 to 16 h.h.*
POPULATION STATUS: *Rare*

A World War I Waler gelding. Photo: Peter Fischer

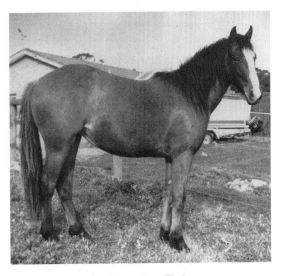

A modern-type Waler. Photo: Peter Fischer

While the Waler was never officially a recognized "breed," this horse was a definite type and received worldwide recognition and fame during the nineteenth century and into the twentieth, through the First World War. "Waler" is a term first used in India in 1846, referring to the outstanding horses from New South Wales.

Some maintain that the Waler is extinct, its blood living on only in the modern Australian Stock Horse and some of the feral brumbies that roam the outback. The issue is confusing at best. The Australian Stock Horse is certainly based on remnants of the old-type Waler, but a great deal of additional Thoroughbred and, in recent decades, American Quarter Horse blood has been infused. The result is a refined sporting-type horse that does not resemble the old Waler of World War I fame. The brumby also does not resemble the famous Waler, having evolved through natural selection into a smaller, chunkier animal, coarser in appearance and shorter in the leg. To further confuse the issue, the term "brumby Waler" is used to mean horses of true Waler type that are feral, for "brumby" is the Australian term for a wild or feral horse. One ponders the question; is *any* horse running loose or feral in Australia a brumby—regardless of its type?

While the Waler, brumby, and Australian Stock Horse have identical early history, through the years there have been forks in the path and today there are three distinct animals which have developed on Australian soil.

To preserve and maintain the last remaining true Waler-type horses, the Waler Horse Society of Australia was formed, and its manager, Peter Fischer, was most helpful in furnishing information.

Horses were first taken to Australia about two hundred years ago by European settlers who needed them to work the land. The newly settled areas of Australia were first given the name New South Wales, and consequently all saddle horses from Australia became known as Walers. Eventually, when Australia was divided into separate states with their own names, the general term Waler was retained for the Australian-bred saddle horse.

The Waler emerged from crossing of English Thoroughbred, Arab, and Anglo-Arab stallions to mares that were imported to Australia—Barb, Thoroughbred, native British ponies, Timor ponies from Indonesia, Clydes-

dale, Percheron, Suffolk Punch, and Cleveland Bay. Bred and reared under harsh conditions and periodically influenced by the addition of new imported blood, these horses gave rise to a distinct type that became famous as a cavalry mount, renowned for its endurance and heart.

Many such horses were shipped to India for purchase by the Indian army or for private purchase as polo ponies; exports to India continued into the 1930s, the animals ranging in size from 14.2 to over 15 hands. Australian Walers were also shipped to South Africa during the Anglo-Boer War and had profound influence on the Boer horse and, subsequently, the Basuto breed. It is recorded that some 16,357 Australian horses were shipped to South Africa between 1899 and 1902. In 1905, 20,808 Walers were exported, some to China and some to India.

The war horses ridden by the men of the Light Horse were standard Australian Walers. They were regarded at first by English officers as "rather on the light side," yet one would note later that their record in the First World War placed them "far above the cavalry horses of any other nation." They won many laurels in Britain's imperial wars, but it was in World War I that the Waler stunned the military world.

During the First World War good chargers were in great demand and Walers were supplied in huge numbers throughout this war. Some 121,000 Australian horses were sent overseas. Australia itself provided a cavalry division made up entirely of Walers. The tremendous stamina and endurance of this breed played a major role in the defeat of the Turks. Due to strict quarantine laws, these valiant horses were not allowed to return to their homeland.

All good, locally bred Australian horses having obvious signs of good blood but of indefinite or unknown origin were usually called Walers. Many cattle stations, particularly those in South Australia and the Northern Territory that bred horses for the Indian trade, used the term "Waler" to describe their best horses. Some of the stations where these horses were bred were as large as the state of Texas. Stallions of good blood were turned loose with the mares, and every year at mustering time they were rounded up for selection and sale. The rest of the year the horses were left free to tend to themselves, the weak not surviving and the strongest remaining to perpetuate the breed. Most people came to think of the Waler as an army remount horse containing about three-fourths Thoroughbred blood, the remainder being station horse, heavy horse, or pony. There was always a variety of types within the range of horses thought of as Walers; some were heavier, some lighter, but all bore an unmistakable resemblance.

The classic Waler was 15 to 16 hands tall, although many were smaller. Usually the sire was a Thoroughbred. They had fine, clean legs with a short back, large barrel, fine neck, and broad head. The Walers could be ridden day after day and were said to be able to travel faster and farther than the heavier, coarser breeds favored by England and other cavalry nations. They ate and drank less, rarely collapsed from exhaustion, and recovered from fatigue quickly. One 16-hand chestnut, known as Bill the Bastard because of his habit of bucking, carried five men to safety at a laboring gallop for nearly a mile over soft sand when the Turks overran their outpost during the battle of Romani. Two of the men were mounted behind the rider in the saddle, and the other two were on either side standing on a stirrup and hanging on. This was the only time that Bill the Bastard was known to gallop without first bucking. This gallant horse was never again required to carry a rider and served out the remainder of the war as an officer's pack horse.

Many records of outstanding endurance by the Australian Waler emerged from the wars in which this horse was used. Often the horses were required to travel at speed on light rations, only receiving water once in a thirty-six-hour period although the heat was intense. One of the batteries of the Australian Mounted Division had been able to water its horses only three times in a nine-day period, the actual intervals between watering being sixty-eight, seventy-two, and seventy-six hours, yet this battery arrived at its destination having lost only eight animals. The weight-carrying English Hunter had to be nursed back to fitness for a long period after such operations, while the little Australian horses, without any special care other than good food and plenty of water, were soon fit to go through another campaign as arduous as the last.

Only one Waler returned to Australia after the war; Sandy, the charter of Major-General Sir William Bridges, Commander-in-Chief of the Australian Air Force. When Bridges was killed at Gallipoli, Sandy was shipped back to Australia to be led with an empty saddle at his funeral. Sandy's head is preserved in the Australian War Memorial in Canberra. Most of the remaining horses overseas were shot, usually by their saddened riders, to ensure that they would not fall into the hands of the Arabs or Egyptians.

Today small numbers of the tenacious Waler horse remain. Some are in the hands of private breeders who are working to preserve this old type. Others can be found in isolated places on some Northern Territory cattle stations. Some groups of brumbies are found to have individuals bearing striking resemblance to the old-time Waler. After some fifty years of running wild in the Harts and Strangways Ranges of central Australia, the remaining feral Walers are being mustered with other brumbies and sent to abattoirs (slaughterhouses) or shot under the federal government's brucellosis and tuberculosis eradication campaign. It is believed that less than 1,000 true Walers remain, and those existing in isolated herds in central Australia are in imminent danger of being lost.

Managed breeding of "pure Walers" was established in the McDonnell ranges of the Northern Territory by 1874 and persisted in the area until 1940, when large numbers were released to the wild. They have lived in feral herds since then in almost complete genetic isolation, so they may be seen to be neither a "brumby-type" horse nor a modern stock horse, but a distinct Waler type.

The old bloodlines of the Waler cannot be reproduced once they are lost. The Waler emerged as a specific type with the combination of many breeds that have themselves changed somewhat in the course of time. It is hoped that more interest will be generated in preserving the remnants of this excellent mount that proved its worth on foreign shores, for the Waler's legacy of death by helicopter shooting is abominable. For additional information see Australian Stock Horse and Brumby.

Wielkopolski
(Great Poland Horse)

ORIGIN: *Poland*
APTITUDES: *Riding, light draft*
AVERAGE HEIGHT: *15.1 to 16.2 h.h.*
POPULATION STATUS: *Common*

Wielkopolski of Poland. Photo: Andrzej Turczanski

Originally called Poznan horses, the Wielkopolski breed began in the nineteenth century in the region of the contemporary Great Poznan County. German halfbreds were used, the first of these being Trakehners, on native mares. Between the World Wars the Thoroughbred, Arab, and Anglo-Arab were used, then later Trakehner stallions again dominated the breeding.

After World War II Trakehner horses were bred primarily in the state stud farms, situated mostly in the Mazury District. Horses bred there were called Mazurian horses.

In 1962 Poznan and Mazurian began to be considered indistinguishable and were combined, the breed being renamed Wielkopolski horse. This is a multipurpose horse, equally good for riding or harness and presently used for all types of sport competition.

Most Wielkopolski horses are bay or chestnut. Others may be brown, black, or grey. The head is nicely formed with a straight profile, lively eyes, and well-opened nostrils; the neck is well shaped and slightly arched; the withers are well pronounced; the back is straight and long; the croup slightly sloped; the chest deep; the shoulder long and sloping and well muscled. The legs of this horse are well muscled with long cannons and good, clearly defined joints and tendons. The pastern is sloped and springy, giving a smooth stride.

This is a good quality riding and driving animal with a strong constitution and calm, sensible attitude. Presently there are about 4,500 registered mares in the breed.

Welera

ORIGIN: *United States*
APTITUDES: *Riding pony*
AVERAGE HEIGHT: *11.2 to 14.2 h.h.*
POPULATION STATUS: *Uncommon*

Welera pony. Photo courtesy American Welera Pony Society

The Welera is a new pony breed established in 1981 by a group of ranchers in southern California. This breed resulted from the cross of the Arabian horse and Welsh pony. The first person known to breed this pony was Lady Wentworth, using the stallion Skowronek on mares imported from the Coed Coch Welsh stud farm in North Wales.

This is a beautiful pony, exhibiting the best qualities of both the Arab and Welsh breeds. Welera ponies are characterized by their beauty, gentle disposition, quick, eager intelligence, and versatility. Pure Arabs are also accepted if they are too small to meet requirements for the Arabian stud book. All colors other than Appaloosa are accepted.

The American Welera Pony Society publishes a breed magazine, the *Welera Journal,* and registers part Welsh, part Arabian ponies 46 to 58 inches in height.

Welsh Cob

ORIGIN: *Wales*
APTITUDES: *Riding horse, light draft*
AVERAGE HEIGHT: *14 to 15 h.h.*
POPULATION STATUS: *Common*

Welsh Cob. Photo: Sally Anne Thompson

The Welsh Cob is registered in Section D of the stud book. Wales has possessed a horse of the cob type for many centuries. In the old days it was this type of horse that did all the farm work, harrowing, hoeing, and raking. The Cob was much faster and a better worker than the heavy horse and was also used for riding.

This is a light horse, active and well coupled, with clean legs, good quality bone, and docile disposition. It is hardy and able to live outside all the year round with only a little hay during bad weather. A natural jumper, strong and able to carry weight, the Welsh Cob is agile and fast. This is one of the most useful breeds of light horse in Europe.

Exact information regarding the origin of this breed has been lost to time. Evidence of the existence of the Welsh Cob in the Middle Ages and even earlier can be found in medieval Welsh literature. According to description it had to be "fleet of foot," a good jumper, a good swimmer, and able to carry substantial weight on its back. It also had to be capable of drawing loads of timber from the forests and doing the general work on the upland farms long before the introduction of heavier animals. Both in times of peace and war the Welsh Cob has done its part. The English throne was gained by Henry Tudor in 1485 with the help of the Welsh militia mounted on Cobs.

Prior to the advent of the automobile, the Welsh Cob was one of the speediest modes of

Welsh Cob stallion. Photo: Evelyn Simak

transportation for the doctor or tradesmen and others eager to get from here to there in the shortest time. Businessmen in South Wales were known to select a Cob by trotting it all the way from Cardiff to Dowlais—some thirty-five miles, uphill all the way. The best could do this in less than three hours, never slackening the pace from start to finish.

For some years the Welsh Cob lost favor, being no longer needed on the farm. Recently, however, with renewed interest in driving competitions, this horse is coming back into its own. It has proven the ideal trekking animal—safe, surefooted, and responsive—and for private or competitive driving it is difficult to rival. Due to its tractable, gentle disposition, this horse is an excellent animal for a disabled rider. The Welsh Cob makes a good cross with the English Thoroughbred, producing jumpers, hunters, and event horses.

All coat colors are allowed except piebald and skewbald. The head is small and well proportioned, of pony type, with a straight profile, large, prominent eyes, and small, pointed ears. The neck is long, muscular, and arched; the withers are well pronounced and long; the back is short, wide, and strong, the croup is long, broad, and rounded; the tail is set high and carried gaily; the chest broad and deep; the shoulder muscular and sloping. The legs are short but strong and well made with strong, clean joints and a bit of feather at the fetlocks; the hooves are nicely shaped and tough.

Welsh Mountain Pony

ORIGIN: *Wales*
APTITUDES: *Riding pony and light draft*
AVERAGE HEIGHT: *Not to exceed 12.2 h.h.*
POPULATION STATUS: *Common*

Welsh Mountain ponies. Photo: Sally Anne Thompson

The Welsh Mountain Pony is an ancient breed, inhabiting the hills and valleys of Wales before the Romans reached the area. Welsh ponies have a wide range of type, so the stud book is divided into four sections. Section A includes the Welsh Mountain Pony, which may not exceed 12.2 hands. Section B is for the slightly larger Welsh pony, limited in size to from 12.2 to 13.2 hands. Section C includes the stockier Welsh Pony of Cob Type, which may not exceed 13.2 hands. Section D is for the Welsh Cob, a horse type standing from 14 to 15.1 hands.

The oldest of the four types is also the smallest—the Welsh Mountain Pony. This pony, which is believed to be a close descendant of the ancient Celtic pony, lived wild in the hills of Wales for centuries. The breed is surefooted, hardy, and courageous as well as intelligent and docile. This pony has strong resemblance to the Arab, and some claim that Julius Caesar encouraged the crossing of Welsh Mountain ponies with Oriental stallions.

Due to the quality of this pony it has been used as foundation stock for other breeds such as the Hackney, Welsh Pony, and Cob, and there is some indication that the breed was in the early foundation stock of the English Thoroughbred. Used in crossbreeding, the Welsh Mountain Pony has added beneficial quality to polo ponies, the Riding pony, and hack and hunter types.

The purebred Welsh Mountain Pony has a friendly personality and kind temperament. This is an excellent first mount for small children as it is trustworthy and intelligent. At the same time, this breed is strong and can easily carry an adult. Welsh Mountain Ponies are good jumpers and show ponies, excellent in harness.

The head is small and handsome with large eyes, small, pricked ears, and a straight or slightly concave profile; the neck is well shaped and arched; the withers are well pronounced; the back is short and straight; the croup slightly sloping; the tail set high and carried gaily; the chest deep and broad; and the shoulder long and sloping with good muscling; The legs are sturdy with plenty of muscle and broad, strong joints, a long forearm, and short cannon; the hooves are rounded and very tough.

The Welsh Mountain Pony may be any color other than piebald or skewbald. Grey, brown, and chestnut are the most common.

Welsh Pony

ORIGIN: *Wales*
APTITUDES: *Riding pony, light draft*
AVERAGE HEIGHT: *12.2 to 13.2 h.h.*
POPULATION STATUS: *Common*

Welsh ponies. Photo: Sally Anne Thompson

Welsh Pony of Cob Type. Photo: Sally Anne Thompson

The Welsh Pony is a larger version of its cousin, the Welsh Mountain Pony, having received some admixture of Hackney blood, and near the end of the nineteenth century a small Thoroughbred stallion named Merlin was used and had significant influence. For this reason, Section B Welsh ponies are often referred to as "Merlins."

At various times the stud book for Welsh Ponies (Section B) has been opened to part-bred stock (at least 50 percent Welsh), producing a more lightly built, taller pony. The Section B pony inherits many of the characteristics of its smaller counterpart, the Mountain pony, and is a riding pony with quality and substance. This is the ideal mount for a child who has outgrown the smaller Mountain Pony. A very versatile pony, the Section B animal can be successful in harness as well as under saddle, is a superb jumper, and has a smooth, flowing action.

All colors are permitted except piebald and skewbald. The conformation is similar to that of the Welsh Mountain Pony.

Welsh Pony of Cob Type

ORIGIN: *Wales*
APTITUDES: *Riding and light draft*
AVERAGE HEIGHT: *Not to exceed 13.2 h.h.*
POPULATION STATUS: *Common*

The Welsh Pony of Cob Type, registered in Section C of the Welsh stud book, is a smaller version of the Welsh Cob. This pony was produced through the mixture of Andalusian and Cob blood with the Welsh Mountain pony. It is a stocky animal, active, strong, and surefooted, considered an all-around, versatile mount. Due to its heavier build, this pony is stronger than the smaller versions of Welsh Pony, equally suitable for riding or driving. Throughout ages this pony has lived and worked on the small farms of Wales.

The true worth of this pony as a dual purpose animal has been fully realized in recent years, with the increase in interest in driving classes. These ponies can be seen in all the numerous trekking centers throughout Wales, carrying vacationers up into the hills; in competitive events; or being driven by tradesmen in the towns. Like all the Welsh breeds, they are natural jumpers.

Colors in the Welsh Pony of Cob Type are the same as in the Welsh Pony. The head is of pony type with a straight profile, the eyes widely spaced and having a lively expression; the ears are small; the neck is muscular and long, well carried, and arched; the withers are slightly pronounced; the back is short and strong; the croup is rounded and gently sloped; and the tail is set high. The chest is deep and wide; the shoulder is well muscled and sloping; the legs are short with some feathering on the fetlocks; the foot is well formed and tough.

Western and Central African Horses

South of the Sahara, small horse breeds are found in addition to the Barb, Dongola, and Arab types. There is some controversy regarding the origin of these, but it is believed that they are descended from various crosses of the above mentioned breeds. Little information is available regarding these animals on an individual basis, but all resemble the Pony Mousseye of Cameroon. They include the Koto-Koli of Togo, M'Bayou of Senegal, Borodi of Niger, Bobo of Côte d'Ivoire, Somali ponies, Kordofani, and others.

Small north-equatorial horses in Africa are probably not true ponies but horses which have diminished in size due to severe climate and poor feeding conditions. These breeds are of different racial types, and it is evident that their ancestors came from various sources and were introduced at different times. Most are characterized by a heavy head, short, thick neck, and generally inharmonious conformation.

Attempts are being made to gather more comprehensive information regarding all of the small African breeds, which are known to be extremely hardy and perfectly adapted to their inhospitable environment.

Westphalian warmblood stallion. Photo: Werner Ernst

Westphalian warmblood. Photo: Werner Ernst

Westphalian
(Westfälisches Warmblut)

ORIGIN: *Germany (Westphalia)*
APTITUDES: *Riding, sport horse*
AVERAGE HEIGHT: *15.2 to 16.2 h.h.*
POPULATION STATUS: *Common*

Throughout Germany all warmblood horses are named according to the state or area from which they hail (with the exception of the Trakehner) and not according to the breed of their sire and dam. This can be confusing to a foreigner, as in the example of a purebred Holstein born in Bavaria. This horse would be branded with the Bavarian brand and entered into the registration book as a "Bavarian Warmblood." While it is indeed a warmblood and was foaled in Bavaria, its ancestry is very different from that of the true Bavarian Warmblood, based on the ancient Rottal horse.

Numerically the Westphalian warmblood is important, as the equine population of Westphalia is very large. Important equestrian centers including the Olympic and National Riding Schools are located there.

At the state stud at Warendorf, there are Thoroughbred and Hanoverian stallions along with the Westphalians. The neighboring Hanoverian breed has played a great part in development of the Westphalian horse. Originally the horses of Westphalia were heavier, used for army and farm work. Hanoverian blood changed the breed into the outstanding sport horse it is today.

The Westphalian is among the world's leading breeds of competition horses. Success of the breed is due to the strict methods of selection in breeding animals. Horses are only permitted to stand at stud after having passed rigorous tests for conformation, pedigree, character, and riding. Stallions are tested first for traction power at three and one-half years of age and are tested for riding and jumping without a rider at the age of four. At four and one-half years they are tested in dressage and jumping and undergo a veterinary examination.

The Westphalian is heavier than the Hanoverian but essentially very similar. The head is attractive with good width between the eyes and a straight or sometimes slightly dished profile; the body is deep and muscular. The croup is often much flatter in the Westphalian than in the Hanoverian horse.

Modern-type Württemberg. Photo: Evelyn Simak

Westphalian horses have proven their ability in competition. An exclusive stock of youngsters, carefully selected and highly decorated at the local breeding shows, are admitted to the Foals Auction of the Westphalian horse breeders association at the Halle Münsterland. This is the top market for future sires and elite mares of tomorrow.

For years stallions stationed in Westphalia have held first places among the most successful German stallions, assessed by the winnings of their progeny in equestrian sports.

Württemberg of Germany. Photo: Werner Ernst

Württemberg (Modern type)

(Württemberg Warmblut)

> ORIGIN: *Germany (Württemberg)*
> APTITUDES: *Riding, sport horse*
> AVERAGE HEIGHT: *16.1 h.h.*
> POPULATION STATUS: *Common*

The modern-day Württemberg horse emerged at Marbach Stud in 1958. The old type of heavy coach horse was no longer as important for the farm, and interest turned to sporting events. The Trakehner breed was used initially, then by 1979 Westphalians and Hanoverians were used to reinforce the constitution of the horses. The most influential sire of the postwar period was the Trakehner stallion Julmond, put into the stud at Marbach in 1960.

The goal today in breeding the Württemberg is to produce an ever-improving riding horse suited to sporting events. This is an elegant, long-lined breed of great quality with swinging, wide, elastic movements suited for all riding purposes. The Württemberg has a friendly temperament and smooth gaits. While still retaining a bit of its cob type, the modern Württemberg is much more refined than the old type. Improvement continues to be made as breeders seek to meet the endless demand for good quality sports horses.

Württemberg (Old type)

(Württemberg Warmblut)

> ORIGIN: *Germany (Württemberg)*
> APTITUDES: *Coach horse, light draft*
> AVERAGE HEIGHT: *16.1 h.h.*
> POPULATION STATUS: *Rare*

Old-type Württemberg of Germany. Photo: Evelyn Simak

The old type of Württemberg horse dates to the sixteenth century at a stud founded by Duke Eberhard V the Bearded, where native mares were crossed to Arab stallions. In 1552 this stud was moved into the main stud, which is now called Haupt-und Landgestüt Marbach, operated at that time by Christoph von Württemberg. In addition to stock from the original stud, the best available horses from a variety of countries were used, including Hungarian, Turkish, Suffolk, and Caucasian stallions. In 1687 the first regulations concerning mating of the horses were issued.

The old type of Württemberg rose from mares with Arab and Thoroughbred blood and from stallions of the Anglo-Norman breed around 1888. When Christoph's son, Ludwig, took over the stud, he introduced Andalusians and Neapolitans.

The Marbach stud was dispersed during the Thirty Years War (1618–48), and it was not until the end of the seventeenth century that horse breeding in Württemberg got back onto solid footing. At that time, Marbach saw the introduction of Barb and Spanish mares and East Friesian stallions.

During the Napoleonic wars the breed suffered a great setback, as thousands of horses were lost to the French army. In the second half of the nineteenth century new blood was brought in, including Anglo-Norman and East Prussian (Trakehner), and by the beginning of the twentieth century the Württemberg was finally established as a recognizable breed.

The breeders' society was founded in 1895 in Stuttgart.

Bred originally for general work purposes, the old type Württemberg is a cobby type of warmblood with great power and a friendly attitude. Common coat colors seen in this breed are bay, brown, black, or chestnut. The head ususally has a straight profile with plentiful space between the eyes; the shoulders are sloping, powerful and muscular; the chest is deep and broad; the withers are well pronounced; the back short and strong; the croup gently sloped. The legs are short and well muscled with plenty of bone and good, well-formed hooves.

As interest in sporting events increased after the Second World War, Württemberg breeders began to breed a lighter, more refined version of this breed, and today the old type of Württemberg is nearly extinct. There are about 100 purebred mares, 400 grade mares, and 5 grade stallions left.

Xilingol

ORIGIN: *China (Xilinggral Meng, Inner Mongolia)*
APTITUDES: *Riding, light draft*
AVERAGE HEIGHT: *14.2 to 15 h.h.*
POPULATION STATUS: *Common*

Xilingol of China. Photo: Tiequan Wang and Youchun Chen

The southeastern part of Xilinggral Meng in Inner Mongolia is a large steppe area. Since the 1960s a new breed has been developed using the blood of Soviet Thoroughbred, Akhal-Teke, and Sanhe on local stock. Later, Soviet "highblood" (improved Thoroughbred), Kabarda, and Don were used to cross with Mongolian horses. The result through crossbreeding and selection has been a new breed, the Xilingol, a very fast horse with exceptional endurance and harmonious appearance.

Yakut

(Yakutskaya)

ORIGIN: *Siberia*
APTITUDES: *Riding, pack horse*
AVERAGE HEIGHT: *13.2 to 13.3 h.h.*
POPULATION STATUS: *Common*

A group of Yakut horses. Photo: Nikiphorov Veniamin Maksimovich

Yakut. Photo: Nikiphorov Veniamin Maksimovich

The Yakut developed by natural selection in the harsh weather conditions of northern and central Siberia. This hardy breed lives further north than any other in the world. The climate is ultracontinental, with long, cold winters and temperatures dropping to as low as −70°C (−90°F). Even during the summer, night temperatures range below freezing. Rainfall in this area is low and snow cover is only about ten to twelve inches, so the horses are able to graze throughout the year.

Compared to horses of similar type and Mongolian origin, the Yakut is larger and more massive. There are three types within the breed; the northern, original Yakut (called the Middle Kolyma or Verkhoyansk horse); the smaller southern type, which has not been crossed with improved breeds; and the larger southern type, which tends toward the breeds used for improvement. This last type is widespread in central Yakutia, including the Namtsi, Yakutsk, Orjonikidze, Megino-Kanglass, and Amga regions, where trotters and heavy draft horses were used for breed improvement.

The Middle Kolyma is the most valuable of the three types, having a greater homogeneity and size. Typically, the head is coarse; the neck straight and average in length; the withers low; the back wide and long; the croup drooping; the chest wide and deep. The legs are short and with hard, solid hooves. The mane and tail are very thick and long. Hair on the body is very thick and, in winter, up to four inches long.

In color the Yakut is usually bay, grey-brown or grey, less often roan or mouse grey. Native horses have a dark dorsal stripe and zebra stripes on the legs. A dark grid-like pattern is often found on the point of the shoulder. Some Yakuts exhibit the *agouti* characteristic, having pigment only on the tips of the hair. In the northern territories the prevailing color is grey, turning nearly pure white by the age of three to four years.

The larger, southern type of Yakut includes descendants of the local Suntar, Megezh, and Olekminsk varieties. They show traces of trotter and, to a lesser extent, draft blood.

The Yakut is a good meat producer and also has a milk yield worthy of note. The horse plays an important part in the life and economy of Yakutia. it is the basic means of transportation and in the northern regions is used for agriculture, milk, meat, and beautiful fur.

A study of the native horses was done in Yakutia in 1944 to 1946. A group of rather large horses were discovered in the Middle

Kolyma region in the northeast, the best of them standing slightly over 14 hands.

Yakut horses are very well adapted to hard work under harsh conditions. Their typical movement is a pace called *tropota,* a running-walk. This gait is suitable for work in forests or on narrow, winding paths and in marshlands. At the *tropota* gait the horses are able to travel forty-five to sixty miles in a day, carrying a pack weighing about 220 pounds. The trotting gait is small (with short steps), and the gallop is good. Racing at the gallop is a popular sport.

The Yakut is slow to mature, reaching full development by the age of five to six years. The breed is correspondingly long-lived, and broodmares or working geldings of advanced age are common. The horses are kept outside throughout the year without protection from harsh weather. Hay is given only to youngsters during spring when snow is too deep for them to reach feed.

Yakut horses are being crossed with the Dzhabe (Jabe) to produce an early maturing meat horse. Stallions of the Yakut breed are also being used in Kazakhstan and Bashkiria to breed Dzhabe and Bashkir mares. The produce of these matings are heavier at six months and are good meat animals. Unlike some horse breeds developed in cold country, the Yakut adapts easily to warmer climates and produces quality offspring.

The outlook for this breed is for pure breeding since, with extensive husbandry and primitive management, crossbreeding fails to yield the desired result. The leading Yakut breeding farms are Leninski state farm and Karl Marx collective farm in the Yakut Autonomous Region.

Yanqi

ORIGIN: *China (Xianjiang Autonomous Region)*
APTITUDES: *Riding, farm work, draft*
AVERAGE HEIGHT: *13 to 14 h.h.*
POPULATION STATUS: *Common*

Found in Yanqi and other counties in the northern Xianjiang Autonomous Region, the

Yanqi stallion. Photo: Weigi Feng and Youchun Chen

Yanqi mare. Photo: Weigi Feng and Youchun Chen

Yanqi breed was based on Mongolian horses brought into the area by the Mongols during the thirteenth century and again in 1771. Later, breeds such as the Don and Orlov Trotter were used to improve the breed.

There are two types found in the Yanqi breed today, known as Basin and Mountain horses. The Basin type is light; the Mountain type tends to be coarser. The Yanqi breed is of medium weight and suitable for riding. The head is dry with a straight profile. Some have the "rabbit-like" convex slope near the nostrils. The eye is large and well opened, the ear is set up, the jaw is wide; the neck is of medium length and straight. The withers are moderate in height and length; the chest is moderately wide and deep; the back is long and straight but somewhat narrow; the loin is long and sometimes convex with poor connection to the croup; the ribs are well sprung. The joints are strong with well-developed tendons; the hind legs are slightly sickle-hocked. Bay, chestnut, and black are the most common colors.

This breed is used on farms mainly for plowing and transport but is also ridden a great deal. The breed shows good speed and endurance and is able to carry a pack weighing about 220 pounds nearly forty miles in a day. The mares are milked, often giving nearly 10 pounds in one day.

There were approximately one million Yanqi horses in 1980.

Yemeni Horses

ORIGIN: *Arabian Peninsula, Yemen*
APTITUDES: *Riding horse*
AVERAGE HEIGHT: *15 h.h.*
POPULATION STATUS: *Uncommon*

Horses in Yemen have declined in quality and are descendants of stock indigenous to countries north of the Arabian Peninsula and of the Arab horse introduced early in development of the Arab breed. Yemeni horses descended from the Arab are divided into three groups; the Shami, found in the northern part of the Arabian Peninsula; the Nagdi, found in the central desert; and Yemeni horses, found in the southwestern region.

The first and second groups are desert descendants and are used in desert areas. The third group represents strong horses, having good legs, less height, and carp-like (curved) backs.

Yemeni horses (including all three groups) number about 4,600 and can be found in Tihama, Gawf, Dhamar, and Taiz. Generally they are grey, brown, or dark bay in color. The head is described as conical with a wide forehead, small, thin ears, and wide nostrils; the chest is wide and strong; the back is short and usually curved; the legs are strong with large joints; the hooves are round in shape and very strong.

Yili

ORIGIN: *China (Yili-Kazakh Autonomous District)*
APTITUDES: *Riding, light draft*
AVERAGE HEIGHT: *14 h.h.*
POPULATION STATUS: *Common*

A Yili stallion of China. Photo: Weigi Feng and Youchun Chen

Yili mare. Photo: Weigi Feng and Youchun Chen

The Yili is a developed breed in China, originating in the Yili-Kazakh Autonomous District, Xinjiang Uygur region. The mountain basin there is good grassland.

Since 1900, when Russian emigrants brought stallions to improve the local Kazakh horses, crossbreeding with native horses has taken place. In 1936 the local construction bureau decided to import Don, Orlov, and Anglo-Don horses to cross with native animals, but good results were not obtained after the second generation upgrade system. In 1963 a draft-riding type was finally confirmed as the main breeding goal for the new breed. Interbreeding of horses with the desired type has established the breed and increased population.

Yili horses are compact and harmonious in conformation with a light head and straight profile. The neck is of average length and slightly arched in males; the withers are well pronounced and blend smoothly into the back. The back is short and strong; the loin is longer than usual; the ribs are well sprung; the chest is deep; the shoulder has good slope. The legs are clean with well-defined tendons; the front legs are correctly set and well muscled; the hind legs tend to be sickled, and some toe out. The body coat is light and fine, often showing a metallic sheen, and is most often bay, with chestnut, black, and grey also seen.

The Yili is a very good horse for long distances at speed. This is a hard horse with outstanding endurance. Crossed to Mongolian horses, the Yili yields offspring that reach a mature size of around 13.1 hands.

Yiwu

ORIGIN: *China (Xinjiang Uygur Autonomous Region)*
APTITUDES: *Riding, packing, draft*
AVERAGE HEIGHT: *13.2 to 14 h.h.*
POPULATION STATUS: *Common*

Yiwu horse of China. Photo: Weigi Feng and Youchun Chen

Yiwu mare. Photo: Weigi Feng and Youchun Chen

The Yiwu horse emerged on the Yiwu horse farm in Hami district, Xinjiang Uygur Autonomous Region. It is located on the east part of the Balikun steppe. The elevation in this area is high, ranging from 6,200 to 6,500 feet, with an average temperature of $-0.3°C$ (31.5 F°). There are only about sixty to eighty frost-free days a year in this area. During the summer, horses are driven to the high valleys of Tian Shan Mountain to pasture.

The Yiwu was developed on the base of Kazakh and Yili horses. In 1955 there were 1,200 horses introduced from the Altai region, and in 1957 to 1958, 600 Yili horses and 1,500 additional Kazakh horses were brought in. Crosses among them with 75 to 87.5 percent of Kazakh blood were interbred.

The head of the Yiwu horse is of medium size, dry, and with a straight or ram profile. The ear is short and thick; the neck is of medium length and straight, with good muscling; the chest is deep and wide; the trunk is full with well-sprung ribs; the limbs are "rough" with strong, well-defined joints. The main colors in the breed are bay and black.

Yiwu horses were tested in 1982 for strength and endurance and the results were impressive, especially at high altitude. The Yiwu is a fast horse and can pull 90 percent of its body weight. Ridden at the pace, this horse travels 1,000 meters in 2 minutes, 17.5 seconds; ridden at speed, it covers 1,000 meters in 1 minute, 23.8 seconds.

The Yiwu performs well in high mountain conditions. Horses are pastured throughout the year and are bred naturally, with a stallion running with twelve to eighteen mares. Improvement of the breed is done through pure breeding and selection of herd stallions.

Yonaguni

ORIGIN: *Japan*
APTITUDES: *Riding, light draft*
AVERAGE HEIGHT: *10 to 12 h.h.*
POPULATION STATUS: *Rare*

Yonaguni pony of Japan. Photo: Kouiche Maefusato

The Yonaguni is the small native pony of the southwest islands of Japan. Today there are about seventy-five living Yonaguni ponies on East and North Ranches on Yonaguni Island, located on the west side of the Yaeyama Islands, from which you can see Taiwan.

Little is known regarding the origin of the Yonaguni, despite several investigations. Horses in Japan can generally be divided into two groups, larger specimens from Hokkaido and smaller individuals from Yonaguni. Many people believe that the small horses were introduced from the southern islands during the Jyomon Period, about two thousand years ago. In 1983, however, Professor Ken Nozawa of Kyoto University claimed that the gene characteristics of the breed indicate relationship to the small breed (Cheju) in Korea.

The Yonaguni pony strongly resembles the Tokara and Miyako ponies of Japan, and is known to be of great antiquity on the islands. The oldest journal of Yonaguni Island, written by the Koreans in 1479, states that the islanders kept and ate both cattle and horses.

The Yonaguni is mostly chestnut in color. The head is large but with well-placed eyes and relatively small ears; the neck is short and thick; the shoulders tend to be straight; the back is long; the croup is often quite level with a high tail-set; the quarters are slight; the legs often tend to be splayed; the hooves are vertically long and very hard. This pony is gentle in nature and very strong and enduring.

In the old days every household had one horse or more for transportation and plowing. In 1939, when local breeds began to be improved to produce larger war horses, the Yonaguni on their remote island were excluded from that plan, and the original breed has been preserved. As technology has improved, horses have become less important and the population of all horse breeds in Japan has been drastically reduced. Once of great importance in the daily lives of the islanders, today the Yonaguni may be seen on only a few ranches and has become a precious cultural asset.

It is hoped that interest in the small but strong native horse will be revived.

Yunnan

ORIGIN: *China (Wuron Mountains)*
APTITUDES: *Riding*
AVERAGE HEIGHT: *11 to 11.2 h.h.*
POPULATION STATUS: *Common*

Yunnan mare of China. Photo: Weigi Feng and Youchun Chen

Yunnan horse of China. Photo: Tiequan Wang and Youchun Chen

There were 70,000 Yunnan horses in China according to a survey done in 1980. This breed is of strictly local animals with no crossing to exotic breeds. The area of breeding varies greatly in topography, and a number of different strains have developed.

The earliest record of the Yunnan breed was written in 285 B.C. Later the Yunnan horse was recorded in pictures on brick. During the period from 25 to 220 A.D., this horse was used to pull chariots or carts.

This is a very small horse, and three types are found within the breed. The head is somewhat heavy with good width between the eyes and a straight but sometimes dished profile. The ears are small and set forward. Compactly built, the Yunnan has a short, straight neck; high withers; and short back, loin, and croup. The croup is sloped; the shoulder rather straight; the legs dry and well muscled; the foot strong and well shaped. Yunnan horses are bay, chestnut, or black.

This small horse is used mostly for mountain riding and can carry 130 to 220 pounds daily for fifteen consecutive days at high altitude before appearing to need a rest. The breed is quite fast. Performance of the Yunnan is impressive in hilly terrain at altitudes above 9,000 feet, in hot, cold, or wet conditions.

Zaniscari

ORIGIN: *India (Jammu and Kashmir)*
APTITUDES: *Riding, pack horse*
AVERAGE HEIGHT: *13 to 14 h.h.*
POPULATION STATUS: *Unknown*

The Zaniskari originated in the Ladakh region of Jammu and Kashmir State. The most common color is grey, and the body is compact with a broad forehead and strong back. The tail is very long, reaching below the fetlock and often touching the ground. The breed has an alert appearance and is spirited and fast.

The Zaniskari is used for transport, riding, and polo.

Details regarding this breed were not made available.

Zhemaichu

(Zhmudskaya, Zhumdka)

ORIGIN: *Lithuania*
APTITUDES: *Riding, draft*
AVERAGE HEIGHT: *13.2 to 14.3 h.h.*
POPULATION STATUS: *Rare*

Zhumudka, or Zhemaichu. Photo: Nikiphorov Veniamin Maksimovich

A native Lithuanian breed of the forest type and belonging to the Konik family, the Zhemaichu or Zhmudka has been known since the sixteenth century. The breed is also known as Zhmud and sometimes spelled Zemaituka. Ancestor to the breed was the native forest horse, well adapted to wet, cold climate. Despite repeated crossbreeding, it has retained its type and suitability for work on small farms.

In the fourteenth century, Lithuania was attacked by Tartars and their horses had some influence on the breed. Zhemaichu horses were also influenced by some native Russian breeds and, during the Polish-Lithuanian Union, by light breeds from Poland. Since the second part of the nineteenth century some Arabian blood has been used. By the beginning of the twentieth century the Zhemaichu breed was represented by two types—a smaller, very elegant but somewhat massive, clean-cut horse and a heavier type without Arab blood. The latter type was more widespread.

The Zhemaichu is undemanding regarding its management and is highly adaptable. The breed has been exported to western Europe for many years.

The old Zhemaichu type was small, averaging around 13 hands with a small, beautiful head. The profile was straight or somewhat dished with a wide forehead. The neck was well developed and muscled; the trunk undersized; the withers not too high; the back flat and solid; and the legs clean and firm. The main deficiency of the breed was small size, resulting in insufficient strength for agricultural or transport work.

The modern Zhemaichu—formed in the post–World War II years by pure breeding and selection of large individuals with high work endurance and through limited crossing with North Swedish stallions—is taller and has a somewhat extended trunk and a very massive body with clean legs. The new breed type has retained the characteristic features of the old Zhemaichu.

The Zhemaichu is a good combination of sporting and utility horse. It is strong, with good endurance and agility. This breed has also shown good results in steeplechases of average difficulty, and it is not accidental that Lithuanian horses were the foundation for breeding the Trakehner.

Their massive, clean build, combined with high sporting potential and strong adaptability, make them valuable breeding material for medium-sized horses for large-scale equestrian sports and tourism. The Zhemaichu is also suitable in development of new meat-producing breeds such as the Altai crossbred.

Predominant colors in the breed are grullo (mouse dun), black, dark brown, and bay. Nearly every individual has a dark dorsal stripe.

Appendix A

Countries of Origin

This section is given as a point of reference for the origins of world horse breeds. Not all of the breeds listed are included in this volume. Some may be extinct, or may be alternate names for existing or extinct breeds. An attempt is currently being made to obtain information on all horse breeds not included in this volume so they can be included in the first revision.

Afghanistan
  Herati
  Mazari
  Qatgani
  Yabu

Albania
  Albanian

Algeria
  Algerian Barb

Argentina
  Argentine Criollo
  Argentine Polo Pony
  Falabella
  Petiso Argentino

Australia
  Australian Pony
  Australian Stock Horse
  Brumby
  Waler

Austria
  Abtenauer
  Austrian Warmblood
  Lipizzan
  Haflinger
  Noriker

Bangladesh
  Rajshahi Pony

Belgium
  Belgian Ardennais
  Belgian Draft
  Belgian Country Bred
  Belgian Halfblood
  Belgian Warmblood

Benin
  Koto-Koli Pony

Bhutan
  Tanghan

Brazil
  Brazilian Sport Horse
  Campeiro
  Campolina
  Mangalarga Marchador
  Mangalarga Paulista
  Marajoara
  Northeastern
  Piquira Pony

Bulgaria
  Bulgarian Heavy Draft
  Bulgarian Mountain Pony
  Danubian

Deli-Orman
Dolny Iskar
East Bulgarian
Karakachan
Pleven
Rila Mountain
Rodupi
Stara Planina

Burkina Faso
Bobo (Bobodi)
Liptako
Mossi
Yagha

Burma
Shan (Burmese)

Cambodia
Cambodian

Cameroon Republic
Dongola
Poney Mousseye

Canada
Canadian
Canadian Cutting Horse
Canadian Pinto
Canadian Rustic Pony
Canadian Sport Horse
Newfoundland Pony (Island of
Newfoundland)
Royal Canadian Mounted Police Horse
Sable Island Horse

Chad
Bahr-el Ghazal
Kirdi Pony
Mbai

Chile
Chilean Criollo
Chilote

China
Baise
Balikun
Buohai
Chakouyi
Chinese Kazakh
Datong
Erlunchun

Guanzhong
Guizhou
Guoxia
Heihe
Heilongkiang
Hequ
Jianchang
Jielin
Jinhong
Jinzhou
Ke-Er-Qin
Lichuan
Lijiang
Mongolian
Sandan
Sanhe
Sini
Tibetan
Tieling
Wuchumutsin
Xilingol
Yanqi
Yili
Yiwu
Yunnan

Colombia
Colombian Criollo
Trote En Gallope

Costa Rica
Costa Rican Saddle Horse

Czechoslovakia
Czech Coldblood
Czech Trotter
Czech Small Riding Horse
Czech Warmblood
Kladruby
Slovak Mountain

Cuba
Cuban Paso
Cuban Pinto
Cuban Trotter
Patibarcina

Denmark
Danish Oldenborg
Danish Sports Pony
Danish Warmblood
Faeroes Island Pony

Frederiksborg
Jutland
Knabstrup

Egypt
Baladi

England
British Appaloosa
British Riding Pony
British Spotted Pony
British Warmblood
Cleveland Bay
Dales Pony
Dartmoor Pony
English Cob
English Hack
English Hunter
English Thoroughbred
Exmoor Pony
Fell Pony
Hackney Pony
Lundy Pony
New Forest Pony
Shire
Suffolk

Ethiopia
Abyssinian/Galla/yellow

Finland
Finnhorse

France
Auxois
Boulonnais
Breton
Camargue
Comtois
Corsican
French Anglo-Arab
French Anglo-Trotter
French Ardennais
French Cob
French Saddle Pony
French Trotter
Landais Pony
Mérens Pony
Northern Ardennais (Trait du Nord)
Percheron
Petite Boulonnais (Mareyeuse)
Poitou

Pottok
Selle Français

Germany
Bavarian Warmblood
Brandenburg
Dülmen Pony
East Friesian (modern type)
East Friesian (old type)
Elegant Warmblood
German Riding Pony
Hanoverian
Hessen
Holstein
Lewitzer Pony
Oldenburg (modern type)
Oldenburg (old type)
Rhineland Heavy Draft
Rottal
Sárvár
Schleswig Heavy Draft
Schwarzwälder Füchse
Senne
South German Coldblood
Tarpan
Trakehner
Westphalian Warmblood
Württemberg (modern type)
Württemberg (old type)

Greece
Cyclades Island Pony
Eleia (Andravida)
Messara
Peneia
Pindos
Skyros Pony
Thessalian

Holland
Dutch Draft
Dutch Tuigpaard
Dutch Warmblood
Friesian
Gelderland
Groningen

Hungary
Furioso
Gidran
Hungarian Coldblood
Hungarian Dun

Hungarian Halfbred
Hungarian Sport Horse
Kisber Halfbred
Mezöhegyes Sport Horse
Murakoz
Nonius
Shagya Arabian

Iceland
Icelandic Horse

India
Bhutia
Chummarti
Deccani
Kabuli
Kathiawari
Manipuri
Marwari
Spiti
Zaniskari Pony

Indonesia
Bali
Batak
Bima (see Sumbawa)
Flores
Gayoe (see Batak)
Java
Kuningan Pony (see Java)
Lombok
Macassar (see Sulawesi)
Periangan
Sandel (Sandelwood)
Sulawesi (Macassar, Lombok)
Sumba
Sumbawa
Timor

Iran
Caspian
Persian Arab
Plateau Persian
Turkoman (Turkmenian)
Yamud

Ireland
Connemara
Irish Cob
Irish Draft
Irish Hunter

Israel
Israeli Local Horse

Italy
Avelignese
Bardigiano
Calabrian (Calabrese)
Italian Heavy Draft
Maremmana
Monterufoli Pony
Murghese
Noric
Salerno (Persano)
Sanfratello
Sardinian Anglo-Arab (Anglo-Arabo
Sarda)
Sardinian Pony (Giara)

Japan
Banei Racing Horse
Hokkaido
Kiso
Misaki
Miyako
Noma
Taishuh
Tokara
Yonaguni

Korea
Cheju

Lesotho
Basuto

Libya
Libyan Barb

Madagascar
Madagascan Pony

Mali
Bandiagara
Banamba
Sahel
Songhai

Mauritania
Hodh

Mexico
Azteca
Galiceño
Native Mexican

Middle East
  Arab
  Oriental Horse

Mongolia People's Republic
  Griffin
  Mongolian
  Poljakoff
  Przewalski's Horse

Morocco
  Moroccan Barb

Namibia
  Namib Desert Horse

Nepal
  Chyanta
  Tanghan
  Tarai Pony
  Tattu

Niger
  Bandiagara
  Djerma
  Hausa
  Songhai
  Sulebawa
  Torodi

Nigeria
  Bhirum
  Bornu
  Nigerian Horse
  Sulebawa

Norway
  Døle Gudbrandsdal
  Fjord (Westland)
  Northlands Pony

Pakistan
  Baluchi
  Hirazi
  Makra
  Unmol
  Waziri

Peru
  Andean
  Chola
  Chumbivilcas
  Costeño

Morochuco
  Peruvian Paso

Philippines
  Philippine Pony

Poland
  Dahoman
  Garwolin
  Gidran
  Kielce
  Konik
  Kopczyk Podlaski
  Krakow-Rzesow
  Lipzbark
  Løwicz
  Malopolski
  Panje
  Polish Draft
  Silesian
  Sokólka (Sokólski)
  Stuhm
  Sztum
  Tarpan
  Wielkopolski

Portugal
  Alter Real
  Azores Pony
  Garrano (Minho)
  Lusitano
  Mira Pony
  Sorraia

Puerto Rico
  Paso Fino

Romania
  Banat
  Carpathian Pony (Hucul)
  Dobrogea
  Moldavian
  Romanian Saddle Horse
  Romanian Traction Horse
  Transylvanian
  Transylvanian Lowland

Senegal
  Fleuve
  Fouta
  M'Bayar
  M'Par

Scotland
  Clydesdale
  Eriskay Pony
  Highland Pony
  Rhum Island Pony
  Shetland Pony

Somalia
  Nogal
  Somali Pony (Dor, Mijertinian, Dorn)

South Africa
  Cape Horse
  Nooitgedacht Pony
  South African Miniature

Former Soviet Union
  Adaev
  Akhal-Teke
  Altai
  Anglo-Kabarda
  Avar
  Azerbaijan
  Bashkir
  Budyonny
  Black Sea Horse (Chernomor)
  Buryat
  Byelorussian Harness
  Caucasian
  Chara (Charysh)
  Dageston Pony
  Deliboz
  Don
  Estonian Draft
  Estonian Native
  Iomud
  Jabe
  Kabarda (Kabardin)
  Kalmyk
  Karabair
  Karabakh
  Karachai (type of Kabarda)
  Kazakh
  Kirgiz
  Kumyk
  Kurdi Pony
  Kushum
  Kustanai
  Kuznet
  Latvian
  Lezgian
  Lithuanian Heavy Draft

Lokai
Megezh
Megrel
Mezen
Minusin
Narym
New Kirgiz
North Russian Pony
Orlov Trotter
Pechora
Polesian
Priob
Russian Heavy Draft
Russian Saddle Horse
Russian Trotter
Soviet Heavy Draft
Tajik
Tavda
Tersk
Tori (Toric)
Turkoman (Turkmenian)
Tushin
Tuva
Ukrainian Saddle Horse
Vladimir
Voronezh Coach
Vyatka
Yakut
Zhemaichu (Zhmudka)

Spain
  Andalusian
  Asturian
  Balearic (Island of Majorca)
  Carthusian
  Galician
  Hispano-Bretona
  Pottok (Basque)
  Sorraia
  Spanish Arab
  Spanish Trotter (Balearic Island)

Sudan
  Western Sudan Pony (Darfur)

Sweden
  Gotland (Skogruss, Gothland, Russ)
  North Swedish Horse
  North Swedish Trotter
  Swedish Ardennes
  Swedish Warmblood

Switzerland
  Freiberg (Franches-Montagnes)
  Swiss Warmblood

Syria
  Syrian Arab

Thailand
  Thai Pony

Tibet
  Chummarti
  Tibetan (Southern Barbarian Pony)

Turkey
  Anadolu
  Araba
  Canik
  Cirit
  Çukurova
  East and Southeast Anadolu
  Hinis
  Karacabey (extinct)
  Karakaçan
  Malakan
  Mytilene Pony (Ege Midillisi)
  Rahvan
  Trakya
  Turkish Arab Race Horse
  Uzunyayla

United States
  Albino
  American Bashkir Curly
  American Indian Horse
  American Miniature
  American Mustang
  American Paint Horse
  American Quarter Horse
  American Saddlebred (Kentucky Saddler)
  American Shetland
  American Walking Pony
  Appaloosa
  AraAppaloosa
  Assateague/Chincoteague
  Banker (Shackleford)
  Buckskin
  Cerbat
  Chickasaw

Colorado Ranger
Florida Cracker Horse
Half Saddlebred
International Striped Horse
Kiger Mustang
Missouri Fox Trotter
Morab
Morgan
Moyle
Mustang (BLM)
National Show Horse
National Spotted Saddle Horse
Palomino
Pinto
Pony of the Americas (POA)
Rocky Mountain Horse
Southwest Spanish Mustang
Spanish Barb
Spanish Colonial
Spanish Mustang
Standardbred
Tennessee Walking Horse
Welera Pony

Venezuela
  Llañero

Vietnam
  Annamese Pony (Annamite, Vietnamese)

Yemen
  Giawf
  Nagdi
  Shami
  Yemen Horse

Yugoslavia
  Bosnian
  Macedonian
  Yugoslav Draft (Croatian)
  Yugoslav Mountain Pony
  Zobnatica Halfblood

Wales
  Welsh Cob
  Welsh Mountain Pony
  Welsh Pony of Cob Type
  Welsh Pony

Appendix B

Horse and Pony Breeds Not Included in This Volume

The following horse and pony breeds are not included in this volume due to lack of response or authoritative information. Some of the listed animals may no longer exist; some may be alternate names for horses that are included herein.

Abyssinian (Ethiopia)
Algerian Barb (Algeria)
Anglo-Karachai (Soviet Union)
Annamese (Vietnam)
Avar (Soviet Union)
Austrian Warmblood (Austria)
Bagual (feral horse of Argentina)
Bahr-el-Ghazal (Chad)
Banamba (Mali)
Banat (Romania)
Bandiagara (Niger)
Belgian Country Bred (Belgium)
Bhirum (Nigeria)
Bobo (Burkina Faso)
Bornu (Nigeria)
British Warmblood (England)
Bulgarian Heavy Draft (Bulgaria)
Bulgarian Mountain Pony (Bulgaria)
Cambodian (Cambodia)
Campeiro (Brazil)
Canadian Pinto (Canada)
Capitanata (Italy)
Chola (Peru)
Chummarti (Tibet)
Chyanta (Nepal)
Czechoslovakian Trotter (Czechoslovakia)
Dabrowa-Tarnowska (Poland)
Dagestan (Soviet Union)
Dahoman (Poland)
Deccani (India)

Deli-Orman (Bulgaria)
Djerma (Niger)
Dobrogea (Romania)
Dolny-Iskar (Bulgaria)
Esperia Pony (Italy)
Fleuve (Senegal)
Fouta (Senegal)
Giawf (Yemen)
Griffin (Mongolia)
Hailar (China)
Hausa (Niger)
Herati (Afghanistan)
Hispano-Bretona (Spain)
Hodh (Mali)
Kabuli (India)
Karakachan (Bulgaria)
Khakassk (Soviet Union)
Kielce (Poland)
Kirdi Pony (Chad)
Koto-Koli Pony (Benin)
Lezgian (Soviet Union)
Libyan Barb (Libya)
Liptako (Burkina Faso)
Ljutomer Trotter (Yugoslavia)
M'Bayar (Senegal)
M'Par (Senegal)
Macedonian (Yugoslavia)
Madagascan Pony (Madagascar)
Makra (Pakistan)
Mazari (Afghanistan)

Mazury (Poland)
Mbai (Chad)
Mijertinian (Somalia)
Mira Pony (Portugal)
Moldavian (Romania)
Mossi (Burkina Faso)
Nagdi (Yemen)
National Appaloosa Pony
    (United States)
Native Mexican (Mexico)
Nefza (Tunisia)
Nogal (Somalia)
Pantaneiro (Brazil)
Petite Boulonnais (France)
Philippine Pony (Philippines)
Poljakoff (Mongolia)
Puno Pony (Chile)
Rajshahi (Bangladesh)
Rila Mountain (Bulgaria)
Rodopi (Bulgaria)
Romanian Traction Horse (Romania)
Sacz (Poland)
Sahel (Mali)
Shami (Yemen)
Slovak Mountain (Czechoslovakia)

Somali Pony (Somali)
Songhai (Mali, Niger)
Spanish Trotter (Majorca, Spain)
Stara Planina (Bulgaria)
Stuhm (Poland)
Sulebawa (Nigeria)
Syrian Arab (Syria)
Sztum (Poland)
Tajik (Soviet Union)
Tanghan (Nepal)
Tarai (Nepal)
Tattu (Nepal)
Tolfetana (Italy)
Torodi (Niger)
Transylvanian Lowland (Romania)
Transylvanian (Romania)
Tunisian Barb (Tunisia)
Waziri (Pakistan)
Western Sudan Pony (Sudan)
Yabu (Afghanistan)
Yagha (Burkina Faso)
Yemen Horse (Yemen)
Yugoslav Draft (Yugoslavia)
Yugoslav Mountain Pony (Yugoslavia)
Zobnatica halfbred (Yugoslavia)

Appendix C

Horse Breed Organization Addresses

The following list of horse breed association or government addresses will be of assistance to those wishing to obtain additional information about certain horse breeds. While most of the addresses are stable, the organizations being housed in permanent quarters, one should bear in mind that addresses are always subject to change.

Albania
  All breeds
    Ministry of Agriculture
    Tirana, Albania

Argentina
  Criollo, Petiso Argentino
    Sociedad Rural Argentina
    Florida 460
    (1005) Buenos Aires, Argentina

    Dr. Luciano Miguens
    Avda. del Libertador 836, 10º P. "A"
    (1001) Capital Federal
    Buenos Aires, Argentina

  Falabella
    Establecimientos Falabella
    Av. Quintana 494 - Piso 6º
    1129 Buenos Aires, Argentina

  Polo Pony
    Asociación Argentina de Criadores de
      Caballos de Polo
    Avda. Pte. Roque Saenz Peña 1160, 9º
    P. "A"
    (1035) Capital Federal
    Buenos Aires, Argentina

Australia
  Australian Pony
    Australian Pony Stud Book Society,
      Inc.
    Box 4317 G.P.O.
    Sydney, N.S.W. 2001, Australia

  Australian Stock Horse
    Australian Stock Horse Society Ltd.
    P.O. Box 288
    Scone, N.S.W. 2337, Australia

  Waler
    The Waler Horse Society of Australia
    "Taree," Harrisons Road
    Dromana, Victoria 3936, Australia

Austria
  Haflinger
    Haflinger Pferdezuchtverband Tirol
    Postfach 640
    Brixnerstrasse 1
    A-6021 Innsbruck, Austria

  Lipizzan
    Bundesgestüt Piber
    8580 Köflach-Piber, Austria

Noriker
Arbeitsgemeinschaft der norischen
Pferdezuchterverbände Österreichs
A-5751 Maishofen 96, Austria

Belgium
Belgian Ardennais
Société Royale Le Cheval de trait
Ardennais
5408 Marloie, LE
Maison De L'Eleveur
Luxembourgeois, Belgium

Belgian Heavy Draft
Société Royale "Le Cheval de trait
Belge"
Av. du Suffrage Universal 49
B-1020 Brussels, Belgium

Belgian Halfblood
Het Belgisch Halfbloedpaard,
V.Z.W.D.
Hamoirlaan 38
B-1180 Brussels, Belgium

Belgian Warmblood
Belgisch Warmbloedpaard
Minderbroedersstraat 8, P.B. III
B-3000 Leuven, Belgium

Brazil
All breeds
Ministerio Da Agricultura
Commissao Coordenadora da Criacao
Do Cavalo Nacional
Av. Almirante Barroso, 139
10º And.-Modulo 1001 - CEP 20051
Rio de Janeiro - RJ, Brazil

Brazilian Sport Horse
Asociação Brasileira dos Criadores do
Cavalo de Hipismo
Av. Francisco Matarzzo, 455
São Paulo - Sp, Brazil

Campolina
Associação Brasileira dos Criadores do
Cavalo Campolina
Rua São Paulo, 824 - 14º
CEP 30170 - Belo Horizonte
Minas Gerais, Brazil

Mangalarga Marchador
Associação Brasileira dos Criadores do
Cavalo Mangalarga Marchador
Rua Goitacases, 14 - 11º e. 13º andares
Ed Bom Destino
Cep 30190, Belo Horizonte, MG,
Brazil

Lauro Antonio Teixeira de Menezes
Haras Boa Luz
B. Siqueira Campos 49.050
Aracaju, Sergipe, Brazil

Mangalarga "Paulista"
Associação Brasileiria de Criadores de
Cavalos da Raça Mangalarga
Avenida Francisco Matarazzo, 455
Pavilhão 04 CEP 05001
Caixa Postal 61016
CEP 05071 - São Paulo, Brazil

Northeastern
Associação Brasileira de Criadores de
Cavallo Nordestino Av. Caxangá
220 Cordeiro
Recife - PE, Brazil

Bulgaria
All Breeds
State Corporation of Cattle and Sheep
Breeding
Malinova Dolina 10
Bistrishko Shosse Str.
1756 Sofia, Bulgaria

Cameroon
All Breeds
Institute of Animal Research
Garoua Sub Station
B.P. 1073, Garoua, Cameroon

Canada
Canadian
Canadian Horse Breeders Association
1747 St-Hubert
Montreal, Quebec, Canada, H2L 3Z1

Canadian Cutting Horse
Canadian Cutting Horse Association
14141 Fox Drive
Edmonton, Alberta, Canada T6H 4P3

Canadian Rustic Pony
Canadian Rustic Pony Association
Box 268
Kelvington, Saskatchewan, Canada
S0A 1W0

Canadian Sport Horse
Canadian Sport Horse Association
Box 520
Gormley, Ontario, Canada L0H 1G0

Newfoundland Pony
The Newfoundland Pony Society
Site 54, Box 33
Suburban Service #2
St. John's, Newfoundland A1C 5H3

Royal Canadian Mounted Police Horse
Royal Canadian Mounted Police
P.O. Box 8900
Ottawa, Canada K1G 3J2

Chile
Chilean
Sociedad Nacional de Agricultura
Tenerini 187
Casilla 40-D
Santiago 1, Chile

China
All breeds
China National Animal Breeding Stock
Import and Export Corporation
Yang Qing, President
Yangyi Hutong No. 10 Dongdan
Beijing, China

Guoxia
Dr. Wang Tiequan, President
Pony Association of China
Animal Science Institute of the Chinese Academy of Agriculture
Ma Lianwa Xijiao
100094 Beijing, China

Colombia
Colombian Criollo (Paso Fino)
Agropecuaria Betania S.A.
Carrera 68 No. 19-52
Bogotã, Colombia

Federación Colombiana de Asociaciones Equinas "Fedequinas"
Transversal 44 No. 99-36
Bogota, D.E., Colombia

Costa Rica
Costa Rican Saddle Horse
Asociacion de Criadores del Caballo
Costarricense de Paso
Apartado 45, Curridabat 2300
San Jose, Costa Rica

Cuba
All breeds
Ministerio de la Agricultura
Centro de Informacion y
Documentacion
Agropecuario CIDA, Ciudad de la
Habana
Gaveta Postal 4149
Habana 4, Cuba

Czechoslovakia
Czech Coldblood, Czech Warmblood
Státní plemenárský podnik, koncern
Nositel Radu práce, Praha (Prague)
Pracovistě Hradištko p. Medníkem
252 09 Hradištko pod Mednikem
Prague, Czechoslovakia

Kladruby
Státní plemenárský podnik
koncern Praha
Vítezného února 64
Post Box 91
120 77 Prague 2, Czechoslovakia

Denmark
Danish Oldenborg
Dansk Oldenborg AVL
v/Sten Marcussen
Østergade 8 - Egense
9280 Storvorde, Tif. 08 31 1447,
Denmark

Danish Sport Pony
Dansk Sports Ponyavl
Tagkaergard—Tif. (04) 561499
6070 Christianfeld, Denmark

Danish Warmblood
Dansk Varmblod
Landskontoret for Heste
Udkærsvej 15, Skejby
8200 Århus N, Denmark

Frederiksborg
Erik Schmidt Arkitekt M.A.A.
Mosgaard
Skovgårdsvej 59, Balle
8300 Odder, Denmark

Jutland
Avlsforenigen Den Jydske Hest
Battrupholt, Kolstrupvej 4
8581 Nimtofte, Denmark

Knabstrup
Ellen Bendtsen
Sdr. Lyngvej 37, Årsballe
3700 Rónne, Denmark

Finland
Finnhorse
Suomen Hippos r.y.
Raviurheilun ja hevoskasvatuksen
keskusjärjestö
Tulkinkuja 3
02600 Espoo, Finland

France
All Breeds
Union Nationale Interprofessionnelle du
Cheval (U.N.I.C.)
22, rue de Penthièvre
75008, Paris, France

French Ministry of Agriculture
National Stud and Equestrian Sports
Department
14 Avenue de la Grande-Armée
75017 Paris, France

Camargue
Association Des Eleveurs de Chevaux
de Race Camargue
Mas du Pont de Rousty
13200 Arles, France

Corsican
Evviva U Cavallu Corsu
2 rue des Bucherons
20000 Ajaccio
Corse du Sud, France

Pottok
Association Nationale Du Pottok
Armanenia, 64540 Espelette Le
France

Germany
All breeds
Deutsche Reiterliche Vereinigung e.V.
Posfach 640
4410 Warendorf 1, Germany

Karl-Marx Universität
Sektion Tierproduction und
Veterinärmedizin
Johannisallee 21 Leipzig
7010, Germany

Bavarian Warmblood
Landesverband
Bayer, Pferdezüchter
Landshamer Str. 11
8000 München 81, Germany

Dülmen
Erbprinz von Croy
Schlosspark 1
D-4408, Dülmen, Germany

German Riding Pony
Rheinisches Pferdestammbuch e.V.
Endenicher Allee 60
5300 Bonn 1, Germany

Hanoverian
Verband hannoverscher Warmblutzüch-
ter e.V.
Postfach 1743, Lindhopper Strasse 92
D-2810 Verden, Germany

Hessen
Verband Hessischer Pferdezüchter e.V.
Kölnische Strasse 48/50
3500 Kassel, Germany

Hessen Pony (various pony breeds)
Verband der Ponyzüchter Hessen e.V.
Rheinstrasse 91
6100 Darmstadt, Germany

Holstein
Verband der züchter des Holstein
pferdes
mit Reit-und Fahrschule Elmshorn e.V.
Westerstrasse 93-95, Postfach 949
2200 Elmshorn, Germany

Oldenburg
Verband der Züchter des Oldenburger
Pferdes
Haarenfeld 52 c.
2900 Oldenburg, Germany

Rare Breeds
(Society for the Conservation of Old
and Endangered Domestic Breeds)
Gesellschaft zur Erhaltung alter und
gefährdeter
Hofbrunnstrasse, 110
8000 Munich, 71, Germany

Rhineland Heavy Draft and Rheineland
Warmblood
PZV Rheinland-Pfalz-Saar
Planiger Strasse 36a
6550 Bad Kreuznach, Germany

Schleswig
Pferdestammbuch Schleswig-
Holstein/Hamburg
Steenbeker Weg 151
2300 Kiel 1, Germany

Schwarzwälder Füchse
PZV Baden-Württemberg
Heinrich-Baumann, Strasse 1-3
D-7000 Stuttgart 1, Germany

South German Coldblood
Landesverband Bayer. Pferdezüchter
Landshamer Strasse 11,
·8000 München 81, Germany

Trakehner
Verband der Zuchter und Freunde des
Ostpreussischen
Warmblutpferdes Trakehner Abstam-
mung e.V.
Grossflecken 68
2350 Neumünster, Germany

Westphalian
Westfälisches Pferdestammbuch e.V.
Sudmühlenstrasse 31-35, Postfach 460
106
D-4400 Münster-Handorf, Germany

Württemberg
Pferdezuchtverband Baden-Württemberg e.V.
Heinrich-Baumann, Strasse 1-3
7000 Stuttgart 1, Germany

Greece
All breeds
Aristotle University of Thessaloniki
Faculty of Earth Sciences and
Technology

School of Agriculture, Department Animal Production
GR-540 06 Thessaloniki, Greece

Skyros
International Society for the Protection
and Preservation of the Skyros Pony
Ktima Litsas
Thermi 57001
Thessaloniki, Greece

Hungary
All breeds
Institute for Agricultural Quality
Control
Mezögazdasági Minísítö Intézet
1024 Budapest, Keleti Károly u. 24
Budapest 114.Pf.30.,90, Hungary

Iceland
Icelandic Horse
Búnadarfélag Islander
Bændahöllinni v/Hagatorg
P.O. Box 7080
IS - 127, Reykjavík, Iceland

India
All Breeds
National Bureau of Animal Genetic Re-
sources and National Institute of An-
imal Genetics
N.D.R.I. Campus,
Karnal - 132001 (Haryana) India

Kathiawari
Kathiawari, Marwari Horse Breed
Association
Bhuvaneswari Pith
Gondal, District Rajkot, India

Marwari
Marwari Horse Breeders Association
Umaid Bhawan
Jodhpur, India

Indonesia
All breeds
Department of Agriculture
Directorate General of Livestock
Services
P.O. Box 402
Jakarta 10002 Indonesia

Iran
All breeds
Mrs. Louise Firouz
No. 10 Damghan
Teheran 15999, Iran

Israel
All breeds
Ministry of Agriculture
Israel Stud Book
P.O. Box 219
New Ziona 70 451, Israel

Italy
Avelignese
Associazione Nazionale Allevatori
Cavallo Razza Avelignese
Via Cavour N. 106
50129 Firenze, Italy

Bardigiano
Associazione Nazionale Allevatori
del Cavallo Bardigiano
Borgo della Salnitrara
3-43100 Parma, Italy

Calabrian (Calabrese)
Associazione Provinciale Allevatori
Cosenza
(ENTE Morale D.P.R.N. 1019 Del
14-10-1958
Via Caloprese 23, c/c Postale 14
866875,
Italy

Italian Heavy Draft
Associazione Nazionale Allevatori del
Cavallo
Agricolo Italiano da Tiro Pesante
Rapido
Via Locatelli 20
37122 Verona, Italy

Maremmana
Associazione Nazionale Allevatori
Cavallo di Razza Maremmana
Via Monterosa 16
58100 Grosseto, Italy

Murghese
Allevatori dell'Asino di Martina Franca
e del Cavallo delle Murge
74015 Martina Franca (Taranto), Italy

Salerno
Associazione Nazionale del Cavallo
Salernitano-Persano
Via Picenza 76 (P.Co. Arbostella)
84100 Salerno, Italy

Sanfratello
Istituto Incremento Ippico
Via Vittoria Emanuele 508
95124 Catania, Italy

Sardinian
Regione Autonoma della Sardegna
Assessorato All'Agriocoltura e Riforma
ARGO-PASTORALE
Istituto Incremento Ippico della
Sardegna
07014 Ozieri, Italy

Japan
All breeds
Japan Equine Affairs Association
1-2 Kanda Surugadai,
Chiyoda-ku
Tokyo, Japan

Ban-ei Race Horse
The National Association of Racing
2-2-1, Azabudai 2-Chome
Minato-Ku
Tokyo 106, Japan

Hokkaido
The Preservation Association of Hok-
kaido Native Horse
In the Hokuno-kaikan Nishi 1 chyome
Kita 4jyo, Chyuku
Sapporo 060, Japan

Kiso
The Preservation Club of Kiso Horse
Suekawa, kaida-mura,
Kiso-gun
Nagano-ken 897-08, Japan

Misaki
The Cooperator Group of Protection
for Misaki Horse
In the Kushima Municipal Office
5550 Nishikata Kushima-shi
Miyazaki-ken 888, Japan

Miyako
The Preservation Club of Miyako
Horse
4F Hirara-shiminkaikan
5-1 Nishisato Hirara-shi
Okinawa-ken 906, Japan

Noma
The Preservation Club of Noma Horse
In the Noma-noukyo 246-1 kou Akata
Imabari-shi
Ehime-ken 794, Japan

Taishuh
The Promotion Club of Taishuh Horse
In the Tushima-noukyo
606-19 Nakamura Izhuara-machi
Shimoga-tumagun
Nagasaki-ken 817, Japan

Tokara
The Preservation Club of Tokara Horse
In the Laboratory of Animal Breeding
Faculty of Agriculture, Kagoshima
University
1-21-24 Kourimoto
Kagoshima-shi 890, Japan

Yonaguni
The Preservation Club of Yonaguni Horse
129 Yonakuni-chiyo, Yaeyame-gun
Okinawa-ken 907-18, Japan

Korea
All breeds
Korea Horse Affairs Association
685 Juan-dong Kwachon
Kyonggi-do
472-070, Korea

Cheju
Professor Dominicus C. Choung
Department of Animal Science
Cheju National University
Cheju-do, Korea

Lesotho
Basuto
Basotho Pony Project
P.O. Box 1027
Maseru 100, Kingdom of Lesotho

Morocco
Moroccan Barb
Ministry of Agriculture
Et De La Reforme Agraire
Kingdom of Morocco

Netherlands
Dutch Draft
Koninklijke Vereniging "Het
Nederlandsche Trekpaard"
Koninginnegracht 49
2514 AE 's-Gravenhage, Netherlands

Dutch Warmblood
Koninklijke Vereniging Warmbloed
Paarden-Stamboek Nederland
(K.W.P.N.)
Postbus 382
3700 AJ Zeist, Netherlands

Friesian
Het Friesch Paarden-Stamboek
Landbouwhuis, Willemskade 11
Postbus 613
8901 BK Leeuwarden, Netherlands

Przewalski's Horse
Foundation for the Preservation and
Protection of the Przewalski Horse
Mathenesserstraat 101a,
3027 PD Rotterdam, Netherlands

Norway
All Breeds
Gudran Gaustad
Sekretaer i Samarbeidsutvalget
Postboks 85, Årvoll
0515 Oslo 5, Norway

Norwegian Fjord
Norges Fjordhestlag
6800 Førde, Norway

Peru
All Breeds
Asociación Nacional de Criadores y de
Caballos Puruanos de Paso
Bellavista No. 546
Lima 18, Peru

Poland
  All breeds
    Ministerstwo Rolnictwa i Gospodarki
      Zywnosciowe
    Ul. Wspolna 30
    00-930 Warsaw, Poland

    National Association of Animal Breed-
      ing & Marketing, Horse Division
    Wspolna 30
    00-519 Warsaw, Poland

  Malopolski, Konik
    Editorial Office (stud books)
    20 Seietokrzyska St.
    00-002 Warsaw, Poland

Portugal
  Alter Real
    Associaçao de Criadores
    de Raças Selectas
    Rua D. Dinis 2
    Lisbon 1200, Portugal

  Lusitano
    Ministerio da Agricultura
    Serviço Nacional Coudélico
    Rua Victor Cordon 2
    1200 Lisbon, Portugal

Puerto Rico
  Paso Fino
    Departamento de Agricultura
    Apartado 10163
    Santurce
    00908 Puerto Rico

Romania
  All breeds
    Republican Center for Horse Breeding
    Bd. Ion Ionescu de la Brad nr. 4
    Sector 1
    71592, Bucharest, Romania

Russia
  (All breeds)
    V. I. Lenin All-Union Academy of Ag-
      ricultural Sciences
    21 Bolshoi Kharitonievsky pereulok
    Moscow 107814 Russia

    Professor E. M. Pern
    391128 c/o Institute for Horses
    Ribnovskii Region
    Ryazanskaya Oblast, Russia

South Africa
  Cape Horse
    Ernest and Lynn Sykes
    Box 197
    Middelburg 5900, South Africa

    Kaapse Boerperd Breeders' Society
    P.O. Box 76
    Graaff-Reinet
    6280, South Africa

    Historiese Boerperd Breeders' Society
    P.O. Box 5
    Davel 2320, South Africa

  Nooitgedacht
    Nooitgedacht Pony Breeders' Society
    P.O. Box 440
    Middelburg, 5900, South Africa

  South African Miniature
    South African Miniature Horse Breed-
      ing Association
    P.O. Box 32362
    Glenstantia 0010, South Africa

Spain
  All Breeds
    Exco. Sr. General Jefe de Cria Caballar
    Paseo de Extremadura 445
    28014 Madrid, Spain

  Andalusian
    Asociacion de Criadores De Caballos
    de Pura Raza Española
    Trajano 2
    41002 Sevilla, Spain

  Spanish Arab and Spanish Anglo-Arab
    Asociación Española de Criadores de
    Caballos Arabes (AECCA)
    Calle Hermosilla 20
    28001 Madrid-1, Spain

Sweden
  All Breeds
    Lantbruksstyrelsen
    (National Board of Agriculture)
    S-551 83 Jönköping, Sweden

North Swedish Horse
Swedish University of Agricultural
  Sciences
Department of Clinical Nutrition
P.O. Box 7023
S-750 07 Uppsala, Sweden

Swedish Ardennes
Avelsföreningen för Svenska
  Ardennerhästen
Hôgaholmen,
570 20 Bodafors, Sweden

Swedish Warmblood
Avelsföreningen för Svenska
Varmblodiga Hästen
240 32 Flyinge, Sweden

Switzerland
Freiberg, Swiss Warmblood
Swiss Horse Breeding Federation
Kramgasse 58
CH-3011 Bern, Switzerland

Thailand
Native Thai Pony
Royal Turf Club of Thailand
183 Pitsanuloke Road
Bangkok 10300, Thailand

Turkey
All Breeds
Tarim Isletmeleri Genel Müdürlügü
Karacabey Tarim Isletmesi Müdürlügü
Posta Kod. 16720
Karacabey, Bursa, Turkey

(Turkey Ministry of Agriculture)
Tarim Orman Ve Köyişleri Bakanligi
Yüksek Komiserler Kurulu
  Başkanligim
Ankara, Turkey

United Kingdom and Ireland
All Breeds
British Horse Society
British Equestrian Centre
Stoneleigh, Kenilworth
Warwickshire CV8 2LR, UK

British Appaloosa
British Appaloosa Society
Mr. Michael Howkins
c/o 2 Frederick Street
Rugby, Warwickshire CV21 2EN, UK

British Riding Pony
British Show Pony Society
Mrs. J. Toynton
124 Green End Road
Sawtry, Huntington
Cambridgeshire PE17 5XA, UK

British Spotted Pony
British Spotted Pony Society
E. M. Williamson
Weston Manor
Corscombe, Dorset DT2 OPR, UK

Caspian
Caspian Pony Society
Brenda Dalton
6, Nums Walk
Virginia Water, Surrey, England

Cleveland Bay
Cleveland Bay Horse Society
York Livestock Centre
Murton
York YO1 3UF, UK

Clydesdale
The Clydesdale Horse Society
24 Beresford Terrace,
Ayr, Ayrshire, Scotland

Connemara
Connemara Pony Breeders' Society
73 Dalysport Road
Salthill, Galway, Ireland

Dales Pony
Dales Pony Society
196 Springvale Road
Walkley
Sheffield S6 3NU, UK

Dartmoor
The Dartmoor Pony Society
Fordans, 17 Clare Court,
Newbiggen Street
Thaxted, Essex CM6 2RN, UK

English Cob/Hack
    The British Show Hack, Cob and
        Riding Horse Association
    Rookwood, Packington Park
    Meriden, Warwickshire CV7 7HF, UK

English Hunter
    The National Light Horse Breeding
        Society
    96 High Street
    Edenbridge, Kent TN8 5AR, UK

English Thoroughbred
    The Thoroughbred Breeders'
        Association
    Stanstead House, The Avenue
    Newmarket, Suffolk, CB8 9AA, UK

Eriskay Pony
    Eriskay Pony Breed Society
    Buskinburn House
    Coldingham, Eyemouth,
    Berwickshire, TD14 5UA, Scotland

Exmoor Pony
    Exmoor Pony Society
    D. Mansell, Esq.
    Glen Fern, Waddicombe
    Dulverton
    Somerset TA22 9RY, UK

Falabella
    Kilverstone Miniature Horse Stud
    Kilverstone, Thetford
    Norfolk, IP24 2RL, UK

Fell Pony
    Fell Pony Society
    19 Dragley Beck, Ulverston
    Cumbria LA12 0HD, UK

Hackney
    The Hackney Horse Society
    Clump Cottage, Chitterne
    Nr. Warminster
    Wiltshire, BA12 0LL UK

Haflinger
    The Haflinger Society of Great Britain
    Mrs. Helen Robbins
    13 Parkfield, Pucklechurch
    Bristol BS17 3NR, UK

Highland Pony
    The Highland Pony Society
    Beechwood, Elie
    Fife KY1 9DH Scotland

Hunter
    Hunter Improvement Society of Ireland
    Mrs. F. Hefferman
    Knockelly, Castle, Fethard,
    Co. Tipperary, Ireland

Irish Draft
    The Irish Draught Horse Society
    Deirdre Quinn, Secretary
    23 Richelieu Park, Ballsbridge,
    Dublin 4, Ireland

Lusitano
    The Lusitano Breed Society of Great
        Britain
    Foxcroft, Bulstrode Lane,
    Felden, Hemel Hempstead
    Herts HP3 0BP, UK

New Forest Pony
    The New Forest Pony Breeding and
        Cattle Society
    Beacon Corner, Burley, Ringwood
    Hampshire BH24 4EW, UK

Shetland
    The Shetland Pony Stud-Book Society
    Pedigree House, 6 King's Place,
    Perth PH2 8AD, Scotland

Shire
    Shire Horse Society
    East of England Showground
    Peterborough PE2 0XE, UK

Suffolk
    Suffolk Horse Society
    6 Church Street
    Woodbridge, Suffolk IP12 1DH, UK

Trakehner
    The British Trakehner Association
    Hallagenna Stud, St Breward,
    Cornwall, England

Welsh
    The Welsh Pony and Cob Society
    6 Chalybeate Street
    Aberystwyth,
    Dyfed SY23 1HS, UK

United States
All breeds
American Horse Council
1700 K Street N.W.
Washington D.C. 20006, USA

Albino
International American Albino
Association
P.O. Box 79
Crabtree, OR 97335, USA

American Bashkir
American Bashkir Curly
Box 453
Ely, NV 89301, USA

The Curly Horse Foundation
P.O. Box 520
Sunman, IN 47041, USA

American Indian Horse
American Indian Horse Registry
Route 3 Box 64
Lockhart, TX 78644, USA

American Miniature
The American Miniature Horse
Association
2908 S.E. Loop 820
Fort Worth, TX 76140, USA

American Mustang
American Mustang Association
P.O. Box 338
Yucaipa, CA 92399, USA

American Paint Horse
American Paint Horse Association
P.O. Box 18519
Fort Worth, TX 76118, USA

American Quarter Horse
American Quarter Horse Association
P.O. Box 200
Amarillo, TX 79168, USA

American Saddlebred
American Saddlebred Horse
Association
Kentucky Horse Park
4093 Iron Works Pike
Lexington, KY 40511, USA

American Shetland
American Shetland Pony Club
P.O. Box 3415
Peoria, IL 61614, USA

American Walking Pony
American Walking Pony Association
Rt. 5 Box 88
Macon, GA 31211, USA

Appaloosa
Appaloosa Horse Club, Inc.
P.O. Box 8403
Moscow, ID 83843, USA

AraAppaloosa
AraAppaloosa & Foundation Breeders
International
Rt. 8 Box 317
Fairmont, WV 26554, USA

Arabian
Arabian Horse Registry of America
12000 Zuni Street
Westminster, CO 80234, USA

International Arabian Horse
Association
P.O. Box 33696
Denver, CO 80233, USA

Assateague/Chincoteague
Assateague State Park
Rt. 2 Box 293
Berlin, MD 21811, USA

Azteca
Azteca Horse Registry of America
P.O. Box 490
Vancouver, WA 98666, USA

Belgian
Belgian Draft Horse Corporation of
America
Box 335
Wabash, IN 46992, USA

Buckskin
American Buckskin Registry
Association
P.O. Box 3850
Redding, CA 96049, USA

International Buckskin Horse Association Inc.
P.O. Box 268
Shelby, IN 46377, USA

Caspian
The Caspian Horse Society of the
Americas
Monastery of St. Clare
Rt. 7, Box 7504
Brenham, TX 77833, USA

Chickasaw
The Chickasaw Horse Association
Box 265
Love Valley, NC 28677, USA

Chincoteague
Chincoteague Island Chamber of
Commerce
P.O. Box 258
Chincoteague, VA 23336, USA

Chincoteague Volunteer Fire Company
Chincoteague, VA 23336, USA

Clydesdale
Clydesdale Breeders of the United States
17378 Kelley Road
Pecatonica, IL 61063, USA

Clydesdale Operations
Anheuser Busch
1 Busch Place
St. Louis,MO 63118, USA

Colorado Ranger
Colorado Ranger Horse Association
RD. #1, Box 1290
Wampum, PA 16157, USA

Dartmoor Pony
Dartmoor Pony Society of America
1005 Pearl Wood Rd.
Albany, OH 45710, USA

Florida Cracker Horse
Florida Cracker Horse Association
Rt. 3, Box 403
Newberry, FL 32669, USA

Friesian
The Friesian Horse Association of
North America
c/o Herman J. Van Reenen
4127 Kentridge Drive S.E.
Grand Rapids, MI 49508, USA

Johnny and Carolyn Sharp
Midnight Valley Friesians
2950 S. County Road 5
Fort Collins, CO 80525, USA

Galiceño
Galiceño Horse Breeders Association
P.O. Box 219
Godley, TX 76044, USA

Hackney
American Hackney Horse Society
P.O. Box 174
Pittsfield, IL 62363, USA

Haflinger
Haflinger Registry of America
14640 State Route 83
Coshocton, OH 43812, USA

Half Saddlebred
The Half Saddlebred Registry of
America
319 South 6th Street
Cochocton, OH 43812, USA

Hanoverian
The American Hanoverian Society
14615 N.E. 190th St., Suite 108
Woodinville, WA 98072, USA

Holstein
The American Holstein Horse
Association
P.O. Box 410
Chelsea, MI 48118-0410, USA

Icelandic Horse
U.S. Icelandic Horse Federation
38 Park Street
Montclair, NJ 07042, USA

Kiger Mustang
Kiger Mesteno Assn.
P.O. Box 452
Burns, OR 97702, USA

Missouri Fox Trotter
Missouri Fox Trotting Horse Breed
Association
Ava, MO 65608, USA

Morgan
The American Morgan Horse
Association
P.O. Box 960, 3 Bostwick Road
Shelburne, VT 05482-0960, USA

Morab
  The North American Morab Horse
    Association
  W3174 Faro Springs Road
  Hilbert, WI 54129, USA

Moyle Horse
  Rex and Marge Moyle
  9495 W. Beacon Light Road
  Star, ID 83669, USA

Mustang
  Headquarters Office
  Bureau of Land Management
  U.S. Department of the Interior
  18th and C Streets N.W.
  Washington, D.C. 20240-0001, USA

National Show Horse
  National Show Horse Registry
  Plainview Triad North, Suite 237
  10401 Linn Station Road
  Louisville, KY 40223, USA

National Spotted Saddle Horse
  National Spotted Saddle Horse
    Association
  P.O. Box 898
  Murfreesboro, TN 37130, USA

Norwegian Fjord
  The Norwegian Fjord Horse Associa-
    tion of North America
  24570 W. Chardon Road
  Grayslake, IL 60030, USA

Palomino
  The Palomino Horse Association
  Box 324
  Jefferson City, MO 65102, USA

  Palomino Horse Breeders of America
  15253 E. Skelly Drive
  Tulsa, OK 74116-2620, USA

Paso Fino
  American Paso Fino Horse Association
  P.O. Box 2363
  Pittsburgh, PA 15230, USA

  Paso Fino Horse Association
  P.O. Box 600
  Bowling Green, FL 33834, USA

Pony of the Americas
  Pony of the Americas
  5240 Elmwood Avenue
  Indianapolis, IN 46203, USA

Rare Breeds
  The American Livestock Breeds
    Conservancy
  Box 477
  Pittsboro, NC 27312, USA

Rocky Mountain Horse
  Rocky Mountain Horse Association
  1140 McCalls Mill Road
  Lexington, KY 40515, USA

Selle Français
  North American Selle Français Horse
    Association
  P.O. Box 646
  Winchester, VA 22601, USA

Shagya Arabian
  North American Shagya-Arabian
    Society
  Adele Furby
  Star Route
  Hall, MT 59837, USA

Shire
  The American Shire Horse Association
  Lowell Wagoner
  Rt. #1, Box 10
  Adel, IA 50003, USA

Southwest Spanish Mustang
  Southwest Spanish Mustang
    Association
  P.O. Box 148
  Finley, OK 74543, USA

Spanish Barb
  Spanish Barb Foundation
  Paulton Spanish-Barb Horses
  Route 1 Box 149
  Hot Springs, SD 57747, USA

Spanish Mustang
  Spanish Mustang Registry
  8328 Stevenson Ave.
  Sacramento, CA 95828, USA

  Marye Ann Thompson, Registrar
  Star Route 3 Box 7670
  Willcox, AZ 85643, USA

Standardbred
  The United States Trotting Association
  750 Michigan Avenue
  Columbus, OH 43215, USA

Striped Horse
  International Striped Horse Association
  Verde Mont Ranch
  4453 Co. Road 182
  Westcliffe, CO 81252, USA

Suffolk
  American Suffolk Horse Association
  Rt. 1 Box 212
  Ledbetter, TX 78946, USA

Tennessee Walking Horse
  Tennessee Walking Horse Breeders and
    Exhibitors Association
  P.O. Box 286
  Lewisburg, TN 37091, USA

Thoroughbred
  The Jockey Club
  380 Madison Avenue
  New York, NY 10017, USA

Trakehner
  North American Trakehner
    Association
  1660 Collier Road
  Akron, Ohio 44320, USA

Trote En Gallope
  Manuel Mendoza
  Malibu Meadow Farms
  715 Crater Camp Road
  Calabasas, CA 91302, USA

Welera
  American Welera Pony Society
  P.O. Box 401
  Yucca Valley, CA 92286, USA

Welsh
  Welsh Pony and Cob Society of
    America
  P.O. Box 2977
  Winchester, VA 22601, USA

Yemen
  All Breeds
    Arab Organization for Agricultural
      Development
    P.O. Box 822
    Sana'a
    Republic of Yemen

Glossary

Aboriginal: Native to an area.

Adaptability: The ability to adapt or adjust to new living conditions such as harsh weather.

Aged: An older horse, usually over ten years. Sometimes applied to any horse over seven years.

Aid: Any of the signals given by a rider to instruct his horse.

Air above the Ground: Any of the various high school movements performed with either forelegs or both fore and hind legs off the ground.

Albino: A name usually applied to a pure white horse with pink skin. The true living albino does not exist in the horse; all incidents of albinism die in utero.

All-round cow horse: A horse skilled at carrying out all duties required by a cowboy.

Also-ran: Any horse not placing in a race.

Alter: To castrate a horse or colt, making it a gelding.

Amble: Old English name given to the lateral, four-beat gait in which the horse's hind and foreleg on the same side move forward simultaneously but strike the ground separately. A four-beat gait.

Artificial aids: Items such as spurs, whips, martingales, etc. to convey commands to the horse.

Band: A group of horses.

Bang-tail: A name often given to feral horses in the American West.

Barrel: The part of the horse's body between the forelegs and the loins.

Bay: A dark-skinned horse with dark brown to red or yellowish brown coat, black mane and tail, and black legs.

Bit: The device, usually of metal or rubber which is placed in the horse's mouth as an aid in control.

Black: A horse with black coat, mane and tail and no other color—except perhaps white markings on the face or legs.

Blacksmith: A craftsman whose medium is iron and in addition to other things, makes horse shoes.

Blood Sweater: The name given to the ancient Turkmenian horse (ancestor to the Akhal-Teke) by the Chinese in the time of Emperor Wu Ti.

Blood horse, or horse of blood: The English Thoroughbred, or a part-Thoroughbred.

Bloodstock: Thoroughbred horses, particularly race and breeding animals.

Bran: A product of grain milling, which when ground fresh and dampened acts as a mild laxative, aiding in digestion.

Break: To break or break in a horse is to give it initial training.

Break down: A term used when laceration occurs to the suspensory ligament or there is fracture to the sesamoid bone.

Breeder: The owner of a mare that produces a foal. The owner of a stud where horses are bred.

Bridle: The part of the saddlery or tack which is placed about the horse's head.

Brio: A Spanish term used to describe controlled energy and sensitivity to the aids of the rider.

Bronc or bronco: A term describing a rodeo bucking horse, or an unbroken horse.

Broomtail: A name often given to American feral horses.

Brumby: Feral horse in Australia.

By: Sired by: Always referring to the sire of a foal.

Camp drafting: An Australian sport in which a rider separates a steer from a group of cattle and drives it at the gallop around a course marked with poles. It is a timed event.

Canter: A pace of three beats in which the hooves strike the ground in the following order: near hind, near fore, and off hind together, then off fore (leading leg); or off hind, off fore, and near hind together, then near fore (leading leg).

Capriole: An air above the ground in which the horse half rears with the hocks drawn under, then leaps forward and high into the air, at the same time kicking the hind legs out with the soles of the feet turning upward, before landing correctly on all four legs.

Carp-like Back: A back that is humped up—opposite of swaybacked.

Cavalletti: A series of small wooden jumps used in the basic training of a riding horse in order to encourage it to lengthen its stride, improve balance and loosen and strengthen the muscles.

Cayuse: An Indian horse or pony.

Chestnut: A horse with a golden to dark reddish brown coat, having a matching or slightly lighter or darker mane and tail; sometimes mare and tail are flaxen (blond.) Also: A horny growth found on the inside of all four legs, about three inches above the knee and on the inner part of hock joint; Not found in all breeds.

Chubary: A Russian word referring to leopard-spotted Appaloosa coloration.

Cob: A type, not a breed. A short-legged animal with a maximum height of 15.1 hands, showing the bone and substance of a hunter, and capable of carrying substantial weight.

Coldblood: A horse of draft breeding.

Collection: Shortened pace by light contact on the bit and steady pressure with the legs to cause a horse to flex the neck, relax the jaw and bring its hocks well under the body so that it is properly balanced.

Colt: An unaltered male horse less than four years of age.

Conformation: A term referring to the horse's shape. The shape of a purebred horse should conform to its type.

Coronary: Coronary band. The area at the top of the hoof between the sensitive laminae or fleshy leaves of the hoof and the perioplic band.

Cowboy: A man who makes his living working cattle. A man specializing in rodeo work.

Cow pony: A term coined in the West for a horse used to work cattle. Not a pony.

Crossbreeding: The breeding to purebred individuals of different breeds, resulting in crossbreds.

Dragon Horse: A ancient Chinese strain possessing great stamina, having small horn-like projections above the eyes.

Entire: A male horse that has not been gelded.

Equidae: Scientific name for the horse family.

Equine: Pertaining to the horse family.

Ergot: A horny growth at the back point of the fetlock joint. Not found in some breeds.

Fetlock: The portion of the horse's leg between the cannon and the pastern.

Filly: A female horse, usually less than four years old.

Foal: A young horse, male or female, up to the age of twelve months. A colt-foal or filly-foal.

Forehand: The part of the horse in front of the rider; head, neck, shoulders, withers, and forelegs.

Foundation: Foundation sire or dam refers to animals influential in the beginning of a breed.

Frog: The "V"-shaped, elastic tissue in the sole of the hoof which expands and contracts when carrying the horse's body weight.

Garron: A native pony of Scotland or Ireland.

Gaskin: The muscular portion of the horse's leg between the thigh and hock joint.

Gelding: A male horse that has been castrated.

Get: The offspring of a stallion.

Girth: (a) The circumference of a horse, measured behind the withers around the deepest part of the body; (b) a band of leather, webbing or nylon which passes under the belly of the horse to hold the saddle in place.

Hand(s): A horse is measured in four-inch increments, called hands. A horse standing 14 hands high (14 h.h.) at the withers is 56 inches high.

Heavenly horse: Only three special ancient breeds were given this name by the Chinese; Dawan horse (ancient Turkmene), Wasun horse, and Sichuan horse.

Herd: A group of horses.

Hinny: The offspring of a stallion on a jenny.

Hock: The large joint on the horse's hind leg between the gaskin muscle and cannon bone; corresponds to the human heel.

Hotblood: A horse of Oriental breeding.

Kosjak: A Russian term for a small breeding group of *taboon* horses.

Koumiss: Fermented mare's milk. Healthful drink throughout Asia since ancient times.

Kurgan: Ancient burial mounds, found in Asia and Siberia.

Lateral gait: A four-beat gait in which the horse moves the front and hind legs on the same side of the body almost simultaneously, striking the ground with the hind foot slightly before the front foot. Also called *tølt*, rack, single-foot, amble, *paso*, running-walk.

Loin: The area of the horse's body between the back and the croup.

Lope: An easy, uncollected canter.

Lymphatic: Lacking physical or mental energy, sluggish; swelling, usually in joints of the hind legs.

Maiden: A horse of either sex that has not won a race of any distance.

Maiden mare: A mare that has not had a foal, even though she may be carrying one.

Maiden race: For horses that have never won a race.

Mare: A female horse aged four years or more.

Oriental horse: A horse breed originating in western or central Asia.

Out of: Out of a mare. Always referring to the dam of a foal.

Outcross: Breeding a purebred animal to one of another breed. Crossbreeding.

Pace: A gait in which the horse moves both front and hind legs on the same side of the body simultaneously.

Pastor: In China, the herdsman watching over a *taboon* of horses.

Polygamous breeding: Natural selection among free-ranging stallions and mares.

Produce: The offspring of a mare.

Purebred: An animal that is free of any blood not that of its own breed.

Rabbit-like profile: A term used in China when referring to a horse with a convex swell in the profile immediately above the nostrils. Often seen on the Mongolian.

Ram profile: Roman-nosed, convex.

Resistant: A term used when referring to ability to withstand fatigue and/or sickness.

Rustic: A horse with native characteristics.

Sound: A horse that is not lame in any way.

Sovchoz: A Russian term meaning state farm.

Spanish Jennet: An extinct breed of Spanish horse, noted for its smooth, ambling gait, speed and stamina.

Stallion: An entire (not gelded) male horse.

Stud: A horse breeding farm. A stallion used for breeding.

Stud horse: A stallion used for breeding on a stud farm.

Taboon: Also spelled *tabun*. A large herd of horses raised on open pasture with an attendant herdsman. Many *taboons* contain as many as 200 horses.

Tebenevka: A Russian term meaning year-round pasture raising without supplemental feeding.

Tropota: A Russian name for a running-walk gait.

Trot: A gait of two time in which the legs moves in diagonal pairs.

Type: Referring to the use for which a horse is best suited; hunter, racer, show pony, cob, etc. Heavy, medium, and light types may all exist within a pure breed.

Warmblood: A horse with some hotblood and some coldblood in its pedigree.

Withers: The highest portion of a horse's back, at the base of the neck between the shoulder blades.

Bibliography

Barclay, Harold. *The Role of the Horse in Man's Culture.* London: J. A. Allen, 1980.

Barmintsev, Yu. *Produktivnoe konevodstvo.* Moscow: Kolos, 1980.

Barmintsev, Yu. N., and Kozhevnicov, Ye. V. *Horse Breeding in the USSR,* Moscow: Kolos, 1983.

Beattie, Gladys Mackie. *The Canadian Horse,* Lennoxville, Quebec: Sun Books, n.d.

Biard, Pierre. *Relation of New France,* edited by R. G. Thwaites. Cleveland, Ohio: 1896–1901.

Bökönyi, Sandor. *Data on the Iron Age Horses in Central and Eastern Europe in the Mecklenburg Collection,* pt. I. Cambridge, Mass: Harvard University Press, 1968.

Bouma, Ir. G. J. A. *Het Friese Paard.* Zwolle, Netherlands: Durkkerij Tulp bv., 1979.

Bruce, Philip H. *Economic History of Virginia in the 17th Century.* 2 vols. New York, 1895.

Budyonny, S. M. *Book on the Horse,* vol. 1. n.p., 1952.

Christie, Barbara J. *The Horses of Sable Island,* Halifax, Canada: Petheric Press 1980.

Encyclopedia Britannica, all vols. Chicago: Encyclopedia Britannica, 1989.

Encyclopedia Britannica Instant Research Service. *Archaeological Finds at Xian, China.*

Exmoor Pony Society. *Official Handbook.* Somerset: Exmoor Pony Society, 1986.

FAO. *Animal Genetic Resources of the USSR.* Animal Production Health paper no. 65. Rome: U.N. Food and Agriculture Organization, 1989.

Firouz, Louise. *The Caspian Miniature Horse of Iran.* Field Research Projects. Privately published, 1972.

Gibson, James R. *Imperial Russia in Frontier America.* New York: Oxford University Press, 1976.

Irish Draught Horse Society. *The Irish Draught Horse.* Dublin: Irish Draught Horse Society, 1989.

Konevodstvo. Monthly journal ("Horse Breeding and Equestrian Sport"), all issues, 1980–1990. Moscow: Kolos.

Loch, Sylvia. *The Royal Horse of Europe.* London: J. A. Allen and Co., 1986.

Manning, Russell J. *Manning's Illustrated Horse Book.* Edgewood Publishing Co., 1882.

Mason, I. L. *A World Dictionary of Livestock Breeds.* United Kingdom: C.A.B. International, 1988.

Merkins, J. *De Paarden—En Runderteelt in Nederlandsch Indie,* Weltevreden, Netherlands: Landrukkerij, 1926.

National Geographic. "China's Incredible Find." Vol. 153, no. 4 (April 1978).

Pecaud. *Les Chevaux de Notre Colonie du Tchad.* S.L.N.D., n.d.

Porter and Coates. *Every Horse Owners' Cyclopedia.* Philadelphia, 1871.

Pridorogin, M. I. *Konskie porody.* Moscow: Karkomzema Novaja Derevnja, 1923.

Saunders, R. N. "The First Introduction of European Plants and Animals into Canada." *Canadian Historical Review* 16 (1935):392.

Schwark, H. J. *Pferde Zucht.* Berlin: VEB Deutscher, Landwirtschaftsverlag, 1987.

Tiequan Wang. *Horse and Ass Breeds in China.* Shanghai: Shanghai Scientific and Technical Publishers, 1986.

The Western Horseman. December 1948–September 1949.

Index